MULTIVARIATE ANALYSIS-III

Proceedings of the Third International Symposium on Multivariate Analysis Held at Wright State University, Dayton, Ohio, June 19–24, 1972

MULTIVARIATE ANALYSIS–III

Proceedings of the Third International Symposium on Multivariate Analysis Held at Wright State University, Dayton, Ohio, June 19–24, 1972

Edited by *PARUCHURI R. KRISHNAIAH*

AEROSPACE RESEARCH LABORATORIES
WRIGHT-PATTERSON AIR FORCE BASE, OHIO

1973

ACADEMIC PRESS New York and London
A Subsidiary of Harcourt Brace Jovanovich, Publishers

ACADEMIC PRESS, INC.
111 Fifth Avenue, New York, New York 10003

United Kingdom Edition published by
ACADEMIC PRESS, INC. (LONDON) LTD.
24/28 Oval Road, London NW1

Library of Congress Cataloging in Publication Data

International Symposium on Multivariate Analysis, 3d, Wright State University, 1972.
Multivariate analysis, volume III; proceedings.

Sponsored by the Aerospace Research Laboratories, Wright-Patterson Air Force Base, Ohio.
Includes bibliographical references.
1. Multivariate analysis–Congresses.
I. Krishnaiah, Paruchuri R., ed. II. United States. Aerospace Research Laboratories, Wright-Patterson Air Force Base, Ohio. III. Wright State University. IV. Title.
QA278.158 1972 519.5'3 73-9579
ISBN 0–12–426653–3

PRINTED IN THE UNITED STATES OF AMERICA

Contents

PART III/Characteristic Functions and Characterizations

PART IV/Design and Analysis of Experiments

The Analysis of Time Series Collected in an Experimental Design
David R. Brillinger

Max-Min Designs in the Analysis of Variance
R. H. Farrell

Analysis of Covariance Structures
K. G. Jöreskog

Optimum Designs for Fitting Biased Multiresponse Surfaces
J. Kiefer

Asymptotic Properties of Some Sequential Nonparametric Estimators in Some Multivariate Linear Models
Pranab Kumar Sen and Malay Ghosh

PART V/Classification, Modelling, and Reliability

List of Contributors

Numbers in parentheses indicate the pages on which the author's contributions begin.

D. F. Andrews, Bell Laboratories, Murray Hill, New Jersey (95)

Richard E. Barlow, Operations Research Center, University of California, Berkeley, California (319)

P. Bickel, University of California, Berkeley, California (3)

David R. Brillinger, Department of Statistics, The University of California, Berkeley, California (241)

A. K. Chattopadhyay, Department of Statistics, Purdue University, Lafayette, Indiana (117)[1]

Herman Chernoff, Stanford University, Stanford, California (337)

Roger Cuppens, Université Paul Sabatier, Toulouse, France (183)

A. P. Dempster, Department of Statistics, Harvard University, Cambridge, Massachusetts (129)

F. Eicker, Abteikung Statistik, Universitat Dortmund (17)

R. H. Farrell, Department of Mathematics, Cornell University, Ithaca, New York (257)

D. A. S. Fraser, Department of Mathematics, University of Toronto, Toronto, Ontario, Canada (143)

Malay Ghosh, Research Training School, Indian Statistical Institute, Calcutta, India (299)

R. Gnanadesikan, Bell Laboratories, Murray Hill, New Jersey, (95)

A. T. James, Department of Statistics, University of Adelaide, Adelaide, South Australia (157)

K. G. Jöreskog, Department of Statistics, University of Uppsala, Uppsala, Sweden (263)

G. Kallianpur, University of Minnesota, Minneapolis, Minnesota (31)

J. Kiefer, Department of Mathematics, Cornell University, Ithaca, New York (287)

[1] Present address: Applied Mathematics Research Laboratory, Aerospace Research Laboratories, Wright-Patterson Air Force Base, Dayton, Ohio.

Eugene Lukacs, The Catholic University of America, Washington, D.C. (199)[2]

P. Masani, Department of Mathematics, University of Pittsburgh, Pittsburgh, Pennsylvania (209)

Kameo Matusita, The Institute of Statistical Mathematics, Tokyo, Japan (345)

K. C. S. Pillai, Department of Statistics, Purdue University, Lafayette, Indiana (117)

M. B. Priestley, University of Manchester, Institute of Science and Technology, Manchester, England (351)

Frank Proschan, Department of Statistics, Florida State University, Tallahassee, Florida (319)

M. M. Rao, The Institute for Advanced Study, Princeton, New Jersey (45)[3]

T. Subba Rao, University of Manchester, Institute of Science and Technology, Manchester, England (351)

W. L. Root, Department of Aerospace Engineering, College of Engineering, The University of Michigan, Ann Arbor, Michigan (61)

M. Rosenblatt, Imperial College and University College, London, and University of California, San Diego, California (3)

Yu. A. Rozanov, Steklov Mathematical Institute, Academy of Sciences of the U.S.S.R., Moscow, U.S.S.R. (79)

Pranab Kumar Sen, Department of Biostatistics, University of North Carolina, Chapel Hill, North Carolina (299)

H. Tong, University of Manchester, Institute of Science and Technology, Manchester, England (351)

J. N. Srivastava, Department of Statistics, Colorado State University, Fort Collins, Colorado (369)

K. Urbanik, Institute of Mathematics, University of Wroclaw, Wroclaw, Poland (225)

J. L. Warner, Bell Laboratories, Murray Hill, New Jersey (95)

R. A. Wijsman, Department of Mathematics, University of Illinois at Urbana-Champaign, Illinois (171)

Herman Wold, Department of Statistics, University of Goteborg, Goteborg, Sweden (383)

[2] Present address: Bowling Green State University, Bowling Green, Ohio.

[3] Present address: Mathematics Department, University of California, Riverside, California.

Preface

Multivariate analysis is well recognized as an important branch of statistics and probability and, moreover, its tools are very useful in drawing inferences from the data that arise in physical, engineering, biological, and behavioral sciences as well as other disciplines. To stimulate research and disseminate knowledge in this area, three international symposia have been organized under the sponsorship of the Aerospace Research Laboratories (ARL). This volume consists of the invited papers presented at the Third International Symposium on Multivariate Analysis held at Wright State University, Dayton, Ohio, June 19–24, 1972. In these papers, prominent statisticians and probabilists from several countries discuss the present state of knowledge in the theory and applications of multivariate analysis. The areas thus covered include time series and stochastic processes, distribution theory and inference, characteristic functions and characterizations, design and analysis of experiments, classification, modeling, and reliability. A great majority of papers present new results in the field and the rest are expository in nature. This volume will be useful to statisticians and probabilists, as well as to scientists in other disciplines broadly interested in the area of multivariate analysis. The material in this volume complements the material in the earlier volumes, *Multivariate Analysis* and *Multivariate Analysis—II*, and to some extent also the papers appearing regularly in the *Journal of Multivariate Analysis*. Unfortunately, contributed papers presented at the Symposium could not be included in this volume because of space limitations. However, the titles of the contributed papers are listed at the end of the volume.

Acknowledgments

While I take the responsibility for any mistakes in the organization of the Symposium and the editing of the Proceedings, I wish to express my gratitude to several persons for their valuable help. Colonel W. J. Hare, Commander of ARL, was kind enough to make the opening remarks. I wish to thank Professors R. A. Bradley, D. R. Brillinger, H. T. David, A. P. Dempster, F. Eicker, D. A. S. Fraser, S. S. Gupta, A. T. James, L. Katz, C. F. Kossack, E. Lukacs, C. Maneri, A. M. Mathai, G. Mudholkar, M. M. Rao, M. Rosenblatt, V. B. Waikar, and R. A. Wijsman and Drs. H. L. Harter, H. M. Hughes, D. A. Lee, and R. Lundegard for presiding over the sessions. I wish to express my appreciation to those who reviewed the papers: G. A. Anderson, C. B. Bell, S. C. Chay, G. Y. H. Chi, P. Flusser, N. Giri, A. K. Gupta, T. Kailath, R. G. Laha, V. Mandrekar, S. J. Press, M. B. Priestley, B. S. Rajput, T. Subba Rao, S. Saunders, J. G. Saw, P. K. Sen, M. Siotani, V. S. Varadhan, and G. Wahba. I am quite grateful to Professor C. Maneri for his valuable help in making the local arrangements; thanks are also due to Professors A. K. Gupta, W. J. Park, and Dr. A. K. Chattopadhyay for assisting with the local arrangements. I am grateful to my colleagues Drs. H. L. Harter and D. A. Lee for their encouragement and advice on various administrative matters.

I am deeply indebted to Professor M. M. Rao for his valuable and unselfish help in the organization of the program, not only of this symposium but also of the earlier two symposia on multivariate analysis. Special thanks are due to the contributors to this volume and to Academic Press for their excellent cooperation. Last but not least, I wish to thank my wife, Indira, for her constant encouragement in the organization of the three symposia.

P. R. Krishnaiah

PART I

Time Series and Stochastic Processes

Two-Dimensional Random Fields

P. BICKEL[1]
UNIVERSITY OF CALIFORNIA, BERKELEY

M. ROSENBLATT[2]
IMPERIAL COLLEGE AND UNIVERSITY COLLEGE, LONDON, AND UNIVERSITY OF CALIFORNIA, SAN DIEGO

The asymptotic behavior of the maximum

$$\sup_{0 \le t \le T} X(t)$$

as $T \to \infty$ of a Gaussian stationary process $\{X(t),\ -\infty < t < \infty\}$ has been studied under a variety of conditions in [1], [2] and [6]. The object of this paper is to obtain similar results for a class of Gaussian processes with a multidimensional time parameter. The results are obtained by an appropriate modification of the techniques employed in the case of a one-dimensional time parameter and are indicated in some detail in the two-dimensional case. The format of the proof is quite similar to that given in [2].

Preliminaries. Let $X(t)$, $t = (t_1, \ldots, t_k), -\infty < t_i < \infty$, be a separable Gaussian stationary process with mean zero and covariance function $r(t)$. Consider

$$Y_T(t) = X(t) + \mu_T(t) \tag{1}$$

with $\mu_T(\cdot)$ a deterministic sequence of functions so that

$$EY_T(\cdot) = \mu_T(\cdot). \tag{2}$$

[1] Research supported by USPHS grant GM-10525-109.
[2] Research carried on while the author was a Guggenheim Fellow and partially supported by the Office of Naval Research.

Our object is to study asymptotic behavior of

$$\begin{aligned} M_T &= \max\{Y_T(t)\colon 0 \le t_i \le T, \quad i = 1, \ldots, k\}, \\ m_T &= \min\{Y_T(t)\colon 0 \le t_i \le T, \quad i = 1, \ldots, k\}, \end{aligned} \tag{3}$$

as $T \to \infty$. Set $b_T(t) = \mu_T(t)\{2k \log T\}^{1/2}$.

The following theorem is a main result of this paper.

Theorem 1. *Assume that*

(i) *$b_T(t)$ is uniformly bounded in t and T for $t \in [0, T]^k$ as $T \to \infty$.*
(ii) *$b_T(t) \to b(t)$ uniformly on $[0, T]^k$ as $T \to \infty$.*
(iii) *$\frac{1}{T^k} m\{t\colon b(t) \le x,\ 0 \le t_i \le T\} \to \psi(x)$ the distribution function of a probability measure as $T \to \infty$ (m denotes Lebesgue measure).*
(iv) *$b(\cdot)$ is uniformly continuous on R^k.*
(v) *The covariance function*

$$r(t) = 1 - |t|^\alpha \int_{S_k} |(t/|t|, \theta)|^\alpha \mu(d\theta) + o(|t|^\alpha), \tag{4}$$

$0 < \alpha \le 2$, as $t \to 0$ with μ a finite measure on the unit k sphere in R^k with at least two distinct rays in its spectrum so that the integral form on the right-hand side of (4) is nonsingular.

(vi) $\int_{-\infty}^{\infty} \cdots \int r^2(t_1, \ldots, t_k)\, dt_1 \cdots dt_k < \infty.$

Let

$$B(t) = \{2k \log t\}^{1/2} + \left\{\frac{1}{2}\left(\frac{2k}{\alpha} - 1\right)\log\log t + \log((2\pi)^{-1/2} H_\alpha (2k)^{[(2k/\alpha)-1]/2})\right\}\Big/(2k \log t)^{1/2} \tag{5}$$

where

$$H_\alpha = \lim_{T\to\infty} T^{-k} \int_0^\infty e^s P\left[\sup_{0 \le t_1, \ldots, t_k \le T} Y(t_1, \ldots, t_k) > s\right] ds \tag{6}$$

with Y a Gaussian process having

$$E(Y(t)) = -|t|^\alpha \int_{S_k} |(t/|t|, \theta)|^\alpha \mu(d\theta) \tag{7}$$

and

$$\operatorname{cov}(Y(t), Y(\tau)) = |t|^\alpha \int_{S_k} |(t/|t|, \theta)|^\alpha \mu(d\theta) + |\tau|^\alpha \int_{S_k} |(\tau/|\tau|, \theta)|^\alpha \mu(d\theta) - |t-\tau|^\alpha \int_{S_k} |(\{t-\tau\}/|t-\tau|, \theta)|^\alpha \mu(d\theta). \tag{8}$$

Then

$$U_T = \{2k \log T\}^{1/2}(M_T - B(T)) \quad \text{and} \quad V_T = -\{2k \log T\}^{1/2}(m_T + B(T)) \tag{9}$$

are asymptotically independent with

$$P[U_T < z] \to \exp[-\lambda_1 \exp(-z)], \qquad P[V_T < z] \to \exp[-\lambda_2 \exp(-z)] \tag{10}$$

where

$$\lambda_1 = \int e^z \psi(dz), \qquad \lambda_2 = \int e^{-z} \psi(dz). \tag{11}$$

Note. The condition (v) of Theorem 1 given by formula (4) is a natural one and can be motivated as follows. Suppose that we are interested in real symmetric covariance functions $r(t)$, that is, $r(t) = r(-t)$ for all t, with the property that for some α, $0 < \alpha \leq 2$,

$$[1 - r(\tau\theta)]/\tau^\alpha \to C(\theta)$$

as the scalar $\tau \downarrow 0$ where θ is any fixed unit vector and C is continuous on the k-sphere S_k. It then follows (see Lévy [4, pp. 221–224]) that $r(t)$ must satisfy (4). Also observe that if μ is taken to be the uniform measure on S_k in (8), the covariance becomes

$$|t|^\alpha + |\tau|^\alpha - |t - \tau|^\alpha,$$

one discussed by Lévy [5].

Corollary 1. *If*

$$\overline{M}_T = \max\{|Y_T(t)|, 0 \leq t_i \leq T, \quad i = 1, \ldots, k\},$$

under the conditions of Theorem 1

$$P[\{2k \log T\}^{1/2}(\overline{M}_T - B(T)) < x] \to \exp[-(\lambda_1 + \lambda_2)\exp(-x)]. \tag{12}$$

Corollary 2. *The results of Theorem* 1 *and Corollary* 1 *hold for*

$$M_T = \sup\{Y_T(t): t/T \in R\} \qquad \textit{and} \qquad m_T = \inf\{Y_T(t): t/T \in R\}$$

when $\mathrm{vol}(R) = 1$ *and* R *is bounded and Jordan measurable.*

The case $\alpha = 2$ is probably that of greatest interest since the random field is then differentiable in mean square. Let Σ be the covariance matrix of the measure μ, that is

$$\Sigma = \{\sigma_{u,v}; \quad u, v = 1, \ldots, k\}, \qquad \sigma_{u,v} = \int \theta_u \theta_v \mu(d\theta).$$

Equation (4) can then be written

$$r(t) = 1 - t\Sigma t' + o(|t|^2)$$

with corresponding representations for formulas (7) and (8).

Corollary 3. *The constant H_α can be evaluated in the following special cases. If $\alpha = 2$ and the covariance matrix of the measure μ is Σ (nonsingular) then*

$$H_2 = \pi^{-k/2}\{\det(\Sigma)\}^{1/2}.$$

If $\alpha = 1$ and

$$r(t) = 1 - C(|t_1| + \cdots + |t_k|) + o\left(\sum_{i=1}^{k} |t_i|\right),$$

then

$$H_1 = C^k.$$

Details of the proof of Theorem 1. Theorem 1 is obtained very much as given in Pickands' Theorem 3.1 [6] or in [2]. We give a sketch of the details by a series of lemmas and do it for the case $k = 2$, since that is representative of the case of a multidimensional time parameter.

Lemma 1. *Let $\psi(x) = \phi(x)/x$ with ϕ the standard normal density. Set $C = 1$, $x = x(T) = B(T) + z_1/(4 \log T)^{1/2}$. Then for $a > 0$*

$$P[\max\{Y_T(t_1 + ajx^{-2/\alpha}, t_2 + akx^{-2/\alpha}), 0 \le j, k \le n\} > x]$$
$$= \psi(x) \exp\{b(t_1, t_2)\}H_\alpha(n, a) + o(\psi(x)), \tag{13}$$

$$P[\min\{Y_T(t_1 + ajx^{-2/\alpha}, t_2 + akx^{-2/\alpha}), 0 \le j, k \le n\} < -x]$$
$$= \psi(x) \exp\{-b(t_1, t_2)\}H_\alpha(n, a) + o(\psi(x)) \tag{14}$$

as $T \to \infty$ uniformly $0 \le t_1, t_2 \le T$ where

$$H_\alpha(n, a) = \int_{-\infty}^{\infty} e^s P[\max\{Y(ja, ka), 0 \le j, k \le n\} > s]\, ds. \tag{15}$$

Also, if $y = y(T) = B(T) + z_2/(4 \log T)^{1/2}$, then

$$P[\max\{Y_T(t_1 + ajx^{-2/\alpha}, t_2 + akx^{-2/\alpha}), 0 \le j, k \le n\} > x,$$
$$\min\{Y_T(t_1 + ajx^{-2/\alpha}, t_2 + akx^{-2/\alpha}), 0 \le j, k \le n\} < -y]$$
$$= o(\psi(x)) = o(\psi(y)) \tag{16}$$

uniformly in $0 \le t_1, t_2 \le T$. Here j, k take on only integer values and the Gaussian process Y is specified in the statement of Theorem 1.

Proof. Let

$$\tilde{Y}_T(u_1, u_2) = x(Y_T(t_1 + u_1x^{-2/\alpha}, t_2 + u_2x^{-2/\alpha}) - \mu_T(t_1, t_2) - x).$$

Then

$$P[\max\{Y_T(t_1 + ajx^{-2/\alpha}, t_2 + akx^{-2/\alpha}): 0 \le j, k \le n\} > x]$$

$$= \int_{-\infty}^{\infty} \gamma(z) P[\max\{\tilde{Y}_T(ja, ka): 0 \le j, k \le n\} > -x\mu_T(t_1, t_2) \mid \tilde{Y}_T(0,0) = z]\, dz$$

with γ the density of $Y_T(0, 0)$

$$\gamma(z) = \frac{1}{x}\phi\left(x + \frac{z}{x}\right) = \psi(x)\exp\left[\left(-z - \frac{z^2}{2x^2}\right]\right). \tag{17}$$

One wants to show that the finite-dimensional conditional distributions of $Y_T(u_1, u_2)$ given $Y_T(0, 0) = z$ converge uniformly in (u_1, u_2) to those of $Y(u_1, u_2) + z$ where the Gaussian process Y is as specified in Theorem 1. The argument is very much like that in the one-dimensional case but we sketch it at the risk of a bit of tedium. It is clear that the conditional distribution is Gaussian, so that it is enough to examine its mean and covariance properties. Let us look at the covariance matrix of $\tilde{Y}_T(u_1, u_2)$, $\tilde{Y}_T(v_1, v_2)$ and $\tilde{Y}_T(0, 0)$. The covariance matrix is of the form

$$x^2\begin{pmatrix} 1 & 1-a & 1-b \\ 1-a & 1 & 1-c \\ 1-b & 1-c & 1 \end{pmatrix} \tag{18}$$

with $a = k_1x^{-2} + o(x^{-2})$, $b = k_2x^{-2} + o(x^{-2})$, $c = k_3x^{-2} + o(x^{-2})$. The determinant is

$$x^6[1 + 2(1-a)(1-b)(1-c) - (1-b)^2 - (1-a)^2 - (1-c)^2]$$
$$= x^6[2ac + 2bc + 2ab - b^2 - a^2 - c^2 - 2abc],$$

which we will only need to consider up to second-order terms in a, b, c. The computation of cofactors for the entries in the upper left-hand 2×2 matrix of (18) is up to first-order terms

$$x^4\begin{pmatrix} 2c & a-b-c \\ a-b-c & 2b \end{pmatrix}.$$

Aside from the factor x^4 this has determinant

$$4bc - (a-b-c)^2 = 2ac + 2bc + 2ab - a^2 - b^2 - c^2,$$

so that asymptotically we get

$$\lim_{x\to\infty} x^2\begin{pmatrix} 2b & b+c-a \\ b+c-a & 2c \end{pmatrix}$$

as covariance matrix, which leads us to the desired covariance function. A simple computation using the conditional mean of one Gaussian variable given another determines the limiting mean.

Now

$$P[\max\{\tilde{Y}_T(ja, ka): 0 \le j, k \le n\} > -x\mu_T(t_1, t_2) \mid \tilde{Y}_T(0, 0) = z]$$
$$\to P[\max\{Y(ja, ka): 0 \le j, k \le n\} > -z - b(t_1, t_2)]$$

and so we have (13). Clearly (14) follows similarly. In looking at (16), let A be the event whose probability is to be estimated. Then

$$\begin{aligned}
&P(A, Y_T(t_1, t_2) > x - (1/\sqrt{x}) + \mu_T(t_1, t_2)) \\
&\quad \le \int_{-\sqrt{x}}^{\infty} \gamma(z) P[\min(\tilde{Y}_T(ja, ka), 0 \le j, k \le n) - z \\
&\qquad\qquad \le z - x(y + x + \mu_T(t_1, t_2)) \mid \tilde{Y}_T(0, 0) = z]\, dz \\
&\quad \le \psi(x) \int_{-\infty}^{\sqrt{x}} e^z P[\min(\tilde{Y}_T(ja, ka) + z, 0 \le j, k \le n) \\
&\qquad\qquad < z - x(y + x + \mu_T(t_1, t_2)) \mid \tilde{Y}_T(0, 0) = -z]\, dz \\
&\quad \le \psi(x) \sum_{j,k=0}^{n} \left\{ \int_{-\infty}^{0} P[\tilde{Y}_T(ja, ka) + z \right. \\
&\qquad\qquad < z - x(y + x + \mu_T(t_1, t_2)) \mid \tilde{Y}_T(0, 0) = z]\, dz \\
&\quad + \sqrt{x} \exp(\sqrt{x}) \max\{P[\tilde{Y}_T(ja, ka) + z \\
&\qquad\qquad \left. < \sqrt{x} - x(y + x + \mu_T(t_1, t_2)) \mid \tilde{Y}_T(0, 0) = -z], 0 \le z \le \sqrt{x}\} \right\}. \qquad (19)
\end{aligned}$$

Standard estimates show that the expression on the right-hand side of (19) is $o(\psi(x))$. Similarly

$$P(A,\ Y_T(t_1, t_2) < -y + (1/\sqrt{y}) + \mu_T(t_1, t_2)) = 0(\psi(y)).$$

Also

$$\begin{aligned}
&P(A, -y + (1/\sqrt{y}) \le Y_T(t_1, t_2) - \mu_T(t_1, t_2) \le x - (1/\sqrt{x})) \\
&\quad \le \int_{-\infty}^{-\sqrt{x}} \gamma(z) P[\max(\tilde{Y}_T(ja, ka), 0 \le j, k \le n) > x\mu_T(t_1, t_2) \mid \tilde{Y}_T(0, 0) = z]\, dz \\
&\quad \le \psi(x) \int_{M}^{\infty} e^z P[\max(Y_T(ja, ka), 0 \le j, k \le n) > z]\, dz
\end{aligned}$$

for each $M < \infty$ if T is sufficiently large. The proof of Lemma 1 is complete.

Lemma 2. *The statement of Lemma* 1 *is valid if* $a = 1$ *and* j, k *are allowed to range continuously over all values in* $[0, n]$ *with* $H_\alpha(n, a)$ *replaced by*

$$\bar{H}_\alpha(n) = \int_{-\infty}^{\infty} e^s P[\max\{Y(u_1, u_2), 0 \le u_1, u_2 \le n\} > s]\, ds. \qquad (20)$$

By arguing as in Lemma A2 of [2] and using Lemma 2 of Kuelbs [3], the result follows.

The following analogue of Pickands' Lemma 2.5 [6] can be shown to hold by an argument like his.

Lemma 3. *If* (4) *holds with* $C = 1$, *then*

$$P\left[\max\left\{Y_T(t_1 + jax^{-2/\alpha}, t_2 + kax^{-2/\alpha}) : 0 \le j, k \le \left[\frac{x^{2/\alpha}\tau}{a}\right]\right\} > x\right]$$

$$= x^{4/\alpha}\psi(x)\,\frac{H_\alpha(a)}{a^2}\int_{t_1}^{t_1+\tau}\int_{t_2}^{t_2+\tau} \exp\{b(u, v)\}\,du\,dv + o(x^{4/\alpha}\psi(x)) \tag{21}$$

uniformly in $0 \le t_1, t_2 \le T$ *with*

$$H_\alpha(a) = \lim_{n\to\infty} \frac{H_\alpha(n, a)}{n^2}. \tag{22}$$

We continue with another result closely following the format given to a corresponding discussion in [2].

Lemma 4. *With* $H_\alpha(n, a)$, $H_\alpha(a)$, *and* $\bar{H}_\alpha(n)$ *defined in* (15), (22), *and* (20) *we have*

$$0 < H_\alpha = \lim_{a\to 0} \frac{H_\alpha(a)}{a^2} = \lim_{n\to\infty} \frac{\bar{H}_\alpha(n)}{n^2}. \tag{23}$$

First let $0 < \alpha < 2$. For $\gamma > 0$ set

$$\begin{aligned}\bar{H}_\alpha(n, \gamma) &= \int_{-\infty}^{\infty} e^s P\left[\max_{0 < t_1, t_2 \le n} Y(t_1, t_2) > s + \gamma\right] ds \\ &= e^{-\gamma}\bar{H}_\alpha(n).\end{aligned}$$

Then

$$\begin{aligned}&\frac{1}{n^2}\,|\bar{H}_\alpha(na, \gamma) - H_\alpha(n, a)| \\ &\quad\le \frac{1}{n^2}\int_{-\infty}^{\infty} e^s P\left[\max_{0 \le t_1, t_2 \le na} Y(t_1, t_2) > s + \gamma,\ \max_{0 \le j, k \le n} Y(ja, ka) \le s\right] ds \\ &\quad+ \frac{1}{n^2}\int_{-\infty}^{\infty} e^s P\left[s < \max_{0 \le t_1, t_2 \le na} \bar{Y}(t_1, t_2) \le s + \gamma\right] ds \\ &\quad\le \frac{1}{n^2}\sum_{j,k=0}^{n-1}\int_{-\infty}^{\infty} P\left[Y(ja, ka) \le s,\ \max_{\substack{ja \le t_1 \le (j+1)a \\ ka \le t_2 \le (k+1)a}} Y(t_1, t_2) > s + \gamma\right] ds \qquad (24) \\ &\quad+ \frac{1}{n^2}\,|\,\bar{H}_\alpha(na) - \bar{H}_\alpha(na, \gamma)\,|\end{aligned}$$

If the summands on the right of (24) are denoted by $A(j, k; \gamma, a)$, then

$$A(j, k; \gamma, a) = \int_{-\infty}^{\infty} e^s \int_{-\infty}^{s} f(z; ja, ka)$$

$$\times P\left[\max_{0 \le t_1, t_2 \le a} Y(t_1 + ja, t_2 + ka) > s + \gamma \,|\, Y(ja, ka) = z\right] dz\, ds$$

where $f(z; ja, ka)$ is the density of $Y(ja, ka)$. For convenience let us write

$$E(Y(t)) = -|t|^\alpha g(t).$$

Then

$$A(j,k; \gamma, a) = \int_{-\infty}^{\infty} \phi(\omega) \int_0^{\infty} P\left[\max_{0 \le t_i \le a} (Y(t_1 + ja, t_2 + ka) - Y(ja, ka)) > s + \gamma \,|\right.$$

$$Y(ja, ka) = 2^{1/2}\{(ja)^2 + (ka)^2\}^{\alpha/4} g(ja, ka)^{1/2} \omega$$

$$\left. + \{(ja)^2 + (ka)^2\}^{\alpha/2} g(ja, ka)\right] ds\, d\omega.$$

As $j, k \to \infty$, the finite-dimensional conditional distributions of

$$Y(t_1 + ja, t_2 + ka) - Y(ja, ka),$$

given

$$Y(ja, ka) = 2^{1/2}[(ja)^2 + (ka)^2]^{\alpha/4} g(ja, ka)^{1/2}\omega + [(ja)^2 + (ka)^2]^{\alpha/2} g(ja, ka),$$

tend for each ω to those of $Y(t_1, t_2)$, $0 \le t_i \le a$.

It is in the step of the argument we have just carried out that the assumption $0 < \alpha < 2$ is required. We argue just as in Lemma 1 to show that

$$\lim_{j,k \to \infty} A(j, k; \gamma, a) = A(\gamma, a) = \int_0^{\infty} e^s P\left[\max_{0 \le t_i < a} Y(t_1, t_2) > s + \gamma\right] ds.$$

We let $Y^*(t_1, t_2) = Y(t_1, t_2) + |(t_1, t_2)|^\alpha g(t_1, t_2)$, so that

$$A(\gamma, \alpha) \le \int_0^{\infty} \exp(s) P\left[\max_{0 \le t_i < a} Y^*(t_1, t_2) > s + \gamma\right] ds$$

$$= \int_0^{\infty} \exp(s) P\left[\max_{0 \le t_i < 1} Y^*(t_1, t_2) > (s + \gamma) a^{-\gamma/2}\right] ds$$

$$= a^{\alpha/2} \exp(-\gamma) \int_{\gamma a^{-\alpha/2}}^{\infty} \exp(\omega a^{\alpha/2}) P\left[\max_{0 \le t_i \le 1} Y^*(t_1, t_2) > \omega\right] d\omega \qquad (25)$$

and using Fernique's estimate [2], the right-hand side of (25) is $O(\exp(-a^{-\alpha/2}))$ for every $\gamma > 0$. Thus

$$\lim_a \lim_n \frac{1}{n^2 a^2} [\bar{H}_\alpha(na, \gamma) - H_\alpha(n, a)] = 0$$

for every $\gamma > 0$. Since

$$P\left[\max_{0 \le t_i \le n} Y(t_1, t_2) > s\right]$$

$$\le \sum_{j,k=0}^{n-1} P\left[\max_{\substack{j \le t_1 < j+1 \\ k \le t_2 < k+1}} Y(t_1, t_2) > s\right]$$

$$\le \sum_{j,k=0}^{n-1} \left\{P\left[Y(j,k) \le s, \max_{\substack{j \le t_1 < j+1 \\ k \le t_2 < k+1}} Y(t_1, t_2) > s\right] + P[Y(j,k) > s]\right\},$$

one can see that

$$\sup_n \frac{\bar{H}_\alpha(n)}{n^2} < \infty.$$

It follows that

$$\lim_{a \to 0} \overline{\lim_{n \to \infty}} (n^2 a^2)^{-1} |H_\alpha(n, a) - \bar{H}_\alpha(na)| = 0.$$

Since

$$H_\alpha(a) = \lim_{n \to \infty} (1/n^2) H_\alpha(n, a)$$

is well defined and finite for each $a > 0$, it follows that

$$\underline{\lim}_{n \to \infty} \frac{\bar{H}_\alpha(na)}{n^2 a^2} = \overline{\lim_{n \to \infty}} \frac{\bar{H}_\alpha(na)}{n^2 a^2}$$

is well defined and finite. We call this limit H_α. But then

$$\lim_{a \to 0} \frac{H_\alpha(a)}{a^2} = H_\alpha \qquad \text{for} \quad 0 < \alpha < 2.$$

It is clear that $H(n, a) > 0$ if $a > 0$. However, since $H(nm, a/m) \ge H(n, a)$, it follows that

$$H_\alpha(a/m) \ge (1/m^2) H_\alpha(a) \tag{26}$$

with $H_\alpha(a) > 0$ for a sufficiently large by an argument like that given by Pickands' Lemma 24 [6]. But (26) then implies that $H_\alpha(a) > 0$ for all $a > 0$ and further that

$$0 < H_\alpha = \lim_{a \to \infty} \frac{H_\alpha(a)}{a^2}.$$

Lemmas 3 and 4 can be used to obtain the following result very much as in [2].

Lemma 5. *If* (4) *holds with* $C = 1$, *then*

$$P[\max\{Y_T(t_1+u, t_2+v)\colon 0 \le u, v \le \tau\} > x]$$
$$= x^{4/\alpha}\psi(x) \int_{t_1}^{t_1+\tau} \int_{t_2}^{t_2+\tau} \exp\{b(u,v)\}\, du\, dv\, H_\alpha + o(x^{4/\alpha}\psi(x))$$

uniformly in $0 \le t_1, t_2 \le T$. *A similar assertion holds for*

$$P[\min\{Y_T(t_1+u, t_2+v)\colon 0 \le u, v \le \tau\} < -x]$$

with $-b$ *replacing* b. *Also,*

$$P[\max\{Y_T(t_1+u, t_2+v)\colon 0 \le u, v \le \tau\} > x,$$
$$\min\{Y_T(t_1+u, t_2+v)\colon 0 \le u, v \le \tau\} < -y] = o(x^{4/\alpha}\psi(x)).$$

We now continue with the proof of Theorem 1. Break the interval $[0, T]$ up into $2N$ intervals. N of these $W_1, \ldots, W_N$ are of length τ and the remaining N $V_1, \ldots, V_N$ of length ε. They alternate, with V_i following W_i, which in turn follows V_{i-1}. Now $N \sim T/(\tau + \varepsilon)$. If x is as given in Lemma 1,

$$x^{4/\alpha}\psi(x)H_\alpha \sim (1/T^2)\exp(-z_1).$$

By Lemma 5

$$P\left[\max\left\{Y_T(t)\colon t \in \bigcup_{j=1}^{N} V_j \times [0, T]\right\} \ge x\right]$$
$$\le \sum_{j,k=1}^{N} (P[\max\{Y_T(t)\colon t \in V_j \times V_k\} \ge x] + P[\max\{Y_T(t)\colon t \in V_j \times W_k\} \ge x])$$
$$\sim \left[\sum_{j,k=1}^{N} \left(\int_{V_j}\int_{V_k} + \int_{V_j}\int_{W_k}\right) \exp\{b(u,v)\}\, du\, dv\right] \frac{\exp(-z_1)}{T^2}$$
$$= \varepsilon O\left(\frac{N}{T}\right) = \varepsilon O(1)$$

with the O term independent of ε and the V_j. A similar result holds for $\min\{Y_T(t)\colon t \in \bigcup_{j=1}^{N} V_j \times [0, T]\}$. We want to show that

$$\lim_{\varepsilon\to 0} \overline{\lim_{T\to\infty}} P\left[\max\left\{Y_T(t)\colon t \in \bigcup_{j,k=1}^{N} W_j \times W_k\right\} \le x,\right.$$
$$\left.\min\left\{Y_T(t)\colon t \in \bigcup_{j,k=1}^{N} W_j \times W_k\right\} \ge -y\right]$$
$$= \exp[-\lambda_1 \exp(-z_1) - \lambda_2 \exp(-z_2)] \tag{27}$$

where λ_1, λ_2 are as given in (11). Choose an $a > 0$. Let $W_j = (c_j, c_j + t)$,

$j = 1, \ldots, N$. By Lemmas 3 and 5 it follows that

$$\Bigg| P\left[\max\left\{Y_T(t) : t \in \bigcup_{r,s=1}^{N} W_r \times W_s\right\} \le x\right] - P\left[Y_T(c_r + jax^{-2/\alpha}, c_s + kax^{-2/\alpha}) \le x : 0 \le j, k \le \left[\frac{\tau x^{2/\alpha}}{a}\right], 1 \le r, s \le N\right]\Bigg|$$

$$\le \sum_{r,s=1}^{N} \Bigg| P[\max\{Y_T(t) : t \in W_r \times W_s\} \le x] - P\left[Y_T(c_r + jax^{-2/\alpha}, c_s + kax^{-2/\alpha}) \le x : 0 \le j, k \le \left[\frac{\tau x^{2/\alpha}}{a}\right]\right]\Bigg|$$

$$\sim \sum_{r,s=1}^{N} \int_{W_r}\int_{W_s} \exp\{b(u, v)\}\, du\, dv\, x^{4/\alpha}\psi(x) \left| H_\alpha - \frac{H_\alpha(a)}{a^2}\right| \exp(-z_1).$$

A similar result holds for

$$P\left[\min\left\{Y_T(t) : t \in \bigcup_{r,s=1}^{N} W_r \times W_s\right\} \ge -y\right].$$

To show (27) it is enough to verify that

$$\lim_{a \to 0} \lim_{\varepsilon \to 0} \overline{\lim_{T \to \infty}}\, P\Bigg[-y \le Y_T(c_r + jax^{-2/\alpha}, c_s + kax^{-2/\alpha}) \le x : 1 \le r, s \le N, 0 \le j, k \le \left[\frac{\tau x^{2/\alpha}}{a}\right]\Bigg] = \exp[-\lambda_1 \exp(-z_1) - \lambda_2 \exp(-z_2)].$$

From Lemma 5 it follows that

$$\overline{\lim_{T \to \infty}} \sum_{r,s=1}^{N} \left(1 - P\left[-y \le Y_T(c_r + jax^{-2/\alpha}, c_s + kax^{-2/\alpha}) \le x : 0 \le j, k \le \left[\frac{\tau x^{2/\alpha}}{a}\right]\right]\right)$$

$$= \frac{H_\alpha(a)}{H_\alpha a^2} \overline{\lim_{T \to \infty}} \frac{1}{T^2} \sum_{r,s=1}^{N} \int_{W_r}\int_{W_s} \{\exp[b(u, v) - z_1] + \exp[-b(u, v) - z_2]\}\, du\, dv.$$

This implies that

$$\lim_{a \to 0} \lim_{\varepsilon \to 0} \overline{\lim_{T \to \infty}} \sum_{r,s=1}^{N} \left(1 - P\left[-y \le Y_T(c_r + jax^{-2/\alpha}, c_s + kax^{-2/\alpha}) \le x : 0 \le j, k \le \left[\frac{\tau x^{2/\alpha}}{a}\right]\right]\right) = \lambda_1 \exp(-z_1) + \lambda_2 \exp(-z_2). \quad (28)$$

Let $E_{r,s}$, $r, s = 1, \ldots, N$, be the events whose probabilities are summed in (28). If the events $E_{r,s}$ were independent, (27) would follow from (28). Let $\bar{P}$ be the measure which makes random variables $Y_T(t)$ with t belonging to distinct sets $W_r \times W_s$ independent but otherwise agreeing with P. An appropriate modification of the argument given by Berman [1] under the one-dimensional version of condition (vi) gives

$$\lim_{\varepsilon \to 0} \lim_{T \to \infty} \left| (P - \bar{P})\left(\bigcap_{r,s=1}^{N} E_{r,s} \right) \right| = 0.$$

Proof of Corollary 2. The argument goes through as before in the case of Theorem 1 by approximating R within and without by finite unions of k-dimensional cubes.

Proof of Corollary 3. First consider the case $\alpha = 2$ with $\Sigma = I$ the identity matrix. Then

$$Y(t) = Y(t_1, \ldots, t_k) = X_1(t_1) + \cdots + X_k(t_k)$$

where the $X_i(t_i)$, $i = 1, \ldots, k$, are independent Gaussian with

$$E(X_i(t_i)) = -t_i^2, \quad \operatorname{cov}(X_i(t_i), X_i(\tau_i)) = t_i^2 + \tau_i^2 - (t_i - \tau_i)^2.$$

However, one already knows that in the case of processes $X_i(t_i)$ with one-dimensional time parameter $\bar{H}_2 = \pi^{-1/2}$. Since

$$\max_{\substack{0 \le t_i \le T \\ i = 1, \ldots, k}} Y(t) = \sum_{i=1}^{k} \max_{0 \le t_i \le T} X_i(t_i)$$

it follows from (6) that for Y

$$H_2 = (\bar{H}_2)^k = \pi^{-k/2}.$$

A similar argument involving such a reduction to the case of one-dimensional processes shows that $H_1 = C^k$ if

$$r(t) = 1 - C(|t_1| + \cdots + |t_k|) + o\left(\sum_{i=1}^{k} |t_i| \right).$$

Let us now return to the case $\alpha = 2$ with general nonsingular Σ. It is easy to show that for regions such as those specified in Corollary 2

$$\lim_{T \to \infty} \frac{1}{T^n} \int_0^\infty e^s P\left[\max_{u \in TR} \bar{Y}(u) > s \right] ds = H_\alpha.$$

Then

$$H_\alpha = \operatorname{vol}(I_k \sqrt{\Sigma}) \lim_{T \to \infty} [T^k \operatorname{vol}(I_k \sqrt{\Sigma})]^{-1} \int_0^\infty e^s P\left[\max_{u \in TI_k\sqrt{\Sigma}} \bar{Y}(u) > s \right] ds$$

where $\sqrt{\Sigma}$ is a square root of Σ and $\bar{Y}(u) = Y(t)$ with $u = t\sqrt{\Sigma}$ and I_k is the unit k-cube. Here Y is the original process and $\bar{Y}$ a process with the covariance matrix of its measure μ the identity matrix. By using this reduction to the identity matrix we obtain $H_2 = \pi^{-k/2}\{\det(\Sigma)\}^{1/2}$. Notice that by applying a 45 degree rotation in the case $\alpha = 1$ with $k = 2$ as in the manner just indicated one reduces the problem of

$$r(t) = 1 - C\max\{|t_1|, |t_2|\} + o(\max\{|t_1|, |t_2|\})$$

to

$$r(t) = 1 - \frac{C}{\sqrt{2}}\{|t_1| + |t_2|\} + o(|t_1| + |t_2|)$$

and obtains $H_1 = C^2/2$. Linear transformations can generally be used in this way to extend the range of processes for which one can evaluate H_α.

REFERENCES

1. Berman, S. M. (1971). Asymptotic independence of the numbers of high and low level crossings of stationary Gaussian processes. *Ann. Math. Statist.* **42** 927–945.
2. Bickel, P. J. and Rosenblatt, M. (1973). On some global measures of the deviations of density function estimates. *Ann. Math. Statist.*
3. Kuelbs, J. (1968). The invariance principle for a lattice of random variables. *Ann. Math. Statist.* **39** 382–389.
4. Lévy, P. (1937). *Théorie de l'Addition des Variables Aléatoires.* Gauthier-Villars, Paris.
5. Lévy, P. (1948). Processus stochastiques et le mouvement Brownien. Paris.
6. Pickands, III, J. S. (1969). Upcrossing probabilities for stationary Gaussian processes. *Trans. Amer. Math. Soc.* **145** 51–73.

Concepts of Consistency in Spectral Estimation for Multivariate Time Series

F. EICKER

ABTEILUNG STATISTIK
UNIVERSITAT DORTMUND

1. INTRODUCTORY REMARKS

Since the probabilistic theory of multivariate (or multiple) stationary time series (with univariate time parameter) is a straightforward and natural extension of that of univariate stationary time series, most of the basic problems of the statistical inference concerning multivariate processes are almost exactly the ones familiar from the one-dimensional case. Since the methods of inference actually in use are often based on personal beliefs, prior experience, or computational advantages rather than on generally accepted statistically desirable properties (compare, e.g., the discussion of Tukey [20, p. 321] and Parzen [14, p. 396 ff.]), we must explain the attitude assumed in this paper, which is motivated by practical rather than purely theoretical considerations. We concentrate on the basic property of consistency and shall primarily contribute some comments on the alternative of estimating the spectral distribution function (df) or the spectral density function. We do not consider here, e.g., the practically relevant questions of how to obtain asymptotic distributions or higher-order moments of the estimators (from among the many papers we draw attention only to the note by Leppink [13]). The aim here is to obtain widely applicable (robust) and good methods; i.e., the assumptions should be kept as general as possible.

The main conclusion reached is that it is the spectral df F, rather than its derivative f (if existent), that one should estimate. This meets the needs of the applier in the best way, for various reasons which are explained later, although it seems not to be the usually practiced procedure. While the functions F and f (if F is absolutely continuous) are completely equivalent probabilistically, their standard estimates behave quite differently statistically. In order to judge the statistical quality of an estimate and to select some appropriate optimality criterion, we must first ponder carefully the probabilistic reason why spectral analysis is done at all in a particular situation. Some of the arguments are evaluated in Section 2.

Statistical analysis of numerical time series data in many cases consists in answering primarily two principally different questions. The first one is to find a satisfactory probabilistic model (model building); the subsequent second one is to specify its parameters (statistical inference). The interplay between both is close, of course: step two is simply impossible if step one has not been done; on the other hand, the suitability of a model cannot be judged without data fitting resulting in the statistical determination of the parameters of the model. The solution of the more experimental step one may depend on nonmathematical, even subjective, reasons varying widely from case to case (e.g., choice of time domain or frequency domain description), so that a systematic treatment, not to speak of a general theory of model building, is impossible. Step two, however, is a matter of statistical theory and we therefore can give a general treatment. Yet because of its strong causal interplay with model building it is not completely free from nonstatistical arguments. Since they may be contradictory in part, statistical inference has to treat different procedures side by side.

It seems interesting, however, that regarding the special problem of this note, namely, the comparison of spectral distribution with spectral density estimation, in the general case the first seems preferable to the second. One of the facts stressing the practical preferability of estimating F is the relative simplicity and the unambiguity of the standard estimator and the generality of the assumptions needed to establish desirable properties. Thus, e.g., the delicate choice of a spectral window is completely eliminated (compare, e.g., Tukey [20], and Jenkins and Watts [12]).

In order to keep the paper concise and to avoid some technicalities, we restrict ourselves to the discrete time case. But as is well known, the continuous time case is treated principally in the same way under a few new assumptions (such as mean square continuity).

The list at the end of the paper gives only some of the relevant papers; it does not aim at completeness.

2. SHOULD ONE ESTIMATE THE SPECTRAL DISTRIBUTION FUNCTION OR ITS DERIVATIVE?

Throughout this paper let $x_n := (x_{n1}, \ldots, x_{np})'$ ($n \in \mathbb{Z}$ = the set of integers) be a weakly stationary p-dimensional discrete-parameter stochastic process with real-valued square integrable components x_{nj} (for detailed definitions and the basic theory compare, e.g., Hannan [9]). The $p \times p$ covariance matrix R of this process allows a Fourier–Stieltjes transformation: for all $k, n \in \mathbb{Z}$ there holds

$$R_n := E(x_{k+n}x_k') = \int_{-\pi}^{\pi} e^{in\lambda}\, dF(\lambda)$$

where the $p \times p$ spectral distribution matrix F defined on $[-\pi, \pi]$ satisfies $F(-\pi) = 0$-matrix, $\Delta_{uv}F := F(v) - F(u)$ is a Hermitian nonnegative definite matrix for all $-\pi \leq u < v \leq \pi$, $\operatorname{tr}(F(\pi) - F(-\pi)) < \infty$. Hence the elements of F are complex-valued measures of bounded variation. F may be taken to be continuous from the right (componentwise). Integrals like the one above are to be understood as matrices of integrals. The proof of this and similar theorems is by reduction to the scalar stationary process $a'x_n$ ($a \in R^p$ any constant vector) and by application of the corresponding univariate theorem (for a lucid presentation of the latter compare, e.g., Doob [3, Chapter X]).

The inversion of the above Fourier–Stieltjes transformation is

$$F(v) - F(u) = (v - u)R_0 - (2\pi i)^{-1} \lim_{N\to\infty} \sum_{\substack{-N \\ n\neq 0}}^{N} R_n(e^{-inv} - e^{-inu})/n \tag{2.1}$$

for continuity points $v > u$ of F. At a jump point y, $F(y)$ is to be replaced by the arithmetic mean of the left and right limit of F at y. A simple and natural estimator of F is obtained by essentially replacing the true covariances by sample covariances in (2.1). Some of its basic properties in the univariate case are pointed out by Doob [3, p. 496]. His theorem immediately generalizes to the multivariate case (Theorem 3.1 below).

For the physical interpretation of the spectral function as well as for the (probabilistic and practical) understanding of a stationary process its spectral representation is crucial

$$x_n = \int_{-\pi}^{\pi} e^{in\lambda}\, dy(\lambda) \tag{2.2}$$

where $\{y(\lambda), -\pi \leq \lambda \leq \pi\}$ is a p-dimensional complex-valued vector process, determined through an inversion formula by the x_n, with orthogonal increments and

$$E(y(\lambda)y^H(\lambda)) = F(\lambda), \qquad E(dy(\lambda)\, dy^H(\lambda)) = dF(\lambda);$$

the integral is defined (componentwise) as the L^2-limit of the approximating Riemann–Stieltjes sums. (We choose the symbol λ rather than ω as the spectral variable since ω is the standard symbol for a point in the sample space, i.e., ω stands for a realization of the process.) The latter give the simple basic reason why spectral analysis is done for stationary processes and why, consequently, statisticians want to estimate the spectral df: for with

$$\lambda_0 = -\pi < \lambda_1 < \cdots < \lambda_N = \pi$$

and

$$\Delta_{uv}y = y(v) - y(u), \qquad \Delta_k y = y(\lambda_k -) - y(\lambda_{k-1} -),$$

the sums

$$\sum_{k=1}^{N} \exp(in\lambda_k{}') \, \Delta_k y, \; E(\Delta_k y(\Delta y_k)^H) = \Delta_k F \qquad (\lambda_{k-1} \le \lambda_k{}' \le \lambda_k) \qquad (2.3)$$

approximate in a sensible way the $\{x_n\}$ process in terms of the orthogonal $\Delta_k y$ random variables for $N \to \infty$. The n-range of approximated x_n's and the approximation of an individual x_k for k fixed becomes better the larger N and the finer the division $\{\lambda_k\}$ is. The "total mean squared amplitude" of the kth summand is given by $\Delta_k F$. They all add up to

$$E\|x_n\|^2 = \operatorname{tr} R_0 = \operatorname{tr} \int_{-\pi}^{\pi} dF(\lambda).$$

In many applications one is interested in detecting periodicities in the observed realization $x.(\omega)$ of the stationary process. In mathematical terms this means that a Fourier–Stieltjes representation of the particular path is wanted. It is given by (2.2) taken at ω (if the integral is interpreted appropriately). A useful sequence of approximations to this representation is given by the random trigonometric sums (2.3). They show immediately the approximate periods $\lambda_k{}'$ involved and the associated random amplitudes $\Delta_k y$. A measure for their magnitude is easily obtained from the increments $\Delta_k F$ of F. Thus it is clear that it is primarily the spectral df that is of practical importance and not the spectral density.

This remains true also if two closely neighboring relative maxima of f are to be distinguished, in which case the statistical analysis should have high resolution in a certain interval of the spectrum. Of course, the maxima of the graph of the true f are depicted equivalently by large increments in the graph of F. However, the latter picture may appear less dramatic to some, and this may be the reason why it is traditional to use the graph of f.

In this context it should be recalled that it is rarely if ever the peak frequencies of the true density that are of interest as long as jumps of F (or delta function components of f) are excluded. (The absence of the latter should always be assumed in order to eliminate clearly regression analysis. If necessary, a regression analysis should precede the spectral analysis.) The peak frequencies themselves in a spectrum of an absolutely continuous F are of hardly any practical interest. The interesting thing is rather to separate (if possible) two disjoint intervals of strong increase of F. A good estimate of F shows such separation early and reliably.

Returning briefly again to formula (2.3), one notes that it is even possible to control the approximation error due to the isometry of two Hilbert spaces

H' and H^* which underlies the definition of the above stochastic integral (compare Doob [3, pp. 426–429]); restricting ourselves for the moment to the scalar case, we have $H' \subset L^2(\Omega, \mathscr{A}, P)$, where $(\Omega, \mathscr{A}, P)$ is the probability space underlying the x_n and $y(\lambda)$ processes. H' is spanned by the x_n, and $H^* = L^2\ (\mathbb{R}^1, \mathscr{B}^1, \mu_F)$ where $(\mathbb{R}^1, \mathscr{B}^1)$ is the Borel line and μ_F the matrix-valued signed measure induced by F on it. In the isometry we have the correspondences

$$x_n \leftrightarrow e^{ins}, \qquad y(\lambda) \leftrightarrow 1(-\pi \leq s \leq \lambda) \qquad (-\pi \leq s \leq \pi), \tag{2.4}$$

$1(A)$ being the indicator function of the set A.

Hence $\sum_k \exp(in\lambda_k')\Delta_k y \leftrightarrow \sum_k \exp(in\lambda_k')1(\lambda_{k-1} < s \leq \lambda_k)$ and by isometry

$$\begin{aligned} E\Big|x_n - \sum_k \exp(in\lambda_k')\Delta_k y\Big|^2 &= \int_{-\pi}^{\pi} \Big|\exp(ins) - \sum_k \exp(in\lambda_k')1(\lambda_{k-1} < s \leq \lambda_k)\Big|^2\, dF(s) \\ &= 2\left(R_0 - \operatorname{Re}\left\{\sum_k \int_{\lambda_{k-1}}^{\lambda_k} \exp[in(s - \lambda_k')]\, dF(s)\right\}\right) \\ &= 2\left(R_0 - \sum_k \int_{\lambda_{k-1}}^{\lambda_k} \cos(n(s - \lambda_k'))\, dF(s)\right), \end{aligned} \tag{2.5}$$

which tends to zero due to Lebesgue's dominated convergence theorem and the equicontinuity and uniform convergence to zero of the integrands.

A rough upper bound for the approximation error is (using $\cos x \geq 1 - x^2/2$)

$$n^2 \sum_k \int_{\lambda_{k-1}}^{\lambda_k} (s - \lambda_k')^2\, dF(s) \leq (2\pi n/N)^2$$

if the λ_k are chosen equidistant: $\lambda_k - \lambda_{k-1} = 2\pi/N$. On a limited range of n this can be made small if N is chosen large enough.

Continuing the analysis of the beginning of this section, one obtains additional insight into the role of the function F after splitting it up into its three components:

$F_1 = \int f$ the absolutely continuous part with spectral density f;
$F_2 =$ the pure jump (discontinuous) part; and
$F_3 =$ the continuous singular part (with respect to Lebesgue measure).

The last does not seem to have any practical importance, and we shall henceforth assume that it vanishes. Correspondingly, x_n can be represented as a sum of orthogonal processes $x_n^{(1)}$ and $x_n^{(2)}$. If $\lambda_1, \lambda_2, \ldots$ is an enumeration of the jump points of F_2, so that the jumps $\Delta_k F$ at λ_k form a nonincreasing sequence, then in analogy to (2.2)

$$x_n^{(2)} = \sum_{k=1}^{\infty} \exp(in\lambda_k)\, \Delta_k y \qquad (\Delta_k y := y(\lambda_k +) - y(\lambda_k -)) \tag{2.6}$$

with $\{\Delta_k y\}$ an orthogonal sequence and

$$E[\Delta_k y(\Delta_k y)^H] = \Delta_k F. \tag{2.7}$$

The approximation error between a finite sum and $x_n^{(2)}$ is easily computed to be

$$E\left\|x_n^{(2)} - \sum_{k=1}^{N} \exp(in\lambda_k)\, \Delta_k y\right\|^2 = \operatorname{tr} \sum_{k=N+1}^{\infty} \Delta_k F \to 0 \tag{2.8}$$

as $N \to \infty$.

Obviously, both the finite and the infinite sum may be considered as a trigonometric regression with time-independent random coefficients $\Delta_k y$ and regressor functions $(\exp(in\lambda_k))_{n \in N}$, $k = 1, 2, \ldots$, the infinite sum being the best possible such regression. It is the "deterministic" part of the process x_n. In certain applications it might be thought of as the signal while $x_n^{(1)}$ is the noise, a purely indeterministic process in the sense of prediction theory.

The $x_n^{(1)}$ can be approximated similarly to (2.3) by approximating f by a matrix-valued finite step function. However, a more enlightening representation of $x_n^{(1)}$ is by a moving average process

$$x_n^{(1)} = \underset{N \to \infty}{\text{l.i.m.}} \sum_{j=-N}^{N} C_j z_{n+j} \tag{2.9}$$

where the $\{z_n\}$ form an orthonormal p-dimensional random sequence (i.e., $E(z_n z_m^H) = \delta_{nm} I_p$) and the C_j are $p \times p$ matrices satisfying

$$\operatorname{tr} \sum_{j=-\infty}^{\infty} C_j C_k^H < \infty \tag{2.10}$$

and

$$f(t) = \left(\underset{N\to\infty}{\text{l.i.m.}} \sum_{j=-N}^{N} C_j e^{ij\lambda}\right)\left(\underset{N\to\infty}{\text{l.i.m.}} \sum_{j=-N}^{N} C_j^H e^{-ij\lambda}\right), \tag{2.11}$$

i.e., the matrix elements $C_j^{k,m}$ (k, m fixed, $j \in Z$) are the Fourier coefficients of the (k, m)-element of $\sqrt{f}$. The C_j, like f and the unique positive semidefinite square root of f, are Hermitian (comp. Appendix Hannan [9]). Since $\sum_{j=0}^{+N} C_j z_{n+j}$ are both Cauchy sequences, (2.9) can be replaced unambiguously by

$$x_n = \sum_{j=-\infty}^{\infty} C_k z_{n+j}, \qquad n \in N. \tag{2.12}$$

The important practical meaning of the above moving average representation —two-sided infinite in general—is that its finite partial sums obviously average successively segments of an infinite random sequence of a very simple covariance structure; moreover, these partial sums approximate the x_n uniformly well in the L^2 sense, and the mean square error is bounded by

$$\operatorname{tr} \sum_{|j|>N} C_j C_j^H \qquad \text{for all } n.$$

To summarize, spectral analysis allows simply structured, uniformly good approximations to the observed process, and it seems to be principally this structure that the applier is looking for (compare, e.g., Gikhman and Shorohod [6a, p. 29], and Tukey [20], p. 300]). However, the Fourier representation of the covariance sequence as given in the first formula of Section 2, although much more vague than the representation (2.2), may also prove sufficiently informative concerning the periodicity behavior of the process. The question, however, whether the direct analysis of the R_n-sequence (time domain analysis) is always or in some special cases preferable to the frequency domain analysis probably does not have a general answer. It may have to be decided upon by experimenting with the time series data.

In prediction theory, spectral analysis plays an important role which, however, is not being discussed here. Actually it may be that the spectral density f itself, is needed here.

Another vague motivation for working in practice with f instead of F may be seen in the following. The representation (2.3) of a general stationary process may be written in case $F'(\lambda) = C^2(\lambda)$ exists (compare the proof of Lemma A1) as

$$x_n = \int_{-\pi}^{\pi} e^{in\lambda} C(\lambda)\, d\tilde{y}(\lambda) \tag{2.13}$$

where the $\tilde{y}(\lambda)$ process has the simplest possible covariance structure:

$$E[d\tilde{y}(\mu)(d\tilde{y}(\lambda))^H] = \delta_{\mu\lambda} I_p\, d\lambda. \tag{2.14}$$

Using this formula, approximation (2.3) becomes particularly suggestive, and here, if anywhere, may lie a valid reason for estimating F'. This is especially true in the multivariate case because of (2.14). However, any spectral representation like (2.13) in practice is interesting only regarding the qualitative structure of the process, i.e., it is impossible to obtain quantitative (L^2) approximations from limited sets of data unless very specific additional assumptions about the probability law governing the process are made (despite the available inversion formula expressing $y(\lambda)$ in terms of the x_n). For computation of the approximation error would require the (unavailable) knowledge of the complete covariance sequence of $\{x_n\}$.

Similarly, the practical value of the moving average representation seems to lie in the structural statement rather than in the possibility of quantitative utilization. It should also be pointed out that in case specification of the functions in formula (2.13) is the aim of the statistical analysis, it is the square root of the spectral density function F' that must be estimated rather than F'. This fact influences, e.g., the choice of the distance between F' and its estimator.

Regarding linear filters there does not seem to be any necessity to work with spectral densities, since the product of the transfer function with a spectral df already gives the spectral df of the filtered process.

Although in principle jumps of F could be admitted, we shall *assume throughout the paper that F does not have jumps*, and this assumption seems to be called for in any statistical spectral analysis (as distinct from regression analysis) for a stationary process based on one single path only, for the following simple reason. If $\Delta F > 0$ is a jump at some frequency, then according to representation (2.12) of the x_n process, there is in general exactly one random variable z_k carrying information about ΔF, according to $\Delta F = E(z_k\, z_k^H)$. Consequently, if z_k could indeed be somehow filtered out, the single path yields just one observation on z_k, no matter how long the path, on the basis of which ΔF would have to be estimated.

Of course, in practically every case no estimate whatsoever could tell much about ΔF in this situation (compare also Eicker [5]). Thus on the basis of one single realization it is impossible to distinguish between a trigonometric regression variable and the presence of a random variable in the infinitely remote past giving rise to a jump of F.

As hinted at before, the amplitude $z_k(\omega)$ as well as the jump frequency λ_k in the particular realization $x_{.}(\omega)$ could of course be estimated by standard regression techniques. But this is something else than estimating ΔF.

Note also that in the presence of jumps of F the usual covariance estimators, and hence the spectral estimators, are in no sense consistent. For simplicity, consider the scalar pure jump case only. Then by (2.2)

$$x_n = \sum_{k=1}^{K} \exp(in\lambda_k) z_k \qquad (k = 1, 2, \ldots, +\infty)$$

($z_k := \Delta_k y$). Consequently

$$(N+1)^{-1} \sum_{j=0}^{N} x_{j+n} \bar{x}_j = (N+1)^{-1} \sum_{k,\, m=1}^{K} z_k\, \bar{z}_m \sum_{j=0}^{N} \exp[i(j+n)\lambda_k - ij\lambda_m]$$

$$(N \in \mathbb{Z}). \quad (2.15)$$

Since $|\lambda_k| \leq \pi$, $\lambda_k - \lambda_m \not\equiv 0 \pmod{2\pi}$,

$$(N+1)^{-1} \sum_{k=1}^{N} \exp[ij(\lambda_k - \lambda_m)] \to \delta_{0,\, \lambda_k - \lambda_m}$$

(δ = the Kronecker function), (2.15) equals asymptotically

$$\sum_{k=1}^{K} |z_k|^2 \exp(in\lambda_k)$$

while $R_n = \sum_{k=1}^{N} \exp(in\lambda_k)\, \Delta F(\lambda_k)$.

A reason for the traditional preference for the spectral density over the spectral df may lie in the fact that in the first decades of time series analysis mainly finitely parametric models were considered (compare Grenander and Rosenblatt [8, p. 115]). But with the increased use of nonparametric models the scope had not been changed.

3. ESTIMATION OF THE SPECTRAL DISTRIBUTION FUNCTION F FOR A MULTIVARIATE STATIONARY RANDOM SEQUENCE

Having advocated the use of F rather than f in the preceding section from probabilistic viewpoints, we now present strong reasons for this use because of simpler estimation. Since generally F is much smoother than f, much weaker assumptions (implying, say, consistency) are expected. This is not surprising if one recalls that, e.g., an estimate for F is obtained by integrating an estimate of f over a frequency interval which thereby is smoothed. Actually, even the periodogram, which is not a useful estimator for f at all, after integration estimates F quite well (see Theorem 3.1). Already this remark shows that the integration eliminates some of the delicate questions (such as choosing a spectral window) raised by the asymptotic good behavior of estimates. (The integration over the frequency is not to be confused with the integration over the product of a weight function with the periodogram, often used to construct point estimates of the spectral density. However, there are similarities: in (3.2) below the periodogram may be said to be integrated over a rectangular window to yield an estimate of the increment $F(\lambda_2) - F(\lambda_1)$ for λ_1, λ_2 fixed. The same expression after division by $\lambda_2 - \lambda_1$ is used to estimate f at the point $(\lambda_1 + \lambda_2)/2$, if λ_1, λ_2 are chosen dependent on the sample size and $\lambda_1, \lambda_2 \to (\lambda_1 + \lambda_2)/2$. Compare, e.g., Grenander and Rosenblatt [8, p. 147].)

The following result is a straightforward generalization of the corresponding fundmental theorem for the scalar case given in Doob [3, p. 496], giving the necessary and sufficient conditions for the process that the natural estimator for F is (strongly or weakly) consistent. The proof uses the standard technique of reduction of x_n to the scalar stationary process $\alpha^H x_n$ ($\alpha \in \mathbb{C}^q$) which is applied also in the spectral representation theorems.

Theorem 3.1. *With the above notations and assumptions*

$$\lim_{N\to\infty} N^{-1} \sum_{m=1}^{N} x_{m+n} x_m{}^H = R_n \qquad a.s. \tag{3.1}$$

for all n iff

$$\lim_{N\to\infty} N^{-1} \int_{\lambda_1}^{\lambda_2} \left(\sum_{m=1}^{N} x_m e^{-im\lambda}\right)\left(\sum_{n=1}^{N} x_n e^{-in\lambda}\right)^H d\lambda = F(\lambda_1) - F(\lambda_2) \qquad a.s. \tag{3.2}$$

for all continuity points of F. If (3.1) *holds only in probability* (*i.p.*), *then* (3.2) *holds i.p.*

For the scalar case Doob states conditions implying (3.1) which amount essentially to fourth-order stationarity and whose multivariate analogue could be found without major effort.

However, since the existence of F' is assumed (which assumption means little loss of generality, according to the remark at the end of the preceding section), it can and will be used in order to obtain very transparent necessary and sufficient conditions for consistent covariance and (by Theorem 3.1) spectral estimation. For the univariate case some results in this direction are given by Eicker [4, 5].

As an estimate of F we shall use the one occurring in Theorem 3.1. We denote it by

$$\begin{aligned} F_N(s) &= N^{-1} \int_{-\pi}^{s} \left(\sum_{m=1}^{N} x_m e^{-i\lambda m}\right)\left(\sum_{n=1}^{N} x_n e^{-i\lambda n}\right)^H d\lambda \\ &= \int_{-\pi}^{s} \sum_{k=-N+1}^{N-1} e^{-i\lambda k}(1 - |k|/N) R_{kN}\, d\lambda \\ &= (s+\pi) R_{ON} + i \sum_{\substack{k=-N+1 \\ k\neq 0}}^{N-1} (\operatorname{sign} k) \frac{N-|k|}{N|k|} (e^{-isk} - (-1)^k) R_{kN}; \end{aligned} \tag{3.3}$$

here

$$\begin{aligned} R_{kN} &= (N-k)^{-1} \sum_{m=k+1}^{N} x_m x_{m-k}^H \qquad \text{for} \quad k = 0, 1, 2, \ldots, N-1, \\ R_{kN} &= R_{|k|,N}^H \qquad \text{for} \quad k = -1, -2, \ldots, -N+1. \end{aligned} \tag{3.4}$$

The R_{kN} are used as covariance estimates. Both these F_N and R_{kN} have long been selected because of their algebraic simplicity, which allows—fortunately—the derivation of some desirable statistical properties on the basis of fairly simple and general hypotheses.

Expression (3.3) shows, by the way, the reason why the integrated periodogram I_N is a good estimator for F whereas I_N is not a good estimate for f

(compare Grenander and Rosenblatt [8, p. 151]): the instability of the R_{kN} for large N as estimates for R_k is attenuated by the factor $(N - |k|)/|k|$.

Since F has no jumps, by assumption, by (2.12) x_n has the L^2-representation

$$x_n = \sum_{j=-\infty}^{\infty} C_j z_{n+j}. \tag{3.5}$$

Exactly because this is an L^2-limit it was shown in Eicker [5], and it generalizes to the multivariate case, that the most natural consistency concept to be used for R_{kN} and, consequently, F_N estimators, is that of L^1-convergence:

$$E[\mathrm{mod}(R_{kN} - R_k)] \to 0 \qquad \text{for} \quad N \to \infty \tag{3.6}$$

(mod of a matrix means the matrix with elements equal to the moduli of the elements of the original matrix). This concept is natural in the sense that it does not require in particular existence and assumptions on moments of order higher than two.

Theorem 3.2. *Under* (3.5) *the* R_{kN} *are* L_1*-consistent in the sense of* (3.6) *for all* k *and all matrix sequences* (C_n) *satisfying* (2.10) *iff the covariance estimators of the* (z_n) *process,*

$$\begin{aligned} r_{kN} &= (N-k)^{-1} \sum_{m=k+1}^{N} z_m z_{m-k}^H \qquad (k = 0, 1, \ldots, N-1), \\ r_{kN} &= r_{|k|,N}^H \qquad (k = -1, -2, \ldots, -N+1), \end{aligned} \tag{3.7}$$

are L_1*-consistent:*

$$E[\mathrm{mod}(r_{kN} - \delta_{k0} I_p)] \to 0 \qquad \text{for} \quad N \to \infty \tag{3.8}$$

for all k. *Here* δ_{k0} *is the Kronecker symbol and* I_p *is the pth unit matrix. The above equivalence also holds if* L_1*-consistency is replaced both times by weak consistency (convergence in probability).*

The proof can be given elementwise and is, except for formal changes, almost exactly the same as the univariate one given by Eicker [5, pp. 269, 270].

Corollary 3.3. *Under* (3.5) *and* (3.8) *we have*

$$F_N(\lambda) \xrightarrow[N \to \infty]{} F(\lambda) \qquad \text{in probability} \quad (|\lambda| \le \pi). \tag{3.9}$$

Since now F *is continuous, we even have*

$$\sup_{|\lambda| \le \pi} |F_N(\lambda) - F(\lambda)| \to 0 \qquad \text{i.p.} \tag{3.10}$$

The proof of this is completely analogous to that of the last sentence of Theorem 3.1.

There is no general and simple answer to the question whether a condition on the generating process $\{z_n\}$ such as (3.8) is a disadvantage in practice as compared with an equivalent condition on the original process $\{x_n\}$. It seems advantageous, though, that the conditions of Theorem 3.2 and Corollary 3.3 do not concern the unknown sequence $\{C_n\}$ of constants in a very specific way, since they are actually to be estimated.

APPENDIX

Lemma A1. *Let A be a positive semidefinite (psd) matrix, whose elements are complex-valued measurable functions. Then there exists a matrix C of the same type such that $C^2 = A$.*

Proof. Enumerate the $2p^2$ real and imaginary parts of the elements of A (where A is a $p \times p$ matrix) in some fixed fashion. Then A can be considered as a $2p^2$-dimensional real-valued measurable function vector $a_1, \ldots, a_{2p^2}$. Subdivide the range space (which is R^1) of the jth function into the intervals (the inner ones of length 2^{-n}) $[-\infty, -2^n]$, $[-2^n, -2^n + 2^{-n}]$, $\ldots$, $[2^n - 2^{-n}, 2^n]$, $[2^n, \infty]$ to be denoted by $B_{n,j,k}$ ($k = 0, 1, \ldots, 2^{n+1} + 1$; $j = 1, \ldots, 2p^2$; $n = 1, 2, \ldots$).

Denote and enumerate in some way the finite set of direct products (rectangles in R^{2p^2})

$$\mathop{\times}_{j=1}^{2p^2} B_{n,j,k_j} =: B_{n,m}$$

where the vectors $(k_1, \ldots, k_{2p^2})$ assume all allowed values.

Let $m = 1, 2, \ldots, M_n$, so that $|M_n|$ = number of the k-vectors. Let $D_{n,m}$ be the measurable set on which the functions of A assume values in $B_{n,m}$, and let $A_{n,m}$ be one value of A assumed at $\omega_{n,m}$ on $B_{n,m}$. Then each $A_{n,m}$ is a psd complex-valued matrix (of constants), and

$$\sum_{m=1}^{M_n} A_{n,m} 1(B_{n,m}) \xrightarrow[n\to\infty]{} A \qquad \text{pointwise}$$

(here $1(B)$ is the indicator function of B).

Choose the psd square root C of A, and square roots $C_{n,m}$ of $A_{n,m}$ such that $C_{n,m} = C(\omega_{n,m})$. Then $\sum_m C_{n,m} 1(B_{n,m}) \xrightarrow[n\to\infty]{} C$, so that C is measurable.

Lemma A2. *Let A and C be as in the preceding lemma. If the elements of A are integrable functions, then those of C are square integrable.*

Proof. Note that $A = (a_{ij})$ and $C = (c_{ij})$ are Hermitian. Hence

$$a_{jj} = \sum_{k=1}^{p} |c_{jk}|^2.$$

The following multivariate generalization of a linear process holds (compare, e.g., Hannan [9, pp. 209, 210], and Rozanov [10]).

Theorem A3. *A weakly stationary multivariate random sequence* $\ldots, x_{-1}, x_0, x_1, \ldots$ *is a two-sided infinite moving average (a linear process)*

$$x_n = \underset{N\to\infty}{\text{l.i.m.}} \sum_{j=-N}^{N} C_j z_{n+j} \qquad (\textit{limit in } L^2 \textit{ sense})$$

if F is absolutely continuous with density

$$f(\lambda) = \left| \lim_{N\to\infty} \sum_{j=-N}^{N} C_j e^{ij\lambda} \right|^2 \tag{A1}$$

where

$$\operatorname{tr} \sum_{j=-\infty} C_j C_j^H < \infty \tag{A2}$$

and $\{z_n\}_{n\in N}$ *is an orthonormal random sequence in the sense that*

$$E(z_n z_m{}^H) = \delta_{nm} I_p.$$

Proof. According to the preceding lemmas, the p.s.d. square root $f^{1/2} =: C$ exists uniquely as a $p \times p$ Hermitian matrix of square integrable functions c_{jk}. Thus each one possesses a Fourier expansion which L^2-converges to c_{ij}. In matrix notation

$$C(\lambda) = \underset{N\to\infty}{\text{l.i.m.}} \sum_{j=-N}^{N} C_j e^{ij\lambda}, \qquad C_j = C_j^H = (c_{ik})$$

and (A1) and (A2) hold.

If $x_n = \int_{-\pi}^{\pi} e^{in\lambda}\, dy(\lambda)$, define $\tilde{y}(\lambda)$ by $d\tilde{y}(\lambda) = C^{-1}(\lambda)\, dy(\lambda)$ (with the usual artifice if $C(\lambda)$ is singular; see Doob [3, p. 498]). Then $E(d\tilde{y}(\lambda)(d\tilde{y}(s))^H) = \delta_{\lambda s} I_p\, d\lambda$. Putting $z_n := \int_{-\pi}^{\pi} e^{\lambda i n}\, d\tilde{y}(\lambda)$, the assertions follow as in the scalar case.

REFERENCES

1. Bartlett, M. S. (1966). *An Introduction to Stochastic Processes.* Cambridge Univ. Press, London and New York.
2. Box, G. E. P. and Jenkins, G. M. (1970). *Time Series Analysis, Forecasting and Control.* Holden-Day, San Francisco, California.
3. Doob, J. L. (1953). *Stochastic Processes.* Wiley, New York.
4. Eicker, F. (1964). Konsistente Spektraldichteschätzungen für lineare Zufallsfolgen bei schwachen Voraussetzungen. *Z. Angew. Math. Mech.* **44** T12-T14.
5. Eicker, F. (1966). Consistent covariance estimation for general classes of stationary discrete stochastic processes. *Z. Wahrscheinlichkeitstheorie und Verw. Gebiete* **5** 265–278.
6. Freiberger, W. (1963). Approximate distributions of cross-spectral estimates for Gaussian processes. *Proc. Symp. Time Ser. Anal. Brown Univ.* (M. Rosenblatt, ed.). Wiley, New York.

6a. Gikhman, I. I. and Skorokhod, A. V. (1969). *Introduction to the Theory of Random Processes*. W. B. Saunders Co., Philadelphia (translated from the Russian in 1965).
7. Goodmann, N. R. (1963). Spectral analysis of multiple time series. *Proc. Symp. Time Ser., Anal., Brown Univ.* (M. Rosenblatt, ed.). Wiley, New York.
8. Grenander, U. and Rosenblatt, M. (1957). *Statistical Analysis of Stationary Time Series*. Wiley, New York.
9. Hannan, E. J. (1970). *Multiple Time Series*. Wiley, New York.
10. Ibragimov, I. A. (1963). On estimation of the spectral function of a stationary Gaussian process. *Theor. Probability Appli.* **8** 366–401.
11. Jenkins, G. M. (1963). Cross-spectral analysis and the estimation of linear open loop transfer functions. *Proc. Symp. Time Ser. Anal., Brown Univ.* (M. Rosenblatt, ed.). Wiley, New York.
12. Jenkins, G. M. and Watts, D. G. (1968). *Spectral Analysis and Its Applications*. Holden-Day, San Francisco, California.
13. Leppink, G. J. (1970). Efficient estimators in spectral analysis. *Proc. Twelfth Canad. Math. Congr. Time Ser. Stochastic Processes, Convexity and Combinatorics* (R. Pyke, ed.). Canadian Math. Congr. Montreal, Canada.
14. Parzen, E. (1969). Multiple time series modeling. *Multivariate Analysis II* (P. R. Krishnaiah, ed.). Academic Press, New York.
15. Priestley, M. B. (1964). The analysis of two-dimensional stationary processes with discontinuous spectra. *Biometrika* **51** 195–217.
16. Quenouille, M. H. (1957). *The Analysis of Multiple Time Series*. Griffin, London.
17. Rosenblatt, M. (1959). Statistical analysis of stochastic processes with stationary residuals. *Probability and Statistics* (U. Grenander, ed.). Wiley, New York.
18. Rosenblatt, M. (1966). Remarks on higher order spectra. *Multivariate Analysis I* (P. R. Krishnaiah, ed.). Academic Press, New York.
19. Rozanov, Yu. A. (1963). *Stationary Random Processes*. Holden-Day, San Francisco, California.
20. Tukey, J. W. (1959). An introduction to the measurement of spectra. *Probability and Statistics* (U. Grenander, ed.). Wiley, New York.

Non-Anticipative Canonical Representations of Equivalent Gaussian Processes[1]

G. KALLIANPUR
UNIVERSITY OF MINNESOTA

1. INTRODUCTION

The purpose of this paper is to study in greater detail non-anticipative representations of equivalent Gaussian processes whose existence has been established by Kallianpur and Oodaira [7] and by Kailath and Duttweiler [6].

The representation given by Kallianpur and Oodaira ([7], Theorem 4.1) is further analyzed in Section 2 using the theory of canonical representations. In Section 3 Hitsuda's approach [5] is extended to obtain an essentially martingale theoretic proof and derivation of non-anticipative representations for Gaussian processes whose canonical representations have continuous multiplicity one and discrete multiplicity zero.

2. THE GENERAL FORM OF THE NON-ANTICIPATIVE REPRESENTATION

Let $\{X(t), P\}$ and $\{X(t), Q\}$ $(t \in [0, 1])$ be equivalent Gaussian processes given on some space $(\Omega, \mathscr{A})$ which have zero mean functions and continuous covariance functions Γ_P and Γ_Q. The term "equivalent" is used in the sense that the probability measures P and Q are mutually absolutely continuous with respect to the σ-field $\mathscr{F}$ generated by the random variables $\{X(t)\}$. We recall the definition of a non-anticipative representation given by Kallianpur and Oodaira [7]. The process $\{X(t), P\}$ has such a representation with respect to $\{X(t), Q\}$ if there is a Gaussian process $\{Y(t), Q\}$ which is a version of $\{X(t), P\}$, i.e., has zero mean and covariance Γ_P, and $Y(t) \in L_Q(X; t)$ for each $t \in [0, 1]$. In Kallianpur and Oodaira [7] we have shown that every Gaussian process $\{X(t), P\}$ equivalent to a given Gaussian process $\{X(t), Q\}$ has a non-anticipative representation with respect to the latter.

[1] Work supported in part by NSF grant GP–30694X.

The starting point of the work of this section is the representation given in Theorem 4.1 of our paper cited above. We need to recapitulate briefly the necessary notation and terminology. (The notation has been slightly changed for typographical convenience.) For details the reader is referred to the work by Kallianpur and Oodaira [7] or to the book of Gohberg and Krein [3].

$L_Q(X; t)$ $(0 \leq t \leq 1)$ denotes the linear subspace of $L^2(\Omega, Q)$ and $P(t)$ the orthoprojector on $L_Q(X; 1)$ with range $L_Q(X; t)$. Let π be a maximal chain of orthoprojectors containing $\{P(t), t \in [0, 1]\}$. It is well known that since P and Q are equivalent, the operator S defined by Γ_P on the reproducing kernel Hilbert space $H(\Gamma_Q)$ is a self-adjoint, positive, invertible operator such that $T = I - S$ is Hilbert–Schmidt. By applying the special factorization theorem of Gohberg and Krein to S (see Theorem 2.2 of [7]), it was shown that the desired nonanticipative representation is given by

$$Y(t) = \Delta(I + W_+)X(t), \tag{2.1}$$

where W_+ is a Hilbert–Schmidt Volterra operator having π as an eigenchain, $\Delta = D^{-1/2}$, and

$$D = I + \sum (P^+ - P^-)[(I - P^+ \tilde{T} P^+)^{-1} - I](P^+ - P^-). \tag{2.2}$$

The summation in (2.2) is over all the gaps of π and $\tilde{T}$ is the operator on $L_Q(X; 1)$ which corresponds to T under the familiar congruence between $L_Q(X; 1)$ and $H(\Gamma_Q)$. D has the following additional properties. It is self-adjoint, positive, invertible,

$$DP = PD \qquad (P \in \pi), \tag{2.3}$$

and

$$D - I \qquad \text{is Hilbert–Schmidt.} \tag{2.4}$$

We shall now study representation (2.1) by using the theory of canonical representations of Lévy and Hida [4]. The following lemma is similar to that of [7], Theorem 4.3, but seems better suited to our purpose.

Lemma 1. *$Y(t)$ defined by* (2.1) *can be put in the form*

$$Y(t) = \Delta X(t) + VX(t) \tag{2.5}$$

where V is the Volterra operator.

$$V = \int_\pi P(\Delta H)\, dP. \tag{2.6}$$

The right-hand side of (2.6) *is an operator-valued integral over the chain π as defined in the book of Gohberg and Krein, and H is the self-adjoint Hilbert–Schmidt operator appearing in the triangular integral of the Volterra operator W_+.*

Proof. From (2.1) we have $V = \Delta W_+$. Since W_+ is Hilbert–Schmidt, Volterra, and has π as an eigenchain, it follows from the work of Gohberg–Krein [3, Chapter I, Theorem 5.1] that W_+ has a triangular integral representation

$$W_+ = \int_\pi PH\,dP. \tag{2.7}$$

The operator integral converges in uniform norm and H is a Hilbert–Schmidt operator uniquely determined by W_+. (Further precision of H is unnecessary for us at the moment.) Since Δ commutes with all the orthoprojectors P, it is easy to see that $\Delta W_+ = \int_\pi P(\Delta H)\,dP$. Clearly ΔW_+ is a Volterra operator with π as an eigenchain and ΔH is Hilbert–Schmidt.

Let the Gaussian process $(X(t), Q))$ $(0 \leq t \leq 1)$ have a canonical representation which we may take to be proper in the sense of Kallianpur and Mandrekar [10]. Let

$$X(t) = \sum_{i=1}^{N} \int_0^t F_i(t, u)\,d\xi_i(u) + \sum_{t_j \leq t} \sum_{q=1}^{N_j} b_j^q(t)\beta_{t_j}^q + \sum_j a_j(t)\eta_j. \tag{2.8}$$

The ξ_i's are mutually independent additive Gaussian processes with $E\xi_i(u) = 0$ and variance functions $\rho_i(u)$. The $\beta_{t_j}^q$ and the η_j are mutually independent standard Gaussian random variables independent of the processes $\xi_i(u)$. The functions $F_i(t, \cdot)$, $b_j^q(\cdot)$, $a_j(\cdot)$ satisfy the condition

$$\sum_i \int_0^t F_i^2(t, u)\,d\rho_i(u) + \sum_{t_j \leq t} \sum_{q=1}^{N_j} [b_j^q(t)]^2 + \sum_j a_j^2(t) < \infty. \tag{2.9}$$

Let us write

$$M_1 = \sum_{i=1}^{N} \oplus L_Q(\xi_i; 1), \tag{2.10}$$

$$M_0' = \sum_j [P(t_j+) - P(t_j)]L_Q(X; 1), \tag{2.11}$$

$$L_Q(X; 0+) = \bigcap_{t>0} L_Q(X; t), \tag{2.12}$$

$$M_0 = L_Q(X; 0+) \oplus M_0'. \tag{2.13}$$

For each j, $\{\beta_{t_j}^q\}$ is an orthonormal (ON) basis in $[P(t_j+) - P(t_j)]L_Q(X; 1)$ and $\{\eta_j\}$ is an ON basis in $L_Q(X; 0+)$. In (2.8) $P_{M_1}X(t) = \sum_{i=1}^N \int_0^t F_i(t, u)\,d\xi_i(u)$ and the sum of the remaining terms on the right-hand side equals $P_{M_0}X(t)$. We may call $P_{M_1}X(t)$ the "continuous" part of $X(t)$ and $P_{M_0}X(t)$ the "discontinuous" part in the representation. To consider the most general situation we shall assume that $L_Q(X; 0+) \neq 0$, so that 0 is a discontinuity of $P(t)$; i.e., π has a gap at 0. We shall incorporate it in our notation and write

$$M_0 = \sum_{j=1}^{m} \oplus (R_j - R_{j-1})L_Q(X; 1) \oplus \sum_{j \geq 1} \oplus [P(t_j+) - P(t_j)]L_Q(X; 1) \tag{2.14}$$

where $R_0 = 0$, $\{R_j\}$ is any set of orthoprojectors with the property $\dim(R_j - R_{j-1}) = 1$ and $\sum_{j=1}^{m} (R_j - R_{j-1}) = P(0+)$ (m may be $=\infty$). In the sequel (P^-, P^+) shall refer to either (R_{j-1}, R_j) or $(P(t_j), P(t_j+))$.

From (2.2) it is easy to see that M_0 and M_1 are invariant subspaces of D and hence also of the self-adjoint, positive operator Δ. Moreover, $DP_{M_1} = P_{M_1}$. Thus by direct verification we have $D^{-1} = D^{-1}P_{M_0} + P_{M_1}$, so that

$$\Delta = \Delta P_{M_0} + P_{M_1}. \tag{2.15}$$

Let us now consider VP_{M_0} and VP_{M_1}. To evaluate the former we need to consider $V(P^+ - P^-)$ where (P^-, P^+) is a typical gap of π. By Lemma 1, V is the limit in uniform norm of sums of the type

$$S_\zeta = \sum_i P_{j-1}(\Delta H)(P_j - P_{j-1}) \tag{2.16}$$

where $\zeta = \{P_j\}$ is a finite partition of π.

Let P_{k-1}, P_k be the orthoprojectors in ζ such that $P_{k-1} \leq P^- < P^+ \leq P_k$. (If equality holds in both places, P^-, P^+ themselves belong to ζ.) It suffices to discuss the case when the inequality is strict. Let $\zeta' \supset \zeta$ be the augmented partition formed by including P^-, P^+ as the new elements. By definition of the operator integral, to every positive ε there is a partition ζ_ε such that $\|V - S_{\zeta'}\| < \varepsilon$ for every partition $\zeta' \supset \zeta_\varepsilon'$. Taking $\zeta_\varepsilon = \zeta$ and for ζ' the augmented partition, we have

$$\|V(P^+ - P^-) - S_{\zeta'}(P^+ - P^-)\| < 2\varepsilon. \tag{2.17}$$

It is easy to check directly that

$$S_{\zeta'}(P^+ - P^-) = P^-(\Delta H)(P^+ - P^-).$$

From (2.17) it follows that $V(P^+ - P^-) = P^-(\Delta H)(P^+ - P^-)$. Thus we have shown that

$$VP_{M_0} = \sum_i P_i^-(\Delta H)(P_i^+ - P_i^-). \tag{2.18}$$

Next, if $\zeta = \{P_j\}$ is a partition of π,

$$VP_{M_1} = \lim \sum_j P_{j-1}(\Delta H)(P_j - P_{j-1})P_{M_1}.$$

The sum on the right-hand side equals

$$\sum_i P_{j-1}\, \Delta P_{M_0} H P_{M_1}(P_j - P_{j-1}) + \sum_j P_{j-1}(\Delta P_{M_1} H P_{M_1})(P_j - P_{j-1}).$$

Each sum in the above expression obviously converges to a limit in uniform norm and we obtain

$$VP_{M_1} = \int_\pi P(\Delta P_{M_0} H P_{M_1})\, dP + \int_\pi P(\Delta P_{M_1} H P_{M_1})\, dP. \tag{2.19}$$

From (2.15), $\Delta P_{M_1} = P_{M_1}$, so that (2.19) becomes

$$VP_{M_1} = \int_{\pi} P(\Delta P_{M_0} HP_{M_1})\, dP + \int_{\pi} P(P_{M_1} HP_{M_1})\, dP. \tag{2.20}$$

Now it is easy to verify by inspecting an approximating sum to the second integral on the right-hand side of (2.20) that the gaps, if any, in a partition ζ of π do not contribute to S_ζ and hence that we may write

$$\int_{\pi} P(P_{M_1} HP_{M_1})\, dP = \int_0^1 P(u)(P_{M_1} HP_{M_1})\, dP(u) \tag{2.21}$$

where the latter integral is the limit in uniform norm of sums $\sum_i P(u_{i-1})$ $[P_{M_1} HP_{M_1}](P(u_i) - P(u_{i-1}))$where $\{0, (P(u_i))\}$ is a partition of π. Thus from (2.5), (2.15), (2.18), (2.20), and (2.21) we finally obtain the non-anticipative representation $Y(t)$ in a form which separates its "continuous" and "discontinuous" parts.

Theorem 2.1. *The non-anticipative representation of $\{X(t), P\}$ with respect to the $\{X(t), Q\}$ process obtained via the Gohberg–Krein factorization theorem is given by the process $\{Y(t), Q\}$ where*

$$Y(t) = \left[\Delta P_{M_0} + \sum_i P_i^-(\Delta H)(P_i^+ - P_i^-) + \int_{\pi} P(\Delta P_{M_0} HP_{M_1})\, dP\right] X(t)$$
$$+ \left[I + \int_0^1 P(u)(P_{M_1} HP_{M_1})\, dP(u)\right] P_{M_1} X(t). \tag{2.22}$$

It is interesting to note that the operator in the second square brackets on the right-hand side of (2.22) is of the form $I + V_1$ where V_1 is a Volterra operator on the Hilbert space $M_1 = \sum_{i=1}^N \oplus L_Q(\xi_i; 1)$. Under the congruence between $L_Q(X; 1)$ and $H(\Gamma_Q)$, M_1 is congruent to $\sum_{i=1}^N \oplus L^2([0, 1], \rho_i)$, so it is possible to give a more concrete expression for V_1. In fact, since

$$\left[\int_0^1 P(u)(P_{M_1} HP_{M_1})\, dP(u)\right] P_{M_1} X(t)$$
$$= \sum_{i=1}^N \left[\int_0^1 P(u)(P_{L_i} HP_{L_i})\, dP(u)\right]\left(\int_0^t F_i(t, u)\, d\xi_i(u)\right)$$

(where L_i stands for $L_Q(\xi_i; 1)$), it suffices to consider each term on the right-hand side separately. We have reduced the problem to a relatively routine calculation, the details of which are left to the reader. We obtain

$$\left[\int_0^1 P(u)(P_{L_i} HP_{L_i})\, dP(u)\right]\left(\int_0^t F_i(t, u)\, d\xi_i(u)\right)$$
$$= \int_0^t F_i(t, u)\left\{\int_0^u \hat{H}_i(s, u)\, d\xi_i(s)\right\} d\rho_i(u) \tag{2.23}$$

when $\hat{H}_i(s, u)$ is a square integrable Volterra kernel in $L^2([0, 1], \rho_i)$, $\hat{H}_i(s, u) = 0$ if $s > u$. We thus have

Theorem 2.2. *Suppose that the process* $(X(t), Q)$ $(0 \leq t \leq 1)$ *has a canonical representation which consists only of the continuous part (i.e., such that its multiplicity* $= N$) *and suppose that* $L_Q(X; 0+) = 0$. *Then the non-anticipative representation* $Y(t)$ *is given by*

$$Y(t) = \sum_{i=1}^{N} \left[\int_0^t F_i(t, u)\, d\xi_i(u) + \int_0^t F_i(t, u) \left\{ \int_0^u \hat{H}_i(s, u)\, d\xi_i(s) \right\} d\rho_i(u) \right] \tag{2.24}$$

where the $\hat{H}_i$ *are Volterra kernels in* $L^2([0, 1], \rho_i)$ *as described above and the usual square integrability conditions are satisfied.*

As another example which is a special case of the most general representation obtained in Theorem 2.1 we let the $(X(t), Q)$ have a canonical representation of continuous multiplicity one (i.e., $N = 1$). The only gap of π is at 0. We then get

Theorem 2.3. *Let* 0 *be the only gap of* π *and let* $N = 1$ *in* (2.8). *Then*

$$Y(t) = \sum_{i=1}^{m} \left\{ C_i(t) + \int_0^t F(t, u) A^{(i)}(u)\, d\rho(u) \right\} \eta_i$$
$$+ \int_0^t F(t, u)\, d\xi(u) + \int_0^t F(t, u) \left\{ \int_0^u \hat{H}(s, u)\, d\xi(s) \right\} d\rho(u). \tag{2.25}$$

$\hat{H}$ *is a square integrable Volterra kernel in* $L^2([0, 1], \rho)$, $\sum_{i=1}^{m} (C_i(t))^2 < \infty$ *and* $\sum_i \int_0^1 \{A^{(i)}(u)\}^2\, d\rho(u) < \infty$.

As special cases of Theorem 2.3 we mention the non-standard Wiener process over [0, 1] and the n-ple Gaussian Markov process in the role of $(X(t), Q)$ both considered in our earlier paper [7].

3. A DERIVATION USING MARTINGALE THEORY

The representation obtained in Theorem 2.1 is based on a purely operator theoretic result: the special factorization of the operator S along the maximal chain π. It would be interesting to obtain a somewhat more illuminating probabilistic proof of this result. Such a proof when $(X(t), Q)$ is a standard Wiener process was given by Hitsuda [5]. In this section we shall reexamine his arguments and show that his approach yields a proof of Theorem 2.1 when $(X(t), Q)$ has a canonical representation of multiplicity one and discrete multiplicity zero. In this case the chain π has no gaps and has rank one [3]. The main ideas of this proof are derived from martingale theory, the representation of square integrable martingales, and Girsanov's theorem [2].

In presenting the proof of our result, we shall frequently refer to Hitsuda's paper [5] and shall omit those details in the proof which can be found in Hitsuda's paper. Also, in order to facilitate comparison of this section with the work of Hitsuda, we have adopted the notation of his paper.

The space $(\Omega, \mathscr{A}, P)$ is a complete probability space. The stochastic process $\{Y(t, \omega), P\}$ $(\omega \in \Omega,\ t \in [0, 1])$ is Gaussian with zero mean and continuous covariance Γ_P. $\tilde{P}$ is a second probability measure on $(\Omega, \mathscr{A})$ with respect to which $\xi(t, \omega)$ is a sample continuous, additive Gaussian process. We shall assume that

$$Y(t, \omega) = \int_0^t F(t, u)\, d\xi(u, \omega) \tag{3.1}$$

with respect to $\tilde{P}$. Here $\xi(u, \omega)$ is sample continuous, hence, (u, ω) measureable, $E[\xi(u)]^2 = \rho(u)$, $F(t, u)$ is (t, u) measurable with

$$\int_0^t F^2(t, u)\, d\rho(u) < \infty, \tag{3.2}$$

and (3.1) is a proper canonical representation (w.r.t. $\tilde{P}$) of $Y(t)$. It means that for each t

$$L_{\tilde{P}}(Y; t) = L_{\tilde{P}}(\xi; t). \tag{3.3}$$

It is well known that the stochastic integral on the right-hand side of (3.1) can be so defined as to render $Y(t, \omega)$ a measurable process. Let $\mathscr{F}_t^{\,0} = \sigma\{Y(s), s \le t\}$. Let $\mathscr{F}_t^{\,P}$ be the σ-field obtained by adjoining to $\mathscr{F}_t^{\,0}$ all P-null sets. Next, we shall make our basic assumption that the measures $\tilde{P}$ and P are equivalent relative to $\mathscr{F}_1^{\,0}$:

$$\tilde{P} \equiv P[\mathscr{F}_1^{\,0}]. \tag{3.4}$$

Clearly (3.4) implies $\mathscr{F}_t^{\,P} = \mathscr{F}_t^{\,\tilde{P}}$. Write $\varphi = d\tilde{P}/dP$. From (3.3) it is easy to see that

$$\mathscr{F}_t^{\,\tilde{P}} = \mathscr{B}^{\tilde{P}}(\xi; t) \tag{3.5}$$

for each t, where $\mathscr{B}^{\tilde{P}}(\xi; t)$ is the σ-field generated by the family $\{\xi(s), s \le t\}$ to which are adjoined all $\tilde{P}$-null sets.

It should be remarked that the setup just described says that under $\tilde{P}$ the Gaussian process $Y(t)$ has a proper canonical representation of continuous multiplicity one. In fact, we have chosen the space on which the canonical representation is given to be $(\Omega, \mathscr{A}, \tilde{P})$ itself. The measurability of ξ has been added for convenience. The continuous covariance of $(Y(t), \tilde{P})$ will be denoted by $\Gamma_{\tilde{P}}$. The mean is seen to be zero from (3.1).

Lemma 1. *Let* $(Z(t), \mathscr{B}^{\tilde{P}}(\xi; t), \tilde{P})$ *be a square integrable martingale with* $E_{\tilde{P}}(Z(t)) = 0$ *for all* t. *Then*

$$Z(t) = \int_0^t f(s)\, d\xi(s) \tag{3.6}$$

where $f(s, \omega)$ *is jointly measurable,* $f(s)$ *is* $\mathscr{B}^{\tilde{P}}(\xi; s)$ *measurable for each* s, *and*

$$E_{\tilde{P}}\left(\int_0^1 f^2(s)\, d\rho(s)\right) < \infty. \tag{3.7}$$

This lemma is well known. For example, see Kunita and Watanabe [8] for the case $\rho(s) = s$.

Let M_t denote a right continuous modification of the martingale $\tilde{E}(\varphi^{-1} | \mathscr{F}_t)$ with respect to $(\mathscr{F}_t, \tilde{P})$. (We write $\tilde{E}$ for $E_{\tilde{P}}$, $\mathscr{F}_t$ for $\mathscr{F}_t^{\tilde{P}}$ or $\mathscr{F}_t^P$, and recall that $\varphi = d\tilde{P}/dP$.) Such a modification exists in view of (3.5).

Lemma 2.

$$P[M_t > 0 \quad \text{for all} \quad t \in [0, 1]] = 1; \tag{3.8}$$

$$M_t = \exp\left[\int_0^t f(s)\, d\xi(s) - \frac{1}{2}\int_0^t f^2(s)\, d\rho(s)\right] \tag{3.9}$$

$(0 \le t \le 1)$, *where* $f(s, \omega)$ *is* (s, ω) *measurable,* $f(s)$ *is adapted to* $(\mathscr{F}_s)$, *and*

$$P\left[\int_0^t f^2(s)\, d\rho(s) < \infty\right] = 1. \tag{3.10}$$

Proof. The above lemma is almost exactly that of Hitsuda ([5], Lemma 1), and the proof too is almost the same. Equation (3.8) is shown in the same was as in the work of Hitsuda [5].

First, defining

$$M_t^n = \tilde{E}\left[\frac{1}{\varphi} \wedge n | \mathscr{F}_t\right], \qquad (n = 1, 2. \ldots),$$

which is a sequence of square integrable martingales, hence continuous by Lemma 1, we show (by an application of Doob's martingale inequality [5]) that M_t is sample continuous. Next, choosing an increasing sequence (T_n) of $(\mathscr{F}_t)$-stopping times, we have the following formula for the square integrable martingale $M_{t \wedge T_n}$.

$$M_{t \wedge T_n} = 1 + \int_0^{t \wedge T_n} g^n(s)\, d\xi(s). \tag{3.11}$$

It follows immediately upon making $n \to \infty$ from the uniqueness of (3.11) that

$$M_t = 1 + \int_0^t g(s)\, d\xi(s). \tag{3.12}$$

Then $g(s, \omega)$ is (s, ω) measurable, $g(s)$ is adapted to $(\mathscr{F}_s)$, and

$$P\left[\int_0^1 g^2(s)\, d\rho(s) < \infty\right] = 1. \tag{3.13}$$

Applying Ito's formula to $\log M_t$ (permissible because of (3.8)), we get

$$\log M_t = \int_0^t f(s)\, d\xi(s) - \frac{1}{2}\int_0^t f^2(s)\, d\rho(s)$$

where $f(s) = (1/M_s)g(s)$. $f(s, \omega)$ satisfies (3.13) with g replaced by f and the same measurability conditions as g. This completes the proof.

The next lemma is a variant of Girsanov's result [2] with the only change that $\xi(t)$ is a sample continuous additive Gaussian process instead of a Wiener process. Alternatively (as we prefer to consider it), it is a special case of Neveu's generalization to continuous martingales of Girsanov's theorem [9].

Lemma 3. *Let $f(s)$ be the process of Lemma 2 and let*

$$\eta(t) = \xi(t) - \int_0^t f(s)\, d\rho(s). \tag{3.14}$$

Then $(\eta(t), \mathscr{F}_t, P)$ is a sample continuous martingale with

$$\langle\eta\rangle(t, \omega) = \rho(t) \qquad \text{a.s. } (P) \quad \text{for all } t. \tag{3.15}$$

In (3.15) $\langle\eta\rangle(t)$ is the increasing process associated with the continuous martingale $(\eta(t), \mathscr{F}_t, P)$.

The next two lemmas are the same as Hitsuda's Lemmas 3 and 4 in [5].

Lemma 4. *Under the assumptions of Lemma 2 we have*

$$E\left[\int_0^1 f^2(s)\, d\rho(s)\right] < \infty, \qquad \tilde{E}\left[\int_0^1 f^2(s)\, d\rho(s)\right] < \infty. \tag{3.16}$$

Lemma 5. *If $f(s)$ is the process of Lemma 2, then*

$$F_t = \int_0^t f(s)\, d\rho(s) \tag{3.17}$$

belongs to $L_{\tilde{P}}(\xi; t)$ $(=L_P(\xi; t))$.

It is to be noted that the proof of Lemma 4 does not need the fact that the $(\eta(t), P)$ process is Gaussian. However, it follows immediately from Lemmas 3 and 5 that $(\eta(t), \mathscr{F}_t, P)$ is a Gaussian process of independent increments and $E[\eta^2(t)] = \rho(t)$. Now the argument in the first part of the proof of Hitsuda's theorem yields the following:

$$f(s) = \int_0^s k(s, u)\, d\xi(u) \qquad \text{a.s. } (P) \tag{3.18}$$

where $k(s, u)$, $(s, u) \in [0, 1]^2$ is a measurable kernel such that for each s, $k(s, \cdot) \in L^2([0, 1], \rho)$. Furthermore, k is a Volterra kernel, $k(s, u) = 0$ (a.e.) if $u > s$, and $\int_0^1 \int_0^s k^2(s, u)\, d\rho(s)\, d\rho(u) < \infty$.

The process $\eta(t)$ defined by (3.14) of Lemma 3 now takes the form

$$\begin{aligned} \eta(t) &= \xi(t) - \int_0^t \left[\int_0^s k(s, u)\, d\xi(u)\right] d\rho(s) \\ &= \xi(t) - \int_0^1 \left[\int_0^t k(s, u)\, d\rho(s)\right] d\xi(u). \end{aligned} \tag{3.19}$$

Let $\hat{K}$ be the Volterra integral operator in $L^2([0, 1], \rho)$ defined by the kernel k and let K be the operator on $L_{\bar{P}}(\xi; 1)$ determined by $\hat{K}$ and the natural congruence between $L^2([0, 1], \rho)$ and $L_{\bar{P}}(\xi; 1)$. Then (3.19) can be written in the form

$$\eta(t) = (I - K)\xi(t). \tag{3.20}$$

K is also a Volterra operator whose invariant subspaces are given by $L_{\bar{P}}(\xi; t)$ $(0 \leq t \leq 1)$. Hence, if $P(t)$ denotes the orthoprojector on $L_{\bar{P}}(Y; 1)$ with range $L_{\bar{P}}(Y; t)$, it follows from (3.3) that $\pi = \{P(t)\}$ $(0 \leq t \leq 1)$ is a maximal eigenchain for K. Clearly π has no gaps.

At this point it is important to comment on the scope of our notation. Equation (3.20) is a statement about elements in Hilbert space. The $\xi(t)$, $\eta(t)$ are regarded as elements (i.e., equivalence classes of random variables) of $L_{\bar{P}}(\xi; 1)$. However, from our measurability condition on $\xi(u, \omega)$ we may assume without loss of generality that the process $\eta(t)$ given by (3.19) is (t, ω) measurable. In other words, we have shown the existence of a measurable process $\eta(t, \omega)$ such that for each t, $\eta(t, \cdot)$ is measurable with respect to $\mathscr{B}^{\bar{P}}(\xi; t)$ and hence with respect to $\mathscr{F}_t$, and such that $\eta(t; \cdot)$ is a member of the equivalence class $\eta(t)$ in (3.20). Strictly speaking, we should use a different notation that brings out this distinction but we have avoided doing so in order to keep the exposition from getting too cumbersome.

Let us now set

$$V = (I - K)^{-1} - I. \tag{3.21}$$

It follows quite simply that V is also a Volterra operator with π as an eigenchain see Lemma 2.4. of [7]. Both K and V are Hilbert–Schmidt operators. From (3.20) and (3.21) we obtain

$$\xi(t) = (I + V)\eta(t). \tag{3.22}$$

Define

$$X(t) = \int_0^t F(t, u)\, d\eta(u). \tag{3.23}$$

From the construction of $\eta(t)$ the following facts emerge. The stochastic integral in (3.23) can be taken in a sense which makes $X(t, \omega)$ (t, ω) measurable. Also for each t, $X(t, \cdot)$ is $\mathscr{F}_t$ measurable and $(X(t), \mathscr{F}_t, P)$ is a version of $(Y(t), \mathscr{F}_t, \tilde{P})$, i.e., $X(t)$ under P has zero mean and covariance $\Gamma_{\tilde{P}}$. Definition (3.23) thus is a canonical representation of $(X(t), \mathscr{F}_t, P)$ which, in fact, is proper canonical in the sense of Hida [4]. Hence

$$L_P(X; t) = L_P(\eta; t) \tag{3.24}$$

for each t. The last property holds because $(Y(t), \mathscr{F}_t, \tilde{P})$ has a proper canonical representation given by (3.1). A necessary and sufficient condition for a canonical representation to be proper involves only the kernel F and the function $\rho(t)$ [4], Theorem I.7. Since it has been shown that $\eta(t)$ is an additive Gaussian process under P with $E[\eta(t)]^2 = \rho(t)$, it follows that (3.23) is proper canonical. From (3.1), (3.20), and (3.21) we have

$$Y(t) = (I + V) \int_0^t F(t, u)\, d\eta(u) \qquad (\text{a.s. } \tilde{P}). \tag{3.25}$$

We recall that in (3.25) $Y(t)$ stands for the equivalence class $(Y(t))^{\tilde{P}}$ to which $Y(t)$ belongs; $\eta(u)$ similarly stands for $(\eta(u))^{\tilde{P}}$, and V is defined on $L_{\tilde{P}}(\xi; 1)$. Let $\mathscr{L}$ be the linear manifold consisting of all random variables Z which are finite linear combinations of the form $\sum_i c_i\, Y(t_i)$ ($t_i \in [0, 1]$ and c_i real) and let $Z^{\tilde{P}}(Z^P)$ be the equivalence class in $L_{\tilde{P}}(Y; 1)$ (resp. $L_P(Y; 1)$) to which Z belongs. Since P and $\tilde{P}$ are equivalent relative to the σ-field $\mathscr{F}_1{}^0$, there exists an equivalence operator, say J, in the sense of Feldman [1] between $L_{\tilde{P}}(Y; 1)$ and $L_P(Y; 1)$ such that $JZ^{\tilde{P}} = Z^P$ for each Z in $\mathscr{L}$. In particular, J is one-to-one, onto, bounded, and has a bounded inverse. Apply J to both sides of (3.25) and again denote the corresponding equivalence classes $(Y(t))^P$, $(\eta(u))^P$ by $Y(t)$ and $\eta(u)$. We obtain

$$Y(t) = (I + W) \int_0^t F(t, u)\, d\eta(u) \qquad (\text{a.s. } P) \tag{3.26}$$

where

$$W = J V J^{-1}. \tag{3.27}$$

W is a Volterra operator. V is the Fredholm resolvent operator of K, i.e., $(I+V)(I-K)^{-1} = (I-K)^{-1}(I+V) = I$, so that writing (3.22) using the kernel of V, it is easy to see that (3.20) and (3.22) together imply

$$L_{\tilde{P}}(\xi; t) = L_{\tilde{P}}(\eta; t) \tag{3.28}$$

for each t. Now $JL_{\tilde{P}}(\eta; t) = L_P(\eta; t) = L_P(X; t)$, the last equality from (3.24). Hence it follows that the subspaces $\{L_P(X; t), 0 \le t \le 1\}$ are invariant subspaces of W. The chain of orthoprojectors $\mathscr{P}(t)$ defined on $L_P(X; 1)$ with range $L_P(X; t)$ is an eigenchain of W. From (3.26) we have

$$Y(t) = (I + W)X(t) \qquad (\text{a.s.} \quad P) \tag{3.29}$$

where $WL_P(X; t) \subset L_P(X; t)$.

It is also clear that for each t

$$L_P(Y; t) = L_P(\eta; t), \tag{3.30}$$

showing that representation (3.29) or (3.26) is proper canonical. We state these results in the following theorem.

Theorem 3.1. *Let the Gaussian process* $\{Y(t), \tilde{P}\}$, $t \in [0, 1]$, *with mean zero and continuous covariance* $\Gamma_{\tilde{P}}$ *have a proper canonical representation of continuous multiplicity one and discrete multiplicity zero. Let* $(\Omega, \mathscr{A}, \tilde{P})$ *be a complete probability space on which a measurable version of the canonical representation* (3.1) *is given. Suppose that* P *is another probability measure on* $(\Omega, \mathscr{A})$ *such that* $\{Y(t), P\}$ *is Gaussian with zero mean and continuous covariance* Γ_P. *Let the measures* P *and* $\tilde{P}$ *be equivalent relative to* $\sigma\{Y(t), 0 \le t \le 1\}$.

Then we can construct a process $X(t)$ *on* $(\Omega, \mathscr{A}, P)$ *such that* $(X(t), \mathscr{F}_t, P)$ *is a version of* $(Y(t), \mathscr{F}_t, \tilde{P})$ *and is defined by the proper canonical representation*

$$X(t) = \int_0^t F(t, u)\, d\eta(u) \tag{3.31}$$

where $(\eta(t), \mathscr{F}_t, P)$ *is a continuous, additive, Gaussian process with* $E[\eta^2(t)] = \rho(t)$. *The process* $\{Y(t), \mathscr{F}_t, P\}$ *is given by the following non-anticipative representation:*

$$Y(t) = (I + W) \int_0^t F(t, u)\, d\eta(u) = (I + W)X(t), \tag{3.32}$$

where W *is a Hilbert–Schmidt Volterra operator on* $L_P(Y; 1)$ *satisfying*

$$WL_P(\eta; t) \subseteq L_P(\eta; t) \tag{3.33}$$

(or equivalently, $WL_P(X; t) \subseteq L_P(X; t)$*).*

Finally, (3.32) *is a proper canonical representation of the* $\{Y(t), P\}$ *process.*

It is easy to see that the converse also holds in the following sense.

Converse of Theorem 3.1. *Suppose* $(Y(t), \mathscr{F}_t, P)$ *is a Gaussian process given by* (3.32) *where* $(X(t), \mathscr{F}_t, P)$ *is defined by* (3.31). *Then* $(Y(t), P)$ *is equivalent to* $(Y(t), \tilde{P})$.

Finally, the proof of Theorem 3.1 and its converse just stated show the following. If $(Y(t), \tilde{P})$ has the canonical representation (3.1), the equivalent Gaussian process $(Y(t), P)$ has a nonanticipative representation with respect to the first process if and only if there is a Gaussian process $(\eta(t), P)$ equivalent to the additive process $(\xi(t), \tilde{P})$ occurring in (3.1) so that (3.22) holds, and such that $(Y(t), P)$ has the canonical representation (3.32).

4. CONCLUDING REMARKS

The ideas and techniques of the preceding section generalize to the case when the canonical representation of $(Y(t), \tilde{P})$ has continuous multiplicity N (where $1 \leq N \leq \infty$) and discrete multiplicity zero. In terms of the chain of orthoprojectors, this is equivalent to saying that $\{P(t), t \in [0, 1]\}$ is itself a maximal chain with no gaps and whose rank is N (see Gohberg and Krein [3]). The generalization involves appropriate extensions of Lemmas 1–5 of Section 3 and will be deferred to a later paper. There is also the question of obtaining a purely probabilistic derivation of the most general case (Theorem 2.1), i.e., when the chain π has gaps. It has independent interest since such a derivation would provide an alternative proof of Gohberg and Krein's special factorization theorem, at least when $I - S$ is positive. We hope to return to these problems soon.

The derivation of the Radon–Nikodym derivative and related questions have been left out of this paper for lack of space but can be treated in the same manner.

REFERENCES

1. Feldman, J. (1958). Equivalence and perpendicularity of Gaussian processes. *Pacific J. Math.* **9** 699–708.
2. Girsanov, I. V. (1960). On transforming a certain class of stochastic processes by absolutely continuous substitution of measures. *Theor. Probability Appl.* **5** 285–301.
3. Gohberg, I. C. and Krein, M. G. (1970). *Theory and Applications of Volterra Operators in Hilbert Space.* (English translation. American Mathematical Society, Providence, R.I.)
4. Hida, T. (1960). Canonical representations of Gaussian processes and their applications. *Mem. Coll. Sci., Univ. Kyoto Ser. A* **33** 109–155.
5. Hitsuda, M. (1968). Representation of Gaussian processes equivalent to Wiener process. *Osaka J. Math.* **5** 299–312.

6. Kailath, T. and Duttweiler, D. (1972). An RKHS approach to detection and estimation problems, Part III: Generalized innovations, representations and a likelihood-ratio formula. *IEEE Trans. Inform. Theor.* **18** 730–745.
7. Kallianpur, G. and Oodaira, H. (1973). Non-anticipative representations of equivalent Gaussian processes. *Annals of Probability* **1** 104–122.
8. Kunita, H. and Watanabe, S. (1967). On square integrable martingales. *Nagoya Math. J.* **30** 209–245.
9. Neveu, J. (1971). Unpublished Lecture Notes.
10. Kallianpur, G. and Mandrekar, V. (1965). Multiplicity and representation theory of purely non-deterministic stochastic processes. *Theor. Probability Appl.* **10** 614–644.

Abstract Martingales and Ergodic Theory[1]

M. M. RAO[2]

THE INSTITUTE FOR ADVANCED STUDY
PRINCETON, NEW JERSEY

INTRODUCTION

It was felt for a long time that martingales and ergodic theory, being essentially theories of integration in infinitely many variables, should be obtainable from a single structure. In fact there are many similarities in form as well as in proofs of the main theorems in both cases for this expectation (cf. [13, p. 135], [6, p. 342]). However, this hope has not yet been completely realized. In this paper, the main attempts at a unification and the results thus obtained will be discussed so as to clarify the problem and illuminate the difficulties involved. Thus, in part, the paper surveys the work on these questions, since this aspect of the theory seems to be lacking in the literature.

In the past there have been three different approaches in attempting to solve the above-mentioned problem. The earliest one, due to Jerison [12], is a formulation of the key ergodic sequences as (decreasing) martingales on suitable measure spaces which are typically nonfinite (and σ-finite). The next approach, due to Rota [16], is to devise a family of operators on certain function spaces, such as the L^p-spaces, for which the desired limit theorems can be proved and from which one can deduce both the martingale and ergodic theorems by an appropriate specialization of these operators. The third approach is the observation that the proofs of both the martingale convergence and ergodic theorems depend on a maximal theorem (called the upcrossings inequality in the former and the maximal inequality in the latter). Thus a general inequality, capable of including both these results by specialization, has been established by Tulcea and Tulcea [11]; and now the main theorems can be obtained with its help. This description indicates the diverse nature of these methods and the difficulties present at a unification of these two seemingly intimate problems.

[1] Supported under NSF grant GP-30672.

[2] Present address: Mathematics Department, University of California-Riverside, Riverside, California.

Thus after introducing some terminology, motivation, and the statement of the problem in the next section, the three methods will be discussed in the following three sections. The final section is devoted to a few remarks. It is hoped that these viewpoints, outlined in this paper, will give a better appreciation of the problem and that their comparison leads to a possibly more satisfactory solution. New results are contained mainly in Section 3.

1. THE PROBLEM

If (Ω, Σ, P) is a probability space and $\tilde{T}: \Omega \mapsto \Omega$ is a mapping which is measurable and measure preserving (i.e., $\tilde{T}^{-1}(\Sigma) \subset \Sigma$, $P(\tilde{T}^{-1}(E))P(E) =$ for all $E \in \Sigma$), let $\tilde{T}^n = \tilde{T}(\tilde{T}^{n-1})$ with $\tilde{T}^0 =$ identity. Then a classical result, due to Poincaré[13b], says that for almost every $\omega \in E \in \Sigma$, $\tilde{T}^n(\omega) \in E$ for infinitely many n. An important problem in statistical mechanics is thus to determine the mean sojourn time (if any) for the points of E visited by $\tilde{T}$. More precisely, to find out whether

$$\lim_n \frac{1}{n} \sum_{i=0}^{n-1} \chi_E(\tilde{T}^i \omega)$$

exists in some sense. This is a special case of finding the class of measurable mappings $f: \Omega \mapsto \mathbb{R}$ for which $\lim_n (1/n) \sum_{i=0}^{n-1} f(\tilde{T}^i \omega)$ exists in some sense. But then the transformation $U: f \mapsto f \circ \tilde{T}$ is a positive linear mapping on the (linear) class of such functions, and the above problem is subsumed under the generalization: Find conditions on $U: \mathscr{C} \mapsto \mathscr{C}$ where the class $\mathscr{C} \subset L^p(\Omega, \Sigma, P)$, say (the Lebesgue space for $1 \le p < \infty$), such that for each $f \in \mathscr{C}$,

$$\lim_n \frac{1}{n} \sum_{i=0}^{n-1} U^i f \qquad \text{exists in some sense.} \tag{1}$$

Thus ergodic theory concerns the convergence of the sequences, such as (1), involving continuous (in fact, contractive) linear operators U, the type of convergence being pointwise a.e., norm, and such others. If U arises from a measure preserving mapping $\tilde{T}$, as described above, then the pointwise a.e. convergence of (1) with $\mathscr{C} = L^1(\Omega, \Sigma, P)$ is the celebrated individual ergodic theorem, due to Birkhoff [2a], and if $\tilde{T}$ is not necessarily measure preserving, then the corresponding result is due to Hurewicz [10]. If U is any linear operator which is a contraction on both L^1 and L^∞, then a very general result on the pointwise a.e. convergence of (1) is due to Dunford and Schwartz (1955) [7, p. 675]. These major results can be formulated as martingales on a suitable measure space. That possibility, and the proof of the corresponding convergence, constitute a central problem of the present study. For this, a general martingale must be defined. It may be remarked that the pointwise a.e. convergence of (1) is a form of the strong law of large numbers in probability theory, and can also be regarded as a generalization of the latter.

Let $(S, \mathscr{A}, \mu)$ be a measure space and $\mathscr{B} \subset \mathscr{A}$ be a σ-field such that the restriction $\mu|\mathscr{B}$ is σ-finite. If $f: S \mapsto \overline{\mathbb{R}}$ is a measurable function such that either $\int_S f^+ \, d\mu < \infty$ or $\int_S f^- \, d\mu < \infty$, where $f^+ = \max(f, 0)$ and $f^- = f^+ - f$, let $v_f(A) = \int_A f \, d\mu$, $A \in \Sigma$. Then $v_f: \mathscr{A} \mapsto \overline{\mathbb{R}}$ is a σ-additive semibounded (i.e., $+\infty$ or $-\infty$ is *not* in the range of v_f) set function, and since $v|\mathscr{B}$ is $\mu|\mathscr{B}$-continuous, by the Radon–Nikodým theorem (cf. [7, p. 176]), there is a μ-unique $\mathscr{B}$-measurable function $\bar{f}: S \mapsto \overline{\mathbb{R}}$, such that

$$\int_B f \, d\mu = v_f(B) = \int_B \bar{f} \, d\mu_{\mathscr{B}}, \qquad B \in \mathscr{B} \tag{2}$$

where $\mu_{\mathscr{B}} = \mu|\mathscr{B}$. Hence the operator $E^{\mathscr{B}}: f \mapsto \bar{f}$ is well defined and is a positive linear mapping on such f's. In case $f \in L^p(S, \mathscr{A}, \mu)$, $1 \leq p < \infty$, then $\mathscr{B} \subset \mathscr{A}$ can be any σ-field and μ any measure for the truth of this statement and, moreover, $E^{\mathscr{B}}: L^p(S, \mathscr{A}, \mu) \mapsto L^p(S, \mathscr{B}, \mu)$ is a contraction. All these properties are proved, for instance, in [5]. Further, if $\mathscr{B}_1 \subset \mathscr{B} \subset \mathscr{A}$ are σ-fields, then $E^{\mathscr{B}_1}E^{\mathscr{B}}(f) = E^{\mathscr{B}}E^{\mathscr{B}_1}(f) = E^{\mathscr{B}_1}(f)$, a.e. $[\mu]$. In this notation, a sequence of functions $\{f_n, \mathscr{B}_n, n \geq 1\}$ is said to be an *increasing* (or *decreasing*) *martingale* if (i) f_n is $\mathscr{B}_n$-measurable for each n; (ii) the $\mathscr{B}_n \subset \mathscr{A}$ are filtering to the right (left); (iii) each f_n is semiintegrable (i.e., the corresponding v_{f_n} is semibounded); and (iv)

$$E^{\mathscr{B}_m}(f_n) = f_m \quad \text{a.e. if} \quad \mathscr{B}_m \subset \mathscr{B}_n \quad \text{for} \quad m < n, \tag{3}$$

$$(E^{\mathscr{B}_n}(f_m) = f_n \quad \text{a.e. if} \quad \mathscr{B}_m \supset \mathscr{B}_n \quad \text{for} \quad m < n.) \tag{4}$$

The problem here is to express the sequence of (1), when U's are as given in the general Dunford–Schwartz formulation, as a martingale for suitable $\mathscr{B}_n$'s on an appropriate measure space $(S, \mathscr{A}, \mu)$. It will be shown, in the next section, that this is comparatively easy, but that the limit theorem is not easily obtainable.

2. MARTINGALE FORMULATION

Let $\mathbb{N}$ be the set of all natural integers, $\mathscr{N}$ the power set of $\mathbb{N}$, and $\zeta(\cdot): \mathscr{N} \mapsto \mathbb{R}^+$ the counting measure. Let (Ω, Σ, P) be a probability space (or even a σ-finite space) and set $(S, \mathscr{A}, \mu) = (\mathbb{N}, \mathscr{N}\ \zeta) \times (\Omega, \Sigma, P)$, their Cartesian product, so that it is only σ-finite (and nonfinite) if $P(\Omega) > 0$. Let $\mathscr{F}_n = \sigma[\{0, 1, \ldots, n-1\}, \{k\}: k \geq n]$ be the σ-field generated by the sets shown, so that the sets $\{k\}$, $k \geq n$, and $\{0, 1, \ldots, n-1\}$ are atoms of $\mathscr{F}_n(\supset \mathscr{F}_{n+1})$. Let $\mathscr{B}_n = \mathscr{F}_n \times \Sigma \subset \mathscr{A}$. Then $\mu|\mathscr{B}_n$ is σ-finite for each n $(\mu(S) = +\infty)$ and $\mathscr{B}_n \supset \mathscr{B}_{n+1}$. To see that the sequence of (1) can be expressed as a martingale on $(S, \mathscr{A}, \mu)$, let $T\colon L^1(\Sigma)\mapsto L^1(\Sigma)$ $(=L^1(\Omega, \Sigma, P))$ be a continuous linear operator and $f \in L^1(\Sigma)$. For each $(k, \omega) \in S$, define $h(k, \omega) = (T^k f)(\omega)$, and

$$h_n(k, \omega) = \begin{cases} \dfrac{1}{n}\displaystyle\sum_{i=0}^{n-1}(T^i f)(\omega) & \text{for} \quad (k, \omega) \in \{0, 1, \ldots, n-1\} \times \Omega, \\ h(k, \omega) & \text{for} \quad (k, \omega) \in \{k\colon k \geq n\} \times \Omega. \end{cases} \tag{5}$$

Here and in what follows $(Tf)(\omega)$ stands for $(T\tilde{f})(\omega)$ where $\tilde{f}$ is a member of the equivalence class of f. As usual, this will be understood in similar situations, and not be pointed out separately.

The following result, due to Jerison [12], can now be established.

Theorem 1. *With the assumptions and notation of* (5), *h is semiintegrable on S, and $\{h_n, \mathscr{B}_n, n \geq 1\}$ is a decreasing martingale. In fact, $E^{\mathscr{B}_n}(h) = h_n$, a.e. $[\mu]$ and the a.e. convergence of the martingale is equivalent to that of the sequence* (1).

Proof. The simple proof will be given here to illustrate the problem. First note that if $\nu_h(A) = \int_A h\, d\mu$, $A \in \mathscr{A}$, then ν_h is semibounded. In fact, the variation measure $|\nu_h|(\cdot)$ is σ-finite, which trivially implies the asserted property of ν_h. To see the former property, h (and h_n) being clearly measurable, let $S_k = \{k\} \times \Omega \in \mathscr{A}$, and $S = \bigcup_{k=1}^{\infty} S_k$. Then $|\nu_h|(S_k) = \int_{S_k}|h|\, d\mu = \int_\Omega |T^k f|\, dP \leq \|T\|^k \int_\Omega |f|\, dP < \infty$, so that $|\nu_h|$ is σ-finite. Since $\mu(S_k) < \infty$ and $\mu|\mathscr{B}_n$ is σ-finite (and if P is σ-finite, a trivial modification is needed) $E^{\mathscr{B}_n}$ is well defined for each n, by (2) of the preceding section. From the fact that $\mathscr{B}_n \subset \mathscr{B}_{n-1}$, if it is shown that $E^{\mathscr{B}_n}(h) = h_n$ a.e. $[\mu]$, it will follow that $\{h_n, \mathscr{B}_n, n \geq 1\}$ is a decreasing martingale. Indeed, since h_{n+1} is $\mathscr{B}_{n+1}$-measurable (hence, $\mathscr{B}_n$-measurable), one has

$$E^{\mathscr{B}_{n+1}}(h_n) = E^{\mathscr{B}_{n+1}}(E^{\mathscr{B}_n}(h)) = E^{\mathscr{B}_n}(E^{\mathscr{B}_{n+1}}(h)) = E^{\mathscr{B}_n}(h_{n+1}) = h_{n+1} \qquad \text{a.e.} \tag{6}$$

To prove that $E^{\mathscr{B}_n}(h) = h_n$ a.e., it suffices to show, by (2) that

$$\int_A h_n\, d\mu_{\mathscr{B}_n} = \int_A E^{\mathscr{B}_n}(h)\, d\mu_{\mathscr{B}_n} = \int_A h\, d\mu, \qquad A \in \mathscr{B}_n. \tag{7}$$

Since $S = \bigcup_{k=1}^{\infty} S_k$, as above, and $S_k \notin \mathscr{B}_n$ for $0 \leq k \leq n-1$, it follows, from the definition of $\mathscr{B}_n$, that $\bigcup_{k=0}^{n-1} S_k \in \mathscr{B}_n$. Thus for any $A \in \mathscr{B}_n$, if $A_1 = A \cap \bigcup_{k=0}^{n-1} S_k$, $A_2 = A - A_1$, then $A_i \in \mathscr{B}_n$, $i = 1, 2$. By (5), $h_n = h$ on A_2, so that (7) is true when A is replaced by A_2. On A_1, h_n is a constant (in the first component). Since $A_1 \subset \bigcup_{k=0}^{n-1} S_k$, it can be expressed as $A_1 = \mathbb{N}_n \times E \in \mathscr{B}_n$ where $\mathbb{N}_n = \{0, 1, \ldots, n-1\} \in \mathscr{F}_n$, $E \in \Sigma$, because $\mathbb{N}_n$ is an atom of $\mathscr{F}_n$. Thus

$$\begin{aligned} \int_{A_1} h_n\, d\mu &= \int_{\mathbb{N}_n}\int_E h_n(k, \omega)\, dP\, d\zeta = n\int_E h_n(k, \omega)\, dP, \qquad k \in \mathbb{N}_n, \\ &= \int_E \sum_{i=0}^{n-1}(T^i f)(\omega)\, dP \qquad \text{by (5).} \end{aligned} \tag{8}$$

On the other hand,

$$\int_{A_1} h(k, \omega)\, d\mu = \int_{\mathbb{N}_n} \int_E (T^k f)(\omega)\, dP\, d\zeta = \int_E \sum_{k=0}^{n-1} (T^i f)(\omega)\, dP. \tag{9}$$

Thus (8) and (9) imply the truth of (7) on A_1 and hence in general. This proves the martingale property.

If $h_n \to h_0$ a.e. $[\mu]$, then h_0 is $\mathscr{B}(= \bigcap_{n=1}^{\infty} \mathscr{B}_n)$-measurable. Since every set in $\mathscr{B}$ is of the form $\mathbb{N} \times E$ for some $E \in \Sigma$, h_0 must be independent of the first variable and hence is a function of ω only. This is equivalent to the statement that $(1/n) \sum_{i=0}^{n-1} (T^i f)(\omega) \to h_0(\omega)$, a.e. $[P]$, as asserted.

A similar formulation can be given for abstract Banach (or B-) space-valued ergodic averages of the type considered by Yosida and Kakutani. (For an account, see [7, p. 661].) Here is an outline of this formulation.

Let $\mathscr{X}$ be a B-space and $T\colon \mathscr{X} \mapsto \mathscr{X}$ be a bounded linear operator. Then the ergodic averages in this case are $\{A_n x\}$, where $A_n = (1/n) \sum_{i=0}^{n-1} T^i$ and $x \in \mathscr{X}$. This can be viewed as an abstract martingale in much the same way as the L^p-case. Thus let $\mathbb{N}$ be as above and $\tau_n = \{0, 1, \ldots, n-1\}$ so that $\tau_n \in \mathscr{N}$ and the latter is ordered by inclusion. Thus $\tau_n < \tau_{n+1}$ if $\tau_n \subset \tau_{n+1}$. Let $f\colon \mathbb{N} \mapsto \mathscr{X}$ be a mapping defined by $f(k) = T^k x$, $x \in \mathscr{X}$, and let

$$h_{\tau_n} = (A_n x)\chi_{\tau_n} + f \cdot \chi_{\mathbb{N} - \tau_n}. \tag{10}$$

Then the mapping $E_{\tau_n}\colon f \mapsto h_{\tau_n}$ is an idempotent linear operator and, as a simple computation shows, $E_{\tau_n} E_{\tau_{n+1}} = E_{\tau_{n+1}} E_{\tau_n} = E_{\tau_{n+1}}$. So $\{h_{\tau_n}, E_{\tau_n}, n \geq 1\}$ is a decreasing martingale sequence in the sense that $E_{\tau_{n+1}}(h_{\tau_n}) = h_{\tau_{n+1}}$. Then the Yosida–Kakutani (mean ergodic) sequence is equivalent to the above abstract martingale and one converges if and only if the other does.

It must be noted that neither h, h_n, nor h_{τ_n} belongs to the spaces $L^1(S\, \mathscr{A}, \mu)$ or $\mathscr{X}$ if f or x is nontrivial. Hence no known martingale convergence theorem is applicable to conclude the convergence of the corresponding ergodic sequences. On the other hand, the known Dunford–Schwartz and Yosida–Kakutani ergodic theorems give general conditions for the convergence of the ergodic sequences, and hence, under these conditions, there appear some "new" martingale convergence theorems. Thus the preceding results tell very little beyond indicating the connections between these theories. This disappointment was strongly expressed by Jerison: "It seems most unlikely that a proof of an individual ergodic theorem might be based, in a straightforward way, upon the martingale convergence theorem," (see, e.g. [12, p. 536]). However, a special ergodic theorem, when T is "metrically transitive" coincides with a strong law of large numbers for independent random variables with a common distribution function. A proof of the latter *can* be based, in a straightforward way, upon the martingale convergence theorem (see, e.g. [6, pp. 341–342]). This again sustains the hope that a general theorem including both these results can be discovered in the framework of martingale theory. The theory of stopping-time transformations may be useful here, although no such proof has ever been given. Clearly a deeper analysis of the problem is in order, and this is still an open question.

In the next two sections, two different approaches will be described, as they yield positive solutions to some questions at hand.

3. AN OPERATOR THEORETIC APPROACH

The connections discussed in the preceding section show that both the martingale and ergodic theories are related through an appropriate family of mutually commuting operators $\{E^{\mathscr{B}_n}\}$. Amplifying this point, one may try to find a family of linear operators on a function space (for the pointwise convergence), such as L^p, which can be sepecialized to yield both these theorems. The following is one approach, advanced by Rota in [16], and since it looks somewhat involved, it will be considered here at some length. This should be of interest to many readers since the details of some points in [16] have not been published before.

Recall that for a sequence of scalars $\{a_n, n \geq 0\}$, $\sum_{n=0}^{\infty} a_n$ converges in the sense of $(C, 1)$, or Cesàro mean, if the sequence of averages $s_n = 1/n \sum_{i=0}^{n-1} a_i$ converges, and it A-converges, or converges in the sense of Abel mean, if $\lim_{x\to 1-0} (1-x) \sum_{n=0}^{\infty} a_n x^n$ exists. While the $(C, 1)$ convergence implies A-convergence, the converse implication is false in general, but holds true provided the sequence $\{s_n, n \geq 1\}$ is bounded in addition to the A-convergence. This is an ancient result of Littlewood. (See Hille [9, p. 251] for a discussion of these concepts.) If $\{a_n\}$ is replaced by $\{(T^n f)(\omega)\}$, the (C, 1) means are then the ergodic averages (1). Looking at the A-convergence part, this becomes formally

$$\lim_{x\to 1-0} (1-x)\sum_{n=0}^{\infty} x^n (T^n f)(\omega) = \lim_{x\to 1-0} (1-x)[(I-xT)^{-1}f](\omega)$$
$$= \lim_{y\to 1+0} (y-1)[(y-T)^{-1}f](\omega), \qquad (11)$$

on replacing x by $1/y$. Now noting that $(y-T)^{-1} = R_y(T)$ is the resolvent of the operator T, it follows from what precedes that the pointwise a.e. limit of (1) exists whenever

$$\sup_n \left| \frac{1}{n} \sum_{i=0}^{n-1} (T^i f)(\omega) \right| < \infty \qquad \text{a.e.}$$

and $\lim_{y\to 1+0} (y-1)(R_y f)(\omega)$ exists a.e. But R_y can be expressed in the following more suggestive form, at first formally.

$$\lim_{x\to 1-0} (1-x)R_x = \lim_{x\to 1-0} \sum_{n=0}^{\infty} (1-x)x^n T^n$$

$$= \lim_{\lambda\to 0^+} \sum_{n=0}^{\infty} [1-\exp(-\lambda)] \exp(-n\lambda) T^n. \tag{12}$$

Define $U(t) = T^n$ for $n \le t < n+1$, $n \ge 0$, so that (12) becomes

$$\lim_{\lambda\to 0+} \lambda R_\lambda f = \lim_{\lambda\to 0^+} \lambda \int_0^\infty e^{-t\lambda} U(t) f\, dt, \qquad f \in L^p(\Sigma), \tag{13}$$

where the integral is an abstract Riemann integral. Thus $\{T^n f\}$ has a $(C, 1)$ limit when the $A - \lim_{\lambda\to 0^+} \lambda R_\lambda f$ exists and a boundedness condition, stated above, holds. But $\{U(t), t \ge 0\}$ is a family of commuting linear operators such that $U(t+s) = U(t)U(s)$ for $s, t \ge 0$ and $\|U(t)f - f\|_p \to 0$ as $t \to 0^+$. Thus $\{U(t), t \ge 0\}$ is a family of strongly continuous semigroup of operators which moreover has the property, when T arises from a measure preserving transformation τ (so $U(t): f \mapsto f \circ \tau^n$ for $n \le t < n+1$, and the T here is the same as the U leading to (1)), that $U(t)(fg) = (U(t)f)(U(t)g)$, for $f, g \in L^1 \cap L^\infty$, i.e., *it is also a homomorphism*. Thus in (13), $\{U(t), t \ge 0\}$ is a strongly continuous semigroup of operators on L^p such that for each $t \ge 0$, $U(t)$ is a homomorphism. The above computations in (11)–(13), which are formal thus far, can now be justified rigorously by use of the operational calculus (cf. [7, Ch. VII]).

The equation of (13) defining R_λ for the particular semigroup $\{U(t), t \ge 0\}$ is the key connecting link between the ergodic theory, from which it sprang in the first place as seen above, and a class of well-behaved transformations called Reynolds operators, studied in great detail by Kampe de Fériet and others (cf. [17] for an extensive bibliography). They occur in the statistical theory of turbulence and this class includes the conditional expectation operators which are crucial for the martingale theory. What is more, this class will now be shown to include the operators defined by (13). Thus there is a strong possibility of realizing a general enough theory which includes both the results under consideration.

In order to show that $\{R_\lambda, \lambda > 0\}$, defined by (13) are also related to Reynolds operators, it is necessary to recall the latter, which have been defined axiomatically, based on experimental evidence, in turbulence work. Thus a continuous linear mapping $R: L^p \mapsto L^p$ is a *Reynolds operator* if the following algebraic identity holds.

$$R(fg) = (Rf)(Rg) + R[(f - Rf)(g - Rg)], \qquad f, g \in L^p \cap L^\infty(\Sigma). \tag{14}$$

Since a conditional operator $E^{\mathscr{B}}$ is a projection and an averaging (i.e., $E^{\mathscr{B}}E^{\mathscr{B}} = E^{\mathscr{B}}$ and $E^{\mathscr{B}}(fE^{\mathscr{B}}g) = (E^{\mathscr{B}}f)(E^{\mathscr{B}}g)$ for $f, g \in L^p \cap L^\infty$), it is trivial that $R = E^{\mathscr{B}}$ satisfies (14). So it is only nontrivial to show that for each $\lambda > 0$, R_λ

of (13), or rather λR_λ, also satisfies (14). This was noticed in [17], and an inclusion of a proof here will exhibit the depth of the problem and illuminate the discussion.

Since $\{U(t),\ t \geq 0\}$ is a strongly continuous semigroup on L^p, i.e., $\|U(t)f - f\|_p \to 0$, as $t \to 0$ for each $f \in L^p$, the function $t \mapsto U(t)f$ is (strongly) differentiable and one has

$$\frac{d}{dt}U(t)f = \lim_{h\to 0}\frac{U(t+h)f - U(t)f}{h} = \lim_{h\to 0} U(t)\cdot\frac{U(h)f - f}{h}$$

$$= U(t)Df = DU(t)f, \tag{15}$$

for f in a dense set $\mathscr{D}$ of L^p, and D is the so-called infinitesimal generator of the semigroup with domain $\mathscr{D}$. (D is the "derivative" of the operators at the origin when the limits are taken in the strong operator topology, and it is an unbounded but closed operator in general.) Moreover, R_λ and $U(t)$ commute and that $R_\lambda = (y - D)^{-1}$. These are consequences of the operational calculus and related results (cf. [7, VIII. 1.11]) for semigroups. Replacing f by $R_\lambda f = (\lambda - D)^{-1}f$ in (15), one gets

$$\frac{d}{dt}U(t)R_\lambda f = U(t)[-I + \lambda R_\lambda]f. \tag{16}$$

This and the homomorphism property of $U(t)$ yield the following identities when $R_\lambda f$ is replaced by $R_\lambda f \cdot R_\lambda g$ in (16), $f, g \in L^1 \cap L^\infty$:

$$\frac{d}{dt}U(t)(R_\lambda f\cdot R_\lambda g) = \frac{d}{dt}(U(t)R_\lambda f)\cdot U(t)R_\lambda g + U(t)R_\lambda(f)\cdot\frac{d}{dt}(U(t)R_\lambda g)$$

$$= U(t)[(\lambda R_\lambda - I)fR_\lambda g + R_\lambda f\cdot(\lambda R_\lambda - I)g]. \tag{17}$$

On the other hand, by (13), R_λ is given by an integral. Hence integration by parts (cf. [7, p. 154]) and the fact that $U(0) = I$ together imply:

$$R_\lambda(R_\lambda f\cdot R_\lambda g) = \int_0^\infty e^{-t\lambda}U(t)(R_\lambda f\cdot R_\lambda g)\,dt$$

$$= -\frac{e^{-t\lambda}}{\lambda}\cdot U(t)R_\lambda f\cdot R_\lambda g\Big|_{t=0}^{t=\infty} + \frac{1}{\lambda}\int_0^\infty e^{-t\lambda}\frac{d}{dt}U(t)(R_\lambda f\cdot R_\lambda g)\,dt$$

$$= \frac{1}{\lambda}R_\lambda f\cdot R_\lambda g + 2\int_0^\infty e^{-t\lambda}U(t)(R_\lambda f\cdot R_\lambda g)\,dt$$

$$-\frac{1}{\lambda}\left[\int_0^\infty e^{-t\lambda}U(t)(fR_\lambda g + g\cdot R_\lambda f)\,dt\right], \quad \text{by (17)}$$

Using the expression in (13) again for R_λ, this simplifies to:

$$\lambda R_\lambda(R_\lambda f \cdot R_\lambda g) = R_\lambda f \cdot R_\lambda g + \lambda R_\lambda \left[2R_\lambda f \cdot R_\lambda g - \frac{fR_\lambda g}{\lambda} - \frac{gR_\lambda f}{\lambda}\right]. \quad (18)$$

Setting $\lambda R_\lambda = V_\lambda$ in (18), and adding $V_\lambda(fg)$ to both sides, this becomes

$$V_\lambda(fg) = V_\lambda f \cdot V_\lambda g + V_\lambda[(f - V_\lambda f)(g - V_\lambda g)], \qquad f, g \in L^1 \cap L^\infty. \quad (19)$$

Thus $V_\lambda = \lambda R_\lambda$ is a Reynolds operator for *each* $\lambda > 0$, as asserted. Since R_λ and R_μ commute for λ, $\mu > 0$, it follows that the resolvent operator R_λ for $\lambda = 1$, and more generally λR_λ for all $\lambda > 0$, is included in the class of Reynolds operators, and hence it is a natural class to consider for a unified theory.

Since R_λ is also known, from the semigroup theory, to be the resolvent operator of D, the infinitesimal generator of $\{U(t), t \geq 0\}$ on L^p, $p \geq 1$, it satisfies another algebraic equation: $R_\lambda - R_\mu = (\lambda - \mu)R_\lambda R_\mu$ which is immediately verified by a direct substitution of $R_\lambda = (\lambda - D)^{-1}$. Now replacing R_λ by $(1/\lambda)V_\lambda$ in this equation, one obtains an identity that must be satisfied in addition to (15) for the problem at hand. This can be stated thus.

Following [16], a family $\{V_\lambda, \lambda > 0\}$ of mutually commuting Reynolds operators on $L^p(\Sigma)$ is called a *generalized martingale* if the following algebraic identity is satisfied:

$$(\mu V_\lambda - \lambda V_\mu)V_\mu f = (\mu - \lambda)V_\lambda V_\mu^2 f, \qquad 0 < \lambda < \mu, \qquad f \in L^1 \cap L^\infty. \quad (20)$$

From the preceding discussion, it follows that with $V_\lambda = E^{\mathscr{B}_\lambda}$, where $\Sigma \supset \mathscr{B}_\lambda \supset \mathscr{B}_\mu$ for $\lambda < \mu$, or $V_\lambda f = \lambda R_\lambda f = \lambda \int_0^\infty e^{-\lambda t}U(t)f\,dt$, the sequence $\{V_\lambda f, \lambda > 0\}$ represents a martingale or an ergodic average, respectively. Hence a limit theorem for this sequence is the desired result. However, there is now a new problem to resolve before the limit theorem can be considered. Namely, while a martingale or an ergodic sequence is also a generalized martingale, there may be other objects that are included in the latter. Hence it is desirable to characterize the class V_λ of operators (and this turns out to be essential for a proof of the limit theorem later) defining a generalized martingale. The results are as follows.

The next result, on the structure of Reynolds operators, shows why the unification of the ergodic and martingale problems is contained in this class. Since ergodic theory is based on general measure spaces (Ω, Σ, μ), it is important that $\mu(\Omega) = \infty$ is allowed in the treatment here.

Theorem 2 (Structure). *Let (Ω, Σ, μ) be a strictly localizable measure space; i.e., there exists a disjoint family $\{A_\alpha, \alpha \in I\} \subset \Sigma$ such that $\bigcup_{\alpha \in I} A_\alpha \doteq \Omega$ (i.e., except for a null set), $0 < \mu(A_\alpha) < \infty$ and each $A \in \Sigma$, $0 < \mu(A) < \infty$ satisfies $A \doteq \bigcup_{\alpha \in J}(A \cap A_\alpha)$ where $J \subset I$ is countable. (If μ is σ-finite, this means that $I = J$, a countable set.) Let $V_\lambda: L^p(\Sigma) \mapsto L^p(\Sigma)$, $1 \leq p < \infty$, be a contractive* *linear operator satisfying* (14) *and such that:* (i) *in case* $p = 1$, *it is also weakly compact* (*i.e., maps the unit ball of* L^1 *into a relatively weakly compact set and this is automatic in* L^p *for* $1 < p < \infty$), (ii) $V_\lambda(L^p(\Sigma) \cap L^\infty) \subset L^p(\Sigma) \cap L^\infty$, *and* (iii) $V_\lambda \chi_{A_\alpha} = \chi_{A_\alpha}$, $\alpha \in I$ *for a distinguished sequence in the definition of strict localizability. Then there exist uniquely* (a) *a* σ*-field* $\mathscr{B}_\lambda \subset \Sigma$ *such that* $\{A_\alpha, \alpha \in I\} \subset \mathscr{B}_\lambda$, (b) $E^{\mathscr{B}_\lambda}: L^p(\Sigma) \mapsto L^p(\mathscr{B}_\lambda)$, *the conditional expectation operator, and* (c) *a strongly continuous semigroup* $\{U(t), t \geq 0\}$ *of operators on* $L^p(\mathscr{B}_\lambda)$ *induced by a measure preserving transformation* (*so they have the desired homomorphism property!*), *in terms of which one has*

$$V_\lambda f = \lambda \int_0^\infty e^{-\lambda t} U(t) E^{\mathscr{B}_\lambda}(f)\, dt, \qquad f \in L^p(\Sigma), \tag{21}$$

the integral being a "vector Lebesgue" or Bochner integral; moreover, (d) $V_\lambda(L^p(\mathscr{B}_\lambda)) \subset L^p(\mathscr{B}_\lambda)$, *the former being dense in the latter, and* (e) V_λ *and* $E^{\mathscr{B}_\lambda}$ *commute, and* $V_\lambda E^{\mathscr{B}_\lambda} = E^{\mathscr{B}_\lambda} V_\lambda = V_\lambda$.

The proof of this result is long and involved. The important case when $p = 2$, $\mu(\Omega) < \infty$ (so that $I = \{1\}$ and $A_1 = \Omega$ in the above), was first proved by Rota [17]. The general case, which is a nontrivial extension of [17], was proved by the author in [14]. The argument is not simplified even if μ is σ-finite instead of being localizable. An interesting point of Theorem 2 is that V_λ is a "composition" of $\{U(t)\}$ and $E^{\mathscr{B}_\lambda}$ operators and hence it is no "surprise" that both ergodic and martingale theories are included in this generalization.

The main result on convergence of the sequences can now be given as follows.

Theorem 3. *Let* $\{V_\lambda f, \lambda > 0\}$ *be a generalized martingale for* $f \in L^p(\Omega, \Sigma, \mu)$, $1 \leq p < \infty$. *Suppose that for each* $\lambda > 0$, V_λ *satisfies the hypothesis of Theorem* 2 *relative to a fixed family* $\{A_\alpha, \alpha \in I\}$. *Then for each* $f \in L^p(\Sigma)$, $\lim_{\lambda \to 0^+} V_\lambda f$ *exists both in norm and pointwise a.e.,* $[\mu]$. (*A similar statement holds if* $\lambda \to +\infty$ *also.*)

Proof. If $\mathscr{B}_0 = \sigma(\bigcup_{\lambda > 0} \mathscr{B}_\lambda)$, then $L^p(\mathscr{B}_0) = \overline{\mathrm{sp}}\{\bigcup_{\lambda > 0} L^p(\mathscr{B}_\lambda)\}$ and thus the latter union is (norm) dense in $L^p(\mathscr{B}_0)$. If $f \in L^p(\mathscr{B}_\lambda)$, then (21) implies

$$R_\lambda f = \lambda \int_0^\infty e^{-\lambda t} U(t) f\, dt,$$

and, since, by the Structure Theorem above (cf. [14] and [17]), R_λ is the resolvent of the infinitesimal generator D of the semigroup $\{U(t), t \geq 0\}$, it follows [7, VIII. 9.21] that $\lim_{\lambda \to 0^+} R_\lambda f$ exists a.e. (In [7], "λ" is $1/\lambda$ in the present notation.) Thus $\lim_{\lambda \to 0^+} R_\lambda f$ exists a.e. for f in a dense set of elements in $L^p(\mathscr{B}_0)$.

Also, R_λ is a positive contraction (in both L^1 and L^∞ norms) by Theorem 2. For this proof, it suffices to consider the case that $\mu(\Omega) < \infty$, since the general case can be deduced from this. If $1 < p < \infty$, both R_λ and $E^{\mathscr{B}_\lambda}$ are weakly compact and for $p = 1$, this is implied by the hypothesis. So $\{E^{\mathscr{B}_\lambda}(f), \lambda > 0\}$ is a uniformly integrable (separable) martingale, and $f^* = \sup_{\lambda>0} E^{\mathscr{B}_\lambda}(|f|) \in L^p$ by a standard result in martingale theory. Hence $|E^{\mathscr{B}_\lambda}(|f|)| \le f^*$ and one has a.e.,

$$|R_\lambda f| \le R_\lambda |f| \le \lambda \int_0^\infty e^{-\lambda t} U(t) E^{\mathscr{B}_\lambda}(|f|)\, dt \le \int_0^\infty e^{-\lambda t} U(t) f^*\, dt, \tag{21$'$}$$

for $f \in L^p(\mathscr{B}_0)$ since $U(t)$ is a positive operator. Thus

$$R_\lambda |f| \le \int_0^\infty U(t) f^* \beta_\lambda(t)\, dt$$

where $\beta_\lambda(t) = \lambda e^{-\lambda t}$ defines a decreasing positive function $\beta_\lambda\colon \mathbb{R}^+ \mapsto \mathbb{R}^+$, for each $\lambda > 0$. But $\{U(t), t \ge 0\}$ is a positive contraction (strongly continuous) semigroup on L^1 and L^∞, so that by [7, VII. 9.3] there exists an a.e. finite function $f_0{}^*$ such that

$$R_\lambda |f| \le \int_0^\infty U(t) f^* \beta_\lambda(t)\, dt \le f_0{}^* \int_0^\infty \beta_\lambda(t)\, dt = f_0{}^* \qquad \text{a.e.} \tag{*}$$

Consequently $\sup_{\lambda>0} |R_\lambda f| \le f_0{}^* < \infty$ a.e. for each $f \in L^p(\mathscr{B}_0)$. Thus by Banach's theorem [7, p. 332], $\lim_{\lambda\to 0^+} R_\lambda f$ exists a.e. for each $f \in L^p(\mathscr{B}_0)$. Since, by Theorem 2(e), $R_\lambda = R_\lambda E^{\mathscr{B}_\lambda} = E^{\mathscr{B}_\lambda} R_\lambda$, and since $E^{\mathscr{B}_\lambda}(f) = E^{\mathscr{B}_\lambda} E^{\mathscr{B}_0}(f)$ a.e. for all $f \in L^p(\Sigma)$, (*) is true for all $f \in L^p(\Sigma)$. This proves the result in the pointwise case and the parenthetical statement is analogous.

The norm convergence of this result (when L^p spaces are replaced by a more general space—the so-called Banach function space L^ρ) was proved [14], and the pointwise case was indicated there very briefly. When $\mu(\Omega) = 1$, $1 < p < \infty$, the pointwise convergence was first obtained by Rota [16]. The key step in the pointwise result, as always, is to have a maximal theorem. Independently of [14], and almost simultaneously, this maximal theorem was published by Edwards [8], which generalizes an earlier result of Rota. All of this is for the case of Abel limits.

The following proposition on $(C, 1)$ averages can now be deduced (non-trivially) from the preceding theorem and that is the sought for result.

Proposition 4. *The abstract ergodic theorem for measure preserving transformations (i.e., G. D. Birkhoff's case) and the martingale convergence (in the general case) are consequences of Theorem 3. More explicitly,* $\lim_{\lambda\to 0^+} E^{\mathscr{B}_\lambda}(f)$ *exists a.e. and in norm*, and *lim*$_{t\to\infty}$ $(1/t)\int_0^t U(s)f\, ds$ *exists a.e. and in norm for each* $f \in L^p(\Sigma)$, $1 \le p < \infty$, *where* $\{U(s), s \ge 0\}$ *is the semigroup given in Theorem 2.*

Proof. As noted earlier (cf. also [9, p. 254]), the result on Abel averages implies the $(C, 1)$-convergence provided the following two conditions hold:

(i) $\sup_{t>0} \left| \frac{1}{t} \int_0^t (U(s)f)(\omega)\, ds \right| < \infty$ for almost all ω;

(ii) $\left\{ \frac{1}{t} \int_0^t (U(s)f)(\omega)\, ds \right\}$ is feebly oscillating as $t \to \infty$ for almost all ω (the definition recalled below);

where $\{U(s)\ s \geq 0\}$ is the (strongly) continuous semigroup of linear operators on $L^p(\Sigma)$ induced by a measurable and measure preserving transformation $\tau\colon \Omega \mapsto \Omega$.

To prove (i), note that $U(t)\colon f \mapsto f \circ \tau^t$ is a positive linear operator such that $\|U(t)\|_1 \leq 1$, and $\|U(t)\|_\infty \leq 1$ for all $t \geq 0$, i.e., is a contraction on L^1 and L^∞. By [7, VIII, 9.3], for such a family, for each $f \in L^p$ there exists an f^* which is finite a.e. $[\mu]$ such that

$$\int_0^\infty |(U(t)f)(\omega)|\alpha(t)\, dt \leq f^*(\omega) \int_0^\infty \alpha(t)\, dt, \tag{22}$$

where $\alpha(\cdot)\colon (0, \infty) \mapsto \mathbb{R}^+$ is a decreasing function and $f^* \in L^p$ if $p > 1$. Now let $\alpha_s(\cdot) = (1/s)\chi_{(0,s)}$ for any fixed but arbitrary $s > 0$. Then (22) becomes

$$\left| \frac{1}{s} \int_0^s (U(t)f)(\omega)\, dt \right| \leq \frac{1}{s} \int_0^s |(U(t)f)(\omega)|\, dt \leq f^*(\omega), \qquad \text{a.e. } [\mu]. \tag{23}$$

Thus (23) implies (i) at once.

As regards (ii), recall a function $g\colon (0, \infty) \mapsto \mathbb{R}$ is *feebly oscillating* as $t \to \infty$, if $\lim_{t\to\infty} |g(t) - g(r)| = 0$ when $t/r \to 1$. In the present case, let $g(t) = (1/t)\int_0^t (U(s)f)(\omega)\, ds$. Then for $r < t$

$$\begin{aligned} |g(t) - g(r)| &\leq \left| \frac{1}{t} - \frac{1}{r} \right| \int_0^r |(U(s)f)(\omega)|\, ds + \frac{1}{t} \int_r^t |(U(s)f)(\omega)|\, ds \\ &\leq \frac{t-r}{t} \cdot \frac{1}{r} \int_0^r |(U(s)f)(\omega)|\, ds + \frac{1}{t} \int_0^{t-r} |(U(\tau + r)f)(\omega)|\, d\tau \\ &\leq \frac{t-r}{t} f^*(\omega) + \frac{t-r}{t} f^*(\omega), \end{aligned} \tag{24}$$

since $\{W_r(t),\ t \geq 0\}$ satisfies the hypothesis leading to (22)–(23), where $W_r(t) = U(t + r)$. As $t \to \infty$ with $t/r \to 1$, it follows from (24) that $g(\cdot)$ is feebly oscillating, so that (ii) is true.

From (i) and (ii) and the Abel convergence established in Theorem 3 for the semigroup $\{U(t), t \geq 0\}$, it follows that $\lim_{t\to\infty} (1/t) \int_0^t U(s)f\, ds$ exists a.e.

$[\mu]$ and in L^p-norm for each $f \in L^p(\Sigma)$. This implies the ergodic theorem for the continuous parameter (i.e., flows).

The martingale convergence is a direct consequence. In fact, if $V_t = E^{\mathscr{B}_t}$, then $\mathscr{B}_t \uparrow$ as $t \downarrow 0$, and $\{E^{\mathscr{B}_t}(f), \mathscr{B}_t, t > 0\}$ is a martingale "closed on the left." Hence $E^{\mathscr{B}_t}(f) = V_t f \to f_0$ a.e. and in norm as $t \to 0^+$ by Theorem 3. This is the key statement. A more general statement for martingales which are not closed and for which the mean convergence is false can then be deduced from this case by well-known techniques (cf., e.g., [15, p. 149]). The above result is the continuous parameter version of [1]. This completes the proof of the proposition.

The results of this section may be considered as one solution of a unified theory. It is clear, however, that this does not tell the whole story. In particular, if $\{U^n f\}$ does not correspond to the measure preserving case, but is more general, as in the Dunford–Schwartz case (or even Hurewicz's case), then the corresponding V_λ need not satisfy (14), since U need not be a homomorphism (or U need not be isometric), so that the computations between (16)–(19) (or those for the proof of Theorem 2) break down. Thus other methods are desirable and the next section is devoted to one such which takes care of some of the above problems.

4. A MAXIMAL INEQUALITY

Since the crucial part of the proofs of the pointwise convergence theory, of both the ergodic and martingale sequences, is to establish certain maximal inequalities (with essentially the same ideas and methods), it is natural to look for a global maximal theorem which yields both these results. Such a theorem was proved by Tulcea and Tulcea [11], generalizing some earlier ideas of Chacón's [3] and Oxtoby's [13a]. This is the third approach mentioned earlier.

The desired result can be stated as follows.

Theorem 5 (Maximal inequality). *Let* $T_i: L^1 \cap L^\infty(\Sigma) \mapsto L^1 \cap L^\infty(\Sigma)$, $i = 0, 1, \ldots, n_0$, *be linear operators such that* $\|T_i\|_1 \le 1$, *and* $\|T_i\|_\infty \le 1$. *Suppose that the following two conditions also hold.* (a) $T_i T_{i-1} = T_i$, $i = 1, \ldots, n_0$, *and* (*b*) $T_0 =$ *identity. If for* $f \in L^p(\Sigma)$, *and* $a > 0$,

$$E_f(a) = \left\{\omega\colon \sup_{\substack{j \le n_0 \\ 0 \le n < \infty}} \frac{1}{n+1} \left| \sum_{i=0}^{n} (T_j^i f)(\omega) \right| > a \right\}, \tag{25}$$

then

$$\mu(E_f(2a)) \le \frac{1}{a} \|f\|_1. \tag{26}$$

The fact that the T_i are contractions on these two spaces (and hence are defined and are contractions on each $L^p(\Sigma)$, $1 \leq p \leq \infty$, by the Riesz convexity theorem) is crucial for the proof here. Such operators are called *Dunford–Schwartz operators*. Taking $n_0 = 1$, $T_1 = T$, the inequality (26) implies the maximal theorem for the Dunford–Schwartz ergodic theorem. In fact letting

$$\tilde{E}_f(a) = \left\{\omega\colon \sup_n \left|\frac{1}{n}\sum_{i=0}^{n-1}(T^i f)(\omega)\right| > a\right\} \dot{\subset} E_f(a),$$

one gets $\mu(\tilde{E}_f(2a) \leq (1/a)\|f\|_1$, and from this the usual proof [7, p. 676] of the individual ergodic theorem, stating that $\lim_{n\to\infty} (1/n)\sum_{i=0}^{n-1} (T^i f)(\omega)$ exists a.e. $[\mu]$ can be quickly given. The details may be found in [11].

If $T_j^2 = T_j$, then

$$\frac{1}{n}\left|\sum_{i=0}^{n-1}(T_j^i f)(\omega)\right| = \left|\frac{f(\omega)}{n} + \frac{n-1}{n}(T_j f)(\omega)\right|,$$

and hence $\lim_n (1/n)|\sum_{i=0}^{n-1} (T_j^i f)(\omega)| = |(T_j f)(\omega)|$. Consequently, if

$$G_f(a) = \left\{\omega\colon \sup_{0\leq j\leq n_0} |(T_j f)(\omega)| > a\right\}(\dot{\subset} E_f(a)),$$

then $\mu(G_f(a)) \leq (1/a)\|f\|_1$. This is sufficient to prove the martingale convergence theorem even somewhat more generally as follows.

Proposition 6. *Let $\{T_n,\ n \geq 0\}$ be a decreasing sequence of projection operators on $L^p(\Sigma)$, which are also Dunford–Schwartz operators. Then for each $f \in L^p(\Sigma)$, $1 \leq p < \infty$, $\lim_n (T_n f)(\omega) = (T_\infty f)(\omega)$ exists a.e. $[\mu]$, (and in norm, and similarly for the increasing case).*

The point here is that (for the usual martingale theorem, $T_n = E^{\mathscr{B}_n}$ where $\mathscr{B}_n \supset \mathscr{B}_{n+1}$ (or $\mathscr{B}_n \subset \mathscr{B}_{n+1}$)) the T_n can be more general projections than conditional operators. This result and the preceding one are proved quickly with the above maximal theorem and hence both theories are unified by it. On the other hand, the first observation indicates that Proposition 6 is really more general than the martingale theorem. How much more general is it? This is best answered by giving an independent proof of Proposition 6. In fact, a slightly more general version of it can be proved at no extra effort. For simplicity (Ω, Σ, μ) will be taken to be a probability space.

Proposition 7. *Let $\{T_n,\ n \geq 0\}$ be an increasing sequence of projection operators on an $L^p(\Omega, \Sigma, \mu)$ such that for $1 < p < 2$, $\|T_n\|_p \leq 1$ and $\|T_n\|_{p'} \leq 1$ where $p' = p/p - 1$. Then for each $f \in L^q(\Sigma)$ $1 \leq q < \infty$, $\lim_n (T_n f)(\omega)$ exists a.e. $[\mu]$, (and in norm and similarly for decreasing sequences).*

Proof. The structure of contractive projections on L^p-spaces was not fully known in 1962, when [11] was written, but it is later worked out and fairly well understood from 1965 on. The assumption that μ is a probability measure is not essential since the structure theory, used in the proof below, is now known to be true for arbitrary measure spaces.

Each contractive projection, with the hypothesis of the proposition, can be expressed as follows (cf. Andô [2, p. 402, Cor. 3]): If $\mathcal{M}_n = T_n(L^p)$, then there exists a set B_n, called the support of $\mathcal{M}_n$ (i.e., every function in $\mathcal{M}_n$ lives on B_n and it is the smallest such set, modulo null sets, with this property) and a σ-field $\mathcal{B}_n \subset \Sigma$ such that $B_n \in \mathcal{B}_n$, and a (complex) function φ_n with $|\varphi_n| = 1$ a.e. on B_n (set $\varphi_n = 1$ on $\Omega - B_n$), in terms of which (uniquely) a.e.,

$$T_n f = \varphi_n \chi_{B_n} E^{\mathcal{B}_n}(\bar{\varphi}_n f), \qquad f \in L^p. \tag{27}$$

As usual, $E^{\mathcal{B}_n}$ is the conditional expectation operator. From (27) it is immediate that T_n is also defined on L^1 and L^∞ and that $\|T_n\|_1 \leq 1$, $\|T_n\|_\infty \leq 1$, so that the present hypothesis already implies the stronger condition that each T_n is in fact a Dunford–Schwartz operator!

In ([2], Lemma 3), $f_{0n} = \varphi_n \chi_{B_n}$ was constructed as a pointwise limit of a sequence of functions in $\mathcal{M}_n$. But $\mathcal{M}_n \subset \mathcal{M}_{n+1}$. So the inductive construction of these functions, given in the proof of the above-quoted lemma of Andô [2], can be continued from $\mathcal{M}_n$ to $\mathcal{M}_{n+1}$ in such a way that $\varphi_{n+1}\chi_{B_{n+1}}$ coincides with φ_n on B_n. (This is quite similar to the construction of orthonormal sequences in expanding subspaces of a Hilbert space.) But it is well known that every uniformly bounded monotone sequence of projections in a reflexive B-space converges (strongly) to a projection. Hence $T_n \to T_\infty$ $(= T_\infty{}^2)$ and if $\mathcal{M}_\infty = T_\infty(L^p)$, then $\mathcal{M}_\infty = \overline{\text{sp}}\,(\bigcup_{n=1}^\infty \mathcal{M}_n)$. If B_∞, $\mathcal{B}_\infty$, and φ_∞ are the corresponding elements (as in (27)), then $T_\infty = \varphi_\infty \chi_{B_\infty} E^{\mathcal{B}_\infty}(\bar{\varphi}_\infty)$ and by the preceding observation $\varphi_\infty \chi_{B_n} = \varphi_n \chi_{B_n}$ for all n.

With this reduction, the sequence under consideration becomes $\{T_n f\} = \{\varphi_\infty \chi_{B_n} E^{\mathcal{B}_n}(\bar{\varphi}_\infty f)\} \subset L^p(\Sigma)$, $1 \leq p < \infty$, where $|\varphi_\infty| = 1$ a.e., $B_n \uparrow B_\infty$ a.e., and $\mathcal{B}_n \uparrow \mathcal{B}_\infty$. Then it follows from a classical martingale theorem [6, p. 331] that $E^{\mathcal{B}_n}(\bar{\varphi}_\infty f) \to E^{\mathcal{B}}(\bar{\varphi}_\infty f)$ a.e. (and in norm). Since $\chi_{B_n} \uparrow \chi_{B_\infty}$ a.e., it follows immediately that $T_n f \to T_\infty f$, a.e. and in norm, and $T_\infty{}^2 = T_\infty$. This completes the proof of this proposition and hence that of Proposition 6.

It may be remarked at this point that, for the ergodic theorems obtainable from Theorem 5, not only the proofs but the result itself can be false if one only assumes that $\|T\|_1 \leq 1$. (Even if $\|T\|_\infty \leq 1 + \varepsilon$ for some $\varepsilon > 0$, then also the result of the Dunford–Schwartz ergodic theorem may be false. T being a positive operator, in addition, does not save the situation.) A counterexample may be found in [4] to this effect.

5. FINAL REMARKS

The preceding three approaches have shown that, while very general ergodic sequences can be formulated as certain martingales, there is at present no convergence theory of the latter to include the former. It is desirable that this gap be filled in the theory of martingales. On the other hand, the last two methods indicate that it is profitable to find a "superstructure" including both the theories. There are reasons to believe that the theory of projective limits of measure spaces, which is known to include the martingale part, can be modified to yield the ergodic theory at least of the type of [10]. Also, the operator approach of Section 3 may be modified for inclusion of [10]. A discussion of these methods will not be entered into at this point in order to keep the paper within the bounds set for these Proceedings.

REFERENCES

1. Andersen, E. S. and Jessen, B. (1948). On the introduction of measures in infinite product sets. *Danske Vid. Selsk. Mat.-Fys. Medd.* **25** No. 5, 8 pp.
2. Andô, T. (1966). Contractive projections in L^p-spaces. *Pacific J. Math.* **17** 391–405.
2a. Birkhoff, G. D. (1931). Proof of the ergodic theorem. *Proc. Nat. Acad. Sci. U.S.A.* **17** 656–660.
3. Chacón, R. V. (1961). On the ergodic theorem without the assumption of positivity. *Bull. Amer. Math. Soc.* **67** 186–190.
4. Chacón, R. V. (1964). A class of linear transformations. *Proc. Amer. Math. Soc.* **15** 560–564.
5. Dinculeanu, N. (1971). Conditional expectations for general measure spaces. *J. Multivariate Anal.* **1** 347–364.
6. Doob, J. L. (1953). *Stochastic Processes.* Wiley, New York.
7. Dunford, N. and Schwartz, J. T. (1958). *Linear Operators. Part I. General Theory.* Wiley (Interscience), New York.
8. Edwards, D. A. (1971). A maximal ergodic theorem for Abel means of continuous parameter operator semi-gropus. *J. Functional Analysis* **7** 61–70.
9. Hille, E. (1945). Remarks on ergodic theorems. *Trans. Amer. Math. Soc.* **57** 246–269.
10. Hurewicz, W. (1944). Ergodic theorem without invariant measure. *Ann. of Math.* (2), **45** 192–206.
11. Tulcea, A. I. and Tulcea, C. I (1963). Abstract ergodic theorems. *Trans. Amer. Math. Soc.* **107** 107–124.
12. Jerison, M. (1959). Martingale formulation of ergodic theorems. *Proc. Amer. Math. Soc.* **10** 531–539.
13. Kakutani, S. (1952). Ergodic theory. *Int. Congr. Math., 1950,* Proc. **2** 128–142.
13a. Oxtoby, J. C. (1962). Unpublished work, referred to in [11].
13b. Poincaré, H. (1890). Sur le problème des trois corps et les équations de la dynamique. *Acta Math.* **13** 1–270.
14. Rao, M. M. (1970). Generalized martingales. *Contrib. Ergodic Theory and Prob.* (Springer Lecture Notes in Math.) **160** 241–261.
15. Rao, M. M. (1971). Abstract non linear prediction and operator martingales. *J. Multivariate Anal.* **1** 129–157.
16. Rota, G.-C. (1961). Une théorie unifiée martingales et des moyennes ergodiques. *C. R. Acad. Sci. Paris, Ser. A-B* **252** 2064–2066.
17. Rota, G.-C. (1964). Reynolds operators. *Proc. Symp. Appl. Math., Amer. Math. Soc.* **16** 70–83.

On the Modelling and Estimation of Communication Channels[1]

W. L. ROOT

UNIVERSITY OF MICHIGAN

The identification of communication channels characterized by a signal transformation and additive output noise is discussed briefly and qualitatively. An abstract mathematical structure, called an ε-representation, in which channel identification problems can be set, is described, and some general results stated. With the use of this structure, the identification problems reduce in a certain way to estimation of parameters in a linear model. There is a review of certain facts about linear estimators, and a modification of linear unbiased minimum-variance estimation is given. Channel identification procedures are described using ε-representations and linear estimation.

1. INTRODUCTION AND PRELIMINARY DISCUSSION

The term "communication channel" is somewhat flexible in its common usage in radio engineering, and a short discussion about what it can mean follows a little later in this section. Abstractly, however, a communication channel will always be a "black box," a system that accepts inputs and produces outputs. It may provide a functional relationship between inputs and outputs, or a stochastic relationship, or something that can conveniently be thought of as a little bit of both. Quite often a real communication channel is not very well described by prior information; sometimes it may change with time so that even if a description is available it goes out of date and no longer applies. However, roughly speaking, the more that is known about the communication channel, the more than can be done in the signal processing to make the overall communication system effective. Consequently the communicators may be presented the possibility of improving communications

[1] Research supported in part by the Air Force Office of Scientific Research under research grant AFOSR-72-2328, and in part by the Willow Run Laboratories, University of Michigan.

by testing the channel, so as to determine as far as possible its characteristics, and thus infer as far as possible its behavior. The determination of the characteristics of any input-output system from data obtained by observing actual corresponding inputs and outputs is called system identification. The problems of identifying communication channels are special problems in system identification.

A point-to-point communication system, as opposed to a communication network, is usually thought of as being made up of three separate successive components: transmitter, channel, and receiver. Where the transmitter leaves off and the channel begins, and where the channel leaves off and the receiver begins, is somewhat arbitrary; it might be reasonable to break down a given communication system in different ways, so that the channel of one description properly includes the channel of a second description. Conventionally, however, the channel always includes those parts of the system that have to do with actually transporting the signal—the physical medium for electromagnetic wave propagation, or the electric cables. The term is used two ways, to denote a certain physical part of a communication system, and to characterize in some way how a signal is transformed and distorted in that part of the system. It is convenient to retain this ambiguous usage here.

A typical model for a radio communication channel, and a good one in many instances, is given by

$$z(t) = \int_A h(t, s)u(s)\, ds + w(t), \qquad t \in T, \tag{1.1}$$

where $w(t)$ is a stationary, real-valued stochastic process with zero mean, u is a real-valued function representing the signal into the channel, z is a real-valued stochastic process representing the signal out of the channel, and A, T are appropriate time intervals, not necessarily finite. Further conditions that sometimes apply are that w be Gaussian, $h(t, s)$ be a function of $t - s$ only, and $h(t, s) = 0$ whenever $t < s$. The linear integral operator might model electromagnetic wave transmission, particularly when there is scattering, or it might model the action of electrical filters, or both. The stochastic process w represents chiefly the sum of additional random voltages appearing at or near the input from the receiving antenna due to thermal noise in resistors, electromagnetic noise radiated from the sky, and shot noise and similar effects in electron tubes and solid-state devices.

There is a point here to be noted, one which distinguishes typical communication channel identification problems from many other system identification problems. Thermal noise and noise in electron devices are added throughout the system; however, they are almost always very small, so they cause a significant distortion of the signal only at points in the system where the intensity of the signal is small. Usually the only place where the signal is weak enough to be distorted appreciably by such noise is at or "near" the terminals of the receiving antenna, before amplification has taken place. This means that one can often model a communication channel and include additive noise only at its output, whereas in other kinds of systems non-negligible additive noise, due to different causes, may appear at various points in the system.

In the model (1.1) one would want to know the kernel $h(t, s)$, and the covariance function $R(t)$ of $w(t)$. The latter could be estimated empirically by standard methods from preliminary measurements, if indeed it could not be computed from physical principles. The kernel $h(t, s)$ could then perhaps be estimated by tests made on the channel; of course, the observations for these tests would have to be made in the presence of the noise, $w(t)$. If we suppose that $R(t)$ is given as prior information, the problem of identifying the channel becomes the problem of estimating $h(t, s)$.

We now formalize a class of problems, of which those described informally by (1.1) and the accompanying discussion may be considered as prototypical. We allow, however, a much greater level of generality; most notably, the linear integral operator of (1.1) may be replaced by transformations that are not linear or are not even integral operators. The basic notations to be used are as follows: u denotes the input signal to the channel; y and z denote, respectively, noise-free and noisy output signals. The relationship among u, y, and z is given by

$$z = H(u) + w = y + w \tag{1.2}$$

where w represents additive noise at the output. The symbols $\mathscr{U}$, $\mathscr{Y}$, $\mathscr{H}$ denote, respectively, the classes of input signals u, output signals y and z, and mappings H under consideration. $\mathscr{U}$ and $\mathscr{Y}$ will be treated abstractly, but in a specific model they are spaces of real-valued or vector-valued functions of time. $\mathscr{Y}$ is sometimes a Hilbert space, and $\mathscr{U}$ is sometimes a subset of a Hilbert space; however, unless otherwise specified it will be required only that $\mathscr{U}$ be a metric space and $\mathscr{Y}$ be a Banach space. In any case the mappings $H \in \mathscr{H}$ are always to be bounded and continuous. The noise, w, is a $\mathscr{Y}$-valued random variable with mean zero.

If there is no noise present, so that Eq. (1.2) reduces to $z = y = H(u)$, we say the channel is deterministic, and define the channel to be the mapping H. If there is noise present, we say the channel is deterministic with additive output noise, and define the channel to be the pair (H, w). The problem of identifying the channel will always be taken to be the problem of finding or estimating $H \in \mathscr{H}$; any statistical facts about w that are desired are assumed to be known a priori. Some mention is made of what may be called a stochastic channel in the concluding remarks, but channels in which randomness enters other than in the form of additive noise are not otherwise discussed.

The identification of a deterministic channel with additive output noise is a regression problem that is often inherently infinite dimensional. However, it appears that a meaningful constraint to be imposed, if there is to be any hope of practicability, is that the identification procedure to be devised reduce in the end to the estimation of a finite number of parameters. Further, it appears that identification should be treated as approximation, entirely apart from the statistical aspects, since it is never necessary (and really impossible) to get an exact solution, even in the noise-free case. These two considerations are consistent, of course, and they lead to the point of view that one should allow originally for an infinite-dimensional model, and then reduce it systematically to finite-dimensional ones that are uniformly good in some reasonable sense. Obviously this may not always be possible, but if it is not possible, it is reasonable to say the channel is not identifiable. A structure that can be used for setting up channel identification problems in this way is summarized in the next section. Most of this material appeared in an earlier paper [6], and full proofs are given by Root [9]; the word "channel" here replaces the word "system" in those references. From a theorem to be stated in the next section it follows that under certain compactness conditions, arbitrarily good, finite-dimensional, *linear* regression models can nearly always be obtained. This same theorem provides, under the same conditions, a solution in principle to the problem of identifying deterministic channels without noise.

In Section 3 there is a summary of certain facts about least squares and linear unbiased minimum-variance estimators, stated in a form convenient for our purposes. There is also described a modification of linear unbiased minimum-variance estimation that is appropriate for channel identification; this appeared originally in the paper [7].

Channel identification is discussed in Section 4 in terms of the models provided in Section 2 and the estimation procedures described in Section 3.

2. CLASSES OF CHANNELS AND REPRESENTATIONS

We define a class of channels $\mathscr{C}$ to be $\mathscr{C} = \{\mathscr{Y}, f, \mathscr{X}, \mathscr{U}\}$ where $\mathscr{Y}$ is a Banach space, $\mathscr{X}$ and $\mathscr{U}$ are metric spaces, and f is a bounded, continuous mapping from the topological product of $\mathscr{X}$ and $\mathscr{U}$, $\mathscr{X} \times \mathscr{U}$, into $\mathscr{Y}$. $\mathscr{U}$ is the channel input space, $\mathscr{Y}$ is the channel output space, and $\mathscr{X}$ is the parameter space. Fixing $x \in \mathscr{X}$ determines one particular channel, $c = \{\mathscr{Y}, f, x, \mathscr{U}\}$, and that channel relates inputs to noise-free outputs according to $y = f(x, u)$, $u \in \mathscr{U}$, $y \in \mathscr{Y}$. If there is additive output noise, the total received signal is $z = y + w$. In this section we do not consider further the $\mathscr{Y}$-valued random variable w, but concern ourselves solely with properties of classes $\mathscr{C}$ and with relations between different classes.

A channel $c = (\mathscr{Y}, f, x, \mathscr{U})$ is *linear* if $\mathscr{U}$ is a linear space and $f(x, u)$ is linear in u. A class of channels $\mathscr{C}$ is a *linear* class if $\mathscr{X}$ is a linear metric space and $f(x, u)$ is linear in x for each u. We need to deal with situations where $\mathscr{X}$ is only a subset of a linear space, however, so we introduce the definition: $\mathscr{C}$ is *prelinear* if $f(\alpha x_1 + \beta x_2, u) = \alpha f(x_1, u) + \beta f(x_2, u)$ for all scalars α and β and all x_1 and $x_2 \in \mathscr{X}$ for which all three terms are defined.

$\mathscr{C} = (\mathscr{Y}, f, \mathscr{X}, \mathscr{U})$ and $\mathscr{C}' = (\mathscr{Y}', f', \mathscr{X}', \mathscr{U}')$ are *equivalent* if there exist homeomorphisms of $\mathscr{X}$ onto $\mathscr{X}'$ and $\mathscr{U}$ onto $\mathscr{U}'$, and a linear homeomorphism from $\mathscr{Y}$ onto $\mathscr{Y}'$ such that if x', u', y' are the images of x, u, y, then $y' = f'(x', u')$ if and only if $y = f(x, u)$.

The concept of natural representation is central in what follows. It is defined in terms of a space of mappings from $\mathscr{U}$ into $\mathscr{Y}$. Let $\mathscr{F} = \mathscr{F}(\mathscr{U}, \mathscr{Y})$ be the class of all bounded continuous functions from $\mathscr{U}$ into $\mathscr{Y}$ made into a Banach space with the linear structure and norm defined as in (i) and (ii):

(i) $(\alpha H_1 + \alpha H_2)(u) \overset{d}{=} \alpha H_1(u) + \beta H_2(u)$ for all $u \in \mathscr{U}$, where α and β are real numbers and $H_1, H_2 \in \mathscr{F}$.

(ii) $\|H\| \overset{d}{=} \sup_{\mathscr{U}} \|H(u)\|$ where $\|H(u)\|$ is the norm of $H(u)$ in $\mathscr{Y}$.

Observe that $y = H(u)$ can be written as $y = g(H, u)$, where g is defined by $H(u) = g(H, u)$ for all $u \in \mathscr{U}$, all $H \in \mathscr{F}$. It is easy to prove:

Lemma 2.1. *$\mathscr{C}_{\mathscr{F}} = (\mathscr{Y}, g, \mathscr{F}, \mathscr{U})$ is a linear class of channels.*

Now let $\mathscr{C} = (\mathscr{Y}, f, \mathscr{X}, \mathscr{U})$ be any class of channels. We assign to it a subclass of $\mathscr{C}_{\mathscr{F}}$ as follows. Since $f(x, \cdot)$ is a bounded continuous function for any fixed $x \in \mathscr{X}$, it can be written $H(\cdot)$, $H \in \mathscr{F}$. Let ψ denote the mapping from $\mathscr{X}$ into $\mathscr{F}$ defined by $\psi(x) = H$, where $H(u) = f(x, u)$ for all $u \in \mathscr{U}$. Let $\mathscr{H} = \psi(\mathscr{X}) \subset \mathscr{F}$. Then $\mathscr{C}_0 = (\mathscr{Y}, g, \mathscr{H}, \mathscr{U})$ is called the *natural representation* of $\mathscr{C}$ and ψ is called the natural mapping. $\mathscr{C}_0$ is obviously a prelinear class of channels. One can show

Lemma 2.2. *If $\mathscr{C} = (\mathscr{Y}, f, \mathscr{X}, \mathscr{U})$ is a class of channels with compact input space $\mathscr{U}$, then the mapping ψ as defined above is continuous.*

An immediate consequence of this lemma is

Theorem 2.1. *Let $\mathscr{C}$ have the properties:* (i) *$\mathscr{U}$ is compact,* (ii) *$\mathscr{X}$ is compact,* (iii) *$f(x_1, u) = f(x_2, u)$ for all $u \in \mathscr{U}$ implies that $x_1 = x_2$. Then $\mathscr{C}$ is equivalent to its natural representation $\mathscr{C}_0$.*

It should be noticed that $\mathscr{C}_0$ is prelinear always, even when there is no linear structure defined in $\mathscr{X}$. However, if $\mathscr{C}$ itself is prelinear, it is easy to verify that ψ preserves the (partial) linear structure of $\mathscr{X}$.

We desire to approximate an arbitrary $\mathscr{C}$ by other classes of channels with certain desirable properties. The natural representation has been introduced as an intermediate step; it provides a sort of coordinate-free version of the original class which is convenient to deal with. Now, let $\mathscr{C}_0 = (\mathscr{Y}, g, \mathscr{H}, \mathscr{U})$ be a subclass of $\mathscr{C}_{\mathscr{F}}$, and let $\mathscr{C}_1 = (\mathscr{Y}, f_1, \mathscr{X}_1, \mathscr{U})$ be a class of channels. Let $\varepsilon > 0$ be fixed. $\mathscr{C}_1$ is an *ε-approximation* to $\mathscr{C}_0$ if for every $H \in \mathscr{H}$ there is an $x_1 \in \mathscr{X}_1$ such that $\|H - \psi_1(x_1)\| \leqq \varepsilon$, where ψ_1 is the natural mapping from $\mathscr{X}_1$ into $\mathscr{F}$. Note that it is not required that $\psi_1(x_1) \in \mathscr{H}$. Let ϕ_1 be a mapping from $\mathscr{H}$ into $\mathscr{X}_1$, then $(\mathscr{C}_1, \phi_1)$ is an *ε-representation* of $\mathscr{C}_0$ if $\|H - \psi_1 \circ \phi_1(H)\| \leqq \varepsilon$ for all $H \in \mathscr{H}$. Again it is not required that $\psi_1 \circ \phi(H)$ belong to $\mathscr{H}$. If $\mathscr{C}_0$ is the natural representation of $\mathscr{C}$, then we say also that $\mathscr{C}_1$ is an ε-approximation or that $(\mathscr{C}_1, \phi_1)$ is an ε-representation of $\mathscr{C}$, respectively, according as the above conditions are satisfied. If $(\mathscr{C}_1, \phi_1)$ is an ε-representation of $\mathscr{C}_0$, then clearly $\mathscr{C}_1$ is an ε-approximation to $\mathscr{C}_0$. A limited converse is stated in Theorem 2.3.

An ε-approximation $\mathscr{C}_1$ is said to be *linear* if its parameter space $\mathscr{X}_1$ is prelinear (this implies that ψ_1 is a restriction of a linear map), and *finite-dimensional* if $\mathscr{X}_1$ is a subset of a Euclidean space. An ε-representation $(\mathscr{C}_1, \phi_1)$ is said to be *continuous* if ϕ_1 and ψ_1 are continuous, and *determined by* $\mathscr{U}_0$, $\mathscr{U}_0 \subset \mathscr{U}$, if the mapping ϕ_1 depends only on the functions H, $H \in \mathscr{H}$, restricted to $\mathscr{U}_0$. Finally, an ε-representation $(\mathscr{C}_1, \phi_1)$ is *linear* if $\mathscr{C}_1$ is linear as an ε-approximation and ϕ_1 is a restriction of a linear mapping,[2] and is *finite-dimensional* if $\mathscr{C}_1$ is. The basic result is

Theorem 2.2. *Let $\mathscr{Y}$ be a Banach space and let $\mathscr{X}$ and $\mathscr{U}$ be compact metric spaces. Let $\mathscr{C} = (\mathscr{Y}, f, \mathscr{X}, \mathscr{U})$ be a class of channels with $\mathscr{X}$, $\mathscr{U}$, $\mathscr{Y}$ as prescribed. Then, given $\varepsilon > 0$, there exists an ε-representation $(\mathscr{C}_1, \phi_1)$ of $\mathscr{C}$ that is continuous, finite-dimensional, and determined by a finite subset $\mathscr{U}_0 \subset \mathscr{U}$. $\mathscr{C}_1$ is a linear ε-approximation. If $\mathscr{Y}$ is a Hilbert space, then there exists an ε-representation which, in addition to possessing the other properties listed, is linear; i.e., ϕ_1 is also a restriction of a linear mapping.*

The full details of the proof are not given (see Root [9]), but the idea of the construction which yields the proof is perfectly straightforward. A finite set of "input-output" pairs (u_i, y_i) for the actual channel are obtained, i.e. $y_i = H(u_i)$, where the u_i, $i = 1, \ldots, N$, are chosen so that any $u \in \mathscr{U}$ is within a specified distance of one of the u_i. Since $\mathscr{H}(\mathscr{U}) \subset \mathscr{Y}$ is compact, there is a finite-dimensional subspace $\mathscr{Y}'$ of $\mathscr{Y}$ such that every point in $\mathscr{H}(\mathscr{U})$ is close to $\mathscr{Y}'$. A continuous mapping π from $\mathscr{Y}$ into $\mathscr{Y}'$ is devised such that if $y \in \mathscr{H}(\mathscr{U})$, $y' = \pi(y) \in \mathscr{Y}'$ is close to y. If $\mathscr{Y}$ is a Hilbert space, π is taken to be the orthogonal projection on $\mathscr{Y}'$. A basis $(e_1, \ldots, e_M)$ is established in $\mathscr{Y}'$, and y' is written

$$y' = \sum_{k=1}^{M} e_k^*(y')e_k. \tag{2.1}$$

[2] This definition is more restrictive than that given by Root [6].

where the $e_k{}^*$ are continuous linear functionals. The parameter space $\mathscr{X}_1$ is taken to be NM-dimensional Euclidean space, and a fixed but arbitrary basis is determined. The element $x \in \mathscr{X}$ corresponding to H is given as the point with coordinates a_{kj}, $k = 1, \ldots, M, j = 1, \ldots, N$, where

$$a_{kj} = e_k{}^* \circ \pi \circ H(u_j). \tag{2.2}$$

A set of continuous interpolation functionals, $\gamma_n(u)$, $n = 1, \ldots, N$, $u \in \mathscr{U}$, are constructed. These functionals depend on $\{u_1, \ldots, u_N\}$, and they have the properties: $\gamma_n(u) \geqq 0$, $\Sigma\gamma_n(u) = 1$, $\gamma_n(u_k) = \delta_{nk}$. Finally, f_1 is defined by

$$f_1(x, u) = \sum_{k=1}^{M} \left(\sum_{n=1}^{N} \gamma_n(u) a_{kn} \right) e_k . \tag{2.3}$$

The formulas for a_{kj} and $f_1(x, u)$ are used later. There is nothing unique about the mapping π, and if a π can be found that is linear as well as continuous, Eq. (2.2) shows that ϕ is a restriction of a continuous linear mapping carrying $\mathscr{H}$ to $\mathscr{X}$. In this case ϕ can be extended to a continuous linear mapping on the closed linear subspace spanned by $\mathscr{H}$, and since its range is finite-dimensional it can be extended without increase of norm to all of $\mathscr{F}$.

An ε-representation of the form just described will be called a *standard ε-representation*. It depends on the particular set $\{u_1, \ldots, u_N\}$, and that set will be called its *determining set*. Any finite set $\{u_i\}$ that is δ-dense in $\mathscr{U}$ for sufficiently small δ can be a determining set for a standard ε-representation Reduced to its basic idea, this theorem says that under suitable compactness conditions, each function in $\mathscr{H}$ can be approximated uniformly well by a finite table of input-output pairs. Not surprisingly, it is possible, at least under certain conditions, to get ε-representations which have some or all of the desirable properties of a standard ε-representation, but which are constructed differently. In fact we have:

Theorem 2.3. *Let $\mathscr{C}_0 = (\mathscr{Y}, g, \mathscr{H}, \mathscr{U})$, and let $\mathscr{C}_1 = (\mathscr{Y}, f_1, \mathscr{X}_1, \mathscr{U})$ be a linear ε-approximation to $\mathscr{C}_0$. Let $\mathscr{C}_1$ satisfy the conditions* (i), (ii), *and* (iii) *of Theorem* 2.1 *and let $\mathscr{X}_1$ be convex. Then, for any $\eta > \varepsilon$, there is a ϕ_1 such that $(\mathscr{C}_1, \phi_1)$ is a continuous η-representation of $\mathscr{C}_0$ determined by a finite set $\{u_1, \ldots, u_N\}$.*

For example, when $\mathscr{Y}$ is an L_2-space and $\mathscr{U}$ is a compact subset of an L_2-space, certain classes of Volterra integral polynomials of degree less than or equal to some positive integer can be used to provide finite-dimensional linear ε-approximations to any compact $\mathscr{H}$ (see Root [6, 8] for details). By the above theorem they can be used also to provide ε-representations. Unfortunately, although it is not very hard to construct a satisfactory ϕ_1 in principle, it appears to be awkward to construct a ϕ_1 in usable form unless $\mathscr{H}$ is rather special. The implications of this for identification are touched on in Section 4.

3. ESTIMATION OF PARAMETERS IN A LINEAR MODEL

The kinds of estimation procedures that are to be admitted in the discussion of channel identification in the next section are least-squares, linear-unbiased-minimum-variance (LUMV), and a modification of the latter. A summary of certain basic facts about least-squares and LUMV estimators is presented (with some apology) because the emphasis and form desired seem to be not quite standard, and because some background is necessary in order to describe the modified LUMV estimator.

Consider the linear model

$$z = Ax + w \tag{3.1}$$

where z is a vector of observations, x is the vector of parameters to be estimated, A is a linear transformation, and w is an error vector. The vector w may represent either model error or noise or both; so it may be a random variable, or it may be a sum of an unknown and a random variable. We suppose that $\mathscr{Y}$, $\mathscr{X}$ are either finite-dimensional Euclidean spaces or separable Hilbert spaces, and that z and w are vectors in $\mathscr{Y}$ and x is a vector in $\mathscr{X}$. Let A^* denote the adjoint of A, $\mathscr{R}(A)$ the range of A, $\mathscr{D}(A)$ the domain of A, and $\mathscr{N}(A)$ the null space of A. If $\mathscr{E}$ is a linear manifold in, say, $\mathscr{X}$, let $\mathscr{E}^\perp$ denote its orthogonal complement in $\mathscr{X}$.

In the finite-dimensional case where $\mathscr{X}$ and $\mathscr{Y}$ are Euclidean spaces, $\mathscr{R}(A^*A) = \mathscr{R}(A^*)$, A^*A is a 1:1 mapping from $\mathscr{R}(A^*)$ onto $\mathscr{R}(A^*)$, and $(A^*A)^{-1}$ is well defined as an operator from $\mathscr{R}(A^*)$ to $\mathscr{R}(A^*)$. Also, of course, $\mathscr{R}(A^*)^\perp = \mathscr{N}(A)$. The formula

$$\tilde{x}(z) = (A^*A)^{-1}A^*z \tag{3.2}$$

which gives the least-squares estimate of x, i.e., which minimizes $\|z - A\hat{x}(z)\|^2$, is always defined. In what might be regarded as the conventional situation, at least from the statistical point of view, $\mathscr{N}(A) = 0$, and $\mathscr{R}(A) = \mathscr{N}(A^*)^\perp$ is a proper subspace of $\mathscr{Y}$, usually of much lower dimension. In other words, there is redundancy in the observations, and the least squares formula uses this redundancy to minimize the errors. At the other extreme, the dimension of $\mathscr{X}$ exceeds the dimension of $\mathscr{Y}$, and $\mathscr{R}(A) = \mathscr{Y}$. There is then not only no redundancy in the observations, but part of x cannot be estimated because $\mathscr{N}(A)$ is necessarily nonzero. In this case the projection of x on $\mathscr{N}(A)$ is treated as error and its norm is set to zero by the estimator $\tilde{x}$. The formula (3.2) thus gives a pseudo-inverse. In general, of course, neither $\mathscr{N}(A)$ nor $\mathscr{N}(A^*)$ are zero, so a part of x cannot be estimated, while on the other hand there is redundancy available for reducing the error in estimating the projection of x on $\mathcal{N}(A)^{\perp}$.

If now $\mathcal{X}$ and $\mathcal{Y}$ are allowed to be separable Hilbert spaces and A is a bounded operator, $\mathcal{R}(A)$ and $\mathcal{R}(A^*)$ need not be closed, $\tilde{x}$ may not be defined, and exact minima may not exist. However, if $\mathcal{R}(A)$ is closed, so are $\mathcal{R}(A^*)$ and $\mathcal{R}(A^*A)$; they are equal, and the preceding statements are still true, except for the references to dimension. The kind of channel identification problem to be discussed is initially infinite-dimensional, though there always can be a reduction to finite-dimensional spaces. If w is a $\mathcal{Y}$-valued random variable with mean zero and known covariance operator R, one may consider LUMV estimators, given under certain circumstances by the formula

$$\hat{x}(z) = (A^*R^{-1}A)^{-1}A^*R^{-1}z. \tag{3.3}$$

Again, temporarily, let $\mathcal{X}$, $\mathcal{Y}$ be finite-dimensional Euclidean spaces and require that R be strictly positive definite, so that it is 1:1 on $\mathcal{Y}$. Then $\mathcal{R}(A^*) = \mathcal{R}(A^*R^{-1}) = \mathcal{R}(A^*R^{-1/2}) = \mathcal{R}(A^*R^{-1}A), \mathcal{N}(A^*R^{-1}A) = \mathcal{N}(A), (A^*R^{-1}A)^{-1}$ is defined on $\mathcal{R}(A^*) = \mathcal{N}(A)^{\perp}$, and the formula (3.3) is meaningful and defines the LUMV estimator for the orthogonal projection of x on $\mathcal{N}(A)^{\perp}$. If now $\mathcal{N}(A^*R^{-1/2}) \neq 0$, which is equivalent to $\mathcal{N}(A^*) \neq 0$, there is redundancy; if $\mathcal{N}(A) \neq 0$ but $\mathcal{N}(A^*) = 0$, $\hat{x}$ is again some kind of pseudo-inverse. In fact, the comments made above about the interpretation of the least-squares estimator $\tilde{x}$ can be repeated for $\hat{x}$.

The solution to the LUMV estimation problem where $\mathcal{X}$ and $\mathcal{Y}$ are separable Hilbert spaces has been known for some time (see Parzen [2]), and I suspect it has appeared in various different forms. A partial solution is stated here (see Root and Fiske [10] for a proof) which emphasizes the formula given by (3.3). It naturally includes the finite-dimensional case.

Theorem 3.1. *Let $z = Ax + w$, where $z \in \mathcal{Y}$, $x \in \mathcal{X}$, and $\mathcal{Y}$ and $\mathcal{X}$ are separable Hilbert spaces. Let w be a $\mathcal{Y}$-valued random variable for which a mean exists and is equal to zero, and for which a bounded cocvariance operator R exists. In addition, let the following conditions hold:*

(i) *R is strictly definite and $R^{1/2}$ is Hilbert–Schmidt.* (*So R^{-1} exists and is densely defined, and R has finite trace,* $\mathrm{Tr}(R)$.)

(ii) *A is a bounded linear transformation from $\mathcal{X}$ into $\mathcal{Y}$.*

(iii) *$\mathcal{D}(R^{-1}A)$ is dense in $\mathcal{X}$.*

(iv) *Let $L = R^{-1/2}A$. $(L^*L)^{-1}$ considered as an operator from $\overline{\mathcal{R}(L^*L)}$ into itself necessarily exists and is densely defined in that subspace. Require that $(L^*L)^{-1}$ be bounded and let B be its continuous extension to $\overline{\mathcal{R}(L^*L)}$.* (*Note that $\overline{\mathcal{R}(L^*L)} = \overline{\mathcal{R}(A^*)}$.*)

(v) *BA^*R^{-1} is bounded. Since it is necessarily densely defined on $\mathscr{Y}$, it has a continuous extension S, $\mathscr{D}(S) = \mathscr{Y}$.*

Then an LUMV estimator for the projection of x on $\mathscr{N}(A)^\perp$ exists and is given by $\hat{x}(z) = Sz$. The variance of $\hat{x}$, i.e., $E\|x - \hat{x}(z)\|^2$, exists and is equal to $\mathrm{Tr}(SRS^*) = \mathrm{Tr}(B) < \infty$. (Observe that $\mathrm{Tr}(SRS^*)$ is finite by the conditions on the operators S and R.)

It is useful for our purposes to note two special cases of the linear model, even though the comments to be made about them are essentially trivial. We shall say the linear model is in *repeated form* if it can be written

$$z_1 = Ax + w_1, \ldots, z_N = Ax + w_N \tag{3.4}$$

where $x \in \mathscr{X}$, $z_n \in \mathscr{Y}$ for $n = 1, 2, \ldots, N$. The total observation is $z = (z_1, \ldots, z_N) \in \mathscr{Y}^N$, the N-fold direct sum of N copies of $\mathscr{Y}$. The least-squares estimate $\tilde{x}(y)$ reduces to

$$\tilde{x} = (A^*A)^{-1}A^*\left(\frac{1}{N}\sum_{n=1}^{N} z_n\right). \tag{3.5}$$

If the conditions of Theorem 3.1 are satisfied for the individual equations $y_i = Ax + w_i$, and if the w_i are uncorrelated and have identical covariance operators R, then also

$$\hat{x}(z) = \frac{1}{N}\sum_{n=1}^{N} Sz_n \tag{3.6}$$

and

$$E\|x - \hat{x}\|^2 = \frac{1}{N}\,\mathrm{Tr}(B) \tag{3.7}$$

where B is as defined above.

We shall say the linear model (3.1) is *simple* if A is a continuous 1:1 linear transformation with $\mathscr{D}(A) = \mathscr{X}$, $\mathscr{R}(A) = \mathscr{Y}$. If the model is simple, $\tilde{x}(y) = A^{-1}y$, even when $\mathscr{X}$ and $\mathscr{Y}$ are Hilbert spaces, since $\mathscr{R}(A)$ is closed and A^{-1} is bounded. The LUMV estimate also exists and is given by $\hat{x}(z) = A^{-1}z$ if the conditions of Theorem 3.1 are satisfied (only (i) and (iii) need special verification if the preliminary conditions are satisfied and the model is simple).

In general, as one has more prior information available about a quantity to be estimated, one expects to be able to improve the quality of estimates. Let us suppose that a linear model is supplemented with the prior information that the unknown x belongs to (1) a known bounded subset, or (2) a known compact subset of $\mathscr{X}$. We use this information to improve on the LUMV estimator $\hat{x}$, in the sense that new linear estimators will be constructed with mean-squared-error uniformly smaller than the (constant) mean-squared-error of $\hat{x}$. The new estimators are biased, but that fact is really of no concern.

Let $\mathscr{X}$ be a separable Hilbert space or a finite-dimensional Euclidean space and let $\{\psi_i\}$ be a complete orthonormal system (cons) in $\mathscr{X}$. Any subset $\mathscr{B}$ of $\mathscr{X}$ of the form $\mathscr{B} = \{x \in \mathscr{X}\colon |(x, \psi_i)| \leqq b_i,\ i = 1, 2, \ldots\}$ where $\{b_i\}$ is a bounded sequence of positive numbers will be called a rectangular parallelepiped (r.p.) with respect to the ψ_i.

Suppose, to start with, that the conditions of Theorem 3.1 are satisfied. Then B, the continuous extension of $(A^*R^{-1}A)^{-1}$ to $\overline{\mathscr{R}(A^*)}$, is compact and self-adjoint. Hence we may write

$$B\psi_i = \sigma_i^2\psi_i, \qquad i = 1, 2, \ldots,$$

where the eigenvectors ψ_i form a cons and the eigenvalues σ_i^2 are real non-negative numbers (zero eigenvalues are allowed if necessary, so that the set $\{\psi_i\}$ can be complete). For convenience suppose that $x \in \mathscr{N}(A)^{\perp}$; then the LUMV estimate is

$$\hat{x}(z) = Sz = SAx + Sw = x + Sw = \sum_i (\hat{x}(z), \psi_i)\psi_i. \tag{3.8}$$

One can think of Eq. (3.8) as describing the trivial case of the simple linear model when $A = I$, with $\hat{x}(z)$ the new "observation." The covariance operator of the "noise" Sw is the operator Γ satisfying $(\Gamma u, v) = E(Sw, u)(Sw, v) = E(w, S^*u)(w, S^*v) = (RS^*u, S^*v) = (SRS^*u, v)$. Hence $\Gamma = SRS^* = B$. It follows then that

$$E(Sw, \psi_i)(Sw, \psi_j) = (B\psi_i, \psi_j) = \sigma_i^2\delta_{ij}. \tag{3.9}$$

Consider an arbitrary linear estimate $x(z)$ of x in Eq. (3.8). Such an estimate can be written

$$x(z) = \sum_{ij} a_{ij}(\hat{x}(z), \psi_j)\psi_i. \tag{3.10}$$

The error in the estimate is given by

$$x(z) - x = \sum_i \Big[(a_{ii} - 1)(x, \psi_i) + \sum_{j \neq i} a_{ij}(x, \psi_j) + \sum_j a_{ij}(Sw, \psi_j)\Big]\psi_i,$$

and, taking account of (3.9), the mean-squared-error is given by

$$E\|x(z) - x\|^2 = \sum_i \Big|(a_{ii} - 1)(x, \psi_i) + \sum_{j \neq i} a_{ij}(x, \psi_j)\Big|^2 + \sum_{ij} a_{ij}\sigma_j^2. \tag{3.11}$$

Now impose the condition that x belongs to a bounded subset of $\mathscr{X}$, and hence to an r.p. $\mathscr{B}$ with respect to the ψ_i. Then $|(x, \psi_i)| \leqq b_i$, and

$$E\|x(z) - x\|^2 \leqq \sum_i \Big(|a_{ii} - 1|b_i + \sum_{j \neq i} |a_{ij}|b_j\Big)^2 + \sum_{ij} a_{ij}^2\sigma_j^2 \tag{3.12}$$

for all $x \in \mathscr{B}$. This upper bound is minimized by putting $a_{ij} = 0$, $i \neq j$, and $a_{ii} = b_i^2/b_i^2 + \sigma_i^2$. The linear estimator formed by using these a_{ij} is written $\hat{\hat{x}}$ and is given by

$$\hat{\hat{x}}(z) = \sum_j \frac{b_j^2}{b_j^2 + \sigma_j^2} (Sz, \psi_j)\psi_j . \tag{3.13}$$

From the inequality (3.12), it follows that

$$E\|\hat{\hat{x}}(z) - x\|^2 \leqq \sum_i \frac{\sigma_i^2 b_i^2}{\sigma_i^2 + b_i^2} . \tag{3.14}$$

Since $E\|\hat{x}(z) - x\|^2 = \mathrm{Tr}(B) = \sum_i \sigma_i^2$ for all x,

$$\sup_{\mathscr{B}} E\|\hat{\hat{x}}(z) - x\|^2 < E\|\hat{x}(z) - x\|^2.$$

The estimate $\hat{\hat{x}}(z)$ may not belong to $\mathscr{B}$. By truncating the components and thereby sacrificing linearity one can form an estimate $\hat{\hat{h}}_b(z)$ that does belong to $\mathscr{B}$ and has necessarily smaller mean-squared error:

$$\hat{\hat{h}}_b(z) = \sum_j \alpha_j \psi_j \tag{3.15}$$

where

$$\begin{aligned}
\alpha_j &= b_j \qquad \text{if} \quad \frac{b_j^2}{b_j^2 + \sigma_j^2} (Sz, \psi_j) \geqq b_j \\
&= -b_j \qquad \text{if} \quad \frac{b_j^2}{b_j^2 + \sigma_j^2} (Sz, \psi_j) \leqq -b_j \\
&= \frac{b_j^2}{b_j^2 + \sigma_j^2} (Sz, \psi_j) \qquad \text{otherwise.}
\end{aligned}$$

Now if the prior restriction on x is made more stringent by the requirement that x belong to a known compact set $\mathscr{K} \subset \mathscr{X}$, the condition that Theorem 3.1 hold can be waived, and we can still use the above estimator in the context of an arbitrarily good finite-dimensional approximation. Since $\mathscr{K}$ is compact, given $\varepsilon > 0$ there exists a finite-dimensional linear subspace $\mathscr{M}_\varepsilon$ of $\mathscr{X}$ such that $\|x - P_\varepsilon x\| \leqq \varepsilon$ for all $x \in \mathscr{K}$, where P_ε is the orthogonal projection on $\mathscr{M}_\varepsilon$. Let Q_ε be the orthogonal projection on $\mathscr{R}(AP_\varepsilon)$, which is a finite-dimensional subspace of $\mathscr{Y}$. We replace the original linear model, where $\mathscr{X}$ and $\mathscr{Y}$ are separable Hilbert spaces, by the approximate finite-dimensional linear model

$$z = AP_\varepsilon x + Q_\varepsilon w, \qquad y \in \mathscr{R}(AP_\varepsilon) = \mathscr{R}(Q_\varepsilon). \tag{3.16}$$

Put $x' = P_\varepsilon x$ and $w' = Q_\varepsilon w$, and let A' denote the restriction of A to $\mathscr{M}_\varepsilon$, so that $\mathscr{R}(A') = \mathscr{R}(Q_\varepsilon)$. Then Eq. (3.16) becomes

$$z = A'x' + w', \qquad y \in \mathscr{R}(Q_\varepsilon). \tag{3.17}$$

The covariance operator for w' is $Q_\varepsilon \mathscr{R} Q_\varepsilon$, and it is strictly positive definite on $\mathscr{R}(Q_\varepsilon)$ if R is strictly positive definite on $\mathscr{Y}$. An LUMV estimator $\hat{x}'$ now exists for the projection of x' on $\mathscr{N}(A')^\perp$ and is given by the standard formula. Furthermore, the modified linear estimator $\hat{\hat{x}}$ developed above exists for x' since the problem is now entirely finite-dimensional.

4. CHANNEL IDENTIFICATION

Let us consider first the perhaps idealized situation that a standard ε-representation is available. Let $\mathscr{C}_0 = (\mathscr{Y}, g, \mathscr{H}, \mathscr{U})$ be the natural representation of the class of unknown channels, where there are no restrictions on $\mathscr{Y}$, $\mathscr{H}$, $\mathscr{U}$ except that $\mathscr{H}$ and $\mathscr{U}$ are compact. Let $(\mathscr{C}, \phi)$ be a standard ε-representation with $\mathscr{C} = (\mathscr{Y}, f, \mathscr{X}, \mathscr{U})$ and determining set $\{u_1, \ldots, n_N\}$. We then have

$$x = \phi(H) = (\ldots, a_{kj}, \ldots)^T,$$

a vector in R^{NM} with respect to an arbitrary fixed basis, where, repeating (2.2) and (2.3),

$$a_{kj} = e_k^* \circ \pi \circ H(u_j) \tag{4.1}$$

and

$$f(x, u) = \sum_{k=1}^{M} \left(\sum_{n=1}^{N} \gamma_n(u) a_{kn} \right) e_k \tag{4.2}$$

and where $\{e_1, \ldots, e_M\}$ is a basis in $\mathscr{Y}'$, which we call the space of processed observations. The noise-free channel identification is given by $x = \phi(H)$ in the representing class $\mathscr{C}$. This identification is accomplished in principle by successively applying the input signals $u_1, \ldots, u_N$, observing the corresponding outputs $y_1 = H(u_1), \ldots, y_N = H(u_N)$, and substituting these in Eq. (4.1). In the case that is probably of most practical interest $\mathscr{Y}$ is an L_2-space. Then π can be chosen to be the orthogonal projection on $\mathscr{Y}'$, and the transformation $e_k^* \circ \pi(y_j)$ can be readily implemented.

In order that the identification be possible it is necessary that the channel can be tested N successive times, and that the tests not interfere. Thus the channel must not change perceptibly over a sufficiently long period of time; or else—and these are hardly likely for a physical transmission medium—it must be periodic or nearly so, or it must be capable of being reset to a base state. Because the whole identification is an approximation, some slight time-variation is permitted. How rapid the time variation may be and still permit identification depends on how much time is required for each test (the "ringing time" of the channels) and on how many tests are required, that is, on N, the number of elements in a determining set. But N depends on ε and on how "large" the class of unknown channels may be, or in other words, on the amount and usefulness of the prior information. These comments are obvious, but they suggest that one can give precise meaning to the concept of "slowly varying channel." Something close to this is done by Prosser and Root [3].

Now suppose that output noise w, with mean zero, is added so that the raw or unprocessed observations are given by

$$z = y + w = H(u) + w. \tag{4.3}$$

Consider first the case that $\mathcal{Y}$ is a Hilbert space and π is an orthogonal projection. Then the determination of x, that is, the determination of the a_{kn}, becomes a problem of estimating parameters in a linear model, as follows. For any u_i in the determining set, $f(x, u_i) = \pi \circ H(u_i)$, so we have

$$\begin{aligned} \pi z_i &= \pi[H(u_i)] + \pi w_i \\ &= f(x, u_i) + \pi w_i, \qquad i = 1, \ldots, N, \end{aligned} \tag{4.4}$$

where w_i and z_i are the noise and the raw observation, respectively, corresponding to the input u_i. Let w_i' and z_i' be πw_i and πz_i, respectively, represented as column vectors in $\mathcal{Y}'$ with respect to the basis $\{e_1, \ldots, e_M\}$, and let

$$\tilde{z} = \begin{bmatrix} z_1' \\ \cdots \\ \vdots \\ \cdots \\ z_N' \end{bmatrix}, \qquad \tilde{w} = \begin{bmatrix} w_1' \\ \cdots \\ \vdots \\ \cdots \\ w_N' \end{bmatrix}.$$

Since f is linear in x, which can be represented as the column vector $(a_{11}, \ldots, a_{M1}; \ldots; a_{1N}, \ldots, a_{MN})^T$, the set of equations (4.4) can be written

$$z_i' = U_i x + w_i', \qquad i = 1, \ldots, N,$$

or as the single equation

$$\tilde{z} = \tilde{U}x + \tilde{w} \tag{4.5}$$

where

$$\tilde{U} = \begin{bmatrix} U_1 \\ \cdots \\ \vdots \\ \cdots \\ U_N \end{bmatrix}$$

is an $NM \times NM$ matrix. Actually $\tilde{U}$ is the identity matrix, as may be seen from (4.2) and the properties of the γ_n's.

Thus, finally

$$\tilde{z} = x + \tilde{w}. \tag{4.6}$$

Let the covariance matrix for $\tilde{w}$ be $\tilde{R}$. Assume, as would usually be the case, that each w_i has the same covariance operator R and that the noise is uncorrelated from one test to the next. Then the covariance operator for each w_i' is $\pi R \pi$ and $Ew_i'w_j'^T = 0$. If we denote the matrix of $\pi R \pi$ with respect to $\{e_1, \ldots, e_M\}$ by $[\pi R \pi]$, then

$$\tilde{R} = \begin{bmatrix} [\pi R\pi] & 0 & \cdots & 0 \\ 0 & [\pi R\pi] & & \\ \vdots & & \ddots & \\ 0 & \cdots & & [\pi R\pi] \end{bmatrix}.$$

Since the linear model (4.6) is simple, there is no statistical redundancy, and in fact $\hat{x}(\tilde{z}) = \tilde{z}$. However, even in this case $\hat{\hat{x}}$ gives an improvement on $\hat{x}$. $\hat{\hat{x}}$ is defined since $\mathscr{X}$ is bounded, and it is given by Eq. (3.13) with the ψ_i a set of orthonormal eigenvectors of $\tilde{R}$, and the b_i the half-edges of the smallest r.p. with respect to the ψ_i that contains $\mathscr{X}$.

To improve the estimate of x, redundancy can be introduced by repeating the measurements *with the same* u_i. In fact, since the standard ε-representation depends on the determining set, the parameters to be estimated depend on the u_i, so the u_i must be repeated or new parameters are introduced. A repetition of measurements leads to a repeated linear model:

$$\begin{aligned} \tilde{z}_1 &= x + \tilde{w}_1 \\ &\vdots \\ \tilde{z}_K &= x + \tilde{w}_K. \end{aligned} \tag{4.7}$$

With the assumptions on the noises made above, one has that $\hat{x}$ is simply the average of the $\tilde{z}_i$, and that $\hat{\hat{x}}$ is given by the same expression as in the simple model except that each σ_i^2 is replaced by σ_i^2/K.

If we go back to the original model (4.3) and allow for testing the channel with each u_i in the determining set and then repeating the entire procedure K times, we have the equations

$$z_{ik} = H(u_i) + w_{ik}, \qquad i = 1, \ldots, N; \quad k = 1, \ldots, K. \tag{4.8}$$

Put

$$\bar{z}_i = \frac{1}{K} \sum_{k=1}^{K} z_{ik} \qquad \text{and} \qquad \bar{w}_i = \frac{1}{K} \sum_{k=1}^{K} w_{ik};$$

then $\bar{z}_i = H(u_i) + \bar{w}_i$, $i = 1, \ldots, N$. The LUMV estimator $\hat{x}$ just obtained can also be expressed, in terms of the parameters a_{kj}, by

$$a_{mi} = e_m^* \circ \pi(\bar{z}_i), \qquad m = 1, \ldots, M; \quad i = 1, \ldots, N. \tag{4.9}$$

If, now, π is nonlinear, the estimator defined by (4.9) is no longer linear, nor is there reason to suppose it is optimum in any sense. However, it is consistent, requires the evaluation of the nonlinear mapping π only once, and is relatively simple.

In general, if certain conditions are satisfied, one may use any ε-representation to reduce a channel identification problem of the type being considered to the estimation of finitely many parameters in a linear model, and to do this in such fashion that if a good estimate is made of the parameters, then a good identification will have been achieved. More specifically, suppose $\mathscr{C}_0 = (\mathscr{Y}, g, \mathscr{H}, \mathscr{U})$ is the natural representation of a class of channels and $(\mathscr{C}_1, \phi)$, where $\mathscr{C}_1 = (\mathscr{Y}, f_1, \mathscr{X}_1, \mathscr{U})$, is an ε-representation for $\mathscr{C}_0$. Suppose that $\mathscr{U}$ and $\mathscr{H}$ are compact and that the ε-representation is continuous, linear, finite dimensional, and determined by a finite set $\{u_i\}$, $i = 1, \ldots, N$. Let $\mathscr{Y}^N$ denote the direct sum of N copies of $\mathscr{Y}$. We define a function $\tilde{\phi}$ from $\mathscr{Y}^N$ into the linear space spanned by $\mathscr{X}_1$ as follows. Put

$$\tilde{\phi}(y_1, \ldots, y_N) = \tilde{\phi}(H(u_1), \ldots, H(u_N)) = \phi(H)$$

for all $(y_1, \ldots, y_N) \in \mathscr{Y}^N$ for which $y_i = H(u_i)$, $i = 1, \ldots, N$, for some H. This definition is meaningful, since if $y_i = H'(u_i) = H''(u_i)$, $i = 1, \ldots, N$, $\phi(H') = \phi(H'')$. Extend $\tilde{\phi}$ linearly to the linear manifold spanned by $\{\mathscr{H}(u_1, \ldots, u_N)\}$; this can be done since ϕ is linear. If this function is continuous, it is defined by continuity on the closed linear manifold $\mathscr{M}$ spanned by $\{\mathscr{H}(u_1, \ldots, u_N)\}$. Finally, if $\mathscr{M}$ has a closed complementary manifold in $\mathscr{Y}^N$, $\tilde{\phi}$ can be extended to a continuous linear operator on all of $\mathscr{Y}^N$ by defining it to be zero in the closed complementary manifold. If such a $\tilde{\phi}$ can be defined, then, for the noise-free case $x = \tilde{\phi}(y_1, \ldots, y_N)$ where the y_i are the outputs corresponding to inputs u_i, $i = 1, \ldots, N$, from the determining set. With additive output noise, $z = y + w$, and

$$\tilde{z} = \tilde{\phi}(z_1, \ldots, z_N) = \tilde{\phi}(y_1, \ldots, y_N) + \tilde{\phi}(w_1, \ldots, w_N) = x + \tilde{\phi}(w_1, \ldots, w_N). \tag{4.10}$$

It appears that neither the continuity of $\tilde{\phi}$ nor its final extendability to all of $\mathscr{Y}^N$ can be guaranteed in general. However, if continuity can be verified from a knowledge of $\mathscr{H}$ and ϕ, and if $\mathscr{Y}$ is a Hilbert space, then a continuous linear $\tilde{\phi}$ on $\mathscr{Y}^N$ does exist, of course. Also, if the linear span of $\{\mathscr{H}(u_1, \ldots, u_N)\}$ is finite-dimensional, then the entire procedure goes through without any further hypotheses, and there is a continuous linear $\tilde{\phi}$ on $\mathscr{Y}^N$.

This last condition clearly holds if $\mathscr{H}(\mathscr{U})$ is finite-dimensional. Although this is perhaps an unreasonable assumption if $y = H(u)$ is taken to be a raw channel output, it is quite reasonable to imagine that there is fixed pre-processing at the output of the channel which has the effect of mapping $\mathscr{Y}$ into a finite-dimensional $\mathscr{Y}'$. Then if the entire channel is conceived of as the mapping from $\mathscr{U}$ into $\mathscr{Y}'$, $\mathscr{H}(\mathscr{U})$ is finite dimensional

Now, although in principle any prelinear class of channels $\mathscr{C}_1$ which is an ε'-approximation to $\mathscr{C}_0$ and which satisfies the conditions of Theorem 2.2 can be used to form an ε-representation, $\varepsilon > \varepsilon'$, in fact it may not be feasible to specify ϕ_1 properly. However, the identification of the unknown channel from input-output data within the set of models provided by $\mathscr{C}_1$ requires the use of some rule for assigning to each set of input-output pairs an $x \in \mathscr{X}$. That is, in the language of this paper, a $\tilde{\phi}$ must be specified, whether it actually yields an ε-representation or not. A common-sense approach might be to choose a reasonable, computationally convenient rule, call it ϕ'; see how many input-output pairs are needed to make the rule work and what restrictions it imposes on the admissibility of the u_i, and then choose a set $\{u_i\}$ that meets the admissibility conditions and is sufficient in number. See Root [8] for an example of this approach using Volterra polynomials. If ϕ' is linear, a linear model is obtained for the estimation, as in (4.10), but the x may not satisfy the condition that $f(x, u)$ is a good approximation to $H(u)$ uniformly in u and H. If extra tests are made with different inputs to provide redundant input-output pairs, it may well be that a least-squares estimate, say, of x gives an improved value, even in the noise-free case. However, none of the theory presented here says anything about this situation. If there is additive output noise and measurements are repeated using the same minimal determining set $\{u_1, \ldots, u_N\}$, the error terms are due to noise alone and therefore are random variables. One can then have a good satistical estimate of x, but still not have a satisfactory identification because x defines a bad model. If there are redundant measurements, some or all of which are made with different sets $\{u_i\}$, unknown model error is in general superimposed on the additive noise, and no statistical assertions can be made about whatever estimate is used.

5. REMARKS

The assumption that the input space $\mathscr{U}$ is compact, which has been made throughout, is not only a convenient simplifying assumption, but is realistic in the modelling of communications channels. If the input space is a subset of an L_2-space of real-valued or vector-valued functions, for example, compactness means, in engineering language, that the energy in the high-frequency components must trail off uniformly for the class of admissible signals, i.e., for signals in $\mathscr{U}$. It appears that any device that produces electrical signals will inherently impose such a constraint.

The assumption that the class of unknown channels,be totally bounded in the sense that $\mathscr{X}$ (or $\mathscr{H}$) be totally bounded is obviously almost necessary, except when $\mathscr{H}$ is finite dimensional, to allow for uniformly good approximations involving only finitely many parameters. Whether the totally bounded sets $\mathscr{H}$ are closed and hence compact is irrelevant, except in the investigation of what happens when $\varepsilon \to 0$. This question of necessity has not been pursued here, but it is by Prosser and Root [3], Varaiya [11], and by Rajput [4, 5] in studies of determinable classes, which have been defined for special input and output spaces and special classes of mappings, and are fairly near the same thing as linear, continuous, finite-dimensional ε-representations.

The choice of the Banach space $\mathscr{F}$, with the sup norm, is arbitrary, but it seems justifiable because it so clearly relates the channel transformations to what can be observed. However, other normed spaces can be used, at least in special cases (see the literature [3, 4, 11]).

A different kind of channel identification problem is posed if, e.g., the kernel $h(t, s)$ in the model (1.1) is taken to be stochastic. Such a model is entirely appropriate for radio communications in some circumstances where the radio wave propagation is effected by apparently random, time-varying scattering from parts of the ionosphere. A thing that one can hope to do in this situation if $k(t, u) = h(t, t - u)$ is stationary in t, is to identify the autocorrelation function, $Ek(t, u)k(t', u') = \Gamma(t - t'; u, u')$. A procedure is given by Bello [1]. At least some problems of this type can be formulated and solved in principle by using ε-representations coupled with classical linear estimation procedures.

REFERENCES

1. Bello, P. (1964). On the measurement of a channel correlation function. *IEEE Trans. Information Theory* **10** 381–383.
2. Parzen, E. (1961). An approach to time series analysis. *Ann. Math. Statist.* **32** 951–989.
3. Prosser, R. T. and Root, W. L. (1966). Determinable classes of channels. *J. Math. Mech.* **16** 365–397.
4. Rajput, B. S. (1971). Bounded determinable subsets of Banach spaces. *SIAM J. Appl. Math.* **20** 753–748.
5. Rajput, B. S. (1971). Unbounded determinable subsets of Banach spaces. *SIAM J. Appl. Math.* **21** 19–29.
6. Root, W. L. (1971). Some general structure theory of systems to be used in identification and measurement. *Proc. Fifth Annu. Princeton Conf. Info. Sci. Systs.* 13–19.
7. Root, W. L. (1971). Estimation in identification theory. *Proc. Ninth Allerton Conf. Circuit Syst. Theory* 1–10.
8. Root, W. L. (1971). On the structure of a class of system identification problems. *Automatica—J. IFAC* **7** 219–231.
9. Root, W. L. (1972). On the modelling of systems for identification. Part 1. Report, Dept. of Aerospace Engineering, Univ. of Michigan.
10. Root, W. L. and Fiske, P. (1973). Linear estimation for use in identification of systems. (Report in preparation.)
11. Varaiya, P. P. (1967). On determinable classes of signals and linear channels. *SIAM J. Appl. Math.* **15** 440–449.

Innovation and Nonanticipative Processes

YU. A. ROZANOV

STEKLOV MATHEMATICAL INSTITUTE, MOSCOW

1. INNOVATION PROCESSES AND REGULARITY

Let $\xi(t) = \{\xi_i(t)\}_1^m$, $t > t_0$, be a multivariate random process with mean zero and $E|\xi_i(t)|^2$ finite. Denote by $H(\xi)$ a subspace generated by all values $\xi_i(t)$ in a separable Hilbert space of random variables η, $E|\eta|^2 < \infty$, with the inner product

$$\langle \eta_1, \eta_2 \rangle = E\eta_1\bar{\eta}_2 .$$

Suppose that all univariate components $\xi_i(t)$, $t > t_0$, are left-continuous in $H(\xi)$:

$$\lim_{h \to +0} \xi_i(t-h) = \xi_i(t), \qquad t > t_0 .$$

Let $H_t(\xi)$ be a subspace generated by all values $\xi_i(s)$, $t_0 < s \le t$, and P_t be a projection operator onto $H_t(\xi)$ in $H(\xi)$, $t > t_0$. Obviously a "chain" of the projections P_t, $t > t_0$, is left-continuous:

$$\lim_{h \to +0} P_{t-h} = P_t, \qquad t > t_0 .$$

Assuming $P_{t_0} = 0$, we have a projection family P_t, $t \ge t_0$, in the Hilbert space $H(\xi)$. It is well known (see, e.g., Plessnev and Rohlin [8]) that there exist elements $x_1, \ldots, x_M \in H(\xi)$ such that the variables

$$X_j(t) = P_t x_j, \qquad j = 1, \ldots, M, \quad t \ge t_0, \tag{1}$$

generate $H(\xi)$ and that the random processes $X_j(t)$, $t \ge t_0$ with orthogonal increments are orthogonal under different $j = 1, \ldots, M$. In addition, their "time-spectral" functions $F_j(t) = E|X_j(t)|^2$, $t \ge t_0$, are absolutely continuous with respect to the preceding functions $F_1(t), \ldots, F_{j-1}(t)$; that is,

$$dF_1(t) \succcurlyeq dF_2(t) \succcurlyeq \cdots \succcurlyeq dF_M(t). \tag{2}$$

The M-variate process $X(t) = \{X_j(t)\}_1^M$, $t \ge t_0$, is called the *innovation process* of $\xi(t)$, $t > t_0$, and the corresponding number M is called the *multiplicity* of the random process $\xi(t) = \{\xi_i(t)\}_1^m$, $t > t_0$. The process $\{\xi_i(t)\}_1^m$, $t > t_0$, has the so-called *Hida-Cramér representation*

$$\xi_i(t) = \int_{t_0}^{t} \sum_{j=1}^{M} c_{ij}(t, s)\, dX_j(s), \qquad i = 1, \ldots, m. \tag{3}$$

As a consequence of formula (3) we obtain that the projection of $\xi_i(t + u)$, $u > 0$, onto the subspace $H_t(\xi)$ is

$$P_t \xi_i(t + u) = \int_{t_0}^{t} \sum_{j=1}^{M} c_{ij}(t + u, s)\, dX_j(s), \qquad i = 1, \ldots, m. \tag{4}$$

The innovation process $X(t) = \{X_j(t)\}_1^M$, $t \geq t_0$, determined by formula (1) can be described as a left-continuous process with orthogonal increments such that

$$H_t(X) = H_t(\xi), \qquad t > t_0; \tag{5}$$

in addition, the different components $X_j(t)$ are orthogonal and the time-spectral functions $F_j(t) = E|X_j(t)|^2$ satisfy relation (2). All innovation processes of $\xi(t) = \{\xi_i(t)\}_1^m$, $t > t_0$, have *the same type*—the same multiplicity M and equivalent time-spectral functions $F_j(t)$ of the corresponding components $X_j(t)$. As was shown by Cramér [1], there is even a *univariate* random process $\xi = \xi(t)$ with any multiplicity $M = 1, \ldots, \infty$ and any given time-spectral functions $F_1(t), \ldots, F_M(t)$.

For some random processes (say, for wide-sense stationary and Markov processes) the type of corresponding innovation processes are known (see the literature [2, 4, 6, 7]). Our first results concern a class of processes with innovation processes of the same type.

It is more convenient to treat generalized processes. Let $\eta(t) = \{\eta_i(t)\}_1^m$, $t > t_0$, be a random process for which the projection operators Q_t onto the corresponding subspaces $H_t(\eta)$, $t > t_0$, are given. Let $\xi(t) = \{\xi_i(t)\}_1^m$, $t > t_0$, be a random process such that the mapping

$$\mathscr{A} : \eta_i(t) \to \xi_i(t)$$

can be extended to a linear *bounded* operator from $H(\eta)$ onto $H(\xi)$. Instead of $\xi(t)$, $t > t_0$, we can consider the generalized random process

$$\xi(u) = Au, \qquad u \in U, \tag{6}$$

with the correlation operator $B = A^*A$ on the Hilbert space $U = H(\eta)$ in which a "chain" of subspaces $U_t = H_t(\eta)$, $t > t_0$, is given.

Let $\xi(u)$, $u \in U$, be an arbitrary generalized random process with a correlation operator B on Hilbert space U in which some chain of subspaces U_t, $t > t_0$, is given; that is,

$$U_s \subseteq U_t \quad \text{if} \quad s \leq t, \qquad \lim_{h \to +0} U_{t-h} = U_t.$$

Let $H(\xi)$ be a closure of all random variables $\xi(u)$, $u \in U$, and $H_t(\xi)$ be a closure of variables $\xi(u)$, $u \in U_t$. Denote by P_t and Q_t projections onto the corresponding subspaces $H_t(\xi)$ and U_t, $t > t_0$.

We suggest that the process $\xi(u)$, $u \in U$, is *regular* with respect to the chain U_t, $t > t_0$, if the projection families P_t, $t > t_0$, and Q_t, $t > t_0$, are unitary isomorphic; i.e., there is a unitary operator X from U onto $H(\xi)$ such that

$$P_t = XQ_tX^{-1} \qquad \text{for all} \quad t > t_0. \tag{7}$$

The question on the projections P_t is reduced for regular processes to the problem of how to find the corresponding unitary operator X.

It is well known (see, e.g., Plessner and Rohlin [8]) that the projection families are unitary isomorphic if and only if their "spectral types" are the same. Returning to the process $\xi(u)$, $u \in U$, determined by formula (6), we can say that *this process is regular if and only if the original processes* $\xi = \xi(t)$ *and* $\eta = \eta(t)$, $t > t_0$, *have innovation processes of the same type.*

The following general proposition holds true.

Theorem 1. *The process* $\xi(u)$, $u \in U$, *is regular with respect to the chain* U_t, $t > t_0$, *if and only if its correlation operator B admits so-called proper factorization along this chain; i.e.,*

$$B = C^*C \tag{8}$$

where C is some linear operator in the Hilbert space U such that

$$\overline{CU_t} = U_t \qquad \textit{for all} \quad t > t_0. \tag{9}$$

Proof. Let $\mathscr{A}$ be a linear operator in the Hilbert space U such that $\mathscr{A}^*\mathscr{A} = B$ (for instance, it may be $\mathscr{A} = B^{1/2}$). Because the mapping $\xi(u) \leftrightarrow Au$, $u \in U$, is unitary, we can assume that $\xi(u) = Au$, $u \in U$, and then

$$H_t(\xi) = \overline{AU_t}, \qquad t > t_0.$$

The relation (7) can be rewritten in the form

$$P_tX = XQ_t, \qquad t > t_0,$$

where

$$P_tXU = P_tH(\xi) = H_t(\xi), \qquad XQ_tU = XU_t,$$

and the regularity means that

$$H_t(\xi) = XU_t, \qquad t > t_0. \tag{10}$$

If the process $\xi(u) = Au$, $u \in U$, is regular, so that relation (10) holds true for some unitary operator X from U onto $H(\xi)$, then the operator $C = X^{-1}A$ obviously satisfies condition (9) and gives the proper factorization (8).

It is clear that a generalized process of the form $\xi(u) = Au$, $u \in U$, where the operator $A = C$ satisfies conditions (8)–(9), is regular because $H_t(\xi) = U_t$ for all $t > t_0$, since it follows from condition (9).

Example 1. Stationary processes. Let U^0 be some Hilbert space with an inner product (u, v); $u, v \in U$, and $U = L^2(U^0)$ be the Hilbert space for all measurable functions $u = u_\lambda$ of λ with values $u_\lambda \in U^0$, $\int \|u_\lambda\|^2 \, d\lambda < \infty$, in which the inner product of functions $u, v \in U$ is

$$\langle u, v\rangle = \int (u_\lambda, v_\lambda) \, d\lambda$$

where $-\pi \leq \lambda \leq \pi$ in the case of discrete time t and $-\infty < \lambda < \infty$ in the case of continuous time t, $-\infty < t < \infty$. Let U_t, $-\infty < t < \infty$, be a family of the subspaces of all functions $u_\lambda \in U$ for which $\int e^{-i\lambda s} u_\lambda \, d\lambda = 0$ if $s > t$; that is in the case of discrete time

$$U_t = \bigvee_{s \leq t} e^{i\lambda s} U^0.$$

Let f_λ be a *bounded* positive operator function of λ in U^0 (spectral density). According to our scheme (6), let us determine the operator A in U as

$$Au = f_\lambda^{1/2} u_\lambda, \qquad u = u_\lambda \in U.$$

Then for the *stationary process* $\xi(u) = Au$, $u \in U$, the corresponding chain $H_t(\xi)$, $t > -\infty$, is formed by the subspaces

$$H_t(\xi) = \bigvee_{s \leq t} e^{i\lambda s} f_\lambda^{1/2} U^0.$$

The process $\xi(u)$, $u \in U$, *is regular with respect to the chain* U_t, $t > -\infty$, *if and only if*

$$\bigcap_t H_t(\xi) = 0 \tag{11}$$

and besides

$$\dim f_\lambda U^0 = \dim U^0 \qquad \text{a.e.}$$

We see that for stationary processes our notion of regularity coincides with Kolmogorov's well-known definition of regularity, which was given in the case of $\dim U^0 = 1$ and means that condition (11) is satisfied (see, e.g., Rozanov [9]).

This condition (11) is equivalent to the factorization of the correlation operator B ($Bu = f_\lambda u_\lambda$, $u = u_\lambda \in U$) or, that the same, to the factorization of the spectral density f_λ by an "analytical" operator function φ_λ from the so-called Hardy class H^2, i.e.,

$$\varphi_\lambda \sim \begin{cases} \sum_0^\infty \Phi_t e^{i\lambda t} & \text{for discrete time } t, \\ \int_0^\infty \Phi_t e^{i\lambda t}\, dt & \text{for continuous time } t, \end{cases}$$

namely, that

$$f_\lambda = \phi_\lambda \cdot \phi_\lambda{}^* \qquad \text{a.e.}$$

The question on the regularity of stationary processes with discrete time was treated in detail in a recent paper [11] where the effective criteria of regularity were given. The same criteria hold true for continuous time: condition (11) is satisfied if and only if there is an "analytical" operator-valued function ψ_λ of the Hardy class H^2 such that

$$\psi_\lambda U^0 \subseteq f_\lambda^{1/2} U^0, \qquad \overline{f_\lambda^{-1/2}\psi_\lambda U^0} = \overline{f_\lambda^{1/2} U^0} \qquad \text{a.e.}$$

and (11′)

$$\int_{-\infty}^{\infty} \| f_\lambda^{-1/2} \psi_\lambda u^0 \|^2 \, d_\lambda < \infty, \qquad u \in U^0.$$

These criteria very easily give us all conditions of regularity in this or that known case, say, in the case of univariate stationary processes ($\dim U^0 = 1$) as follows from condition (11′)

$$\int_{-\infty}^{\infty} \frac{\log |f_\lambda^{-1/2}\psi_\lambda|^2}{1+\lambda^2}\, d_\lambda = \int_{-\infty}^{\infty} \frac{\log|\psi_\lambda|^2}{1+\lambda^2}\, d_\lambda - \int_{-\infty}^{\infty} \frac{\log f_\lambda}{1+\lambda^2}\, d_\lambda < \infty$$

and because the former integral on the right-hand side with the function ψ_λ of Hardy class H^2 is finite, we obtain the well-known Kolmogorov–Krein condition of regularity

$$\int_{-\infty}^{\infty} \frac{\log f_\lambda}{1+\lambda^2}\, d_\lambda > -\infty$$

(see, e.g., Rozanov [9]). Inversely, if the regularity takes place, so that the spectral density f_λ admits the factorization $f_\lambda = \varphi_\lambda \cdot \varphi_\lambda{}^*$ by an "analytical" function φ_λ of Hardy class H^2, then the function $\psi_\lambda = \varphi_\lambda / 1 + i\lambda$ of the same class H^2 satisfies condition (11)′.

Let us return to the general processes $\xi(u)$, $u \in U$.

Theorem 2. *If the correlation operator B is invertible and*

$$B = I + \Delta,$$

where Δ *is a Hilbert–Schmidt operator in* U, *then the process* $\xi(u)$, $u \in U$, *is regular with respect to any chain of subspaces* U_t, $t > t_0$.

Proof. Note that $B^{-1} = I - F$ and

$$B = (I - F)^{-1}. \tag{12}$$

where $F = B^{-1}\Delta$ is also the Hilbert–Schmidt operator.

Let us consider some "maximal chain" of projection operators in U which include all projections Q_t onto U_t, $t > t_0$ (see, e.g., Gohberg and Krein's book [3]). By means of Theorem 6.2, Chapter IV, and Theorem 10.1, Chapter I, of this book, we can conclude that the correlation operator B of the form (12) admits the so-called special factorization along that maximal chain:

$$B = C^*C, \qquad C = D(I + V) \tag{13}$$

where V is a Volterra operator, D is a positive inverse operator, and $VU_t \subseteq V_t$, $DU_t \subseteq U_t$, for all t. Indeed according to Gohberg and Krein [3, Chapter IV, Theorem 6.2], our operator B can be represented in the form (13) if the following condition is satisfied: for any projection Q the operator $I - QFQ$ is invertible. But the operator F is such that

$$\sup_{\|u\|=1} (Fu, u) \leq 1 - \inf_{\|u\|=1} (B^{-1}u, u) = r < 1$$

and

$$P = I - QFQ \geq 0, \qquad \inf_{\|u\|=1} (Pu, u) \geq 1 - r > 0,$$

so the inverse operator P^{-1} exists.

Recall that a spectrum of a Volterra operator consists only of a single point $\lambda = 0$ and the equation $(I + V)u = v$ has a solution $u \in U_t$ for any element $v \in U_t$, so that $(I + V)U_t = U_t$. Obviously the positive operator D has the bounded inverse operator and thus $DU_t = U_t$. Hence we obtain that the operator $C = D(I + V)$ maps U_t onto the whole U_t for all t, so conditions (8)–(9) are satisfied and $\xi(u)$, $u \in U$, is regular. The theorem is proved.

We wish to note that for the generalized process $\xi(u)$, $u \in U$, it is very interesting to consider a *nuclear* correlation operator B. The question on the proper factorization in this case is still open.

It is worth emphasizing that all questions on innovation processes, regularity, etc., concern only a "linear theory" of random processes, and in order that our results be more clear we can assume without loss of generality that these processes are Gaussian.

We say that the probability distributions of random processes are *in wide-sense equivalent* if Gaussian distributions with the same correlation are equivalent in the usual sense.

As a corollary of Theorem 2 and well-known results on the equivalence of Gaussian distributions (see, e.g., Rozanov [10]) we obtain the following proposition.

Theorem 3. *Random processes with wide-sense equivalent distributions have innovation processes of the same type.*[1]

To conclude this section we consider a *nonanticipative transformation* of some given generalized process $\eta(u)$, $u \in U$, with respect to some chain of subspaces U_t, $t > t_0$:

$$\xi(u) = \eta(Au), \qquad u \in U, \tag{14}$$

where a linear operator A in the Hilbert space U satisfies the condition

$$AU_t \subseteq U_t, \qquad t > t_0. \tag{15}$$

There is a question regarding the relation

$$H_t(\xi) = H_t(\eta), \qquad t > t_0. \tag{16}$$

Obviously *this relation holds true if* $\overline{AU_t} = U_t$ *for all* t *or in other terms the operator* $Q_t A^*$ *is invertible in the subspace* U_t *for all* t (recall that Q_t is the projection onto U_t).

In connection with this proposition we note that, for instance, a non-anticipative transformation

$$\xi(t) = \int_{t_0}^{t} \mathscr{A}(t, s)\, d\eta(s)$$

of the Wiener process $\eta(t)$, $t_0 < t < T$, with a Volterra kernel $A = A(t, s)$, $t_0 < s$, $t < T$, may be a process with arbitrary multiplicity and time-spectral functions; more exactly, for any $\varepsilon > 0$ there is a transformation A such that the corresponding process $\xi(t)$, $t_0 \leq t \leq T$, has any given time-spectral functions $F_1(t), \ldots, F_M(t)$ on the interval (ε, T).

2. CANONICAL REPRESENTATIONS AND FULLY SUBMITTED PROCESSES

Let $\mathscr{T}(t) = \{\mathscr{T}_k(t)\}_1^N$, $t \geq t_0$, be some left-continuous process with orthogonal increments and orthogonal components $\mathscr{T}_k(t)$, $k = 1, \ldots, N$, and

[1] Of course, the type of innovation on some interval $t_0 < t \leq T$ depends only on the corresponding process $\xi = \xi(t)$ on the same time interval $t_0 < t \leq T$.

$\xi(t) = \{\xi_i(t)\}_1^m$, $t > t_0$, be a nonanticipative process with respect to $\mathscr{T}(t)$:

$$\xi_i(t) = \int_{t_0}^{t} \sum_{k=1}^{N} c_{ik}(t, s)\, d\mathscr{T}_k(s), \qquad i = 1, \ldots, m. \tag{17}$$

One says it is *a canonical representation* if for all t a projection of $\xi_i(t+u)$, $u > 0$, onto the subspace $H_t(\xi)$ is

$$P_t\, \xi_i(t+u) = \int_{t_0}^{t} \sum_{k=1}^{N} c_{ik}(t+u, s)\, d\mathscr{T}_k(s), \qquad i = 1, \ldots, m, \tag{18}$$

where P_t denotes the projection operator onto $H_t(\xi)$ in Hilbert space $H(\xi)$—cf. condition (4).

For example, representation (17) is canonical if $\mathscr{T}(t)$ is an innovation process for $\xi(t)$.

We say that a process $\xi(t) = \{\xi_i(t)\}_1^m$, $t > t_0$, is *submitted* to a process $\eta(t) = \{\eta_j(t)\}_1^n$, $t > t_0$, if

$$H_t(\xi) \subseteq H_t(\eta) \qquad \text{for all } t, \tag{19}$$

and $\xi(t)$ is *fully submitted* to $\eta(t)$ if, in addition, these processes satisfy the condition

$$H_t(\xi)^\perp \subseteq H_t(\eta)^\perp \qquad \text{for all } t, \tag{20}$$

where

$$H_t(\xi)^\perp = H(\xi) \ominus H_t(\xi), \qquad H_t(\eta)^\perp = H(\eta) \ominus H_t(\eta).$$

For example, *the process $\xi(t) = \{\xi_i(t)\}_1^m$, $t > t_0$, is fully submitted to any process $\eta(t)$ with the innovation process $\mathscr{T}(t) = \{\mathscr{T}_k(t)\}_1^N$ for which the canonical representation* (17) *holds true, because*

$$H_t(\xi) \subseteq H_t(\mathscr{T}) = H_t(\eta)$$

and $H_t(\xi)^\perp$ generated by variables

$$\xi_i(t+u) - P_t\, \xi_i(t+u) = \int_{t}^{t+u} \sum_{k=1}^{N} c_{ik}(t+u, s)\, d\mathscr{T}_k(s), \qquad u > 0,$$

is contained in $H_t(\mathscr{T})^\perp = H_t(\eta)^\perp$, $t > t_0$.

If the process $\xi(t) = \{\xi_i(t)\}_1^m$, $t > t_0$, is fully submitted to some process $\eta(t)$, $t > t_0$, with innovation process $\mathscr{T}(t) = \{\mathscr{T}_k(t)\}_1^N$, $t > t_0$, then the canonical representation (17) *holds.*

Indeed, if P_t and Q_t are the projection operators onto $H_t(\xi)$ and $H_t(\eta)$, then as follows from relations (19), (20),

$$Q_t\, H_t(\xi) = H_t(\xi), \qquad Q_t\, H_t(\xi)^\perp \subseteq Q_t\, H_t(\eta)^\perp = 0 \tag{21}$$

and we see that *the operator* Q_t *in the invariant subspace* $H(\xi) \subseteq H(\eta)$ *coincides with* P_t so representation (17) for $\xi_i(t) \in H_t(\eta) = H_t(\mathscr{T})$, $t > t_0$, is the canonical representation because

$$P_t \xi_i(t+u) = Q_t \xi_i(t+u) = \int_{t_0}^{t} \sum_{k=1}^{N} c_{ik}(t+u, s)\, d\mathscr{T}_k(s), \qquad u > 0.$$

In connection with representation (17) it is worth noting that we obtained in particular the following result.

The process $\mathscr{T}(t)$ *in the canonical representation* (17) *is the innovation process for* $\xi(t)$, $t > t_0$, *if and only if*

$$H(\xi) = H(\mathscr{T}). \tag{22}$$

As application of this result, let us consider the following example.

Example 2. Let a random process $\xi(t) = \{\xi_i(t)\}_1^m$, $t > t_0$, satisfy a linear stochastic equation

$$\dot{\xi}(t) - A(t)\xi(t) = y(t) \tag{23}$$

where $A(t)$ is a continuous matrix and $y(t) = \{y_i(t)\}_1^m$ is a random process of "white noise" type:

$$Ey_i(s)y_j(t) = f_{ij}(t)\, \delta(t-s),$$

so $\xi(t)$ can be represented in the form

$$\xi(t) = \int_{t_0}^{t} R(t, s)\, d\mathscr{T}(s). \tag{24}$$

where $R(t, s) = \{R_{ik}(t, s)\}_1^m$ is the resolvent of the differential operator $\dot{x} - Ax$ and $\mathscr{T}(t) = \int_{t_0}^{t} y(s)\, ds$ is a Wiener process with the diffusion matrix $f(t) = \{f_{ij}(t)\}_1^m$.

Formula (24) gives us a canonical representation of the process $\xi(t)$ because, owing to the well-known property of the resolvent,

$$R(t+u, s) = R(t+u, t)R(t, s), \qquad u \geq 0,$$

and apparently the components of vectors

$$Q_t \xi(t+u) = \int_{t_0}^{t} R(t+u, s)\, d\mathscr{T}(s)$$

$$= R(t+u, t) \int_{t_0}^{t} R(t, s)\, d\mathscr{T}(s) = R(t+u, t)\xi(t)$$

belong to the subspace $H_t(\xi)$ for all $t > t_0$.

Let us show that Wiener process $\mathscr{T}(t)$ is an innovation process for $\xi(t)$. The space $H_T(\mathscr{T})$ consists of all random variables $y = \int_{t_0}^{T} c(t)\, d\mathscr{T}(t)$ where a vector function $c(t) = \{c_k(t)\}_1^m$ is such that $\int_{t_0}^T c(t)f(t)c^*(t)\,dt < \infty$ (here $c^*(t)$ denotes a column function conjugated to $c(t)$). Suppose that for some $y \in H_T(\mathscr{T})$ we have

$$x_i(t) = \langle \xi_i(t), y \rangle \equiv 0.$$

A vector function $x(t) = \{x_i(t)\}_1^m$ represented in the form

$$x(t) = \int_{t_0}^{t} R(t, s)f(s)c^*(s)\,ds$$

satisfies the equation

$$\dot{x}(t) - A(t)x(t) = f(t)c^*(t) \qquad \text{a.e.}$$

and if $x(t) = 0$, then $f(t)c^*(t) = 0$ a.e. It shows that $\xi(t)$, $t_0 < t \leq T$, is a complete system in $H_T(\eta)$ for all T and that condition (22) holds true, so $\mathscr{T}(t)$, $t > t_0$, is the innovation process for $\xi(t)$, $t > t_0$.

Let us consider a general random process $\xi(t) = \{\xi_i(t)\}_1^m$ which is fully submitted to some process $\eta(t) = \{\eta_j(t)\}_1^n$, $t > t_0$. Like (22), *the condition*

$$H(\xi) = H(\eta) \tag{25}$$

is necessary and sufficient in order that

$$H_t(\xi) = H_t(\eta), \qquad t > t_0,$$

because, as was shown earlier (see also the proof of Theorem 4), the projection Q_t onto $H_t(\eta)$ coincides in $H(\xi)$ with the projection P_t onto $H_t(\xi)$ and under condition (25)

$$H_t(\xi) = Q_t H(\xi) = Q_t H(\eta) = H_t(\eta).$$

When is condition (25) satisfied? In connection with this question we suggest the following result.

Theorem 4. *Let $\xi(t)$, $t > t_0$, be fully submitted to a process $\eta(t)$, $t > t_0$, with the same innovation. Then in the case of finite multiplicity relation* (25) *holds true.*

Recall that earlier we proved that $\xi(t)$, $\eta(t)$ have innovation processes of the same time if, for instance, their probability distributions are in wide-sense equivalent (see Theorem 3).

We wish also to pay attention to the fact that in the case of infinite multiplicity Theorem 4 becomes wrong. For example, let $\eta(t) = \{\eta_k(t)\}_1^\infty$, $t > t_0$, be a process with orthogonal identically distributed components $\eta_k(t)$, $k = 1, 2, \ldots$, and $\xi(t) = \{\eta_k(t)\}_2^\infty$, $t > t_0$; obviously, the process $\xi(t)$, $t > t_0$, is fully submitted to the process $\eta(t)$, $t > t_0$, and their innovation processes are of the same type, but $H(\xi) \neq H(\eta)$ because the first component $\eta_1(t)$, $t > t_0$, does not belong to $H(\xi)$.

Proof of Theorem 4. As was shown above, the projection operator Q_t onto $H_t(\eta)$ coincides in the invariant subspace $H(\xi) \subseteq H(\eta)$ with the projection P_t onto $H_t(\xi)$—see (21). Let

$$X_j(t) = Q_t x_j, \qquad \mathscr{T}_j(t) = Q_t y_j, \qquad j = 1, \ldots, M, \tag{26}$$

be innovation processes of $\xi(t)$, $\eta(t)$ with the same time-spectral functions $F_j(t)$, $j = 1, \ldots, M$. It is more convenient to treat densities

$$f_j(t) = \frac{dF_j(t)}{dF(t)}, \qquad j = 1, \ldots, M,$$

with respect to $F(t) = F_1(t)$, $t > t_0$. If

$$x_j = \int \sum_{k=1}^{M} c_{jk}(s) \mathscr{T}_k(s),$$

then

$$X_j(t) = \int_{t_0}^{t} \sum_{k=1}^{M} c_{jk}(s)\, d\mathscr{T}_k(s), \qquad j = 1, \ldots, M,$$

and for all $t > t_0$

$$\begin{aligned} E|X_j(t)|^2 &= \int_{t_0}^{t} \sum_{k=1}^{M} |c_{jk}(s)|^2 f_k(s)\, dF(s) \\ &= F_j(t) = \int_{t_0}^{t} f_j(s)\, dF(s), \end{aligned}$$

$$EX_i(t)\overline{X_j(t)} = \int_{t_0}^{t} \sum_{k=1}^{M} c_{ik}(s)\overline{c_{jk}(s)} f_k(s)\, dF(s) = 0 \qquad \text{for} \quad i \neq j.$$

So for almost all t with respect to $dF(t)$

$$\sum_{k=1}^{M} |c_{jk}(t)|^2 f_k(t)\, dt = f_j(t),$$

$$\sum_{k=1}^{M} c_{ik}(t)\overline{c_{jk}(t)} f_k(t)\, dt = 0 \qquad \text{for} \quad i \neq j \qquad (i, j = 1, \ldots, M).$$

We see that under fixed t the vectors

$$c_j(t) = \{c_{jk}(t)\}_1^M, \qquad j = 1, \ldots, M,$$

form the orthogonal basis in a unitary space of the vectors $c = \{c_k\}_1^M$ with the inner product

$$(c', c'') = \sum_{k=1}^{M} c_k' \bar{c}_k'' f_k(t).$$

Therefore if some variable

$$\eta = \int \sum_{k=1}^{M} c_k(s)\, d\mathcal{T}_k(s) \in H(\eta)$$

is orthogonal to $H(\xi)$, then

$$\langle \eta, X_j(t) \rangle = \int_{t_0}^{t} \sum_{k=1}^{M} c_k(s)\overline{c_{jk}(s)} f_k(s)\, dF(s) = 0$$

for all $t > t_0$, $j = 1, \ldots, M$,

$$\sum_{k=1}^{M} c_k(t)\overline{c_{jk}(t)} f_k(t) = 0 \qquad \text{a.e.,} \qquad j = 1, \ldots, M,$$

and

$$\sum_{k=1}^{M} |c_k(t)|^2 f_k(t) = 0 \qquad \text{a.e.}$$

So $\eta = 0$. It means that $H(\xi) = H(\eta)$.

Example 3. Let $\eta(t)$, $t_0 < t \leq T$, be some univariate process with orthogonal increments and time-spectral function $F(t)$, and let $\xi(t)$, $t_0 < t \leq T$, be a process of the form

$$\xi(t) = \int_{t_0}^{t} a(s)\, dF(s) + \eta(t) \tag{27}$$

where a random process $a(t)$, $t_0 \leq t \leq T$, is such that

$$\int_{t_0}^{T} E|a(t)|^2\, dt < \infty. \tag{28}$$

We can determine a stochastic integral

$$\int_{t_0}^{T} u(t)\, d\xi(t) = \int_{t_0}^{T} u(t)a(t)\, dF(t) + \int_{t_0}^{T} u(t)\, d\eta(t)$$

for any measurable function $u = u(t)$, $t_0 \leq t \leq T$, which satisfies the condition $\int_{t_0}^{T} |u(t)|^2\, dF(t) < \infty$. Let U be the Hilbert space of all such functions with the inner product $(u, v) = \int_{t_0}^{T} u(t)\overline{v(t)}\, dF(t)$. It is very easy to verify that

$$A\colon \int_{t_0}^{T} u(t)\, d\eta(t) \to \int_{t_0}^{T} u(t)a(t)\, dF(t)$$

is a *Hilbert–Schmidt* operator in $H(\eta)$ and

$$\xi(t) = \eta(t) + A\eta(t), \qquad t_0 < t \leq T.$$

Let the generalized random process

$$\xi(u) = \int_{t_0}^{T} u(t)\, d\xi(t), \qquad u \in U,$$

be nondegenerate in the sense that $\xi(u) = 0$ with probability 1 if and only if $u(t) = 0$ for almost all t with respect to $dF(t)$, $t_0 < t \leq T$. Then probability distributions of the random process $\xi(t)$, $\eta(t)$ are in wide-sense equivalent (sce, e.g., Rozanov [10]). As a corollary of Theorem 3 we obtain that the multiplicity of the random process $\xi(t)$ is equal to 1 and its time-spectral function is the same type with $F(t)$.

Suppose the random process $\eta(t)$, $t_0 < t \leq T$, with orthogonal increments is such that

$$\eta(t+u) - \eta(t) \perp H_t(\xi), \qquad u > 0. \tag{29}$$

Then the corresponding *innovation* process of $\xi(t)$, $t_0 < t \leq T$, can be obtained as

$$X(t) = \xi(t) - \int_{t_0}^{t} \hat{a}(s)\, dF(s), \qquad t_0 < t \leq T, \tag{30}$$

where $\hat{a}(t) = \hat{E}(a(t)/\xi(s),\ s \leq t)$ is the projection of the variable $a(t)$ onto the subspace $H_t(\xi)$. Indeed

$$X(t) = \int_{t_0}^{t} [a(s) - \hat{a}(s)]\, dF(s) + \eta(s), \qquad t > t_0,$$

and increments

$$X(t+u) - X(t) = \int_{t}^{t+u} [a(s) - \hat{a}(s)]\, dF(s) + [\eta(t+u) - \eta(t)]$$

are orthogonal to $H_t(\xi)$ for all $u > 0$; the subspace $H_t(X)^{\perp}$ generated by all variables $X(t+u) - X(t)$, $u > 0$, is orthogonal to the subspace $H_t(\xi)$, so $H_t(X)^{\perp} \subseteq H_t(\xi)^{\perp}$ and the process $X(t)$, $t_0 < t \leq T$, is *fully submitted* to the process $\xi(t)$, $t_0 < t \leq T$. Besides, these processes have probability distributions which are in wide sense equivalent to the distribution of the process $\eta(t)$, $t_0 < t \leq T$; hence, from Theorem 4

$$H_t(X) = H_t(\xi) \qquad \text{for all } t.$$

In conclusion we wish to consider the following proposition.

Theorem 5. *Let a process $\xi(t)$, $t > t_0$, be fully submitted to a process $\eta(t)$, $t > t_0$, and*

$$dF_1(t) \succcurlyeq dF_2(t) \succcurlyeq \cdots \succcurlyeq dF_M(t),$$
$$dG_1(t) \succcurlyeq dG_2(t) \succcurlyeq \cdots \succcurlyeq dG_N(t)$$

be time-spectral functions of such processes. Then

$$M \leq N, \qquad dF_j(t) \preccurlyeq dG_j(t), \qquad j = 1, \ldots, M. \tag{31}$$

Proof. Denote by

$$X_j(t) = Q_t x_j, \qquad j = 1, \ldots, M,$$
$$\mathscr{T}_k(t) = Q_t y_k, \qquad k = 1, \ldots, N,$$

components of innovation processes—see (26). Obviously $F_1(t)$ is absolutely continuous with respect to the "maximal spectral type" $G_1(t) = \langle Q_t y_1, y_1 \rangle$. If $dG_2(t) \sim dG_1(t)$, then $dF_2(t) \preccurlyeq dG_2(t)$. If $dG_1(t) = dG_2(t) \oplus dG(t)$ where $dG(t)$ is orthogonal to $dG_2(t)$, then the *cyclic* subspace of all elements $x \in H(\xi)$ with "spectral types" which are absolutely continuous with respect to $G(t)$, obviously belongs to the *first* cyclic subspace $H(X_1)$ in $H(\xi)$, so $dF_2(t)$ is orthogonal to $dG(t)$ and $dF_2(t) \preccurlyeq dG_2(t)$. By similar arguments we can prove that $dF_j(t) \preccurlyeq dG_j(t)$ for all $j = 1, \ldots, M$ (see, e.g., Plessner and Rohlin [8]).

Note that in the relations (31) may be strictly inequality $M < N$ and besides $dF_k \prec dG_k(t)$. For instance, if $\xi(t) = \mathscr{T}_1(t)$, $t > t_0$, is the first component of the multivariate innovation process $\mathscr{T}(t) = \{\mathscr{T}_k(t)\}_1^N$, $N > 1$, of some random process $\eta(t)$, $t > t_0$, then obviously $\xi(t)$ is fully submitted to $\eta(t)$, $t > t_0$, and $M < N$ (cf. Kallianpur and Mandrekar [6]).

REFERENCES

1. Cramér, H. (1964). Stochastic processes as curves in Hilbert space. *Theor. Probability Appl.* **9** 193–204.
2. Cramér, H. (1966). A contribution to the multiplicity theory of stochastic processes. *Proc. Fifth Berkeley Symp. Math. Statist. Prob.* **2** 215–221.
3. Gohberg, I. C. and Krein, M. G. (1967). *Theory and Applications of Volterra Operators in Hilbert Space.* "Nauka," Moscow (Translated in English by Amer. Math. Soc., 1970).
4. Ivkovich, L. and Rozanov, Yu. A. (1971). On the canonical Hida-Cramer representation of stochastic processes. *Theor. Probability Appl.* **16** 348–353.
5. Kailath, T. (1970). The innovations approach to detection and estimation theory. *Proc. IEEE* **58** 000.
6. Kallianpur, G. and Mandrekar, V. (1965). Multiplicity and representation theory of purely non-deterministic stochastic processes. *Theor. Probability Appl.* **10** 553–581.
7. Mandrekar, V. (1968). On multivariate wide-sense Markov processes. *Nagoya Math. J.* **33** 7–19.
8. Plessner, A. I. and Rohlin, V. A. (1946). Spectral theory of linear operators, II. *Uspehi Mat. Nauk* **1** 71–191.
9. Rozanov, Yu. A. (1967). *Stationary Random Processes* (Engl. transl.). Holden-Day, San Francisco, California.
10. Rozanov, Yu. A. (1968). *Infinite-Dimensional Gaussian Distributions.* "Nauka," Moscow (Proceedings of the Steklov Mathematical Institute, Vol. 108; translated in English by Amer. Math. Soc., 1971).
11. Rozanov, Yu. A. (1972). Some approximation problems in the theory of stationary processes. *J. Multivariate Anal.* **2** 135–144.

PART II

Distribution Theory and Inference

Methods for Assessing Multivariate Normality

D. F. ANDREWS, R. GNANADESIKAN, and J. L. WARNER

BELL LABORATORIES
MURRAY HILL, NEW JERSEY

1. INTRODUCTION

The assumption of multivariate normality underlies much of the standard "classical" multivariate statistical methodology. The effects of departures from normality on the methods are not easily or clearly understood. Moreover, for analyzing multiresponse data, the development of statistical methods that are robust to such distributional departures is still at a rudimentary stage. (Cf. Gnanadesikan and Kettenring, 1972.) Thus, it would be useful to have procedures for verifying the reasonableness of assuming normality for a given body of multiresponse observations. If available, such a check would be helpful in guiding the subsequent analysis of the data, perhaps by suggesting the need for and nature of a transformation of the data to make it more nearly normally distributed, or perhaps by indicating appropriate modifications of the models and methods for analyzing the data.

The class of multivariate non-normal distributions which have been put forth as alternatives to the multivariate normal is not very rich. Most of the alternatives (e.g. the multivariate lognormal and gamma) are defined so as to have properties that are similar to those of the multivariate normal (e.g. that all marginal distributions belong to the same class). Real data will, of course, not necessarily conform to such specified types of multivariate non-normality.

With multiresponse data, it is clear that the possibilities for departure from joint normality are indeed many and varied. One implication of this is the need for a variety of techniques with differing sensitivities to the different types of departures. Seeking a single best method would seem to be neither pragmatically sensible nor necessary. Developing several techniques and enabling an accumulation of experience with, and insight into, their properties is a crucial first step.

One way of seeing the need for a variety of techniques in the multivariate case is in terms of the degree of commitment one wishes to make to the coordinate system for the multiresponse observations. At one extreme is the situation wherein the interest is completely confined to the observed coordinates. In this case, the marginal distributions of each of the observed variables and conditional distributions of certain of these given certain others would be the objects of interest. On the other hand, the interest may be in the original coordinates as well as all possible orthogonal transformations of them, and here the things (such as Euclidean distance) which remain invariant under orthogonal transformations would be the ones of interest. More generally, the class of all nonsingular linear transformations of the observed variables may be the one of interest and then affine invariance would guide the analysis. Aside from linear transformations, sometimes one may be willing to make simple nonlinear transformations (perhaps of each coordinate separately) so as to be able to use simple models and techniques. In this case, the methods used should be cognizant of this degree of flexibility and should attempt to incorporate it statistically. Much of the formal theory of multivariate analysis has been concerned solely with affine invariance, thus limiting the class of available procedures. The present paper considers techniques that are applicable to situations with different degrees of commitment to the observed coordinate system, including the classical one requiring affine invariance.

Another important issue with multivariate techniques is that although some complexity of the methods is to be expected, yet, if possible, they should be kept computationally economical. The feasibility of extensive computing made easily accessible by modern computers still does not imply that every technique is economically tenable. One objective used in developing the methods described in this paper was that they be computationally reasonably economic and efficient.

The specific concerns of this paper are to review some techniques (both old and new) for assessing the normality of multivariate data and to illustrate their performance by some examples. Of necessity, the more familiar techniques will merely be mentioned with appropriate references and even for the newer techniques the emphasis will be on the more promising ones based on preliminary and very limited experience with them. The reader interested in more extensive details is referred to a larger working document (Andrews et al., 1972) from which the material in this paper has been extracted.

More specifically, this paper covers methods under the following headings: (i) univariate techniques for evaluating marginal normality (Section 2); (ii) multivariate techniques for evaluating joint normality (Section 3); and (iii) other procedures based on unidimensional views of the multiresponse data (Section 4). As mentioned earlier, performing an initial transformation on the data and then using "standard" methods of analysis is a prevalent and often useful approach in analyzing data. Hence, as a general approach under each of the three categories of methods mentioned above, the assessment of normality may be made by inquiring about the need for a transformation. However, an approach which is not explicitly dependent on data-based transformations is also possible. Techniques of both types are discussed in this paper.

2. UNIVARIATE TECHNIQUES FOR EVALUATING MARGINAL NORMALITY

In practice, a single overall multivariate analysis of data is seldom sufficient or adequate by itself, and almost always it needs to be augmented by analyses of subsets of the responses, including univariate analyses of each of the original variables. Although marginal normality does not imply joint normality, the presence of many types of non-normality is often reflected in the marginal distributions as well. Hence a natural, simple, and preliminary step in evaluating the normality of multiresponse observations is to study the reasonableness of marginal normality for the data on each of the variables.

For this purpose, one can utilize a variety of methods for assessing univariate normality, including: (i) likelihood ratio tests associated with transformations for enhancing univariate normality, (ii) skewness and kurtosis tests (cf., e.g., Pearson & Hartley, 1966), (iii) omnibus tests for normality (e.g. W-test of Shapiro & Wilk, 1965; D-test of D'Agostino, 1971; and a test based on gaps, Andrews et al., 1972), and (iv) normal probability plots. For the sake of brevity, only the transformation-related method (item (i)) will be described here.

Box and Cox (1964) have proposed methods for estimating a shifted-power transformation, $(X + \xi)^\lambda$, of a single variable X so as to improve normality. This simple but flexible class (cf. Tukey, 1957, 1970; Moore and Tukey, 1954) of transformations includes many commonly used transformations. Specifically, one of the approaches suggested by Box and Cox (1964) is to estimate ξ and λ by maximum likelihood using the observations on X. The logarithm of a likelihood function (initially maximized with respect to the mean and variance for given ξ and λ), $\mathscr{L}_{\max}(\xi, \lambda)$, is maximized to provide estimates $\hat{\xi}$ and $\hat{\lambda}$, and using the approximate asymptotic theory Box and Cox (1964) proposed a $100(1 - \alpha)\%$ confidence region for ξ and λ defined by

$$\mathscr{L}_{\max}(\hat{\xi}, \hat{\lambda}) - \mathscr{L}_{\max}(\xi, \lambda) \leq \tfrac{1}{2}\chi_2{}^2(\alpha),$$

where $\chi_2{}^2(\alpha)$ denotes the upper $100\alpha\%$ point of χ^2 with two degrees of freedom. A simple transformation-related procedure for assessing the normality of the distribution of X consists in not rejecting (at a $100\alpha\%$ level of significance) the hypothesis of normality if the above confidence region overlaps with the line $\lambda = 1$. A related more stringent "likelihood ratio test" would consist of comparing the value of $2\{\mathscr{L}_{\max}(\hat{\xi}, \hat{\lambda}) - \mathscr{L}_{\max}(\hat{\xi}, 1)\}$ to a chi-squared distribution with one degree of freedom. [Note that $\mathscr{L}_{\max}(\xi, 1)$ is independent of ξ so that any value, including $\hat{\xi}$, maximizes $\mathscr{L}_{\max}(\xi, 1)$ as a function of ξ.]

3. MULTIVARIATE TECHNIQUES FOR EVALUATING JOINT NORMALITY

The tests of the previous section are sensitive to forms of multivariate non-normality which result in non-normality of the marginal distributions. In practice, this marginal non-normality will be present whenever joint non-normality is present except perhaps in rare or pathological examples. However, there is a need for tests which explicitly exploit the multivariate nature of the data, hopefully to yield greater sensitivity.

Some methods addressed to this need are (i) likelihood ratio tests associated with transformations for enhancing joint normality, (ii) goodness-of-fit tests such as χ^2 and Kolmogorov–Smirnov, and tests based on local densities (e.g. Weiss, 1958; Anderson, 1966; and a nearest distance test proposed by Andrews et al., 1972), and (iii) an informal graphical method associated with a *radius-and-angles* representation of the data. The three subsections which follow pertain respectively to the above three groups of methods.

3.1. Transformation-Related Methods

The univariate techniques of Box and Cox (1964) referred to in Section 2 may be extended to multivariate data for estimating transformations within a specified class, such as the shifted-power class, the objective being to make the transformed data more amenable than the untransformed data to treatment by simple multivariate normal models (cf. Andrews et al., 1971). There is a test for joint normality corresponding to this estimation problem—evidence that a nonlinear transformation is required to significantly improve joint normality is considered as evidence that the untransformed data is non-normal.

Assume $\boldsymbol{\Lambda} = \{\boldsymbol{\lambda}\}$ is a parametric family of transformations such that the transformed data $\mathbf{y}^{(\boldsymbol{\lambda})}$ may be described by a simple *multivariate normal* model. Then this model yields a distribution for $\mathbf{y}$ depending on the transformation parameters $\boldsymbol{\lambda}$ and the parameters of the multivariate normal model. If the transformations $\boldsymbol{\Lambda}$ are invertible, a log likelihood function of all the parameters may be found. The function is easily maximized over the normal model parameters using common maximum likelihood methods. This maximized function, $\mathscr{L}_{\max}(\boldsymbol{\lambda})$, now depends only on $\boldsymbol{\lambda}$.

The details of this procedure when $\mathbf{\Lambda}$ is the power transformation family and the observations are bivariate are given in Andrews et al. (1971). Even when there are only two variables involved, in order to keep the computational effort down and also be able to display some of the analyses graphically, the transformations actually employed are just power transformations of each variable separately, viz. $Y_j^{\lambda_j}$, with no shift parameters involved. (Cf., however, the discussion of Example 2 in Section 5 for possible limitations imposed by not including shift parameters.)

For the power family, the linear transformation $\boldsymbol{\lambda} = (\lambda_1, \ldots, \lambda_p)' = \mathbf{1}$ is the only transformation consistent with the hypothesis that the data is normally distributed. A likelihood ratio test of the hypothesis $\boldsymbol{\lambda} = \mathbf{1}$ may be based on the asymptotically approximate $\chi_p{}^2$ distribution of

$$2(\mathscr{L}_{\max}(\hat{\boldsymbol{\lambda}}) - \mathscr{L}_{\max}(\mathbf{1})),$$

where $\mathscr{L}_{\max}(\boldsymbol{\lambda})$ is defined in Table I and $\hat{\boldsymbol{\lambda}}$ is the value of $\boldsymbol{\lambda}$ which maximizes $\mathscr{L}_{\max}(\boldsymbol{\lambda})$. This $\chi_p{}^2$ distribution may be used to obtain both a significance level α associated with the observed $\hat{\boldsymbol{\lambda}}$ and a confidence set for $\boldsymbol{\lambda}$. This procedure not only indicates when data is non-normal—which we may be willing to grant for many large samples—but also it suggests data transformations which may be used to enhance normality. Table I presents the steps in applying this test using the power class of transformations.

TABLE I

A Transformation Test for Multivariate Normality

Data $\mathbf{Y} = (y_{ij})$ $i = 1, \ldots, n$; $j = 1, \ldots, p$; $y_{ij} > 0$

(a) Given $\boldsymbol{\lambda} = (\lambda_1, \ldots, \lambda_p)'$ calculate $\mathbf{Y}^{(\boldsymbol{\lambda})} = (y_{ij}^{(\boldsymbol{\lambda})})$ from

$$y_{ij}^{(\boldsymbol{\lambda})} = \begin{cases} (y_{ij}^{\lambda_j} - 1)/\lambda_j & \text{for } \lambda_j \neq 0, \\ \log y_{ij} & \text{for } \lambda_j = 0. \end{cases}$$

(b) Using $\mathbf{Y}^{(\boldsymbol{\lambda})}$ calculate the maximum likelihood estimates of the mean vector and covariance matrix

$$\hat{\boldsymbol{\mu}} = (1/n)\mathbf{Y}^{(\boldsymbol{\lambda})'} \cdot \mathbf{1},$$
$$\hat{\boldsymbol{\Sigma}} = (1/n)(\mathbf{Y}^{(\boldsymbol{\lambda})} - \mathbf{1} \cdot \hat{\boldsymbol{\mu}}')'(\mathbf{Y}^{(\boldsymbol{\lambda})} - \mathbf{1} \cdot \hat{\boldsymbol{\mu}}'),$$

and thence obtain

$$\mathscr{L}_{\max}(\boldsymbol{\lambda}) = -\frac{n}{2}\log|\hat{\boldsymbol{\Sigma}}| + \left\{\sum_{j=1}^{p} (\lambda_j - 1) \sum_{i=1}^{n} \log y_{ij}\right\}.$$

(c) Numerically find $\hat{\boldsymbol{\lambda}}$ to maximize $\mathscr{L}_{\max}(\boldsymbol{\lambda})$.

(d) Obtain the significance level associated with the test statistic

$$2\{\mathscr{L}_{\max}(\hat{\boldsymbol{\lambda}}) - \mathscr{L}_{\max}(\mathbf{1})\}$$

by referring to the chi-squared distribution with p degrees of freedom.

3.2. Tests Based on Distributional Densities

The classical goodness-of-fit tests, such as χ^2 and Kolmogorov–Smirnov, belong in this category. However, the known difficulties and drawbacks of these tests in univariate situations are greatly magnified in the multivariate case, thus severely limiting their usefulness for present purposes (cf. Andrews et al., 1972, for some discussion).

Also, a number of tests have been proposed which are based at least indirectly on the distances between near neighbors in p-space. Two of these, proposed by Weiss (1958) and Anderson (1966), respectively, have been available but have not seen wide application. A third, which attempts to be more sensitive to local density departures by looking explicitly at the behavior of nearest neighbor distances, is proposed by Andrews et al. (1972). The authors' limited experience with all of these procedures seems to suggest that while they can be useful in some circumstances these tests are not uniquely nor sufficiently sensitive to many types of departure from normality.

3.3. Radius and Angles Decompositions

The first step in conceptualizing the method is to obtain the scaled residuals

$$\mathbf{z}_i = \mathbf{S}^{-1/2}(\mathbf{y}_i - \bar{\mathbf{y}}), \qquad i = 1, \ldots, n,$$

where $\mathbf{y}_i$ denotes the observations, $\bar{\mathbf{y}}$ and $\mathbf{S}$ are respectively the sample mean vector and covariance matrix, and $\mathbf{S}^{-1/2}$ is the symmetric square root of $\mathbf{S}^{-1}$. Under the null hypothesis, the scaled residuals are approximately spherically symmetrically distributed. The squared radii, or squared lengths of the $\mathbf{z}_i$,

$$r_i^2 = \mathbf{z}_i'\mathbf{z}_i = (\mathbf{y}_i - \bar{\mathbf{y}})'\mathbf{S}^{-1}(\mathbf{y}_i - \bar{\mathbf{y}}),$$

will have approximately a chi-squared distribution with 2 degrees of freedom in the bivariate case (and p degrees of freedom in the p-variate case). Also, in the bivariate case, the angle θ_i that $\mathbf{z}_i$ makes with say the abscissa direction will be approximately uniformly distributed over $(0, 2\pi)$. All quantities, viz. the r_i^2's and the θ_i's, will be approximately independent for large n. The dependence enters, among other things, via the estimates of the mean and the covariance matrix, and hopefully for large enough samples this dependence will have no serious effects. A further comment which may be in order is that the exact *marginal* distribution of r_i^2 is known to be a constant multiple of a beta distributed variable rather than a chi-squared distribution; but again even for moderate samples (i.e. $n = 20$ or 25 in the bivariate case) the difference between using the beta and the χ^2 approximation appears to be insignificant (cf. Gnanadesikan & Kettenring, 1972).

3.3.1. Plotting Radii and Angles. The above-mentioned properties suggest that summaries in terms of radii and angles may be useful for assessing joint normality. Indeed some authors (e.g. Healy, 1968; Kessel & Fukunaga, 1972) have suggested procedures based purely on the squared radii. Other simple procedures based on both radii and angles are proposed in this section.

The proposed procedures are informal graphical techniques, much in the same spirit as univariate probability plots in their role as aids in assessing the adequacy and appropriateness of univariate distributional assumptions (cf. Wilk & Gnanadesikan, 1968).

In the bivariate case, one can make a $\chi^2_{(2)}$ probability plot of the r_i^2 and a uniform probability plot of the normalized form of θ_i, viz. $\theta_i^* = \theta_i/2\pi$. If the data conform statistically to the null hypothesis of bivariate normality, the configurations on these two probability plots should be reasonably linear. Departures from linearity on either or both of the plots would indicate specific types of departure from null conditions. [Note: The origin on the plot of the θ_i^* is arbitrary. Also, the θ_i^* that correspond to large r_i^2 may be more statistically stable than those with very small r_i^2 and, therefore, one might wish to "trim" the observations with the smallest values of r_i^2 and study an appropriate uniform probability plot of the θ_i^* only for the remaining observations.]

For bivariate data, one can also combine the information in the radii and angles in a single two-dimensional display. Let u_i denote the probability integral transformation of r_i^2 based on a $\chi^2_{(2)}$ distribution of the latter, i.e. $u_i = P\{\chi^2_{(2)} \leq r_i^2\}$ for $i = 1, \ldots, n$. Then a plot of the n points whose coordinates are (u_i, θ_i^*), $i = 1, \ldots, n$, may be made. Under the null hypothesis, one would expect to get a uniform scatter of points on the unit square. Non-uniformity of scatter, or indication of any relationship between the two coordinates in the plot, would suggest departures from the null hypothesis. (Note: Formal tests of uniformity can also be made; however, the main value and appeal of the procedure is its graphical character.)

For higher-dimensional data (say p-dimensional with $p > 2$), there are $(p - 1)$ independent angles involved. In this general case, a $\chi^2_{(p)}$ probability plot of the squared radii (which now would be the squared lengths of p-dimensional scaled residuals) and $(p - 1)$ beta probability plots of the $(p - 1)$ different normalized angles may be made, obtaining p probability plots in all for study. Scatter plots of the unit-square type, described in the preceding paragraph for the bivariate case, may be obtained for subsets of size 2 out of the p quantities involved in the p-variate situation. [Cf. Andrews *et al.*, 1972, for further details when $p > 2$.]

4. TESTS BASED ON UNIDIMENSIONAL VIEWS OF MULTIVARIATE DATA

One attractive property of tests for marginal normality (cf. Section 2) is that the computational effort involved increases only linearly with p, the dimensionality of the data. It is therefore not inappropriate to look into the possibility of using various unidimensional views of the data in addition to just the marginal variables. A study of the r_i^2 of Section 3.3 by themselves, as proposed by Healy (1968) and Kessell & Fukunaga (1972), would be an example. Investigating the linearity of the regression of each variable on all the others is another example of using a collection of unidimensional views of the data.

Another obvious class of techniques to seek would be based on the characterization of the multivariate normal distribution in terms of univariate normality of all linear combinations of the variables. Tests of multivariate normality that look at "all possible" unidimensional projections and utilize the union–intersection principle of Roy (1953) have received some attention recently (cf. Malkovitch & Afifi, 1971; Aitkin, 1972). The computational efforts involved in some of these tests tend to be prohibitive. A different scheme based on looking at unidimensional projections of the multivariate data along specified directions rather than "all possible" directions is described next.

4.1. Directional Normality

Marginal analysis of each of the original variables considers the projections of the data on each of the coordinate axes separately. Other one-dimensional projections may be considered. It is of some interest to use those projections that are likely to exhibit certain types of marked non-normality.

One approach to this problem is to look at projections of the data along directions that are in part determined by the data, but also in part chosen to be sensitive to particular types of non-normality. The work of Andrews et al. (1971) in the context of estimating transformations to enhance directional normality provides a contact point for the testing problem of present concern.

Specifically, once again the scaled residuals $\mathbf{z}_i = \mathbf{S}^{-1/2}(\mathbf{y}_i - \bar{\mathbf{y}})$, which are approximately spherically symmetrically distributed under the null hypothesis, may be obtained. Any non-normal characteristics of the observations $\mathbf{y}_i$ will be reflected in corresponding characteristics of the $\mathbf{z}_i$, and the direction of any non-normal clustering of points, if present, may perhaps be identified by studying a normalized weighted sum of the $\mathbf{z}_i$

$$\mathbf{d}_\alpha = \frac{\sum_{i=1}^n w_i \mathbf{z}_i}{\left\|\sum_{i=1}^n w_i \mathbf{z}_i\right\|}, \qquad w_i = \|\mathbf{z}_i\|^\alpha,$$

where $\|\mathbf{x}\|$ denotes the Euclidean norm, or length, of the vector $\mathbf{x}$ and α is a constant to be chosen.

The vector $\mathbf{d}_\alpha$ provides a parametrization of directions in the z-space (and hence in the y-space of the original observations) in terms of the single parameter α. If $\alpha = -1$, $\mathbf{d}_\alpha$ is a function only of the orientation of the $\mathbf{z}_i$'s, while if $\alpha = +1$, $\mathbf{d}_\alpha$ becomes sensitive primarily to those observations $\mathbf{y}_i$ that are far from the mean $\bar{\mathbf{y}}$. More generally, for $\alpha > 0$ the vector $\mathbf{d}_\alpha$ will tend to point toward any clustering of observations far from the mean, while for $\alpha < 0$ the vector $\mathbf{d}_\alpha$ will point in the direction of any abnormal clustering near the center of gravity of the data.

For a specified α, the vector $\mathbf{d}_\alpha$ (chosen to be sensitive to particular types of non-normal clusterings if any are present) corresponds to the vector $\mathbf{d}_\alpha^* = \mathbf{S}^{1/2}\mathbf{d}_\alpha$ in the space of the original observations. The lengths of the projections of the original observations onto the unidimensional space specified by the "direction" $\mathbf{d}_\alpha^*$ constitute a univariate sample which can be studied by any of the univariate tests for departure from univariate normality mentioned in Section 2. In the examples to be discussed in Section 5, the likelihood ratio test associated with the transformation approach is the one most extensively employed.

Because of the data-dependent, as well as certain other, aspects of the approach the significance levels associated with these tests are probably not formally applicable. However, they do provide useful benchmarks for measuring the non-normality along particular directions (using different values of α would enable one to look in many directions) in the space of the original variables. If this measure is not significant, there is some hope that subsequent methods of analysis will behave as expected. If this measure is highly significant, a further transformation may make the subsequent analysis more meaningful. The transformation test provides an indication of what transformation will ameliorate the abnormalities when the data is viewed in specified directions.

5. EXAMPLES

Many of the methods described heretofore have been used for analyzing several sets of simulated and real bivariate data. Only a few are included here for illustrative purposes.

The simulation included samples from the null (i.e. normal) and several non-null distributions ranging from reasonably close (e.g. χ_ν^2 with large ν) to moderately distant (e.g. Laplace and lognormal) alternatives. Two methods of correlating the variables were employed for each of the alternatives. The first method, leading to so-called *correlated conditional* distributions, consisted of initially generating random samples from one of the alternative distributions independently for each variable, and then introducing a correlation through a linear transformation of the independent coordinates. The second method, leading to so-called *correlated marginal* distributions, conformed to the definitions of many commonly used multivariate non-normal distributions. For this method, observations on each coordinate of a correlated bivariate normal were transformed marginally to the alternative of interest. (Cf. Andrews et al., 1972, for details.) For both methods, a range of values of the correlation was used and the sample size was always 100.

Comfortingly, none of the techniques exhibited significant departures at any reasonable significance level for the few sets of simulated bivariate normal data included in the study. At the other extreme, almost all of the techniques indicated highly significant departures from normality in the lognormal samples. More interesting were the cases wherein some but not all techniques were able to detect the departure from normality and the next two examples illustrate this.

Example 1. This mildly non-normal example consisted of 100 observations from a bivariate correlated conditional distribution with a value of 0.9 for the correlation used in the computer generation of the data.

The techniques described in Section 2 for assessing marginal normality were applied. The transformation-related test indicated highly significant (.0007 for the first variable and .0001 for the second) departures from normality, while the skewness test was moderately significant ($<.01$ and $<.05$, respectively) and the kurtosis, D'Agostino, and gaps tests were all sufficiently nonsignificant ($>.2$). Figures 1a and 1b show normal probability

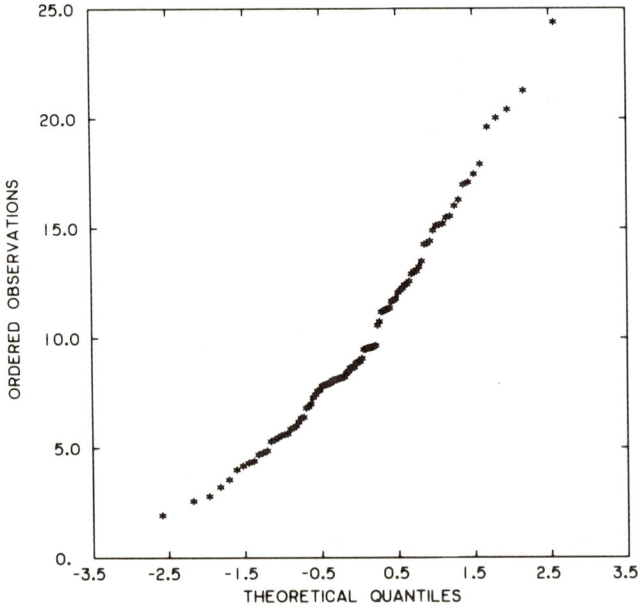

Fig. 1a. Example 1: Normal probability plot of variable 1.

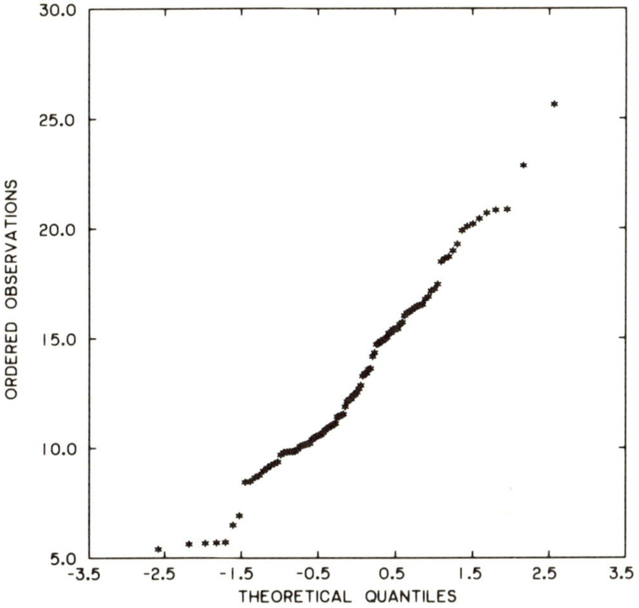

Fig. 1b. Example 1: Normal probability plot of variable 2.

plots of the observations on each variable. The value of this informal tool is quite apparent in this example since both these plots exhibit some curvature and departures from the linearity that would be expected if the data were normal.

Figure 1c is a scatter plot of the data and there is some evidence here (especially for a trained eye) of departures from normality. Next the tests of joint normality discussed in Sections 3.1 and 3.2 were applied. The likelihood ratio test associated with the bivariate power transformation approach yielded a moderately significant (.06) result, whereas Weiss's test and the nearest distance test indicated no significant ($\geq .4$) departures. The techniques of Section 3.3 based on the radius and angle decomposition of the data were then applied. Figure 1d is a scatter plot of the unit-square type described in Section 3.3.1. Departures from a uniform scatter (to be expected under normality) are clear in this plot in that several cells are empty and also several horizontal and vertical strips are sparse in points relative to other strips. Figure 1e is a $\chi^2_{(2)}$ probability plot of the squared radii. This plot is reasonably linear and suggests no marked departures of the squared radii from null expectations. On the other hand, Fig. 1f, which is a uniform probability plot of the normalized angles, exhibits irregularities, especially at the upper end. A χ^2 goodness-of-fit test based on 10 equal cells turns out

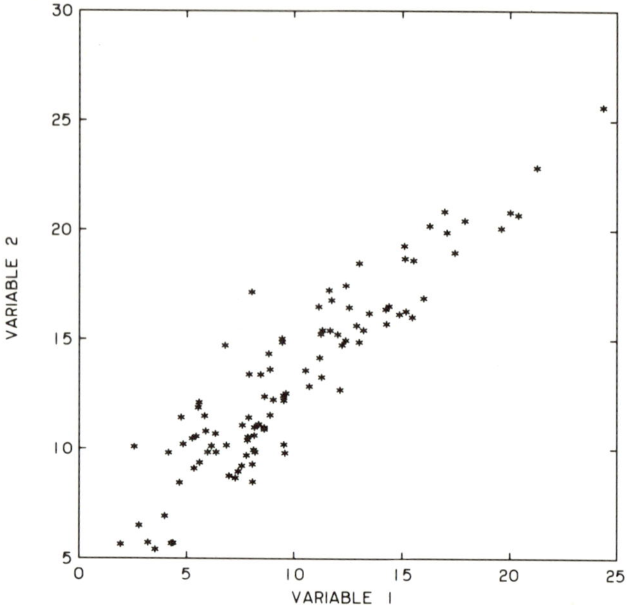

Fig. 1c. Example 1: Scatter plot of data.

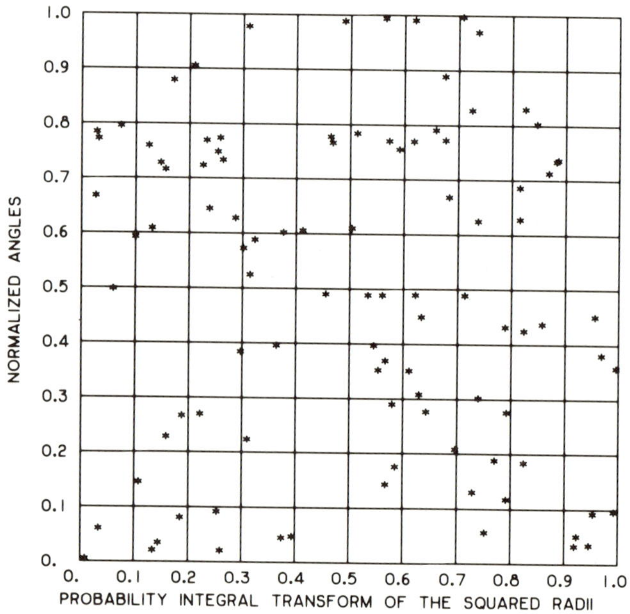

Fig. 1d. Example 1: Scatter plot of normalized angles versus probability integral transform of squared radii.

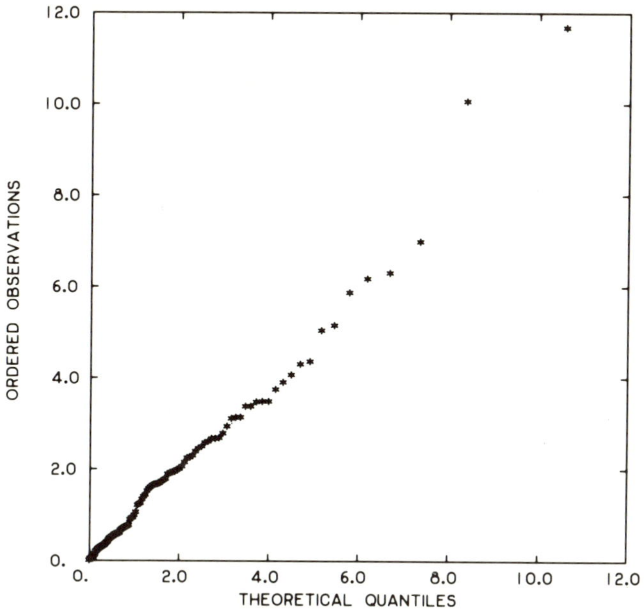

Fig. 1e. Example 1: Chi-squared probability plot of squared radii (degrees of freedom = 2).

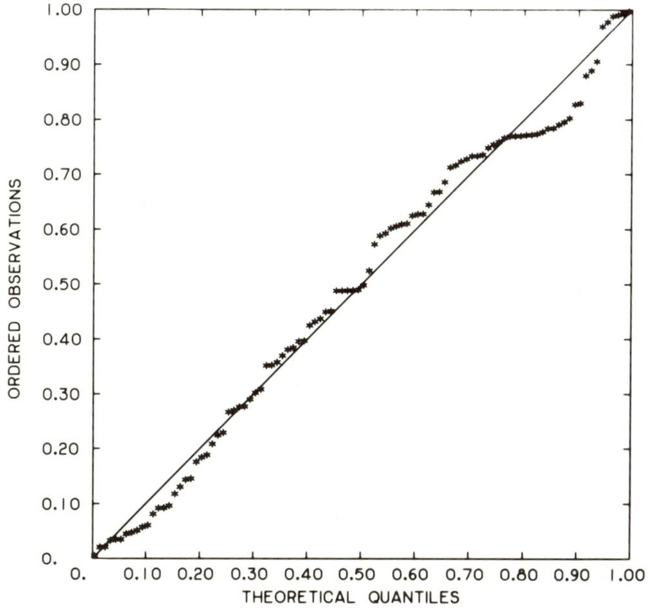

Fig. 1f. Example 1: Uniform probability plot of normalized angles.

to be highly statistically significant (.002), further confirming the departure indicated by Fig. 1f.

Finally, the test for directional normality described in Section 4.1 was used with a value of $\alpha = 1$ in computing the "direction" $\mathbf{d}_\alpha^*$. The transformation test of Section 2 applied to the projections of the data on $\mathbf{d}_{1.0}^*$ led to a highly significant ($\sim$.00001) departure from normality for this data-dependent unidimensional view of the bivariate observations. In the direction chosen, the data departs more noticeably from normality than either of the marginal variables did.

Example 2. The second example of simulated data consisted of 100 observations from a correlated conditional Laplace distribution with an underlying correlation coefficient of 0.9. The distribution here is long tailed but symmetric. Only the results of using the techniques discussed in Sections 3 and 4 are summarized here.

The significance level was quite high (0.4) for the likelihood ratio test associated with the bivariate power transformation method of Section 3.1. This nonsignificance is in part explainable by the fact that the transformation involved only power and no shift parameters and, as such, one should expect sensitivity to skewness but not to long-tailedness in the presence of sym-

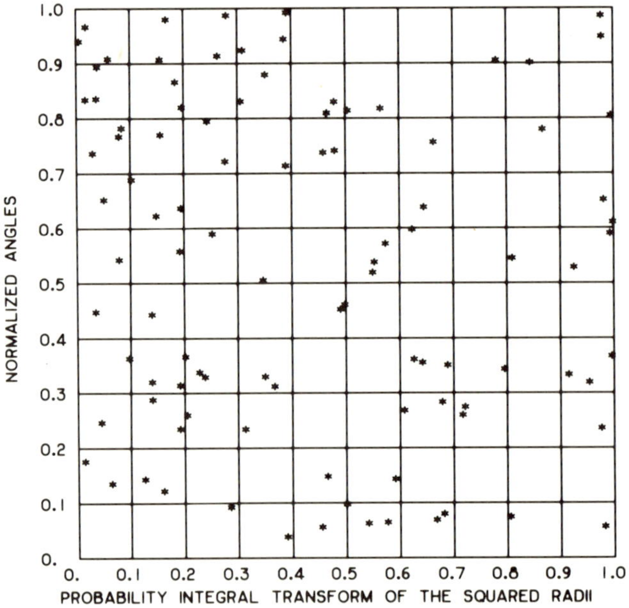

Fig. 2a. Example 2: Scatter plot of normalized angles versus probability integral transform of squared radii.

metry. Including shift parameters would most probably remedy the situation, but then the required computational effort would increase substantially.

Among the procedures of Section 3, the most striking indication of departure from joint normality was provided by the plots derived from the radius and angle decomposition. Figures 2a–c show the combined scatter, and the two probability, plots associated with the radii and angles in this example. The nonuniformity of the scatter of points in Fig. 2a indicates departure from normality, while the long-tailedness of the data is clearly exhibited in the $\chi^2_{(2)}$ probability plot of the squared radii (Fig. 2b). Also, some evidence of the lack of spherical symmetry of the scaled residuals is provided by the uniform probability plot of the normalized angles (Fig. 2c).

Next, the directional normality test described in Section 4.1 was applied with a choice of six different values of α in computing the "directions" $\mathbf{d}_\alpha^*$. Figure 2d shows a scatter plot of the data together with the six "directions" determined in this example, and Table II summarizes the results of the directional normality tests. The directions determined by using $\alpha < 0$, being sensitive to the center of the data, do indicate more significant departures and this is not too surprising in view of the difference between the densities of the Laplace and the normal in the center.

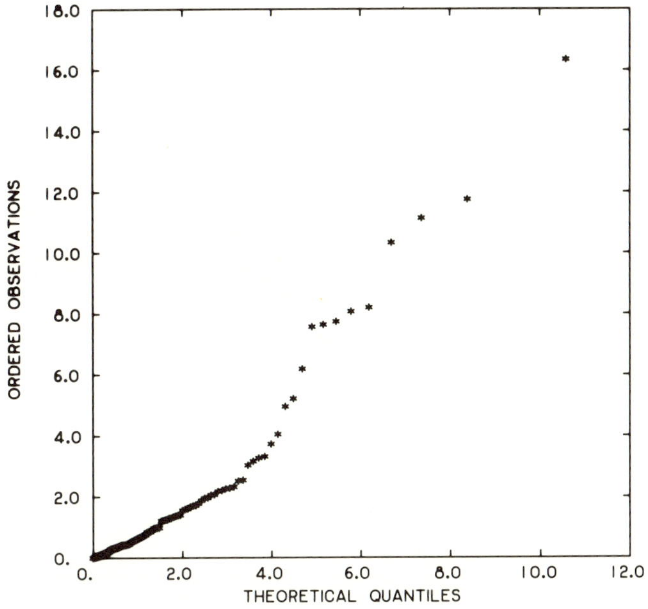

Fig. 2b. Example 2: Chi-squared probability plot of squared radii (degrees of freedom = 2).

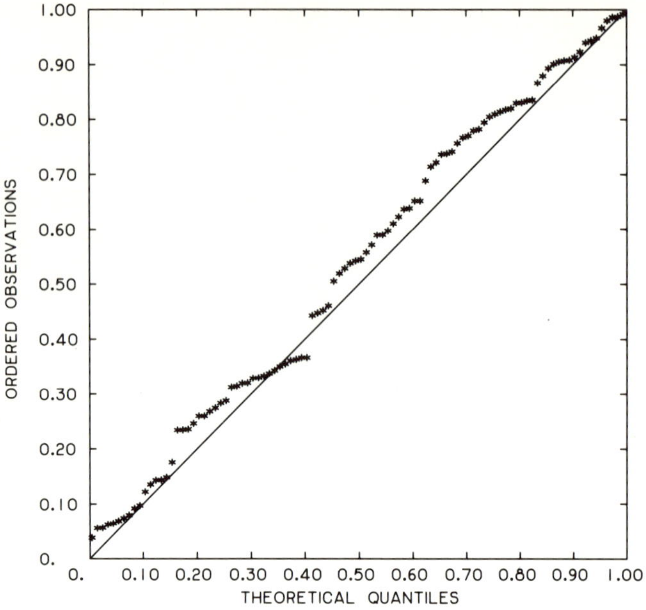

Fig. 2c. Example 2: Uniform probability plot of normalized angles.

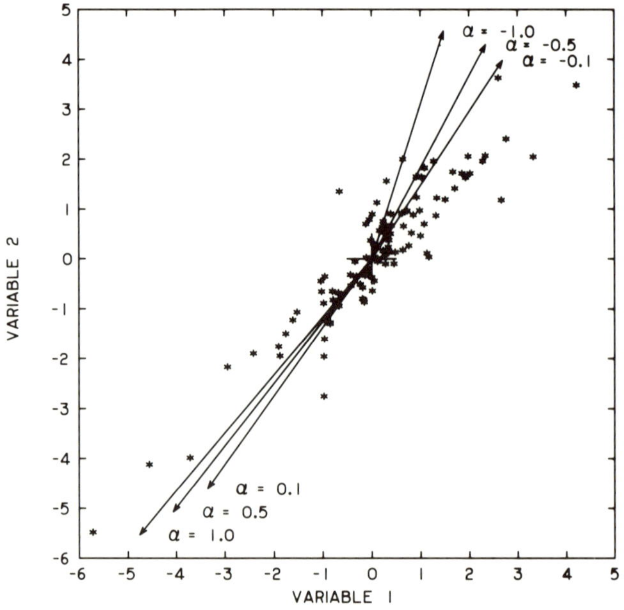

Fig. 2d. Example 2: Scatter plot of data along with various directions derived from directional normality test.

TABLE II

Example 2—Directional Normality

α	$\mathbf{d}_\alpha^*$		$\hat{\xi}$	$\hat{\lambda}$	$\mathscr{L}_{\max}(\hat{\xi}, \hat{\lambda}) - \mathscr{L}_{\max}(\hat{\xi}, 1)$	Approximate Significance Level
1.0	(−0.65	−0.76)	23.2	−1.0	3.2	0.011
0.5	(−0.63	−0.78)	24.5	−1.0	3.2	0.011
0.1	(−0.59	−0.81)	24.5	−1.0	3.1	0.013
−0.1	(0.56	0.83)	10.0	1.8	4.1	0.004
−0.5	(0.48	0.88)	9.7	1.8	4.0	0.005
−1.0	(0.31	0.95)	8.9	1.7	3.9	0.005

Example 3. This example is based on data from Standard & Poor's COMPUSTAT tape and consists of values of the debt ratio and the dividends/price ratio for 94 utility firms in 1969. Figure 3a is a scatter plot of the observations.

The univariate transformation-related test, when applied to the two variables separately, indicated some departure from univariate normality for the

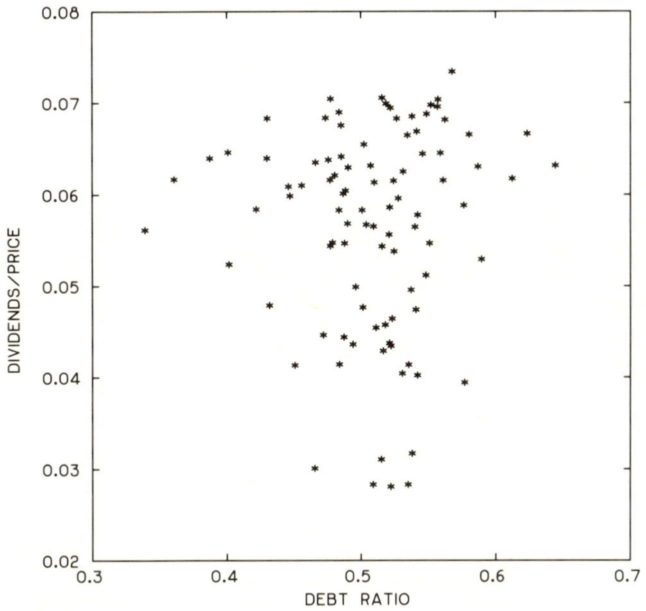

Fig. 3a. Example 3: Scatter plot of debt ratio and dividends/price for 94 utilities in 1969.

debt ratio distribution (significance level $\simeq$.06) and more striking departure for dividends/price (significance level $\simeq$.00007). The D'Agostino test indicated no strikingly significant departures. Figures 3b and 3c are normal probability plots of the values of the two variables, and the departures from normality are very clearly exhibited.

The test for bivariate normality based on power transformations (Section 3.1), led to a highly significant ($\simeq$.0003) value of the likelihood ratio statistic. The nearest distance test of Andrews et al. (1972) indicated only a mildly significant (.03) departure. The radius and angle decomposition, however, proved to be a sensitive indicator of departures again. Figures 3d–f show the combined scatter, and the two probability, plots involved. The nonuniform scatter and the sparseness in several contiguous blocks in Fig. 3d are striking. Also, Fig. 3e indicates peculiarities in the upper tail of the distribution of squared radii, while Fig. 3f shows up some departures from spherical symmetry as judged by the distribution of the normalized angles.

The directional normality test tended to use directions that were highly influenced by dividends/price, a fact that is not surprising in view of the striking non-normality of this variable. The directional test yielded moderately significant ($\simeq$.02) results in this example.

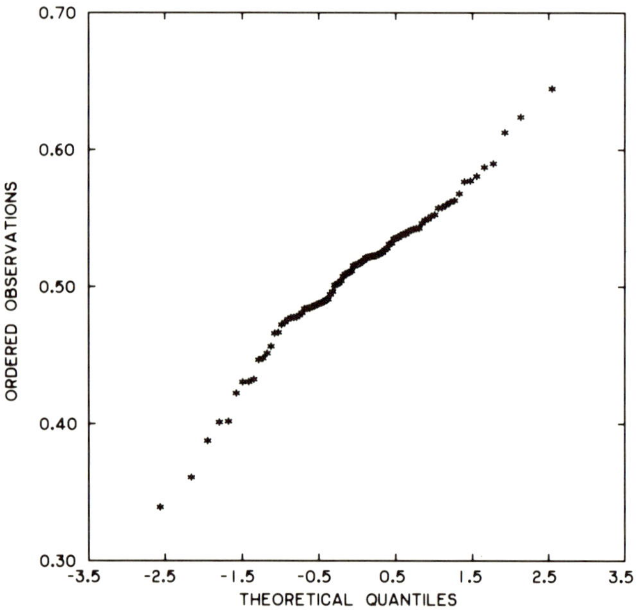

Fig. 3b. Example 3: Normal probability plot of debt ratio.

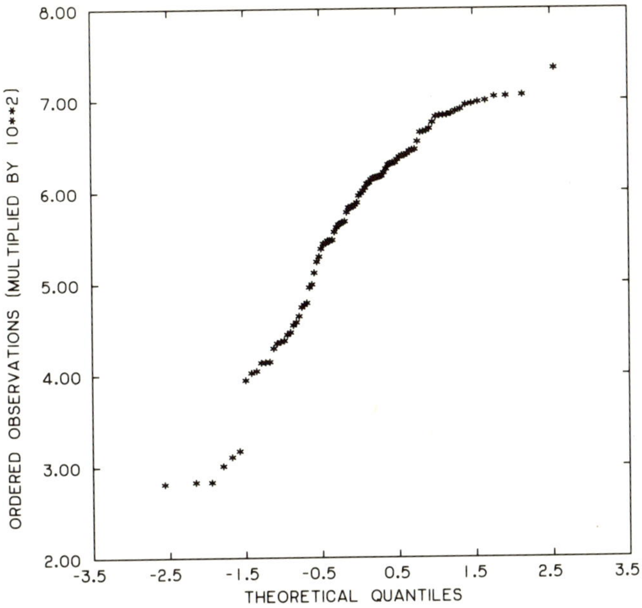

Fig. 3c. Example 3: Normal probability plot of dividends/price.

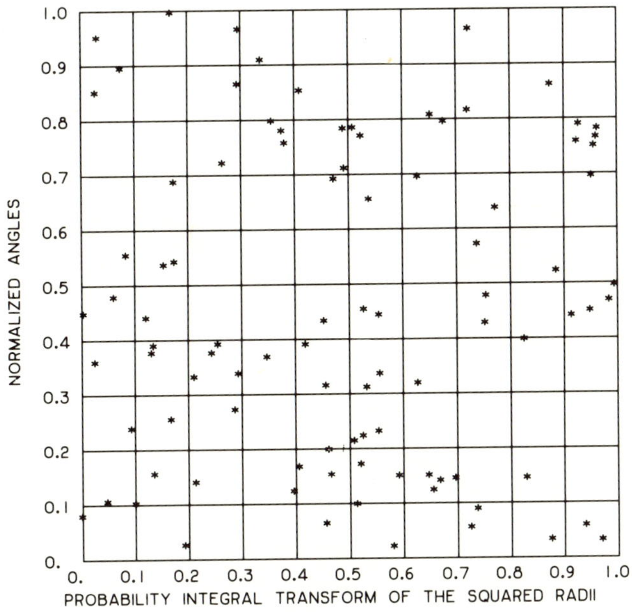

Fig. 3d. Example 3: Scatter plot of normalized angles versus probability integral transform of squared radii.

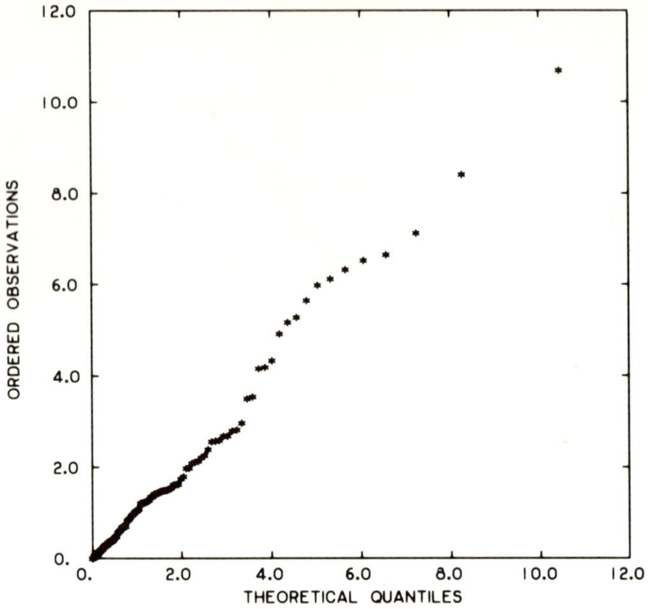

Fig. 3e. Example 3: Chi-squared probability plot of squared radii (degrees of freedom = 2).

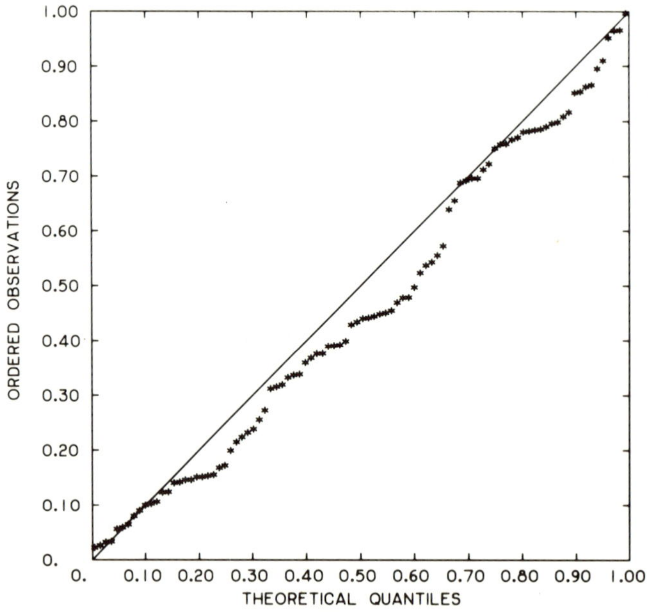

Fig. 3f. Example 3: Uniform probability plot of normalized angles.

6. CONCLUDING REMARKS

This paper has been partly a review and largely a preliminary report. The intention has been to illustrate the reasonably useful character of the methods rather than to provide any precise or exhaustive guidelines for choice among them. Also, for the initial investigations, only moderately large samples (i.e. 50–100 observations) were considered. The issues of more detailed developments pertaining to the relative sensitivities of the methods and their properties for smaller sample sizes are not unimportant but are left for continuing study.

General indications, based on the authors' limited experience, are that the transformation-related techniques (Sections 2, 3.1, and 4.1) and the plotting procedures for the radius and angles representation of the data are likely to be useful tools of multivariate data analysis. Further study of and experience with these techniques, as well as the development of additional methods, would be useful. The motivation for such techniques should perhaps not be influenced as much by issues of formal inferences concerning a distributional hypothesis as it should be by objectives of data analysis. Thus, for example, the usefulness of techniques proposed initially for assessing normality for additional purposes, such as detecting outliers and other peculiarities in data, should not be overlooked.

ACKNOWLEDGMENT

Comments of C. L. Mallows and H. O. Pollak on an earlier version are gratefully acknowledged.

REFERENCES

Aitkin, M. A. (1972). A class of tests for multivariate normality based on linear functions of order statistics. Unpublished manuscript.

Anderson, T. W. (1966). Some nonparametric multivariate procedures based on statistically equivalent blocks. *Multivariate Analysis* (Ed. P. R. Krishnaiah), pp. 5–27. Academic Press, New York.

Andrews, D. F., Gnanadesikan, R. and Warner, J. L. (1971). Transformations of multivariate data. *Biometrics* **27** 825–40.

Andrews, D. F., Gnanadesikan, R. and Warner, J. L. (1972). Methods for assessing multivariate normality. Unpublished memorandum.

Box, G. E. P. and Cox, D. R. (1964). An analysis of transformations. *J. R. Statist. Soc. B* **26** 211–52.

D'Agostino, R. B. (1971). An omnibus test of normality for moderate and large size samples. *Biometrika* **58** 341–8.

Gnanadesikan, R. and Kettenring, J. R. (1972). Robust estimates, residuals, and outlier detection with multiresponse data. *Biometrics* **28** 81–124.

Healy, M. J. R. (1968). Multivariate normal plotting. *Appl. Statist.* **17** 157–61.

Kessel, D. L. and Fukunaga, K. (1972). A test for multivariate normality with unspecified parameters. Unpublished report, Purdue University School of Electrical Engineering.

Malkovitch, J. F. and Afifi, A. A. (1971). On tests for multivariate normality. Unpublished UCLA report.

Moore, P. G. and Tukey, J. W. (1954). Answer to query 112. *Biometrics* **10** 562–8.

Pearson, E. S. and Hartley, H. O. (1966). *Biometrika Tables for Statisticians, Vol. I* pp. 67–9, 207–8. Cambridge Univ. Press.

Roy, S. N. (1953). On a heuristic method of test construction and its use in multivariate analysis. *Ann. Math. Statist.* **24** 220–38.

Shapiro, S. S. and Wilk, M. B. (1965). An analysis of variance test for normality (complete samples). *Biometrika* **52** 591–611.

Tukey, J. W. (1957). On the comparative anatomy of transformations. *Ann. Math. Statist.* **28** 602–32.

Tukey, J. W. (1970). *Exploratory Data Analysis.* Limited preliminary edition, Addison-Wesley, Reading.

Weiss, L. (1958). A test of fit for multivariate distributions. *Ann. Math. Statist.* **29**, 595–9.

Wilk, M. B. and Gnanadesikan, R. (1968). Probability plotting methods for the analysis of data. *Biometrika* **55** 1–17.

Asymptotic Expansions for the Distributions of Characteristic Roots When the Parameter Matrix Has Several Multiple Roots[1]

A. K. CHATTOPADHYAY[2] and *K. C. S. PILLAI*
PURDUE UNIVERSITY

1. INTRODUCTION

In recent studies on asymptotic distributions of characteristic roots in multivariate analysis, the approach has generally been to maximize an integrand of the form

$$I = \int_{O(p)} {}_sF_t(a_1, \ldots, a_s; b_1, \ldots, b_t; \mathbf{LHRH}')\, d(\mathbf{H}),$$

where $O(p)$ is the group of orthogonal matrices $\mathbf{H}(p \times p)$ with respect to which the maximization is carried out, $\mathbf{L} = \operatorname{diag}(l_1, \ldots, l_p)$, $\mathbf{R} = \operatorname{diag}(r_1, \ldots, r_p)$, $d(\mathbf{H})$ is the invariant or Haar measure over the group $O(p)$ normalized so that the measure of the whole group is unity, ${}_sF_t$ is a hypergeometric function of matrix variates [7], and $a_1, \ldots, a_s, b_1, \ldots, b_t$ are functions of degrees of freedom and are positive real numbers. In the one-sample or covariance matrix case, G. Anderson [1] has shown that the maximum of the integrand with $s = t = 0$, $\mathbf{R}$, the sample characteristic root matrix, and $\mathbf{L}^{-1}$, the matrix of population characteristic roots assumed to be distinct, is attained where $\mathbf{H} = \operatorname{diag}(\pm, \pm 1, \ldots, \pm 1)$. James [8] extended the study to the case of an extreme multiple population root. Chang [2] and Li and Pillai [9, 10] found the form for $\mathbf{H}$ the same as in the one-sample case when maximizing the integrand in the two-sample (two covariance matrices) problem with $s = 1$ and $t = 0$. In the complex analogue of both one-sample and two-sample cases, Li and Pillai [9, 10] obtained a similar form of the unitary matrix $\mathbf{U}$. Chattopadhyay and Pillai [4] generalized the results of the previous authors

[1] This research was supported by National Science Foundation grant GP-11473.

[2] Present address: Applied Mathematics Research Laboratory, Aerospace Research Laboratories, Wright-Patterson Air Force Base, Dayton, Ohio.

not only to ${}_sF_t$ hypergeometric function but also when l_i's are equal in each of several sets.

The asymptotic expansion for large degrees of freedom is obtained by evaluating the integral for small neighborhoods of the matrices $H = (\pm 1, \ldots, \pm 1)$. Anderson [1] and Chang [2] have also shown, by invoking a lemma of Hsu [6], that at least they have obtained an asymptotic representation of the integral I. While the earlier authors all considered the asymptotic expansions of a single covariance matrix or two covariance matrices, Chattopadhyay and Pillai [5] extended the study to MANOVA and canonical correlation cases, and from one extreme multiple root situation to that of one multiple root, extreme or intermediate. Again, they [5] extended their results to the complex case as well. However, in extending the work further to the case of several multiple population roots, the method used by James [8] was not found to be suitable in view of the fact that the invariance of a function with respect to the choice of a submatrix in the orthogonal (unitary) matrix used there, does not extend to the simultaneous invariance with respect to choices of several submatrices, as is needed to extend that method. In order to overcome this difficulty we proceed in a different manner without recourse to the invariance property and restate Lemma 3.1 from [4] in a detailed fashion before demonstrating the new approach.

2. THE MAXIMIZATION PROCEDURES

Let us define $\mathbf{R} = \operatorname{diag}(r_1, \ldots, r_p)$; $\infty > r_1 > r_2 > \cdots > r_p > 0$,

$$\mathbf{L} = \operatorname{diag}(l_1, \overset{k_1}{\ldots}, l_1, l_2, \overset{k_2}{\ldots}, l_{k_1+k_2+1}, \ldots, l_p), \qquad (1)$$
$$\infty > l_t > l_2 > l_{k_1+k_2+1} > \cdots > l_p \geq 0,$$

and let $\mathbf{H} \in O(p)$, where $O(p)$ is the group of orthogonal matrices of order p. Then we state the following lemma.

Lemma 2.1. *If* (1) *holds, then for all variations of* $\mathbf{R} > 0$

$$\max_{\mathbf{H} \in O(p)} C_\kappa(\mathbf{H}'\mathbf{LHR}) = \max_{\mathbf{H} \in O(p)} C_\kappa(\mathbf{HRH}'\mathbf{L}) = C_\kappa(\mathbf{LR})$$

and if the ordering of the elements of $\mathbf{L}$ *is reversed, then*

$$\min_{\mathbf{H} \in O(p)} C_\kappa(\mathbf{H}'\mathbf{LHR}) = \min_{\mathbf{H} \in O(p)} C_\kappa(\mathbf{HRH}'\mathbf{L}) = C_\kappa(\mathbf{LR}).$$

The optimum values are attained if $\mathbf{H}$ *has the form*

$$\mathbf{H} = \begin{array}{c} \begin{matrix} k_1 & k_2 & q \end{matrix} \\ \begin{bmatrix} \mathbf{H}_1 & \mathbf{0} & \mathbf{0} \\ \mathbf{0} & \mathbf{H}_2 & \mathbf{0} \\ \mathbf{0} & \mathbf{0} & \mathbf{I}_0 \end{bmatrix} \end{array} \begin{matrix} k_2 \\ k_2 \\ q \end{matrix}$$

where $q = p - k_1 - k_2$ *and* $\mathbf{I}_0(q) = \operatorname{diag}(\pm 1, \ldots, \pm 1)$.

Proof. By Lemma 3.2 of [4] we get **H** must have the form

$$\mathbf{H} = \begin{bmatrix} \mathbf{H}_1 & \mathbf{H}_{11} & \mathbf{H}_{12} \\ \mathbf{0} & \mathbf{H}_2 & \mathbf{H}_{22} \\ \mathbf{0} & \mathbf{0} & \mathbf{I}_0 \end{bmatrix}.$$

But, because of the orthogonality of **H**, we get $\mathbf{H}_{12} = \mathbf{H}_{22} = \mathbf{0}$, which in turn gives $\mathbf{H}_2 \mathbf{H}_2' = \mathbf{H}_2'\mathbf{H}_2 = \mathbf{I}(k_2)$. Thus, $\mathbf{H}_{11} = \mathbf{0}$, and hence the proof.

The proof just outlined also goes through in the complex analogue of this problem when **H** is replaced by **U**, where $\mathbf{U} \in U(p)$, and $U(p)$ is the group of unitary matrices. Thus let us consider the following formalization.

$$\begin{gathered} \mathbf{R} = \operatorname{diag}(r_1, \ldots, r_p), \qquad \infty > r_1 > r_2 > \cdots > r_p > 0, \\ \mathbf{L} = \operatorname{diag}(l_1, \overset{k_1}{\ldots}, l_1, \ldots, l_m, \overset{k_m}{\ldots}, l_m, l_{k_1+\cdots+k_m+1}, \ldots, l_p), \\ \infty > l_1 > l_2 > \cdots > l_m > l_{k_1+\cdots+k_m+1} > \cdots l_p \geq 0, \end{gathered} \tag{2}$$

and let $\mathbf{H} \in O(p)$.

Lemma 2.2. *If* (2) *holds, then for all variations of* $R > 0$

$$\max_{\mathbf{H} \in O(p)} C_\kappa(\mathbf{H}'\mathbf{L}\mathbf{H}\mathbf{R}) = \max_{\mathbf{H} \in O(p)} C_\kappa(\mathbf{H}\mathbf{R}\mathbf{H}'\mathbf{L}) = C_\kappa(\mathbf{L}\mathbf{R})$$

and if the ordering of the elements of L is reversed, then

$$\min_{\mathbf{H} \in O(p)} C_\kappa(\mathbf{H}'\mathbf{L}\mathbf{H}\mathbf{R}) = \min_{\mathbf{H} \in O(p)} C_\kappa(\mathbf{H}\mathbf{R}\mathbf{H}'\mathbf{L}) = C_\kappa(\mathbf{L}\mathbf{R})$$

and the optimum values are attained iff **H** *has the form*

$$\mathbf{H} = \operatorname{diag}(\mathbf{H}_1, \ldots, \mathbf{H}_m, \mathbf{I}_0(p - k_1 - \cdots - k_m)),$$

where $\mathbf{H}_j(k_j \times k_j)$ *is an orthogonal matrix of order* k_j, $j = 1, \ldots, m$, *and*

$$\mathbf{I}_0(p - k_1 - \cdots - k_m) = \operatorname{diag}(\pm 1, \ldots, \pm 1).$$

In order to facilitate the subsequent generalization to the complex case, we state an analogue of Lemma 2.2, the proof being self-evident from previous discussion.

Lemma 2.3. *If* (2) *holds, then for all variations of* $\mathbf{R} > 0$

$$\max_{\mathbf{U} \in U(p)} \tilde{C}_\kappa(\mathbf{U}^*\mathbf{L}\mathbf{U}\mathbf{R}) = \max_{\mathbf{U} \in U(p)} \tilde{C}_\kappa(\mathbf{U}\mathbf{R}\mathbf{U}^*\mathbf{L}) = \tilde{C}_\kappa(\mathbf{L}\mathbf{R}),$$

and if the ordering of the elements of **L** *is reversed, then*

$$\min_{\mathbf{U} \in U(p)} \tilde{C}_\kappa(\mathbf{U}^*\mathbf{L}\mathbf{U}\mathbf{R}) = \min_{\mathbf{U} \in U(p)} \tilde{C}_\kappa(\mathbf{U}\mathbf{R}\mathbf{U}^*\mathbf{L}) = \tilde{C}_\kappa(\mathbf{L}\mathbf{R})$$

and the optimum values are attained iff **U** *has the form* $\mathbf{U} = \operatorname{diag}(\mathbf{U}_1, \ldots, \mathbf{U}_m, \mathbf{U}_{m+1})$, *where* $\mathbf{U}_j(k_j \times k_j)$ *is a unitary matrix of order* k_j, $j = 1, \ldots, m$, *and*

$\mathbf{U}_{m+1} = \text{diag}(\exp(\sqrt{-1}\,\theta_1), \ldots, \exp(\sqrt{-1}\,\theta_q))$, $0 \le \theta_j < 2\pi$, $j = 1, \ldots, q$, *and* $q = p - k_1 - \cdots - k_m$.

Using the above results, we get corresponding results for Theorem 1.2 and its complex analogue Theorem 1.3 in the paper [4].

3. ASYMPTOTIC EXPANSION FOR THE DISTRIBUTION OF THE LATENT ROOTS OF THE ESTIMATED COVARIANCE MATRIX—SEVERAL MULTIPLE POPULATION ROOTS

Let $\mathbf{R} = \text{diag}(r_1, \ldots, r_p)$, $\infty > r_1 > \cdots > r_p > 0$, where r_i's are the latent roots, in descending order, of a sample covariance matrix $\mathbf{C}$ with n degrees of freedom (df) calculated from a sample from a normal population with covariance matrix $\mathbf{\Sigma}$. Let the diagonal matrix of the latent roots of $\mathbf{\Sigma}^{-1}$ be $\mathbf{L}$, and let $\mathbf{L}$ have the form

$$\begin{aligned} \mathbf{L} &= \text{diag}(l_1, \ldots, l_q, l_{q+1}, \overset{k_1}{\ldots}, l_{q+1}, \ldots, l_{q+m}, \overset{k_m}{\ldots}, l_{q+m}), \\ \mathbf{R} &= \text{diag}(r_1, \ldots, r_p), \qquad \infty > r_1 > r_2 > \cdots > r_p > 0, \end{aligned} \tag{3}$$

$$\infty > l_{q+m} > \cdots > l_{q+1} > l_q > \cdots > l_1 \ge 0, \tag{4}$$

where $p = k_1 + \cdots + k_m + q$. Then the joint distribution of $r_1, \ldots, r_p$ is

$$c_1 \int_{O(p)} \exp\left(-\frac{n}{2} \operatorname{tr} \mathbf{H}'\mathbf{LHR}\right) d(\mathbf{H}) \tag{5}$$

where

$$c_1 = \left\{ n^{np/2} \pi^{p^2/2} \Big/ \left[2^{np/2} \Gamma_p\left(\frac{n}{2}\right) \Gamma_p\left(\frac{p}{2}\right) \right] \right\} \prod_{i=1}^{q} l_i^{n/2}$$
$$\times \prod_{j=1}^{m} l_{q+j}^{(nk/2)j} \prod_{i=1}^{p} r_i^{(n-p-1)/2} \prod_{i<j} (r_i - r_j) \prod_{i=1}^{p} dr_i .$$

Now by Lemma 2.2 and as shown by Chattopadhyay and Pillai [4] the integrand in (5) is maximized for all variations of $\mathbf{R} > 0$ when $\mathbf{H}$ has the form

$$\mathbf{H} = \text{diag}(\mathbf{I}_0(q), \mathbf{H}_1, \ldots, \mathbf{H}_m) \tag{6}$$

where

$$\mathbf{H}_i(k_i \times k_i), \qquad i = 1, \ldots, m,$$

are orthogonal matrices.

As stated earlier, we do not resort to the invariance technique as used by earlier authors. Now following Anderson [1], we use the transformation

$$\mathbf{H} = \exp[\mathbf{S}], \tag{7}$$

where $\mathbf{S}$ is a $p \times p$ skew symmetric matrix. Now under (4), the transformation (7) reduces the integrand in (5) to a form which does not yield to direct evaluation. Hence to avoid this difficulty we note that if (4) holds, then for all $\mathbf{R} > 0$ the integrand in (5) is maximized when $\mathbf{H}$ has the form (6). Also, when n is large, the whole integral is concentrated around its unique maximum value. Thus, instead of (7) we use the transformation

$$\mathbf{H} = \exp[\mathbf{S}_1] \tag{8}$$

where $\mathbf{S}_1$ is a $p \times p$ skew symmetric matrix but has the form

$$\mathbf{S}_1 = \begin{bmatrix} \mathbf{S}_0 \\ \mathbf{S}_i \\ \mathbf{S}_m \end{bmatrix},$$

$$\mathbf{S}_0(q \times p); \qquad \mathbf{S}_i = (\mathbf{S}_{i1}, \mathbf{S}_{i2}, \mathbf{S}_{i3});$$
$$\mathbf{S}_{i1}(k_i \times (q + k_1 + \cdots + k_{i-1})); \qquad \mathbf{S}_{i2}(k_i \times k_i) = \mathbf{0};$$
$$\mathbf{S}_{i3}(k_i \times (k_{i+1} + \cdots + k_m)), \qquad i = 1, \quad \ldots, m-1;$$

and

$$\mathbf{S}_m = (\mathbf{S}_{m1}, \mathbf{S}_{m2}); \qquad \mathbf{S}_{m1}(k_m \times (p - k_m)); \qquad \mathbf{S}_{m2}(k_m \times k_m) = \mathbf{0}.$$

This is no loss of generality provided the constant factor is adjusted, as for large n the integrand is concentrated around its unique maximum and at least one maximizing set is covered by this substitution. Let $g = p - k_m$. Then

$$\begin{aligned} \operatorname{tr}(\mathbf{H}'\mathbf{LHR}) &= \operatorname{tr}(\mathbf{LHRH}') \\ &= l_{q+m} \sum_{j=1}^{p} r_j + \sum_{i=1}^{g} \sum_{j=1}^{p} (l_i - l_{q+m}) r_j h_{ij}^2, \end{aligned}$$

since

$$\sum_{i=g+1}^{p} h_{ij}^2 = 1 - \sum_{i=1}^{g} h_{ij}^2 \qquad \text{for} \quad j = 1, \ldots, p.$$

For large n and l_i's and r_j's well spaced, most of the integrand in (5) will be given by small values of $\mathbf{S}_1$. Now under (8)

$$h_{ii} = 1 - \frac{1}{2} \sum_{j=1}^{p} s_{ij}^2 + \text{higher-order terms in } s_{ij}\text{'s}$$

and

$$h_{ij} = s_{ij} + \text{higher-order terms in } s_{ij}\text{'s}.$$

Thus we get

$$\operatorname{tr}(\mathbf{H}'\mathbf{L}\mathbf{H}\mathbf{R}) = l_{q+m}\sum_{j=1}^{p} r_j + \sum_{i=1}^{q}(l_i - l_{q+m})r_i\left(1 - \sum_{j=1}^{p} s_{ij}^2\right)$$
$$+ \sum_{i=1}^{q}\sum_{j=1}^{p}(l_i - l_{q+m})r_j s_{ij}^2 + \text{higher-order terms in } s_{ij}\text{'s}$$
$$= \sum_{i=1}^{p} l_i r_i + \sum_{\substack{i=1\\i<j}}^{q}\sum_{j=1}^{p}(l_i - l_j)(r_j - r_i)s_{ij}^2$$
$$+ \sum_{u=1}^{m}\sum_{i=q_u}^{q_{u+1}-1}\sum_{j=q_{u+1}}^{p}(l_i - l_j)(r_j - r_i)s_{ij}^2$$
$$+ \text{higher-order terms involving } s_{ij}^2, \tag{9}$$

where

$$q_1 = q + 1, \qquad q_i = q + \sum_{j=1}^{i-1} k_j + 1, \qquad i = 2, \ldots, m + 1. \tag{10}$$

Substituting (9) in (5), we note that the integrand tends to zero as each $s_{ij} \to \infty$. Also for large n and for l_i's and r_j's well spaced, we can approximate the integral over $\mathbf{N}$ ($\mathbf{S}_1 = 0$) by varying each s_{ij} over the whole real line, i.e., $-\infty < s_{ij} < \infty$ for each pair (i, j), involved in our representation (8). Thus for large n, noting that the maximum of the integrand in (5) is attained when $\mathbf{H}$ has the form (6), following Anderson [1], and using the notations

$$c_{ij} = (l_i - l_j)(r_j - r_i), \qquad i = 1, \ldots, q, \quad j = 1, \ldots, p, \tag{11}$$

$c_{ij}^0 = (l_i - l_j)(r_j - r_i)$, i and j varying over the indicated set where it is nonzero,

$$\omega_i = \pi^{k_i^2/2}\{\Gamma_{k_i}(k_i/2)\}^{-1}, \qquad i = 1, \ldots, m; \qquad \omega_{m+1} = \pi^{p^2/2}\{\Gamma_p(p/2)\}^{-1},$$

where the factor involving ω_i accounts for the fact that integrand in (5) is maximized when $\mathbf{H}$ has the form (6), we get the following theorem.

Theorem 3.1. *An asymptotic expansion of the distribution of the roots* $r_1, \ldots, r_p$ *of the sample covariance matrix* $\mathbf{C}$ *for large degrees of freedom* n, *when the population roots satisfy* (4), *is given by*

$$\omega_{m+1}^{-1}\prod_{i=1}^{m}\omega_i 2^q c_1 \prod_{\substack{i=1\\i<j}}^{q}\prod_{j=1}^{p}\left(\frac{2\pi}{nc_{ij}}\right)^{1/2}\prod_{u=1}^{m}\prod_{i=q_u}^{q_{u+1}-1}\prod_{j=q_{u+1}}^{p}\left(\frac{2\pi}{nc_{ij}^0}\right)^{1/2}$$
$$\times\left(1 + \frac{1}{2n}\left(\sum_{\substack{i=1\\i<j}}^{q}\sum_{j=1}^{p} c_{ij}^{-1} + \sum_{u=1}^{m}\sum_{i=q_u}^{q_{u+1}-1}\sum_{j=q_{u+1}}^{p} c_{ij}^{0^{-1}}\right) + \cdots\right)\exp\left[-\frac{n}{2}\operatorname{tr}\mathbf{LR}\right]$$

where $q = p - k_1 - \cdots - k_m$, *and* $\mathbf{L}$, $\mathbf{R}$, q_i $(i = 1, \ldots, m + 1)$, c_{ij}^0*'s*, c_{ij}*'s are defined in* (3), (4), (10), *and* (11), *respectively.*

4. ASYMPTOTIC EXPANSION FOR THE DISTRIBUTION OF THE LATENT ROOTS OF $\mathbf{S}_1\mathbf{S}_2^{-1}$—SEVERAL MULTIPLE POPULATION ROOTS

The problem of finding the asymptotic expansion of the roots of $\mathbf{S}_1\mathbf{S}_2^{-1}$ in case of one extreme multiple population root has been studied by Li and Pillai [9, 10]. We here extend their results to the case when there are several multiple population roots.

Let $\mathbf{S}_i$ be independently distributed as Wishart $(n_i, p, \mathbf{\Sigma}_i)$, $i = 1, 2$, and let $r_i = ch_i(\mathbf{S}_1\mathbf{S}_2^{-1})$, $l_i = ch_i(\mathbf{\Sigma}_1\mathbf{\Sigma}_2^{-1})$, $i = 1, \ldots, p$, and let $\mathbf{R} = \operatorname{diag}(r_1, \ldots, r_p)$; $\infty > r_1 > \cdots > r_p > 0$,

$$\mathbf{L} = \operatorname{diag}(l_1, \ldots, l_q, l_{q+1}, \overset{k_1}{\ldots}, l_{q+1}, \ldots, l_{q+m}, \overset{k_m}{\ldots}, l_{q+m}), \tag{12}$$

and

$$\infty > l_{q+m} > l_{q+m-1} > \cdots > l_{q+1} > l_q > \cdots > l_1 \geq 0,$$

where $p = k_1 + \cdots + k_m + q$. Then the joint distribution of the roots $r_1, \ldots, r_p$ is given by

$$c_2 \int_{0(p)} |\mathbf{I} + \mathbf{H}'\mathbf{L}\mathbf{H}\mathbf{R}|^{-n/2} \, d(\mathbf{H}) \tag{13}$$

where

$$c_2 = \pi^{p^2/2}\Gamma_p\left(\frac{n_1+n_2}{2}\right)\left\{\Gamma_p\left(\frac{n_1}{2}\right)\Gamma_p\left(\frac{n_2}{2}\right)\Gamma_p\left(\frac{p}{2}\right)\right\}^{-1} |\mathbf{L}|^{n_1/2} |R|^{(n_1-p-1)/2} \prod_{i<j} (l_i - l_j)$$

and

$$n = n_1 + n_2, \qquad \Gamma_p(x) = \pi^{p(p-1)/4} \prod_{j=1}^{p} \Gamma((x - \tfrac{1}{2}(j-1)).$$

$d(\mathbf{H})$ is the invariant measure on the group $0(p)$. Again, as earlier, by Lemma 2.2 and as is shown in the previous paper [4], the integrand in (13) is maximized for all variation of $\mathbf{R} > 0$ when $\mathbf{H}$ has the form (6).

Again we make a substitution of the form (8) and after lengthy algebra similar to that of Li and Pillai [9, 10], for large n and l_i's and r_j's well spaced $(i, j, = 1, \ldots, p)$, and using the notations

$$\begin{aligned} c_{ij} &= (t_{ji} - t_i t_j r_{ij}) r_{ij}, \\ t_{ij} &= t_i - t_j, \qquad t_i = l_i(1 + l_i r_i)^{-1}, \\ r_{ij} &= r_i - r_j, \qquad i = 1, \ldots, q, \quad j = 1, \ldots, p, \quad i < j, \end{aligned} \tag{14}$$

and $c_{ij}^0 = c_{ij}$ but subscripts varying over the indicated set where it is nonzero,

$$\alpha_1(p, q) = (q/12)\{(q-1)(4q+1) + 6(p^2 - q^2)\}$$

and

$$\alpha_2(p, q, k_1, \ldots, k_m) = \frac{1}{2}\sum_{i=1}^{m} k_i(p - q - k_1 - \cdots - k_i)(p - q - k_1 - \cdots - k_i - 1)$$
$$+ \sum_{i<j<l=3}^{m} k_i k_j k_l + \frac{3}{2}\sum_{i<j=2}^{m} k_i k_j,$$

we have the theorem:

Theorem 4.1. *For large degrees of freedom $n = n_1 + n_2$, an asymptotic expansion for the distribution of the roots $\infty > r_1 > \cdots > r_p > 0$ when the population roots satisfy* (12) *is given by*

$$2^q \prod_{i=1}^{m} \omega_i \omega_{m+1}^{-1} c_2 |\mathbf{I} + \mathbf{LR}|^{n/2} \prod_{\substack{i=1 \\ i<j}}^{q} \prod_{j=1}^{p} \left(\frac{2\pi}{nc_{ij}}\right)^{1/2}$$
$$\times \prod_{u=1}^{m} \prod_{i=q_u}^{q_{u+1}-1} \prod_{j=q_u+1}^{p} \left(\frac{2\pi}{nc_{ij}^0}\right)^{1/2} \left[1 + \frac{1}{2n}\left(\sum_{\substack{i=1 \\ i<j}}^{q} \sum_{j=1}^{p} c_{ij}^{-1}\right.\right.$$
$$\left.\left. \times \sum_{u=1}^{m} \sum_{i=q_u}^{q_{u+1}-1} \sum_{j=q_u+1}^{p} c_{ij}^{0^{-1}} + \alpha_1(p, q) + \alpha_2(p, q, k_1, \ldots, k_m)\right) + \cdots\right],$$

where the constants are defined by (14).

In the following we give the asymptotic expansions for the roots of relevant matrices for MANOVA and canonical correlation cases and for complex analogs of all these problems. Detailed groundwork having already been done in [5, 9, 10] and above, we just state the problems and the corresponding solutions omitting the details.

5. ASYMPTOTIC EXPANSION FOR MANOVA—SEVERAL MULTIPLE POPULATION ROOTS

Let $\mathbf{B}$ be the between SP (sums of squares and cross products) matrix and $\mathbf{W}$ the within SP matrix. Then $\mathbf{B}(p \times p)$ has a noncentral Wishart distribution with s *df* and matrix of noncentrality parameter $\mathbf{A}$, and $\mathbf{W}$ has the central Wishart distribution on t df, the covariance matrix in each case being $\boldsymbol{\Sigma}$. Let $\mathbf{A} = \boldsymbol{\mu\mu}'\boldsymbol{\Sigma}^{-1}$ and $\mathbf{R} = \mathbf{B}(\mathbf{W} + \mathbf{B})^{-1}$ and in terms of the characteristic roots let

$$\mathbf{R} = \mathrm{diag}(r_1, \ldots, r_p), \qquad 1 > r_1 > r_2, \ldots, > r_p > 0, \tag{15}$$

$$\mathbf{L} = \mathrm{diag}(l_1, \ldots, l_q, l_{q+1}, \overset{k_1}{\ldots}, l_{q+1}, \ldots, l_{q+m}, \overset{k_m}{\ldots}, l_{q+m}), \tag{16}$$

where

$$\infty > l_1 > \cdots > l_q > l_{q+1} > \cdots > l_{q+m} \geq 0.$$

Then we have the following theorem.

Theorem 5.1. *For large t (and hence for large sample size) an asymptotic expansion for the distribution of the characteristic roots of* **R** *in* (15), *when the parameter matrix* **L** *satisfies* (16) *is given by*

$$c_3 2^q \prod_{i=1}^{m} \omega_i \omega_{m+1}^{-1} \prod_{\substack{i=1\\i<j}}^{q} \prod_{j=1}^{p} \left(\frac{2\pi}{tc_{ij}}\right)^{1/2} \prod_{u=1}^{m} \prod_{i=q_u}^{q_{u+1}-1} \prod_{j=q_{u+1}}^{p} \left(\frac{2\pi}{tc_{ij}^0}\right)^{1/2}$$

$$\times \left\{1 + \frac{1}{2t}\left(\sum_{\substack{i=1\\i<j}}^{q} \sum_{j=1}^{p} c_{ij}^{-1} + \sum_{u=1}^{m} \sum_{i=q_u}^{q_{u+1}-1} \sum_{j=q_{u+1}}^{p} c_{ij}^{0^{-1}} + \alpha_1(p, q)\right.\right.$$

$$\left.\left. + \alpha_2(p, q, k_1, \ldots, k_m)\right) + \cdots\right\} \exp[\operatorname{tr} \mathbf{LR}]_1F_1\left(-\frac{t}{2}; \frac{s}{2}; -\mathbf{LR}\right) + O(\varepsilon),$$

where

$$c_3 = \pi^{p^2/2}\Gamma_p(\tfrac{1}{2}(s+t))\left\{\Gamma_p\left(\frac{t}{2}\right)\Gamma_p\left(\frac{s}{2}\right)\Gamma_p\left(\frac{p}{2}\right)\right\}^{-1} \exp[\operatorname{tr} \mathbf{L}]|\mathbf{R}|^{(s-p-1)/2}$$

$$\times |\mathbf{I} - \mathbf{R}|^{(t-p-1)/2} \prod_{i<j} (r_i - r_j)$$

and

$$c_{ij} = \frac{(d_i - d_j)(l_j - l_i)}{(1 + l_i d_i)(1 + l_j d_j)}, \qquad i = 1, \ldots, q, \quad j = 1, \ldots, p, \quad i < j.$$

c_{ij}^0's are defined like c_{ij} but the subscripts vary over the indicated set such that c_{ij}^0 is nonzero and $\mathbf{D} = \mathbf{R}^{-1}$, *i.e.*, $d_i = r_i^{-1}$, $i = 1, \ldots, p$.

6. ASYMPTOTIC EXPANSION FOR CANONICAL CORRELATION—SEVERAL MULTIPLE POPULATION ROOTS

Let $x_1, \ldots, x_p, x_{p+1}, \ldots, x_{p+f}$, $p \leq f$, be distributed $N(\mathbf{0}, \mathbf{\Sigma})$, where

$$\mathbf{\Sigma} = \begin{bmatrix} \mathbf{\Sigma}_{11} & \mathbf{\Sigma}_{12} \\ \mathbf{\Sigma}_{12}' & \mathbf{\Sigma}_{22} \end{bmatrix} \begin{matrix} p \\ f \end{matrix}.$$

(with column blocks p, f)

Let $\mathbf{P}^2 = \operatorname{diag}(\rho_1^2, \ldots, \rho_p^2)$, where ρ_i^2, $i = 1, \ldots, p$, be the roots of

$$|\Sigma_{12} \;\; \Sigma_{22}^{-1} \;\; \Sigma_{12}' - \rho^2 \;\; \Sigma_{11}| = 0$$

and let $\tilde{\mathbf{P}}^2 = \text{diag}(\hat{\rho}_1{}^2, \ldots, \hat{\rho}_p{}^2)$, where $\hat{\rho}_i{}^2$, $i = 1, \ldots, p$, be the maximum likelihood estimates. Also let

$$\begin{aligned} \hat{\mathbf{P}}^2 &= \mathbf{R} = \text{diag}(r_1, \ldots, r_p), \\ \mathbf{P}^2 &= \mathbf{L} = \text{diag}(l_1, \ldots, l_q, l_{q+1}, \overset{k_1}{\ldots}, l_{q+1}, \ldots, l_{q+m}, \overset{k_m}{\ldots}, l_{q+m}), \end{aligned} \tag{17}$$

where

$$1 > r_1 > \cdots > r_p > 0, \quad 1 > l_1 > l_2 > \cdots > l_q > l_{q+1} > \cdots > l_q + m \geq 0.$$

Then we have the following theorem.

Theorem 6.1. *For large n, an asymptotic expansion of the distribution of $r_1, \ldots, r_p$ (squares of the canonical correlation coefficients) when population parameters satisfy* (17) *is given by*

$$c_4 2^q \prod_{i=1}^{m} \omega_i \omega_{m+1}^{-1} |\mathbf{I} - \mathbf{LR}|^{-(2n-f)/2} \Bigg\{ 1 + \frac{1}{2(2n-f)} \Bigg(\sum_{\substack{i=1 \\ i<j}}^{q} \sum_{j=1}^{p} c_{ij}^{-1}$$

$$+ \sum_{u=1}^{m} \sum_{i=q_u}^{q_{u+1}-1} \sum_{j=q_{u+1}}^{p} c_{ij}^{0^{-1}} + \alpha_1(p, q) + \alpha_2(p, q, k_1, \ldots, k_m) \Bigg)$$

$$+ \cdots \Bigg\} {}_2F_1(\tfrac{1}{2}(f-n), \tfrac{1}{2}(f-n); \tfrac{1}{2}f, \mathbf{LR}) + O(\varepsilon)$$

where

$$\begin{aligned} c_{ij} &= (t_{ji} - t_i t_j r_{ij}) r_{ij} = c_{ji}, \\ t_{ij} &= t_i - t_j, \qquad r_{ij} = r_i - r_j, \\ t_i &= l_i (1 - r_i l_i)^{-1}, \qquad i = 1, \ldots, q, \quad j = 1, \ldots, p, \quad i < j, \end{aligned}$$

and c_{ij}^0 is defined like c_{ij} but the subscripts vary over the indicated set such that c_{ij}^0 is nonzero, and

$$c_4 = \left\{ \pi^{p^2/2} \Gamma_p\left(\frac{n}{2}\right) / \Gamma_p\left(\frac{f}{2}\right) \Gamma_p\left(\frac{n-f}{2}\right) \Gamma_p\left(\frac{p}{2}\right) \right\} |\mathbf{I} - \mathbf{L}|^{n/2} |\mathbf{R}|^{(f-p-1)/2} \prod_{i<j} (r_i - r_j)$$

$$\times \prod_{\substack{i=1 \\ i<j}}^{q} \prod_{j=1}^{p} \left(\frac{2\pi}{(2n-f) c_{ij}} \right)^{1/2} \prod_{u=1}^{m} \prod_{i=q_u}^{q_{u+1}-1} \prod_{j=q_{u+1}}^{p} \left(\frac{2\pi}{(2n-f) c_{ij}^0} \right)^{1/2}.$$

7. COMPLEX ANALOGUES OF PREVIOUS RESULTS

In the following generalization of the above results to the complex case we refer to Lemma 1.2 and the corresponding results of Theorem 1.3 in the previous paper [4] and proceed as above, the details of algebra obtainable from Li and Pillai [9, 10], with suitable changes. Further details may be found in the work of Chattopadhyay [3].

8. REMARKS

1. As will be seen from the above formulas, they give the already known results of Anderson [1], Chang [2], James [8], Li and Pillai [9, 10] as special cases.

2. Though we have taken the sets with multiple roots in the population parametric matrix at one extreme, actually it would not matter even if they were otherwise. By pre- and postmultiplication by suitable permutation matrix, all multiple roots can be brought to one extreme place without affecting our distribution problem but, of course, care should be taken in defining c_{ij} and c_{ij}^0 coefficients.

3. Since, for all variations of $\mathbf{R} > 0$, the appropriate integral in each case takes the identical maximum when the corresponding orthogonal or unitary matrices take definite special forms, we can take particular transformations like (8) or its complex analogue to approximate the integrand around one such optimum and hence adjust for all such optima.

4. As will be evident, our technique being a generalization of techniques of earlier authors, the restrictions made by earlier authors also apply in our case.

5. As said earlier, we tacitly avoided the "invariance" technique used by James and subsequently followed by others. Moreover, our technique gives their result as a special case and hence gives a different interpretation of their results.

REFERENCES

1. Anderson, G. A. (1965). An asymptotic expansion for the distribution of the latent roots of the estimated co-variance matrix. *Ann. Math. Statist.* **36** 1153–1173.
2. Chang, T. C. (1970). On an asymptotic representation of the distribution of the characteristic roots of $\mathbf{S}_1\mathbf{S}_2^{-1}$ when roots are not all distinct. *Ann. Math. Statist.* **41** 440–445.
3. Chattopadhyay, A. K. (1971). An asymptotic distribution theory and applications in multivariate analysis. Mimeo. Ser. No. 256. Dept. of Statist., Purdue Univ., Lafayette, Indiana.
4. Chattopadhyay, A. K. and Pillai, K. C. S. (1970). On the maximization of an integral of a matrix function over the group of orthogonal matrices. Mimeo. Ser. No. 248. Dept. of Statist., Purdue Univ., Lafayette, Indiana.
5. Chattopadhyay, A. K. and Pillai, K. C. S. (1971). Asymptotic expansions of the distribution of characteristic roots in MANOVA and canonical correlation. Mimeo Ser. No. 249. Dept. of Statist., Purdue Univ., Lafayette, Indiana.
6. Hsu, L. C. (1948). A theorem on the asymptotic behaviour of a multiple integral. *Duke Math. J.* **15** 623–632.
7. James, A. T. (1964). Distribution of matrix variates and latent roots derived from normal samples. *Ann. Math. Statist.* **35** 475–501.
8. James, A. T. (1969). Tests of equality of latent roots of the covariance matrix. *Multivariate Analysis II* (P. R. Krishnaiah, ed.). Academic Press, New York.
9. Li, H. C. and Pillai, K. C. S. (1970). Asymptotic expansions for the distribution of the characteristic roots of $\mathbf{S}_1\mathbf{S}_2^{-1}$ when population roots are not all distinct. Mimeo. Ser. No. 231. Dept. of Statist., Purdue Univ., Lafayette, Indiana.
10. Li, H. C., Pillai, K. C. S. and Chang, T. C. (1970). Asymptotic expansions of the distributions of the roots of two matrices from classical and complex Gaussian populations. *Ann. Math. Statist.* **41** 1541–1556.

Aspects of the Multinomial Logit Model[1]

A. P. DEMPSTER

DEPARTMENT OF STATISTICS
HARVARD UNIVERSITY
CAMBRIDGE, MASSACHUSETTS

1. THE GENERAL LOGIT MODEL

The general model defined below assumes units each observed on three data arrays: **Y** representing a multinomial response variable, **X** representing a set of continuous observables perceived as random, and **Z** representing a set of fixed variables. Where **X** is empty the model will be called *fixed*, while if **Z** is empty the term *random* will be used. The case with both **X** and **Z** present will be called *mixed*. The distributions considered for **X** are limited to multivariate normals given **Y** and **Z**.

The variables $\{Y_a; a = 1, 2, \ldots, A\}$ making up **Y** are indicator variables for A mutually exclusive categories of a multinomial response, so that each Y_a takes only the values 0 or 1 and $\sum_1^A Y_a = 1$. Similarly the elements of **X** will be denoted $\{X_g; g = 1, 2, \ldots, G\}$ and the elements of **Z** will be denoted $\{Z_s; s = 1, 2, \ldots, S\}$. In formulas using matrix notation, **Y**, **X** and **Z** will be treated as arrays of dimensions $1 \times A$, $1 \times G$, and $1 \times S$ in the usual way.

Several different parameterizations of the general model will appear in the course of the discussion. A simple set of parameters for purposes of definition is given by the following five arrays:

$$\begin{aligned} &\boldsymbol{\alpha}: \{\alpha_a; a = 1, 2, \ldots, A\}; \\ &\boldsymbol{\beta}: \{\beta_{as}; a = 1, 2, \ldots, A, s = 1, 2, \ldots, S\}; \\ &\boldsymbol{\xi}: \{\xi_{ag}; a = 1, 2, \ldots, A, g = 1, 2, \ldots, G\}; \\ &\boldsymbol{\eta}: \{\eta_{sg}; s = 1, 2, \ldots, S, g = 1, 2, \ldots, G\}; \\ &\boldsymbol{\Sigma}: \{\sigma_{gh}; g = 1, 2, \ldots, G, h = 1, 2, \ldots, G\}. \end{aligned} \tag{1.1}$$

In matrix formulas, these will be treated as arrays of dimensions $A \times 1$, $A \times S$, $A \times G$, $S \times G$, $G \times G$, respectively, in the obvious way. The entries in

[1] This work was facilitated by National Science Foundation grant GP-31003X.

$\boldsymbol{\alpha}$, $\boldsymbol{\beta}$, $\boldsymbol{\xi}$, $\boldsymbol{\eta}$ are any set of real numbers, while $\boldsymbol{\Sigma}$ is restricted to be a positive definite symmetric matrix. Thus there are only $G(G + 1)/2$ distinct parameters in $\boldsymbol{\Sigma}$. The general logit model specifies a joint distribution for $\mathbf{Y}$ and $\mathbf{X}$ given $\mathbf{Z}$, making use of the factorization

$$h(\mathbf{Y}, \mathbf{X}; \mathbf{Z}) = f(\mathbf{Y}; \mathbf{Z})g(\mathbf{X}; \mathbf{Y}, \mathbf{Z}) \tag{1.2}$$

where $h(\mathbf{Y}, \mathbf{X}; \mathbf{Z})$ denotes the joint density of $\mathbf{Y}$ and $\mathbf{X}$ given $\mathbf{Z}$, $f(\mathbf{Y}; \mathbf{Z})$ denotes the density of $\mathbf{Y}$ given $\mathbf{Z}$, and $g(\mathbf{X}; \mathbf{Y}, \mathbf{Z})$ denotes the density of $\mathbf{X}$ given $\mathbf{Y}$ and $\mathbf{Z}$. These are densities relative to the obvious carrying measures, namely, counting measure over the A categories in the case of $\mathbf{Y}$, and lebesgue measure over R_G in the case of $\mathbf{X}$. The first factor in (1.2) is defined by the fixed logit model

$$\log f(\mathbf{Y}; \mathbf{Z}) = \gamma(\mathbf{Z}; \boldsymbol{\alpha}, \boldsymbol{\beta}) + \mathbf{Y}\boldsymbol{\alpha} + \mathbf{Y}\boldsymbol{\beta}\mathbf{Z}^T \tag{1.3}$$

where γ is the additive constant which guarantees that f sums to 1 over the A possible values of $\mathbf{Y}$. The second factor in (1.2) is obtained by assigning $\mathbf{X}$ the $N(\mathbf{Y}\boldsymbol{\xi} + \mathbf{Z}\boldsymbol{\eta}, \boldsymbol{\Sigma})$ distribution given $\mathbf{Y}$ and $\mathbf{Z}$. Thus

$$\begin{aligned}\log g(\mathbf{X}; \mathbf{Y}, \mathbf{Z}) - [-\tfrac{1}{2}\log\det\boldsymbol{\Sigma}] &+ \mathbf{Z}[\boldsymbol{\eta}\boldsymbol{\Sigma}^{-1}\boldsymbol{\eta}^T]\mathbf{Z}^T + \mathbf{Y}[-\tfrac{1}{2}\operatorname{diag}\boldsymbol{\xi}\boldsymbol{\Sigma}^{-1}\boldsymbol{\xi}^T] \\ &+ \mathbf{Y}[\boldsymbol{\xi}\boldsymbol{\Sigma}^{-1}\boldsymbol{\eta}^T]\mathbf{Z}^T + \mathbf{Y}[\boldsymbol{\xi}\boldsymbol{\Sigma}^{-1}]\mathbf{X}^T + \mathbf{X}[\boldsymbol{\Sigma}^{-1}\boldsymbol{\eta}^T]\mathbf{Z}^T \\ &+ \mathbf{X}[-\tfrac{1}{2}\boldsymbol{\Sigma}^{-1}]\mathbf{X}^T.\end{aligned} \tag{1.4}$$

If the conditional density of $\mathbf{Y}$ given $\mathbf{X}$ and $\mathbf{Z}$ is denoted by $k(\mathbf{Y}; \mathbf{X}, \mathbf{Z})$ it follows from (1.2)–(1.4) that

$$\begin{aligned}\log k(\mathbf{Y}; \mathbf{X}, \mathbf{Z}) = \gamma_1(\mathbf{X}, \mathbf{Z}; \boldsymbol{\alpha}, \boldsymbol{\beta}, \boldsymbol{\xi}, \boldsymbol{\eta}, \boldsymbol{\Sigma}) &+ \mathbf{Y}[\boldsymbol{\alpha} - \tfrac{1}{2}\operatorname{diag}(\boldsymbol{\xi}\mathbf{Z}^{-1}\boldsymbol{\xi}^T)] \\ &+ \mathbf{Y}[\boldsymbol{\beta} + \boldsymbol{\xi}\boldsymbol{\Sigma}^{-1}\boldsymbol{\eta}^T]\mathbf{Z}^T + \mathbf{Y}[\boldsymbol{\xi}\boldsymbol{\Sigma}^{-1}]\mathbf{X}^T.\end{aligned} \tag{1.5}$$

Note that (1.5) has the same general form as the fixed logit model (1.3) with $(\mathbf{X}, \mathbf{Z})$ in place of $\mathbf{Z}$ alone, whence the term logit remains appropriate for the mixed model. In the case where $\mathbf{X}$ is absent, the parameters $\boldsymbol{\xi}$, $\boldsymbol{\eta}$, $\boldsymbol{\Sigma}$ do not appear, and (1.5) reduces to the *fixed* model (1.3). In the case where $\mathbf{Z}$ is absent, both $\boldsymbol{\beta}$ and $\boldsymbol{\eta}$ disappear from the model. This yields the *random* model, which may also be called the *normal mixture* model because the conditional distribution of $\mathbf{X}$ given $\mathbf{Y}$ is simply normal with mean vector given by the appropriate row of $\boldsymbol{\xi}$ and common covariance matrix $\boldsymbol{\Sigma}$, while the marginal distribution of $\mathbf{Y}$ is specified by probabilities proportional to $(e^{\alpha_1}, e^{\alpha_2}, \ldots, e^{\alpha_A})$.

The parameterization (1.1) of the fixed model includes a redundancy in the sense that adding a constant to each element of $\boldsymbol{\alpha}$ or to each element of any column of $\boldsymbol{\beta}$ leaves $f(\mathbf{Y}; \mathbf{Z})$ unchanged. Elimination of the redundancy would introduce a notational asymmetry among the A categories, or would require the introduction of arbitrary constraints on the parameters. The approach taken here is to live with the redundancy, always remembering that only the differences among the elements of the columns of $\boldsymbol{\alpha}$ and $\boldsymbol{\beta}$ are identifiable.

Special cases of the fixed logit model (1.3) have long been used, notably the case $A = 2$, $S = 1$ in bioassay. The general case (1.3) has been presented by Bock (1970) and Press (1972) who illustrate basic procedures for fitting the model to data by maximum likelihood. Multinomial logit models were suggested earlier by Cox (1966) and Mantel (1966), and a restricted version where $\boldsymbol{\beta}$ has rank 1 was introduced by Duncan and Walker (1967). Advocates of logit models, for example Cox (1966), have long recognized that the normal mixture model gives rise to the fixed model after conditioning on $\mathbf{X}$, and have accordingly suggested using maximum likelihood estimates from the random model as starting values for iterative maximum likelihood procedures for a fixed model based on the same variables. The normal mixture model has also appeared in other contexts in the literature of multivariate analysis, for example in the Bayesian treatment of discriminant analysis of Anderson (1957) and in the study of discrete-continuous correlation measures of Olkin and Tate (1961).

The introduction in this paper of the general mixed model is motivated by the idea that in many practical situations it may be acceptable to retain the normal mixture assumption for a subset of the independent variables. Retention of the mixed model, as opposed to the completely fixed model, makes the computation of maximum likelihood estimates a less onerous task, and in general permits increased accuracy of the estimates of the logit parameters.

There is no mystery about the origins of the mixed model. To derive it, one need only start from the normal mixture model, condition on the part $\mathbf{X}_2$ of $\mathbf{X} = (\mathbf{X}_1, \mathbf{X}_2)$, and finally change the notation $(\mathbf{X}_1, \mathbf{X}_2)$ into $(\mathbf{X}, \mathbf{Z})$ with corresponding relabelling of parameter arrays. Since conditioning the normal mixture model on $\mathbf{X}$ is easily seen to produce a fixed logit model for $\mathbf{Y}$, and since conditioning on $\mathbf{X}_1$, after already conditioning on $\mathbf{X}_2$ produces the same result as conditioning on $\mathbf{X}$ in one stroke, it is clear from the start that the mixed model will produce a fixed logit model for $\mathbf{Y}$ after conditioning on $\mathbf{X}$, a result which was directly verified in (1.5).

2. PROPERTIES OF THE LIKELIHOOD

Given n sample observations

$$(\mathbf{Y}^{(l)}, \mathbf{X}^{(l)}, \mathbf{Z}^{(l)}) \tag{2.1}$$

for $l = 1, 2, \ldots, n$ assumed independently drawn from the mixed model (1.1), the log likelihood of the unknown parameters may be expressed as

$$\sum_{l=1}^{n} \log h(\mathbf{Y}^{(l)}, \mathbf{X}^{(l)}; \mathbf{Z}^{(l)}) = \sum_{l=1}^{n} \log f(\mathbf{Y}^{(l)}; \mathbf{Z}^{(l)}) + \sum_{l=1}^{n} \log g(\mathbf{X}^{(l)}; \mathbf{Y}^{(l)}, \mathbf{Z}^{(l)}). \tag{2.2}$$

Since the two terms on the right side of (2.1) depend on disjoint parameter sets, the maximum likelihood estimators of all the parameters can be found by maximizing the two terms separately. The first term is a log likelihood from a fixed logit model. For this term, an iterative computation is generally required. In contrast, maximum likelihood estimates from the second term can be obtained in closed form requiring only a finite number of computing steps, for (1.4) is just a version of the much studied multivariate general linear hypothesis model.

In Section 4 of Dempster (1971), the class of *extended exponential families* was defined by a density f expressed in Eq. (4.7) as

$$\log f = \alpha(\boldsymbol{\phi}, \mathbf{X}) + \mathbf{Y}\boldsymbol{\phi}\mathbf{X}^T \tag{2.3}$$

where $\mathbf{Y}$ denotes a random observed vector, $\mathbf{X}$ denotes a fixed observed vector, and $\boldsymbol{\phi}$ denotes a matrix of parameters which will be called *exponential parameters*. It is easily shown, as in Dempster (1971), that first and second derivatives of $\log f$ with respect to the exponential parameters are expressible in terms of first and second moments of $\mathbf{Y}^T\mathbf{X}$. From (1.2)–(1.4) it follows that each of $\log h(\mathbf{Y}, \mathbf{X}; \mathbf{Z})$, $\log f(\mathbf{Y}; \mathbf{Z})$, and $\log g(\mathbf{X}; \mathbf{Y}, \mathbf{Z})$ defines an extended exponential family, so that first and second derivatives of each with respect to corresponding exponential parameters can be formed by straightforward moment calculations. The desired derivatives of $\log h$ are not simply sums of those for $\log f$ and $\log g$ because the exponential parameters in $\log h$ are not simply the result of concatenating those of $\log f$ and $\log g$. Nevertheless, as will be seen, the derivatives for $\log f$ and $\log g$ do provide key information for the computation of $\log h$ derivatives. After deriving expressions for these derivatives, the main objective of this paper is addressed, namely, to show how the inverse of the second derivative matrix of $\log h$ can be found reasonably simply, through a combination of numerical and analytical manipulation.

The use of the particular matrix representation (2.3) is awkward here because the exponential parameters of the present applications do not appear naturally in the form of rectangular arrays. In a nonspecific notational format, the facts required about extended exponential families are the following:

(I) Apart from the additive constant term depending on parameters and fixed variables only (α in the notation (2.3)), the log likelihood from a single observation has the form

$$\sum_{i \in \mathscr{I}} (\ )_i [\ ]_i \tag{2.4}$$

where the $[\ ]_i$ for $i \in \mathscr{I}$ denote the exponential parameters and the $(\ )_i$ for $i \in \mathscr{I}$ denote a set of functions of both random and fixed observables.

(II) The first derivative of a log likelihood of the form (2.4) with respect to $[\ ]_i$ holding all other $[\ ]_j$ fixed is $(\ )_i - E(\ )_i$ where E denotes expectation with respect to the distribution of the random variables given the values of both parameters and fixed variables. Hence the first derivatives of the log likelihood from n observations are

$$\sum_{l=1}^{n} \{(\ )_i^{(l)} - E^{(l)}(\ )_i^{(l)}\} \qquad \forall \quad i \in \mathscr{I}, \tag{2.5}$$

where the expectation operators $E^{(l)}$ vary because the fixed variables vary across the n sample observation, while the parameter values remain fixed.

(III) The second derivative of a log likelihood of the form (2.4) with respect to $[\ ]_i$ and $[\ ]_j$ holding all other $[\ ]_k$ fixed is $\mathrm{COV}\{(\ )_i, (\ )_j\}$ where COV denotes covariance computed according to the distribution in question. Hence the second derivatives of log likelihood from n observations are

$$\sum_{l=1}^{n} \mathrm{COV}^{(l)}\{(\ )_i^{(l)}, (\ )_j^{(l)}\} \qquad \forall \quad i, j \in \mathscr{I}, \tag{2.6}$$

where $\mathrm{COV}^{(l)}$ varies across the n observations for the same reason that $E^{(l)}$ varies.

The index sets $\mathscr{I}$ required for the three applications to log h, log f, and log g are not the same but can be specified by subcollections of a common collection of index sets. These are

$$\begin{aligned}
\mathscr{I}_1 &= \{a;\ a = 1, 2, \ldots, A\},\\
\mathscr{I}_2 &= \{(a, s);\ a = 1, 2, \ldots, A, s = 1, 2, \ldots, S\},\\
\mathscr{I}_3 &= \{(a, g);\ a = 1, 2, \ldots, A, g = 1, 2, \ldots, G\},\\
\mathscr{I}_4 &= \{(s, g);\ s = 1, 2, \ldots, S, g = 1, 2, \ldots, G\},\\
\mathscr{I}_5 &= \{(g, h);\ g = 1, 2, \ldots, G, h = 1, 2, \ldots, G\},
\end{aligned} \tag{2.7}$$

where in the case of $\mathscr{I}_5$ the pairs (g, h) and (h, g) are considered the same, so that $\mathscr{I}_5$ contains $G(G + 1)/2$ elements. From (1.3) it is seen that the index set for log f is

$$\{\mathscr{I}_1, \mathscr{I}_2\}. \tag{2.8}$$

Since only **X** is random in (1.4), the index set for log g is

$$\{\mathscr{I}_3, \mathscr{I}_4, \mathscr{I}_5\}. \tag{2.9}$$

Summing (1.3) and (1.4) shows that the index set for log h is

$$\{\mathscr{I}_1, \mathscr{I}_2, \mathscr{I}_3, \mathscr{I}_4, \mathscr{I}_5\}. \tag{2.10}$$

The application of (2.5) to the mixed model requires that one find expectations of the two sets of arrays

$$\begin{aligned}
&\{Y_a;\ a \in \mathscr{I}_1\}\\
&\{Y_a Z_s;\ (a, s) \in \mathscr{I}_2\}
\end{aligned} \tag{2.11}$$

and

$$\begin{gathered}\{Y_a X_g; (a, g) \in \mathscr{I}_3\},\\ \{Z_s X_g; (s, g) \in \mathscr{I}_4\},\\ \{X_g X_h; (g, h) \in \mathscr{I}_5\}.\end{gathered} \tag{2.12}$$

The first step is to compute expectations over **X** given **Y** and **Z** according to the distribution (1.4), which leaves the arrays (2.11) unchanged while carrying (2.12) into

$$\begin{gathered}\left\{Y_a \sum_{b=1}^{A} Y_b \xi_{bg} + Y_a \sum_{s=1}^{S} Z_s \eta_{sg}; (a, g) \in \mathscr{I}_3\right\},\\ \left\{Z_s \sum_{a=1}^{A} Y_a \xi_{ag} + Z_s \sum_{t=1}^{S} Z_t \eta_{tg}; (s, g) \in \mathscr{I}_4\right\},\\ \left\{\sigma_{gh} + \left(\sum_{b=1}^{A} Y_b \xi_{bg} + \sum_{s=1}^{S} Z_s \eta_{sg}\right)\left(\sum_{a=1}^{A} Y_a \xi_{ah} + \sum_{t=1}^{S} Z_t \eta_{th}\right); (g, h) \in \mathscr{I}_5\right\}.\end{gathered} \tag{2.13}$$

The second step involves averaging (2.11) and (2.13) over **Y** given **Z** according to the distribution (1.3). Two facts about the second step should be noted. First, because the additive constant term γ in (1.3) has no simple analytical expression in terms of $\boldsymbol{\alpha}$ and $\boldsymbol{\beta}$, the expectations of (2.11) over (1.3) have no simple analytical expressions. Second, the terms in (2.13) which depend on **Y** are expressible as linear combinations of the terms of (2.11) where the coefficients do not depend on **Z**. For numerical calculation of first derivatives of log h from (2.5) with given parameter values, these facts imply (1) that it is necessary to average (2.11) numerically using probability vectors calculated separately from (1.3) for each of the distinct values of $\mathbf{Z}^{(l)}$ in the sample, but (2) that it is unnecessary to carry out the numerical averaging over the array (2.13) separately for each sample observation. Thus the task of finding first derivatives of log h reduces essentially to finding analytically the first derivatives of log g, which appear in (2.13), and finding numerically the first derivatives of log f by averaging (2.11).

To find second derivatives of log likelihood for the mixed model requires that one find covariances among all the elements of the arrays (2.11) and (2.12). Again, a two step procedure is indicated. The first step is to find covariances conditional on **Y** and **Z**. Since the array (2.11) is fixed given **Y** and **Z**, the first step yields nonzero covariances only among the elements of the arrays (2.12), and these are analytically computable as indicated below. The second step is to add to the conditional covariances given **Y** and **Z** the covariances among the conditional expectations given **Y** and **Z**, i.e., the covariances among the elements of the arrays (2.11) and (2.13). Just as the expectations of these conditional expectations fail to have simple analytical expressions, so do their covariances, and numerical computation given specific parameter sets will be required in most applications. Again, however, for numerical calculation of second derivatives of log h from (2.6) it is necessary at the second step to compute only covariances among the members of the arrays (2.11) since the complete linear dependence of the arrays (2.13) on the arrays (2.11) is identical for all n sample observations and therefore is identical for the sum of the $\mathrm{COV}^{(l)}$ in (2.6). From these parameter-dependent but sample-independent linear relations, it is easy to express the covariances among all the elements of (2.11) and (2.13) in terms of the covariances among the elements of (2.11) only. The details are omitted here since they are not needed in the sequel.

Details will now be given for the analytically feasible first step of covariance computation, namely, finding covariances among elements of the arrays (2.12) under the model (1.4). The covariances to be computed fall into six types, namely,

$$\begin{aligned}
\operatorname{cov}(Y_a X_g, Y_b X_h) &= Y_a Y_b \sigma_{gh} && \forall\ (a, g) \in \mathscr{I}_3, \quad (b, h) \in \mathscr{I}_3, \\
\operatorname{cov}(Y_a X_g, Z_s X_h) &= Y_a Z_s \sigma_{gh} && \forall\ (a, g) \in \mathscr{I}_3, \quad (s, h) \in \mathscr{I}_4, \qquad (2.14)\\
\operatorname{cov}(Z_s X_g, Z_t X_h) &= Z_s Z_t \sigma_{gh} && \forall\ (s, g) \in \mathscr{I}_4, \quad (t, h) \in \mathscr{I}_4
\end{aligned}$$

and

$$\begin{aligned}
\operatorname{cov}(Y_a X_g, X_h X_i) &= Y_a E(X_h)\sigma_{gi} + Y_a E(X_i)\sigma_{gh} \\
&\qquad \forall\ (a, g) \in \mathscr{I}_3, \quad (h, i) \in \mathscr{I}_5, \qquad (2.15)\\
\operatorname{cov}(Z_s X_g, X_h X_i) &= Z_s E(X_h)\sigma_{gi} + Z_s E(X_i)\sigma_{gh} \\
&\qquad \forall\ (s, g) \in \mathscr{I}_4, \quad (h, i) \in \mathscr{I}_5,
\end{aligned}$$

and

$$\begin{aligned}
\operatorname{cov}(X_g X_h, X_i X_j) &= \sigma_{gi}\sigma_{hj} + \sigma_{gj}\sigma_{hi} + E(X_g)E(X_i)\sigma_{hj} + E(X_g)E(X_j)\sigma_{hi} \\
&\quad + E(X_h)E(X_j)\sigma_{gi} + E(X_h)E(X_i)\sigma_{gj} \\
&\qquad \forall\ (g, h) \in \mathscr{I}_5, \quad (i, j) \in \mathscr{I}_5, \qquad (2.16)
\end{aligned}$$

where

$$E(X_g) = \sum_{a=1}^{A} Y_a \xi_{ag} + \sum_{s=1}^{S} Z_s \eta_{sg} \qquad \text{for} \quad g = 1, 2, \ldots, G. \qquad (2.17)$$

Applying (2.6) to the model (1.4) yields the following six arrays of second derivatives corresponding to (2.14), (2.15), (2.16):

$$\begin{aligned}
&R_{ab}\sigma_{gh} && \forall\ (a, g) \in \mathscr{I}_3, \quad (b, h) \in \mathscr{I}_3, \\
&S_{as}\sigma_{gh} && \forall\ (a, g) \in \mathscr{I}_3, \quad (s, h) \in \mathscr{I}_4, \qquad (2.18)\\
&T_{st}\sigma_{gh} && \forall\ (s, g) \in \mathscr{I}_4, \quad (t, h) \in \mathscr{I}_4,
\end{aligned}$$

and

$$\left(\sum_{b=1}^{A} R_{ab}\xi_{bh} + \sum_{s=1}^{S} S_{as}\eta_{sh}\right)\sigma_{gi} + \left(\sum_{b=1}^{A} R_{ab}\xi_{bi} + \sum_{s=1}^{S} S_{as}\eta_{si}\right)\sigma_{gh}$$
$$\forall \quad (a, g) \in \mathscr{I}_3, \quad (h, i) \in \mathscr{I}_5 \tag{2.19}$$
$$\left(\sum_{a=1}^{A} S_{as}\xi_{ah} + \sum_{t=1}^{S} T_{st}\eta_{th}\right)\sigma_{gi} + \left(\sum_{a=1}^{A} S_{as}\xi_{ai} + \sum_{t=1}^{S} T_{st}\eta_{ti}\right)\sigma_{gh}$$
$$\forall \quad (s, g) \in \mathscr{I}_4, \quad (h, i) \in \mathscr{I}_5,$$

and

$$n(\sigma_{gi}\sigma_{hj} + \sigma_{gj}\sigma_{hi}) + \left(\sum_{a=1}^{A}\sum_{b=1}^{A} R_{ab}\xi_{ai}\xi_{bg} + \sum_{a=1}^{A}\sum_{s=1}^{S} S_{as}\xi_{di}\eta_{sg}\right.$$
$$\left. + \sum_{a=1}^{A}\sum_{s=1}^{S} S_{as}\xi_{ag}\eta_{si} + \sum_{s=1}^{S}\sum_{t=1}^{S} T_{st}\eta_{sg}\eta_{ti}\right)\sigma_{hj} \tag{2.20}$$
$$+ 3 \text{ sets of similar terms}$$
$$\forall \quad (g, h) \in \mathscr{I}_5, \quad (i, j) \in \mathscr{I}_5,$$

where

$$R_{ab} = \sum_{l=1}^{n} Y_a^{(l)} Y_b^{(l)} \qquad \text{for} \quad a, b = 1, 2, \ldots, A,$$
$$S_{as} = \sum_{l=1}^{n} Y_a^{(l)} Z_s^{(l)} \qquad \text{for} \quad a = 1, 2, \ldots, A, \quad s = 1, 2, \ldots, S, \tag{2.21}$$
$$T_{st} = \sum_{l=1}^{n} Z_s^{(l)} Z_t^{(l)} \qquad \text{for} \quad s, t = 1, 2, \ldots, S.$$

The two preceding paragraphs have described numerical computation of two components of the second derivative matrix of $\sum_l^n \log h(\mathbf{Y}^{(l)}, \mathbf{X}^{(l)}; \mathbf{Z}^{(l)})$ with respect to the exponential parameters of log h, where the first component has contributions from all pairs of indices drawn from $\mathscr{I}_1, \mathscr{I}_2, \mathscr{I}_3, \mathscr{I}_4, \mathscr{I}_5$ while the second component has nonzero contribution only from pairs drawn from $\mathscr{I}_3, \mathscr{I}_4, \mathscr{I}_5$. The results of summing the two components is obviously interpretable as the exact sampling covariance of the maximum likelihood estimators of the parameters of the form

$$\sum_{l=1}^{n} E^{(l)}(\ )_i^{(l)} \qquad \forall \quad i \in \mathscr{I}, \tag{2.21$'$}$$

in the notation of (2.5). These parameters may be called *moment parameters*, and in some applications they are directly interesting quantities. In some other contexts, however, the explicit dependence of the moment parameters on the particular configurations of fixed variables $\mathbf{Z}^{(l)}$ in the sample makes them unnatural for parameterization. In particular, when logit analysis is the center of attention, the exponential parameters which appear in square brackets in (1.7) are of direct interest. Approximate large sample covariances for the exponential parameters are given by the inverse of the second derivative array, either within a sampling distribution framework or perhaps more importantly within a Bayesian framework. Accordingly, the remainder of this section is concerned with the inversion of the second derivative array.

Using the partition notation of formula (A.2) in the Appendix below, the second derivative matrix is expressible in the form

$$\begin{pmatrix} \mathbf{C}_{11} & \mathbf{C}_{12} & \mathbf{C}_{13} & \mathbf{C}_{34} & \mathbf{C}_{35} \\ \mathbf{C}_{21} & \mathbf{C}_{22} & \mathbf{C}_{23} & \mathbf{C}_{35} & \mathbf{C}_{45} \\ \mathbf{C}_{31} & \mathbf{C}_{32} & \mathbf{C}_{33}+\mathbf{D}_{11} & \mathbf{C}_{34}+\mathbf{D}_{12} & \mathbf{C}_{35}+\mathbf{D}_{13} \\ \mathbf{C}_{41} & \mathbf{C}_{42} & \mathbf{C}_{43}+\mathbf{D}_{21} & \mathbf{C}_{44}+\mathbf{D}_{22} & \mathbf{C}_{45}+\mathbf{D}_{23} \\ \mathbf{C}_{51} & \mathbf{C}_{52} & \mathbf{C}_{35}+\mathbf{D}_{31} & \mathbf{C}_{54}+\mathbf{D}_{32} & \mathbf{C}_{55}+\mathbf{D}_{33} \end{pmatrix}, \tag{2.22}$$

where the $\mathbf{C}_{ij}$ denote covariances among the arrays (2.11) and (2.13) under (1.3) and the $\mathbf{D}_{ij}$ denote the covariances given by (2.18), (2.19), and (2.20). The inversion will be carried out in three steps, namely, SWP[$\mathscr{I}_1$, $\mathscr{I}_2$], SWP[$\mathscr{I}_3$, $\mathscr{I}_4$], and SWP[$\mathscr{I}_5$] in terms of the Beaton sweep operator defined in the Appendix.

The result of operating on (2.22) with SWP[$\mathscr{I}_1$, $\mathscr{I}_2$] may be denoted

$$\begin{pmatrix} -\mathbf{C}^{11} & -\mathbf{C}^{12} & \mathbf{C}_3^{\ 1} & \mathbf{C}_4^{\ 1} & \mathbf{C}_5^{\ 1} \\ -\mathbf{C}^{21} & -\mathbf{C}^{22} & \mathbf{C}_3^{\ 2} & \mathbf{C}_4^{\ 2} & \mathbf{C}_5^{\ 2} \\ \mathbf{C}_1^{\ 3} & \mathbf{C}_2^{\ 3} & \mathbf{D}_{11} & \mathbf{D}_{12} & \mathbf{D}_{13} \\ \mathbf{C}_1^{\ 4} & \mathbf{C}_2^{\ 4} & \mathbf{D}_{21} & \mathbf{D}_{22} & \mathbf{D}_{23} \\ \mathbf{C}_1^{\ 5} & \mathbf{C}_2^{\ 5} & \mathbf{D}_{31} & \mathbf{D}_{32} & \mathbf{D}_{33} \end{pmatrix} \tag{2.23}$$

where

$$\begin{pmatrix} \mathbf{C}^{11} & \mathbf{C}^{12} \\ \mathbf{C}^{21} & \mathbf{C}^{22} \end{pmatrix} = \begin{pmatrix} \mathbf{C}_{11} & \mathbf{C}_{12} \\ \mathbf{C}_{21} & \mathbf{C}_{22} \end{pmatrix}^{I} \tag{2.24}$$

and the $\mathbf{C}_j^{\ i} = (\mathbf{C}_i^{\ j})^T$ for $i = 1, 2$ and $j = 3, 4, 5$ denote the regression coefficients of the array (2.13) on the array (2.11). The I superscript in (2.24) denotes generalized inverse, necessary here because the linear relation $\sum Y_a = 1$ forces the upper left $A(1 + S) \times A(1 + S)$ to have rank at most $(A - 1)(1 + S)$. (The maximum is generally achieved if $n \geq 1 + S$.) The definition used here for generalized inverse of a symmetric matrix is any symmetric matrix of the same rank as the original whose matrix product with the original is idempotent, again with the same rank. Such a generalized inverse can be computed by the Beaton sweep applied to any subset of the indices with cardinality equal to the rank of the matrix. Recall that the inversion operation (2.24) must be done numerically rather than analytically. Recall also that the regression coefficients $\mathbf{C}_j{}^i$ are obtainable by direct inspection of (2.13) and (2.11). Specifically,

$$\begin{aligned}
\mathbf{C}_3{}^1 &= \{\delta_{ab}\xi_{bg};\, a \in \mathscr{I}_1, (b, g) \in \mathscr{I}_3\}, \\
\mathbf{C}_4{}^1 &= \{0;\, a \in \mathscr{I}_1, (s, g) \in \mathscr{I}_4\}, \\
\mathbf{C}_5{}^1 &= \{\xi_{ah}\xi_{ag};\, a \in \mathscr{I}_1, (g, h) \in \mathscr{I}_5\}, \\
\mathbf{C}_3{}^2 &= \{\delta_{ab}\eta_{sg};\, (a, s) \in \mathscr{I}_2, (b, g) \in \mathscr{I}_3\}, \\
\mathbf{C}_4{}^2 &= (\xi_{ag}\delta_{st};\, (a, s) \in \mathscr{I}_2, (t, g) \in \mathscr{I}_4\}, \\
\mathbf{C}_5{}^2 &= \{\xi_{ag}\eta_{sh} + \xi_{ah}\eta_{sg};\, (a, s) \in \mathscr{I}_2, (g, h) \in \mathscr{I}_5\}.
\end{aligned} \tag{2.25}$$

Finally, the disappearance of the $\mathbf{C}_{ij}$ for $i, j = 3, 4, 5$ from (2.23) is due to the exact linear dependence of the array (2.13) on the array (2.11).

The result of operating on (2.23) with SWP$[\mathscr{I}_3, \mathscr{I}_4]$ may be denoted

$$\begin{pmatrix}
-\mathbf{C}^{11} - \tilde{\mathbf{C}}^{11} & -\mathbf{C}^{12} - \tilde{\mathbf{C}}^{12} & -\mathbf{C}^{13} & -\mathbf{C}^{14} & \tilde{\mathbf{C}}_5{}^1 \\
-\mathbf{C}^{21} - \tilde{\mathbf{C}}^{21} & -\mathbf{C}^{22} - \tilde{\mathbf{C}}^{22} & -\mathbf{C}^{23} & -\mathbf{C}^{24} & \tilde{\mathbf{C}}_5{}^2 \\
-\mathbf{C}^{31} & -\mathbf{C}^{32} & -\mathbf{D}^{11} & -\mathbf{D}^{12} & \mathbf{D}_3{}^1 \\
-\mathbf{C}^{41} & -\mathbf{C}^{42} & -\mathbf{D}^{21} & -\mathbf{D}^{22} & \mathbf{D}_3{}^2 \\
\tilde{\mathbf{C}}_1{}^5 & \tilde{\mathbf{C}}_2{}^5 & \mathbf{D}_1{}^3 & \mathbf{D}_2{}^3 & \tilde{\mathbf{D}}_{33}
\end{pmatrix} \tag{2.26}$$

where the various pieces are expressible and computable as indicated below. First, the inversion part of SWP$[\mathscr{I}_3, \mathscr{I}_4]$ requires that one compute

$$\begin{pmatrix} \mathbf{D}^{11} & \mathbf{D}^{12} \\ \mathbf{D}^{21} & \mathbf{D}^{22} \end{pmatrix} = \begin{pmatrix} \mathbf{D}_{11} & \mathbf{D}_{12} \\ \mathbf{D}_{21} & \mathbf{D}_{22} \end{pmatrix}^{-1} \tag{2.27}$$

From (2.18) it is seen that

$$\begin{pmatrix} \mathbf{D}_{11} & \mathbf{D}_{12} \\ \mathbf{D}_{21} & \mathbf{D}_{22} \end{pmatrix} = \begin{pmatrix} \mathbf{R} & \mathbf{S} \\ \mathbf{S}^T & \mathbf{T} \end{pmatrix} \otimes \boldsymbol{\Sigma} \tag{2.28}$$

where $\mathbf{R}$, $\mathbf{S}$, and $\mathbf{T}$ are matrices whose elements are defined by (2.21). Since the inverse of a Kronecker product is the Kronecker product of the inverses, it follows that

$$\begin{aligned}
\mathbf{D}^{11} &= \{R^{ab}\sigma^{gh};\, (a, g) \in \mathscr{I}_3, (b, h) \in \mathscr{I}_3\}, \\
\mathbf{D}^{12} &= \{S^{as}\sigma^{gh};\, (a, g) \in \mathscr{I}_3, (s, h) \in \mathscr{I}_4\}, \\
\mathbf{D}^{22} &= \{T^{st}\sigma^{gh};\, (s, g) \in \mathscr{I}_4, (t, h) \in \mathscr{I}_4\}
\end{aligned} \tag{2.29}$$

where the R^{ab}, S^{as}, T^{st} denote typical elements of

$$\begin{pmatrix} \mathbf{R} & \mathbf{S} \\ \mathbf{S}^T & \mathbf{T} \end{pmatrix}^{-1}$$

and the σ^{gh} denote typical elements of $\boldsymbol{\Sigma}^{-1}$. The following parametric functions will simplify subsequent expressions:

$$\begin{gathered}
\xi^{ag} = \sum_{h=1}^{G} \xi_{ah}\sigma^{gh}, \qquad \eta^{sg} = \sum_{h=1}^{G} \eta_{sh}\sigma^{gh}, \qquad \mu^{ab} = \sum_{h=1}^{G} \xi_{ah}\xi^{bh}, \\
\nu^{as} = \sum_{h=1}^{G} \xi_{ah}\eta^{sh}, \qquad \omega^{sr} = \sum_{h=1}^{G} \eta_{sh}\eta^{rh}.
\end{gathered} \tag{2.30}$$

In these terms, the remaining parts of (2.26) may be directly expressed as

$$\begin{aligned}
\mathbf{C}^{13} &= \{-R^{ac}\xi^{ah};\ a \in \mathscr{I}_1, (c, h) \in \mathscr{I}_2\}, \\
\mathbf{C}^{14} &= \{-S^{as}\xi^{ah};\ a \in \mathscr{I}_1, (s, h) \in \mathscr{I}_3\}, \\
\mathbf{C}^{23} &= \{-R^{ca}\eta^{sh} - S^{cs}\xi^{ah};\ (a, s) \in \mathscr{I}_2, (c, h) \in \mathscr{I}_3\}, \\
\mathbf{C}^{24} &= \{-S^{ar}\eta^{sk} - \xi^{ak}T^{sr};\ (a, s) \in \mathscr{I}_2, (r, k) \in \mathscr{I}_4\}, \\
\mathbf{D}_3{}^1 &= \{\delta_{hk}\xi_{ai} + \delta_{hi}\xi_{ak};\ (a, h) \in \mathscr{I}_3, (i, k) \in \mathscr{I}_5\}, \\
\mathbf{D}_3{}^2 &= \{\delta_{hk}\eta_{si} + \delta_{hi}\eta_{sk};\ (s, h) \in \mathscr{I}_4, (i, k) \in \mathscr{I}_5\}, \\
\tilde{\mathbf{D}}_{33} &= \{n(\sigma_{gi}\sigma_{hj} + \sigma_{hj}\sigma_{gi});\ (g, h) \in \mathscr{I}_5, (i, j) \in \mathscr{I}_5\}, \\
\mathbf{C}_5{}^1 &= \{-\xi_{ah}\xi_{ag};\ a \in \mathscr{I}_1, (g, h) \in \mathscr{I}_5\}, \\
\tilde{\mathbf{C}}_5{}^2 &= \{-\xi_{ag}\eta_{sh} - \xi_{ah}\eta_{sg};\ (a, s) \in \mathscr{I}_2; (g, h) \in \mathscr{I}_5\}, \\
\tilde{\mathbf{C}}^{11} &= \{\mu^{ac}R^{ac};\ a \in \mathscr{I}_1, c \in \mathscr{I}_1\}, \\
\tilde{\mathbf{C}}^{12} &= \{\nu^{ar}R^{ac} + \mu^{ac}S^{ar};\ a \in \mathscr{I}_1, (c, r) \in \mathscr{I}_2\}, \\
\tilde{\mathbf{C}}^{22} &= \{\omega^{sr}R^{ac} + \nu^{cs}S^{ra} + \nu^{ar}S^{ct} + \mu^{ac}T^{sr};\ (a, s) \in \mathscr{I}_2, (c, r) \in \mathscr{I}_2\}.
\end{aligned} \tag{2.31}$$

See (2.19) for the calculation of $\mathbf{D}_3{}^1$ and $\mathbf{D}_3{}^2$.

Finally, the result of operating on (2.26) with SWP[$\mathscr{I}_5$] may be denoted

$$\begin{pmatrix}
-\mathbf{C}^{11}-\tilde{\mathbf{C}}^{11}-\tilde{\tilde{\mathbf{C}}}^{11} & -\mathbf{C}^{12}-\tilde{\mathbf{C}}^{12}-\tilde{\tilde{\mathbf{C}}}^{12} & -\mathbf{C}^{13}-\tilde{\mathbf{C}}^{13} & -\mathbf{C}^{14}-\tilde{\mathbf{C}}^{14} & -\tilde{\mathbf{C}}^{15} \\
-\mathbf{C}^{21}-\tilde{\mathbf{C}}^{21}-\tilde{\tilde{\mathbf{C}}}^{21} & -\mathbf{C}^{22}-\tilde{\mathbf{C}}^{22}-\tilde{\tilde{\mathbf{C}}}^{22} & -\mathbf{C}^{23}-\tilde{\mathbf{C}}^{23} & -\mathbf{C}^{24}-\tilde{\mathbf{C}}^{24} & -\tilde{\mathbf{C}}^{25} \\
-\mathbf{C}^{31}-\tilde{\mathbf{C}}^{31} & -\mathbf{C}^{32}-\tilde{\mathbf{C}}^{32} & -\mathbf{D}^{11}-\tilde{\mathbf{D}}^{11} & -\mathbf{D}^{12}-\tilde{\mathbf{D}}^{12} & -\mathbf{D}^{13} \\
-\mathbf{C}^{41}-\tilde{\mathbf{C}}^{41} & -\mathbf{C}^{42}-\tilde{\mathbf{C}}^{42} & -\mathbf{D}^{21}-\tilde{\mathbf{D}}^{21} & -\mathbf{D}^{22}-\tilde{\mathbf{D}}^{22} & -\mathbf{D}^{23} \\
-\tilde{\mathbf{C}}^{51} & -\tilde{\mathbf{C}}^{52} & -\mathbf{D}^{31} & -\mathbf{D}^{32} & -\tilde{\mathbf{D}}^{33}
\end{pmatrix} \tag{2.32}$$

whose negative is the desired inverse of the second derivative matrix, i.e., is an asymptotic covariance matrix for the exponential parameters of the general logit model. The reader may check directly the following formulas, where [ ]** means that an expression [ ] indexed by $(g, h) \in \mathscr{I}_5$ and $(i, j) \in \mathscr{I}_5$ is multiplied by $\frac{1}{2}$ if $g = h$ and by another $\frac{1}{2}$ if $i = j$; similarly [ ]* means that an expression indexed by $(g, h) \in \mathscr{I}_5$ is multiplied by $\frac{1}{2}$ if $g = h$:

$$
\begin{aligned}
n\tilde{\mathbf{D}}^{33} &= \{[\sigma^{gi}\sigma^{hj} + \sigma^{hi}\sigma^{gj}]^{**};\ (g, h) \in \mathscr{I}_5, (i, j) \in \mathscr{I}_5\},\\
n\mathbf{D}^{13} &= \{-[\sigma^{gi}\xi^{ah} + \sigma^{hi}\xi^{ag}]^{*};\ (a, i) \in \mathscr{I}_3, (g, h) \in \mathscr{I}_5\},\\
n\mathbf{D}^{23} &= \{-[\sigma^{gi}\eta^{sh} + \sigma^{hi}\eta^{sg}]^{*};\ (s, i) \in \mathscr{I}_4, (g, h) \in \mathscr{I}_5\},\\
n\tilde{\mathbf{C}}^{15} &= \{[\xi^{ag}\xi^{ah}]^{*};\ a \in \mathscr{I}_1, (g, h) \in \mathscr{I}_5\},\\
n\tilde{\mathbf{C}}^{25} &= \{[\eta^{sg}\xi^{ah} + \eta^{sh}\xi^{ag}]^{*};\ (a, s) \in \mathscr{I}_2, (g, h) \in \mathscr{I}_5\},\\
n\tilde{\mathbf{D}}^{11} &= \{\xi^{ah}\xi^{bg} + \sigma^{gh}\mu^{ab};\ (a, g) \in \mathscr{I}_3, (b, h) \in \mathscr{I}_3\},\\
n\tilde{\mathbf{D}}^{12} &= \{\xi^{ah}\eta^{tg} + \sigma^{gh}\nu^{at};\ (a, g) \in \mathscr{I}_3, (t, h) \in \mathscr{I}_4\},\\
n\tilde{\mathbf{D}}^{22} &= \{\eta^{sh}\eta^{tg} + \sigma^{gh}\omega^{st};\ (s, g) \in \mathscr{I}_4, (t, h) \in \mathscr{I}_4\},\\
n\mathbf{C}^{13} &= \{\xi^{ag}\mu^{ab};\ a \in \mathscr{I}_1, (b, g) \in \mathscr{I}_3\},\\
n\mathbf{C}^{14} &= \{\xi^{ag}\nu^{as};\ a \in \mathscr{I}_1, (s, g) \in \mathscr{I}_4\},\\
n\mathbf{C}^{23} &= \{\xi^{ag}\nu^{bs} + \eta^{sg}\mu^{ab};\ (a, s) \in \mathscr{I}_2, (b, g) \in \mathscr{I}_3\},\\
n\mathbf{C}^{24} &= \{\xi^{ag}\omega^{st} + \eta^{sg}\nu^{at};\ (a, s) \in \mathscr{I}_2, (t, g) \in \mathscr{I}_4\},\\
n\mathbf{C}^{11} &= \{\tfrac{1}{2}(\mu^{ab})^2;\ a \in \mathscr{I}_1, b \in \mathscr{I}_1\},\\
n\mathbf{C}^{12} &= \{\mu^{ab}\nu^{at};\ a \in \mathscr{I}_1, (b, t) \in \mathscr{I}_2\},\\
n\mathbf{C}^{22} &= \{\nu^{at}\nu^{bs} + \mu^{ab}\omega^{st};\ (a, s) \in \mathscr{I}_2, (b, t) \in \mathscr{I}_2\}.
\end{aligned}
\tag{2.33}
$$

It may be noted that the upper left 4 × 4 piece of (2.26) is the negative of the asymptotic covariance of the exponential parameters of the general logit model *under the assumption that* **Σ** *is known*. The addition to the 4 × 4 piece in passing from (2.26) to (2.32) represents the additional variability in estimating these parameters when **Σ** is unknown.

3. COMMENT

The mathematical theory of this paper has been presented without indication of its application. The intention is to use the theory to obtain an understanding of the performance of logit analysis as a practical data analysis tool. The tool itself is being increasingly used, for example in large and sometimes controversial medical studies such as Gordon and Kannel (1970, 1971) and Meinert, Knatterud, Genell, Prout and Klimt (1970). It is therefore important that a clearer theoretical understanding of the behavior of the method be obtained.

For example, it is important to gain a heuristic understanding of the accuracy of estimation under fixed and random models of the exponential parameters appearing in the logit expression (1.5). Thence it will be possible to assess the sample sizes required to disentangle effectively correlated sources of variation among the independent variables. Another interesting subject for study is the effect of measurement error in the independent variables, especially when the logit method is used for successive time periods as in Gordon and Kannel (1970, 1971).

APPENDIX: THE BEATON SWEEP

Suppose that

$$\mathscr{I} = \{\mathscr{I}_1, \mathscr{I}_2, \ldots, \mathscr{I}_t\} \tag{A.1}$$

is an index set made up by concatenating the t index sets $\mathscr{I}_r$ for $r = 1, 2, \ldots, t$. Suppose that $\mathbf{B}$ is a symmetric inner product array defined over all pairs drawn from $\mathscr{I}$. Then $\mathbf{B}$ can be partitioned into $t(t+1)/2$ subarrays $\mathbf{B}_{rs}$ for $r, s = 1, 2, \ldots, t$, where $\mathbf{B}_{rs}$ consists of inner products associated with all pairs having one element in $\mathscr{I}_r$ and one element in $\mathscr{I}_s$.

If the elements of each $\mathscr{I}_r$ are written as a list, then the elements of $\mathscr{I}$ become a list also, according to (A.1). It now becomes natural to view $\mathbf{B}$ as a square symmetric matrix whose rows and columns are associated in order with the list $\mathscr{I}$, and to view $\mathbf{B}_{rs}$ similarly as a matrix of inner products whose rows and columns are associated respectively with the lists $\mathscr{I}_r$ and $\mathscr{I}_s$. Thus the inner product matrix $\mathbf{B}$ is partitioned into

$$\mathbf{B} = \begin{pmatrix} \mathbf{B}_{11} & \mathbf{B}_{12} & \cdots & \mathbf{B}_{1t} \\ \mathbf{B}_{21} & \mathbf{B}_{22} & \cdots & \mathbf{B}_{2t} \\ \mathbf{B}_{t1} & \mathbf{B}_{t2} & \cdots & \mathbf{B}_{tt} \end{pmatrix} \tag{A.2}$$

where $\mathbf{B}_{rs}$ and $\mathbf{B}_{sr} = \mathbf{B}_{rs}^T$ represent the same array.

The Beaton sweep operator which was defined and used extensively in Dempster (1969) applies naturally to multiindexed arrays. Thus one may define

$$\mathrm{SWP}[\mathscr{I}_r]\mathbf{B} = \begin{pmatrix} \mathbf{C}_{11} & \mathbf{C}_{12} & \cdots & \mathbf{C}_{1t} \\ \mathbf{C}_{21} & \mathbf{C}_{22} & \cdots & \mathbf{C}_{2t} \\ \vdots & \vdots & & \vdots \\ \mathbf{C}_{t1} & \mathbf{C}_{t2} & \cdots & \mathbf{C}_{tt} \end{pmatrix} \tag{A.3}$$

where

$$\begin{aligned} \mathbf{C}_{rr} &= -\mathbf{B}_{rr}^{-1}, \\ \mathbf{C}_{rs} &= \mathbf{B}_{rr}^{-1}\mathbf{B}_{rs} = \mathbf{C}_{sr}^T \qquad && \forall \quad s \neq r, \\ \mathbf{C}_{sp} &= \mathbf{B}_{sp} - \mathbf{B}_{sr}\mathbf{B}_{rr}^{-1}\mathbf{B}_{rp} && \forall \quad s, p \neq r. \end{aligned} \tag{A.4}$$

It is obvious that the Beaton sweep depends only on the arrays $\mathscr{I}_1, \mathscr{I}_2, \ldots, \mathscr{I}_t$ and not on any order of the elements within the arrays nor on any ordering of the arrays themselves. In particular, the operator has no special tie to the matrix representation (A.2) in terms of which the definition (A.3) and (A.4) was couched. The key property of the Beaton sweep for present purposes can be expressed as follows in order free language. Suppose that the arrays $\mathscr{I}_r$ for $r \in \mathscr{A}$ are concatenated into a single array $\mathscr{I}^*$. Then the result of applying SWP[$\mathscr{I}_r$] successively for all $r \in \mathscr{A}$ is equivalent to applying SWP[$\mathscr{I}^*$] and does not depend on the order in which the individual operations SWP[$\mathscr{I}_r$] are applied. In particular, the successive application in any order of SWP[$\mathscr{I}_r$] for $r = 1, 2, \ldots, t$ yields SWP[$\mathscr{I}$]. Note from (A.4) that

$$\text{SWP}[\mathscr{I}]\mathbf{B} = -\mathbf{B}^{-1} \tag{A.5}$$

so that the successive application of SWP[$\mathscr{I}_r$] for $r = 1, 2, \ldots, t$ is a step by step procedure for inverting a symmetric inner product matrix. The proofs of the results of this paragraph are given in Dempster (1969).

REFERENCES

Anderson, T. W. (1958). *An Introduction to Multivariate Statistical Analysis.* Wiley, New York.

Bock, R. D. (1970). Estimating multinomial response relation. *Essays in Probability and Statistics* (R. C. Bose, I. M. Chakravarti, P. C. Mahalanobis, C. R. Rao, K. J. C. Smith, eds.), pp. 111–132. Univ. North Carolina Press, Chapel Hill.

Cox, D. R. (1966). Some procedures connected with the logistic qualitative response curve. *Research Papers in Statistics.* (F. N. David, ed.), pp. 55–71. Wiley, London.

Dempster, A. P. (1971). An overview of multivariate analysis. *J. Multivar. Anal.* **1**, 316–46.

Dempster, A. P. (1969). *Elements of Continuous Multivariate Analysis.* Addison Wesley, Reading, Massachusetts.

Gordon, Tavia and Kannel, William B. [Section 26 (1970), Section 27 (1971).] *The Framingham Study. An Epidemiological Study of Heart Disease.* U.S. Government Printing Office.

Mantel, N. (1966). Models for complex contingency tables and polychotomous response curves. *Biometrics* **22**, pp. 83–110.

Meinert, Curtis L., Knatterud, Genell L., Prout, Thaddeus E., and Klimt, Christian R. (1970). A study of the effects of hypoglycemic agents on vascular complications in patients with adult-onset diabetes. Part II. Mortality results. *J. Amer. Diabetes Assoc.* **19**, Suppl. 2, 789–830.

Olkin, I. and Tate, R. F. (1961). Multivariate correlation models with mixed discrete and continuous variables. *Ann. Math. Statist.* **32**, 448–465.

Press, S. James (1972). *Applied Multivariate Analysis.* Holt, New York.

Walker, S. H. and Duncan, D. B. (1967). Estimation of the probability of an event as a function of several independent variables. *Biometrika* **54**, 315–327.

Inference and Redundant Parameters

D. A. S. FRASER
UNIVERSITY OF TORONTO

Various complexities exist in the use of probabilities in applied situations. The most prominent of these has long been recognized and concerns conditional probability given an event having zero probability: a partition is needed to define the necessary limiting process. At another extreme is the bayesian claim that probabilities can describe all unknowns: a counterclaim is discussed in Fraser [5]. This paper examines a complexity that falls somewhere between these and is concerned with a parameter that becomes redundant if symmetrical error is used; information that specifies the value of the symmetry parameter may produce different results if applied formally to the conditional distribution describing the realization or if used realistically to modify the model. The apparent conflict is resolved and establishes that *information cannot necessarily be used to further condition a conditional distribution*. Probability space models are used for the analysis, as they provide the framework in which the complexity can be examined incisively. The complexity exists more generally and it remains unresolved in the framework of the usual bayesian assumptions.

1. INTRODUCTION

Most difficulties and complexities in the applied use of probability are connected with conditional probability. And interestingly most texts are particularly short and evasive in their treatment of this topic.

The prominent complexity involves conditional probability given an event having zero probability. Many examples may be found in the literature, usually in terms of geometrical probability and varied assignments of a uniform distribution. The resolution treats conditional probability as a limit and uses an appropriate and relevant partition on the sample space.

Another complexity is involved with the bayesian claim that all unknowns can be described by probabilities. The magnitude of the claim tends to make the claim elusive to assess. An examination (Fraser [5]) of conditional probability in a well-defined context shows, however, that probabilities are not available in the generality of the bayesian claim.

A large part of the appeal of the bayesian approach rests, it seems, on the inability of the classical methods to produce results in any but the most special cases. For example, with location-scale and linear models the classical methods are essentially restricted to the single case of a normal distribution for the variation; multivariate models seem similarly restricted. In contrast the bayesian approach produces a wealth of answers. To some degree by giving all answers it gives no answer. Indeed the approach does not produce probabilities specific to a particular system being investigated—from the data and the *set* of possible probability descriptions for the system; thus tests of significance are not available.

For many applications an alternative model is available that directly describes the variation in the system under investigation. A *probability space model* consists of a *probability space* ($\mathscr{S}$, $\mathscr{A}$, P) and a *group* G of random variables. The solution for such a model is unique and is obtained directly from probability theory without any need for the reduction and optimality criteria familiar to the traditional model.

The probability space *model* has been criticized by some of traditional persuasion as being too special—restricted to an *invariant* case indicated by the group. This is somewhat surprising, since linear models—group invariant—embrace the great bulk of statistical models used in practice. And it is even more surprising in that the traditional methods as applied to these models only handle the very special case of a normal distribution for variation.

The probability space *results* have been criticized, usually by those who are promoting some other approach to posterior probabilities. One form of criticism refers to a use of ancillary statistics. This reflects an unfamiliarity with the probability space analysis, as ancillary statistics in fact are not used—either explicitly or implicitly. Another form of criticism uses a betting analogy to assess a posterior probability for a real parameter. The parameter can have two values given the data and the tactic is to bet inward towards a value randomly chosen on the real line. For any fixed prior distribution the tactic is shown to have positive gain thus supposedly discrediting the posterior claim. Contrarily, however, with a U-shaped prior and optional termination by the betting opponent, the loss can be as large as desired. The essential element that is overlooked in the second criticism is that a legitimate prior provides additional information concerning a realization and accordingly modifies the posterior.

The citing of invariance as supposedly being a substantial limitation raises the question of a more serious kind of invariance: all the events of a measurable space are typically taken to have the same status and are not treated differentially. This reaches a somewhat serious tone with ancillary statistics.

Basu [2] has shown that a maximal ancillary or a minimal ancillary event is not unique. Barnard and Sprott [1] and Cox [3] have suggested methods of choosing a preferred ancillary. Cox's arguments, however, provide a substantial basis for the conclusion that the ancillary concept itself is basically a mistake (Fraser [6]). The resolving factor is that events are *not* all of the same status. In particular an event can have a fixed probability because it comes from a probability space and it can have a fixed probability as derived from a class of measures on a measurable space; this is a substantial difference in applications and it can have significant effects on meaning and implications.

The complexity examined in this paper concerns a multidimensional parameter and the effect of information that changes the dimension of the parameter. The complexity arises with a probability space model and it arises in the corresponding bayesian analysis based on the conventional right invariant prior. The complexity is resolved for the probability space model. Its resolution for bayesian theory would presumably need to wait until there was a bayesian substantiation of the use of the right invariant prior.

2. THE PROBABILITY SPACE MODEL

Let Z be a variable describing the variation in some physical system; and suppose that Z takes values in a connected open set $\mathscr{S} \subset \mathbb{R}^N$ and has a probability measure P with density f with respect to lebesgue measure on $(\mathscr{S}, \mathscr{B}_N)$. And let $Y = \theta Z$ be a variable describing the response of the physical system; and suppose that the random variable θ (measurable function $\mathscr{S} \to \mathscr{S}$) is an element of a connected open set $G \subset \mathbb{R}^L$ where G is an exact lie group on $\mathscr{S}$. Then $(\mathscr{S}, \mathscr{B}_N, P; G)$ is a probability space model (Fraser, [4]).

In an application the probability space $(\mathscr{S}, \mathscr{B}_N, P)$ describes the variation in the physical system under investigation, and the random variables G present the various possible forms of expression for the response, some one of which is known to provide a reasonable approximation in the particular application.

As a simple illustration consider the linear model $\mathbf{y} = X\boldsymbol{\beta} + \sigma\mathbf{z}$ where $\mathbf{z}$ is a sample from a distribution P with density f and X is a full rank $n \times r$ design matrix. The group multiplication can be expressed by matrix multiplication

$$\begin{pmatrix} X' \\ \mathbf{y}' \end{pmatrix} = \begin{pmatrix} I & \mathbf{0} \\ \boldsymbol{\beta}' & \boldsymbol{\sigma} \end{pmatrix} \begin{pmatrix} X' \\ \mathbf{z}' \end{pmatrix} \qquad \text{or} \qquad Y = \theta Z.$$

The probability space is $(\mathbb{R}^N - \mathscr{L}(X), \mathscr{B}_N, P^N)$ and the group of random variables is $G = \{\theta : \boldsymbol{\beta} \in \mathbb{R}^r, \sigma \in \mathbb{R}^+\}$.

Consider briefly the analysis of this model. Let Y be the observed response value coming from a realization Z on the probability space; and let θ be the unknown parameter value in it. The possible values for Z can be determined by attempting a solution: $Z = \theta^{-1} Y$. Thus Z can be any value in GY and no other values are possible. It follows that the observed Y produces the value GY for the measurable function GZ and gives no identification for Z in GY. An event of known probability has been observed and conditional probability describes the unidentified Z given the event GY.

For convenient notation let $[\cdot]$ be a continuously differentiable function from $\mathscr{S}$ into G such that $[gZ] = g[Z]$ for all g and Z; let $D(Z) = [Z]^{-1}Z$; and let $Q = \{D(Z): Z \in \mathscr{S}\} = \{Z: [Z] = i\}$. It follows that $Z \leftrightarrow ([Z], D(Z))$ is a one-to-one correspondence with $D(Z)$ referring to the observed event and with $[Z]$ giving coordinates within such events.

The change of variable from $Z = gD$ to (g, D) where $g = [Z]$ and $D = D(Z)$ has the following effect on the euclidean volume measure:

$$\begin{aligned} dZ &= J_N(g\colon D) J_L^{-1}(g)\, dg\, J_*(D)\, dD \\ &= J_N(g\colon D)\, d\mu(g)\, J_*(D)\, dD \end{aligned} \tag{1}$$

where

$$J_N(g\colon D) = \left|\frac{\partial gZ}{\partial Z}\right|_{Z=D}, \qquad J_L(g) = \left|\frac{\partial gh}{\partial h}\right|_{h=i}, \qquad J_*(D) = \left|\frac{\partial gZ}{\partial(g, D(Z))}\right|_{g=i},$$

$d\mu(g)$ is the standardized left haar measure on G, and dD is lebesgue measure on Q. Thus the probability differential on $\mathbb{R}^N$ can be factored as

$$\begin{aligned} f(Z)\, dZ &= f(gD) J_N(g\colon D)\, d\mu(g)\, J_*(D)\, dD \\ &= \frac{f(gD) J_N(g\colon D)}{k(D)}\, d\mu(g)\, k(D) J_*(D)\, dD \end{aligned} \tag{2}$$

where

$$k(D) = \int_G f(gD) J_N(g\colon D)\, d\mu(g)$$

is the marginal density for D relative to $J_*(D)\, dD$.

For the observed Y from a realized Z, let $D(Y) = D_0$. Then Y identifies $D(Z) = D_0$ and gives no identification for $g = [Z]$ given $D(Z)$. The distribution has been factored accordingly and

$$k(D_0) J_*(D_0)\, dD \tag{3}$$

is the probability of what has been observed, and

$$\frac{f(gD) J_N(g\colon D)}{k(D)}\, d\mu(g) \tag{4}$$

is the conditional probability for the unidentified g.

Now let $Y = \theta Z = \beta\alpha Z$ where $\alpha \in H_1$, $\beta \in H_2$, and $G = H_2 H_1$ is a semidirect product of continuous groups. In an application the natural inference procedure is to assess the parameter β using the appropriate distribution and then to assess α conditional on a value for β. Note that the linear model example can be expressed in the following form

$$\begin{pmatrix} X' \\ \mathbf{y}' \end{pmatrix} = \begin{pmatrix} I & \mathbf{0} \\ \beta_r 0 \cdots 0 & 1 \end{pmatrix} \begin{pmatrix} I & \mathbf{0} \\ 0\beta_{r-1} \cdots \beta_1 & \sigma \end{pmatrix} \begin{pmatrix} X' \\ \mathbf{z} \end{pmatrix};$$

the inference procedure is then to assess the highest order regression coefficient and then assess the remaining parameters conditional on β_r.

In this paper we consider the problem of assessing the parameters in the reverse order: α first and then β. This is not the natural order as subsequent analysis will show. But it does have certain practical relevance if α belongs to a compact group expressing symmetries of the distribution for variation. For this suppose that the value of α becomes known on the basis of some outside information. As a first option we can proceed *formally* and condition the distribution obtained from the general analysis or (in the symmetry case) even take a marginal distribution that ignores the variable relating to symmetry. Or more realistically, we can analyze the specialized model obtained from the known value for α.

In the next section we record some necessary factorizations of the invariant measures and in later sections we examine the corresponding implications for the inference procedures. The multivariate normal is used as the central example. For the linear model example it happens that the order for inference is not a serious matter; this derives from the normality of the location subgroups.

3. MEASURE FACTORIZATIONS

Consider the parameter factorization $\theta = \beta\alpha$ with α in H_1 and β in H_2. Some coordinate notation is substantially simpler if we choose a function $[\cdot]$ such that Q passes through the observed Y; then $D_0 = D(Y) = Y$. The basic equation then becomes

$$Y = \theta Z = \beta\alpha Z = \beta\alpha g Y = \beta\alpha h_1 h_2 Y \tag{5}$$

where $g = h_1 h_2$ has been factored according to the reverse semidirect product $G = H_1 H_2$. The equation thus produces the identifications $\beta = h_2^{-1}$ and $\alpha = h_1^{-1}$.

The preceding equation shows that the factorization of the left measure should be in the order $G = H_1 H_2$. Let μ_1, ν_1, Δ_1 be the standardized left haar, the corresponding right haar, and the modular function, respectively, for H_1; let μ, ν, Δ be the corresponding functions for G; and let μ_2, ν_2, Δ_2 be the standardized (re euclidean volume orthogonal to H_1 at i in $H_1 H_2$) left haar, the corresponding right haar, and the modular function for H_2. Then

$$d\mu(h_1 h_2) = d\mu_1(h_1) \frac{\Delta(h_2)}{\Delta_2(h_2)} d\mu_2(h_2)$$

$$= d\mu_1(h_1)\, d\mu_{[2]}(h_2) \tag{6}$$

where the support measure

$$d\mu_{[2]}(h_2) = \frac{d\mu(h_1 h_2)}{d\mu_1(h_1)} = \frac{\Delta(h_2)}{\Delta_2(h_2)} d\mu_2(h_2)$$

is the quotient of left haar on $G = H_1 H_2$ and left haar on H_1.

For the right haar measures it is convenient to use some notation that will be convenient later; let $\alpha = h_1^{-1}$, $\beta = h_2^{-1}$, $\theta = (h_1 h_2)^{-1} = \beta\alpha$. Then

$$d\nu(\beta\alpha) = d\nu_1(\alpha) \frac{\Delta^{-1}(\beta)}{\Delta_2^{-1}(\beta)} d\nu_2(\beta)$$

$$= d\nu_1(\alpha)\, d\nu_{[2]}(\beta) \tag{7}$$

where the support measure

$$d\nu_{[2]}(\beta) = \frac{d\nu(\beta\alpha)}{d\nu_1(\alpha)} = \frac{\Delta^{-1}(\beta)}{\Delta_2^{-1}(\beta)} d\nu_2(\beta)$$

is the quotient of right haar on $G = H_2 H_1$ by right haar on H_1.

The further factorization of the probability element in Section 3 can then be obtained by applying the preceding factorization for μ:

$$f(Z)\, dZ = \frac{f(gD) J_N(g : D)}{k(D)} d\mu(g)\, k(D) J_*(D)\, dD$$

$$= \frac{f(h_1 h_2\, D) J_N(h_1 h_2 : D)}{k(D)} d\mu_1(h_1) \frac{\Delta(h_2)}{\Delta_2(h_2)} d\mu_2(h_2)\, k(D) J_*(D)\, dD$$

$$= \frac{f(h_1 h_2\, D) J_N(h_1 h_2 : D)}{k_{[2]}(h_2 : D)} d\mu_1(h_1)$$

$$\times \frac{k_{[2]}(h_2 : D)}{k(D)} \frac{\Delta(h_2)}{\Delta_2(h_2)} d\mu_2(h_2)\, k(D) J_*(D)\, dD; \tag{8}$$

this is the marginal of D preceded by the conditional for h_2 and then preceded by the conditional for h_1. For further details see Fraser and Mackay [7].

The distribution

$$\frac{k_{[2]}(h_2 : D)}{k(D)} \frac{\Delta(h_2)}{\Delta_2(h_2)} d\mu_2(h_2) \tag{9}$$

describes the realized h_2; and by means of the correspondence $h_2 = \beta^{-1}$ this distribution provides the basis for inference (significance, likelihood, or posterior) concerning the parameter β. For example the distribution describing possible values for β is

$$\frac{k_{[2]}(\beta^{-1}: D)}{k(D)} \frac{\Delta(\beta^{-1})}{\Delta_2(\beta^{-1})} d\nu_2(\beta). \tag{10}$$

Then conditionally given a value for $\beta(= h_2^{-1})$ the distribution

$$\frac{f(h_1 h_2\, D) J_N(h_1 h_2 : D)}{k_{[2]}(h_2 : D)} d\mu_1(h_1) \tag{11}$$

describes the realized h_1; and by means of the correspondence $h_1 = \alpha^{-1}$ this distribution provides the basis for inference (significance, likelihood, or posterior) concerning the parameter α. For example the distribution describing possible values for α is

$$\frac{f(\alpha^{-1} h_2\, D) J_N(\alpha^{-1} h_2 : D)}{k_{[2]}(h_2 : D)} d\nu_1(\alpha). \tag{12}$$

4. IF THE INNER PARAMETER BECOMES KNOWN

Now suppose that information becomes available that identifies the value of α and hence the value of $h_1 = \alpha^{-1}$.

As a first option we could consider *formally* conditioning the distribution for $g = h_1 h_2$ in the preceding section. We would then obtain the following distribution for h_2:

$$\frac{f(h_1 h_2\, D) J_N(h_1 h_2 : D)}{k(h_1, D)} \frac{\Delta(h_2)}{\Delta_2(h_2)} d\mu_2(h_2). \tag{13}$$

This distribution would describe the possible values for the realized h_2 and would accordingly provide the basis for inference concerning β. In particular the distribution describing possible values for β would be

$$\frac{f(h_1 \beta^{-1} D) J_N(h_1 \beta^{-1} : D)}{k(h_1, D)} \frac{\Delta(\beta^{-1})}{\Delta_2(\beta^{-1})} d\nu_2(\beta). \tag{14}$$

As a more realistic alternative we can analyze the specialized model obtained by substituting the known value for α. The specialized model has the equation $Y = \beta\alpha Z$ in which a transformation β in H_2 is applied to αZ; or equivalently, the equation $\alpha^{-1} Y = \alpha^{-1}\beta\alpha Z$ in which the transformation $\alpha^{-1}\beta\alpha$ is applied directly to Z—on the *given* probability space. The group $\alpha^{-1} H_2 \alpha$ is a copy of H_2 and the mapping $\alpha^{-1} h_2 \alpha \mapsto h_2$ is an isomorphism.

The measure $d\mu_2(h_2)$ is left invariant for $\alpha^{-1}h_2\alpha$ in $\alpha^{-1}H_2\alpha$ although not necessarily standardized in the conventional way at the identity. The probability element can then be factored as in (2), using the preceding group in place of G:

$$f(Z)\,dZ = \frac{f(gD)J_N(g: D)}{k(D)}\,d\mu(g)\,k(D)J_*(D)\,dD$$

$$= \frac{f(h_1h_2h_1^{-1}\cdot h_1D)J_N(h_1h_2h_1^{-1}\cdot h_1: D)}{k_*(h_1, D)}\,d\mu_2(h_2)\,dw$$

where dw denotes a probability measure on H_1Q, a measure that in most cases depends on the value of α. Thus we obtain the following distribution for h_2:

$$\frac{f(h_1h_2\,D)J_N(h_1h_2: D)}{k_*(h_1, D)}\,d\mu_2(h_2). \tag{15}$$

This distribution describes the possible values for the realized h_2 and accordingly provides the basis for inference concerning β. In particular the distribution describing possible values for β is

$$\frac{f(h_1\beta^{-1}D)J_N(h_1\beta^{-1}: D)}{k_*(h_1, D)}\,d\nu_2(\beta). \tag{16}$$

The direct analysis of the specialized model based on the known value for α produces the distribution (15) for h_2. But the formal analysis that treats the known value α^{-1} as an *observed* value for h_2 produces the distribution (13), which has an additional modulating factor

$$\frac{\Delta(h_2)}{\Delta_2(h_2)}.$$

This extraneous factor arises because the formal conditioning uses the partition

$$\{h_1H_2: h_1 \in H_1\}$$

on G which is not in general a transformation-group partition, whereas the direct analysis uses the partition

$$\{h_1^{-1}H_2h_1g: g \in G\}$$

which *is* a transformation-group $(h_1^{-1}H_2h_1)$ partition; a transformation-group partition is necessary for information to be observational ([5] Fraser).

For the posterior distribution for β given α, note that the direct analysis gives

$$\propto \text{likelihood}\ d\nu_2(\beta),$$

whereas the formal conditioning of the full posterior gives

$$\propto \text{likelihood}\,\frac{\Delta(\beta^{-1})}{\Delta_2(\beta^{-1})}\,d\nu_2(\beta).$$

5. A REDUNDANT PARAMETER

Consider a distribution f for variation which has symmetries that can be expressed by the group H_1. The probability measure

$$\frac{f(gD)J_N(g\colon D)}{k(D)}\,d\mu(g)\,k(D)J_*(D)\,dD$$

is then invariant under transformations in H_1 and it follows that the conditional distribution for h_1

$$\frac{f(h_1h_2\,D)J_N(h_1h_2\colon D)}{k_{[2]}(h_2\colon D)}\,d\mu_1(h_1) = \frac{f(h_2\,D)J_N(h_2\colon D)}{k_{[2]}(h_2\colon D)}\,d\mu_1(h_1) = \frac{d\mu_1(h_1)}{\mu_1(H_1)}$$

is uniform over H_1 with density $1/\mu_1(H_1)$; in particular, H_1 is compact. The joint distribution for h_1h_2 can then be expressed in the form

$$\frac{d\mu_1(h_1)}{\mu_1(H_1)}\,\frac{\mu_1(H_1)f(h_2\,D)J_N(h_2\colon D)}{k(D)}\,\frac{\Delta(h_2)}{\Delta_2(h_2)}\,d\mu_2(h_2)\,k(D)J_*(D)\,dD;$$

this is the marginal of D preceded by the conditional of h_2 and then preceded by the conditional of h_1.

Now suppose that information becomes available that identifies the value of α and hence the value of $h_1 = \alpha^{-1}$.

The direct analysis of the specialized model based on the known value for α produces the following distribution from formula (15):

$$\frac{f(h_2\,D)J_N(h_2\colon D)}{k_*(D)}\,d\mu_2(h_2); \tag{17}$$

the corresponding distribution describing possible values for β is

$$\frac{f(\beta^{-1}D)J_N(\beta^{-1}\colon D)}{k_*(D)}\,d\nu_2(\beta), \tag{18}$$

which is likelihood with respect to the right invariant measure.

On the other hand the *formal* analysis treating the known value of α^{-1} as an *observed* value for h_2 produces the following distribution from formula (13):

$$\frac{\mu_1(H_1)f(h_2\,D)J_N(h_2\colon D)}{k(D)}\,\frac{\Delta(h_2)}{\Delta_2(h_2)}\,d\mu_2(h_2); \tag{19}$$

the corresponding distribution describing possible values for β is

$$\frac{\mu_1(H_1)f(\beta^{-1}D)J_N(\beta^{-1}: D)}{k(D)} \frac{\Delta(\beta^{-1})}{\Delta_2(\beta^{-1})} dv_2(\beta). \tag{20}$$

These formally derived distributions contain the incorrect additional factor involving the modular functions Δ, Δ_2. Note that if we should choose to ignore the parameter α and obtain the marginal distribution of h_2 and correspondingly of β, we would obtain the same expressions found by the formal conditioning procedure (note that the conditional distribution does not depend on the value of the conditioning variable).

6. THE MULTIVARIATE MODEL

Consider the multivariate model and for familiarity suppose that a normal distribution is used for the variation. We examine here the case in which the mean is known; this is notationally easier, and the more general case introduces no new features.

Let Z be a $p \times n$ matrix of independent standard normal variables; let Γ be a $p \times p$ matrix in the positive ($|\Gamma| > 0$) linear group G; and let $Y = \Gamma Z$ be a $p \times n$ response matrix obtained from some element of G applied to a realized value Z on the probability space for variation. The distribution for possible responses is that of a sample from a multivariate normal with mean Σ and covariance matrix $\Sigma = \Gamma\Gamma'$.

The sampled p-variate distribution for variation is rotationally symmetric. Accordingly let

$$\Gamma = \Lambda\Omega = \begin{pmatrix} \sigma_{(1)} & & & 0 \\ \tau_{21} & \sigma_{(2)} & & \\ \vdots & & \ddots & \\ \tau_{p1} & \cdots & \tau_{pp-1} & \sigma_{(p)} \end{pmatrix} \Omega$$

where Λ is positive lower triangular and Ω is positive orthogonal; note that $\Sigma = \Lambda\Lambda'$. Then $G = H_2 H_1$ can be factored as the positive lower triangular group H_2 times the positive orthogonal group H_1.

For the reverse factorization on $G = H_1 H_2$ let

$$g = OT = O\begin{pmatrix} s_{(1)} & & & 0 \\ t_{21} & s_{(2)} & & \\ \vdots & & \ddots & \\ t_{p1} & \cdots & t_{pp-1} & s_{(p)} \end{pmatrix}.$$

And for a convenient factorization on $\mathscr{S} = \mathbb{R}^{np}$ let

$$Y = T(Y)O(Y) = \begin{pmatrix} s_{(1)}(Y) & & & 0 \\ t_{21}(Y) & s_{(2)}(Y) & & \\ \vdots & & \ddots & \\ t_{p1}(Y) & \cdots & t_{pp-1}(Y) & s_{(p)}(Y) \end{pmatrix} O(Y)$$

where $O(Y)$ is semiorthogonal $(p \times n)$. The measure functions on G are

$$d\mu(g) = \frac{dg}{|g|^p}, \qquad dv(g) = \frac{dg}{|g|^p}, \qquad \Delta(g) = 1;$$

the measure functions on H_1 are

$$d\mu_1(O) = dO, \qquad dv_1(O) = dO, \qquad \Delta_1(O) = 1;$$

and the measure functions on H_2 are

$$d\mu_2(T) = \frac{dT}{|T|_\Delta}, \qquad dv_2(T) = \frac{dT}{|T|_\nabla}, \qquad \Delta_2(T) = \frac{|T|_\nabla}{|T|_\Delta}$$

where $|T|_\Delta$ and $|T|_\nabla$ are the decreasing and increasing determinants. The measure $\mu(g)$ can be then factored as

$$d\mu(g) = \frac{dg}{|g|^p} = dO\,\frac{\Delta(T)}{\Delta_2(T)}\,d\mu_2(T) = dO\,dv_2(T).$$

Consider the group G as providing coordinates relative to the point $D = O(Y)$. The probability differential can be factored as

$$f(Z)\,dZ = f(gD)\,|g|^n\,d\mu(g)\,dD$$

where dD denotes euclidean volume orthogonal to GY at $O(Y)$. Then

$$\begin{aligned} f(Z)\,dZ &= (2\pi)^{-np/2}\,\mathrm{etr}\{-\tfrac{1}{2}gg'\}|g|^n\,dO\,dv_2(T)\,dD \\ &= \frac{dO}{A_p \cdots A_2}\,\frac{A_n \cdots A_{n-p+1}}{(2\pi)^{np/2}}\,\exp\{-\tfrac{1}{2}(\Sigma s_{(j)}^2 + \Sigma t_{jj'}^2)\} \\ &\quad \cdot s_{(1)}^{n-p} \cdots s_{(p)}^{n-1} \prod ds_{(j)} \prod t_{jj'}\,\frac{A_p \cdots A_2}{A_n \cdots A_{n-p+1}}\,dD \end{aligned}$$

where $A_d = 2\pi^{d/2}/\Gamma(d/2)$ is the area of a unit sphere in $\mathbb{R}^d$; the A's complete the normalization of the chi variables and complete the normalization of the Euclidean measure over H_1. The conditional distribution for g is thus

$$\frac{dO}{A_p \cdots A_2}\,\frac{A_n \cdots A_{n-p+1}}{(2\pi)^{np/2}}\,\mathrm{etr}\{-\tfrac{1}{2}TT'\}|T|^n\,\frac{dT}{|T|_\nabla}$$

relative to the point $O(Y)$. If coordinates are taken relative to Y itself, then T must be replaced by $TT(Y)$, giving

$$\frac{dO}{A_p \cdots A_2}\,\frac{A_n \cdots A_{n-p+1}}{(2\pi)^{np/2}}\,\mathrm{etr}\{-\tfrac{1}{2}T'TS(Y)\}|T|^n|\mathscr{S}(Y)|^{n/2}\,\frac{dT}{|T|_\nabla} \tag{21}$$

where $\mathscr{S}(Y) = T(Y)T'(Y) = YY'$ is the inner product matrix for the rows of Y. The distribution given by the second factor describes the realized T; and by means of the correspondence $T = \Lambda^{-1}$ it provides the basis for inference concerning Λ. A simple transformation of the distribution for T gives the standard Wishart distribution for $T'T$. And similarly, the transformation $\Sigma = \Lambda\Lambda'$ with $d\Sigma = 2^p|\Lambda|_\nabla\, d\Lambda$ gives the following posterior distribution for Σ:

$$\frac{A_n \cdots A_{n-p+1}}{(2\pi)^{np/2}} \operatorname{etr}\{-\tfrac{1}{2}\Sigma^{-1}\mathscr{S}(Y)\} \frac{|\mathscr{S}(Y)|^{n/2}}{|\Sigma|^{n/2}} \frac{d\Sigma}{2^p|\Sigma|^{(p+1)/2}}. \tag{22}$$

Now suppose that there is information that identifies the value of Ω and hence the value of O (in the special coordinates we have used). The distribution of $Z^* = \Omega Z$ is the same standard normal as for Z. Thus the equation is $Y = \Lambda Z^*$ where Λ is in H_2. The group H_2 can be used to provide coordinates relative to $D = O(Y)$. The probability differential can be factored

$$f(Z)\, dZ = f(TD|T|^n\, d\mu_2(T)\, dD_2$$

and then normalized

$$f(Z)\, dZ = \frac{A_n \cdots A_{n-p+1}}{(2\pi)^{np/2}} \exp\{-\tfrac{1}{2}(\Sigma s_{(j)}^2 + \Sigma t_{jj'}^2)\}$$

$$\cdot\, s_{(1)}^{n-1} \cdots s_{(p)}^{n-p} \prod ds_{(j)} \prod dt_{jj'} \frac{dD_2}{A_n \cdots A_{n-p+1}};$$

note that the A's complete the normalization of the chi variables. The conditional distribution for T is then

$$\frac{A_n \cdots A_{n-p+1}}{(2\pi)^{np/2}} \operatorname{etr}\{-\tfrac{1}{2}TT'\}|T|^n \frac{dT}{|T|_\Delta}$$

relative to the point $O(Y)$ (note the increasing rather than decreasing determinant). If coordinates are taken relative to Y itself, then T must be replaced by $TT(Y)$, giving

$$\frac{A_n \cdots A_{n-p+1}}{(2\pi)^{np/2}} \operatorname{etr}\{-\tfrac{1}{2}T'T\mathscr{S}(Y)\}|T|^n|\mathscr{S}(Y)|^{n/2} \frac{|T(Y)|_\nabla}{|T(Y)|_\Delta} \frac{dT}{|T|_\Delta}. \tag{23}$$

This distribution describes the realized T given the information that specifies Ω; and by means of the correspondence $T = \Lambda^{-1}$ it provides the basis for inference concerning Λ. The corresponding distribution for $\Sigma = \Lambda\Lambda'$ can be obtained from $d\Sigma = 2^p|\Lambda|_\nabla\, d\Lambda$:

$$\frac{A_n \cdots A_{n-p+1}}{(2\pi)^{np/2}} \operatorname{etr}\{-\tfrac{1}{2}\Sigma^{-1}\mathscr{S}(Y)\} \frac{|\mathscr{S}(Y)|^{n/2}}{|\Sigma|^{n/2}}$$

$$\cdot \left(\frac{s_1(Y)}{\sigma_{(1)}}\right)^{p-1} \left(\frac{s_{(2)}(Y)}{\sigma_{(2)}}\right)^{p-3} \cdots \left(\frac{s_{(p)}(Y)}{\sigma_{(p)}}\right)^{1-p} \frac{d\Sigma}{2^p|\Sigma|^{(p+1)/2}}. \tag{24}$$

The analysis based on the full linear group produces the distribution (22) describing Σ.

Alternatively, for the case in which information specifies the rotation Ω the analysis is based essentially on the positive lower triangular group. This produces the distribution (24) describing Σ. A *formal* analysis, however, that treats Ω^{-1} as an *observed* value for O in the full model would produce the distribution (22) for Σ; it has an additional modulating factor that removes

$$\left(\frac{s_1(Y)}{\sigma_{(1)}}\right)^{p-1} \cdots \left(\frac{s_{(p)}(Y)}{\sigma_{(p)}}\right)^{1-p}$$

from the formula (24). And a formal analysis that ignores O would also lead to the incorrect distribution (22).

The specification of the rotation Ω introduces a very special structure in which the first response variable is generated from a first error variable, a second response from the first two error variables, and so on. Correspondingly we find that the distributions (23) and (24) are consistent as we vary the number p of coordinates examined. And this explains why the distributions (21) and (22) based on all the parameters of the positive linear group give differing results as we vary the number p of coordinates examined.

7. THE BAYESIAN RIGHT INVARIANT

It is fairly common practice for bayesian analysis to use a right invariant prior as the support for the likelihood function. For simple problems the routine use of right invariant priors causes no problems. But for more complicated problems the use of right invariant priors can produce contradictory results.

Consider a parameter θ in a group G as described in Section 3. And suppose $\theta = \beta\alpha$ is factored with components β in a group H_2 and α in a group H_1. The right invariant measure then factors as

$$dv(\beta\alpha) = dv_1(\alpha)\, \Delta^{-1}(\beta)\, \Delta_2(\beta)\, dv_2(\beta).$$

Now suppose that a conventional bayesian uses right invariant priors to modulate his likelihood functions. The posterior distribution for $\theta = \beta\alpha$ is then the likelihood function $L(\theta) = L(\beta\alpha)$ supported by the preceding right invariant measure.

If the likelihood function does not depend on α—owing to symmetry, say —then the posterior for β alone is proportional to

$$L(\beta)\, \Delta^{-1}(\beta)\, \Delta_2(\beta)\, dv_2(\beta).$$

Or if the parameter β is being examined conditionally given the parameter α, then the posterior for β is proportional to

$$L(\beta\alpha)\,\Delta^{-1}(\beta)\,\Delta_2(\beta)\,d\nu_2(\beta).$$

These distributions do not in general have the form of likelihood multiplied by the right invariant prior; an additional modulating factor $\Delta^{-1}(\beta)\,\Delta_2(\beta)$ is present. Thus the conventional bayesian use of right invariant priors is self contradictory.

A related problem for bayesian analysis using right invariant priors is the unavailability of marginal likelihood functions.

REFERENCES

1. Barnard, G. A. and Sprott, D. (1971). A note on Basu's examples of anomalous ancillary statistics. *Symp. Foundations Statist. Inference, Waterloo.* Holt, New York.
2. Basu, D. (1959). The family of ancillary statistics. *Sankhyā Ser. A* **21** 247–256.
3. Cox, D. R. (1971). Recovery of ancillary information. *J. Roy. Statist. Soc. Ser. B* **33** 251–255.
4. Fraser, D. A. S. (1968). *The Structure of Inference.* Wiley, New York.
5. Fraser, D. A. S. (1972). Bayes, likelihood, or structural. *Ann. Math. Statist.* **43** 777–790.
6. Fraser, D. A. S. (1972). The elusive ancillary. *Proc. Symp. Multivariate Anal., Dalhousie,* to be published.
7. Fraser, D. A. S. and MacKay, J. (1972). Significance, likelihood, and objective posterior. submitted to *Ann. Math. Statist.*

The Variance Information Manifold and the Functions on It

A. T. JAMES
UNIVERSITY OF ADELAIDE

1. THE VARIANCE INFORMATION MANIFOLD AND ITS BOUNDARY

The $m \times m$ variance matrices Σ form a convex cone in $\frac{1}{2}m(m+1)$-dimensional Euclidean space $R^{1/2\,m(m+1)}$, whose interior consists of the positive definite matrices. Each nonsingular variance matrix Σ is associated with an information matrix $J = \Sigma^{-1}$. We consider the variance information manifold (VIM) as a space in which each interior point has alternative coordinate matrices Σ or $J = \Sigma^{-1}$.

The singular positive semidefinite matrices Σ, which form the surface of the cone of variance matrices in $R^{1/2\,m(m+1)}$, are allotted corresponding points in the variance information manifold to form part of its boundary. Likewise, the singular information matrices J are allotted points in VIM to form a disjoint part of its boundary. The boundary of VIM is completed by a set of points of singular variance and information.

The variance information manifold may be realized by the mapping

$$(\Sigma, J) \to \Sigma(I + \Sigma)^{-1} = (I + J)^{-1} = \Lambda$$

of variance and information matrices into the space of $m \times m$ symmetric matrices whose latent roots λ_i all lie between 0 and 1;

$$0 \leq \lambda_i \leq 1.$$

VIM points whose latent roots λ_i are all less than one correspond to a variance matrix $\Sigma = \Lambda(I - \Lambda)^{-1}$; those with latent roots all greater than zero have an information matrix $J = \Lambda^{-1}(I - \Lambda)$.

If both zero and one occur among the latent roots of a VIM point, then it has neither a variance matrix Σ nor an information matrix J, but it is singular in both variance and information. Nevertheless, one can consider the eigenspace of the zero root, $\lambda = 0$, as a subspace of sample space R^n in which there is zero variance and infinite information, and the eigenspace of the unit root, $\lambda = 1$, as a subspace of infinite variance and zero information.

The Euclidean topology of the space of symmetric matrices with latent roots between 0 and 1 supplies a topology for the variance information manifold. With the inclusion of its boundary, the VIM is compact.

2. THE BIVARIATE CASE

In the case $m = 2$, the VIM is a double cone in R^3, base to base. A slight transformation will put it up on end. We introduce cylindrical polar coordinates (r, θ, z) as the following functions of the elements of

$$\begin{aligned}\Sigma &= \begin{bmatrix} \sigma_{11} & \sigma_{12} \\ \sigma_{21} & \sigma_{22} \end{bmatrix} \\ &= \begin{bmatrix} \cos\varphi & -\sin\varphi \\ \sin\varphi & \cos\varphi \end{bmatrix} \begin{bmatrix} \lambda_1 & 0 \\ 0 & \lambda_2 \end{bmatrix} \begin{bmatrix} \cos\varphi & \sin\varphi \\ -\sin\varphi & \cos\varphi \end{bmatrix} = J^{-1}, \\ &\qquad\qquad\qquad \lambda_1 \geq \lambda_2, \quad 0 \leq \varphi < \pi,\end{aligned}$$

$$t = \operatorname{trace}\Sigma = \sigma_{11} + \sigma_{22}, \qquad \Delta = \det\Sigma,$$

namely,

$$r = \frac{\lambda_1 - \lambda_2}{1 + \lambda_1 + \lambda_2 + \lambda_1\lambda_2} = \frac{(t^2 - 4\Delta)^{1/2}}{1 + t + \Delta},$$

$$\theta = 2\varphi = \arctan\frac{2\sigma_{12}}{\sigma_{11} - \sigma_{22}} = \arctan\frac{2j_{12}}{j_{11} - j_{22}},$$

$$z = \frac{\lambda_1\lambda_2 - 1}{1 + \lambda_1 + \lambda_2 + \lambda_1\lambda_2} = \frac{\Delta - 1}{1 + t + \Delta}.$$

VIM is represented as a double cone with singular variance as the lower surface and singular information as the upper. Points on the circle where the upper and lower surfaces intersect have both singular information and variance. In terms of the distribution of a variate

$$\mathbf{x} = \begin{bmatrix} x_1 \\ x_2 \end{bmatrix}$$

in sample space R^2, singular information implies that probability statements only apply to some linear function

$$l_1 x_1 + l_2 x_2 .$$

If this were to have nonzero variance σ^2, the information matrix would be

$$J = \frac{1}{\sigma^2}\begin{bmatrix} l_1{}^2 & l_1 l_2 \\ l_2 l_1 & l_2{}^2 \end{bmatrix}.$$

However, if there is singular variance as well as singular information, then the linear function is constant

$$l_1 x_1 + l_2 x_2 = c$$

and hence a distribution in R^2 of singular variance and information is equivalent to a linear constraint (Fig. 1). We have

$$\tan\theta = \tan 2\varphi = \frac{2l_1 l_2}{l_1^2 - l_2^2}$$

hence

$$\tan\varphi = -\frac{l_1}{l_2}.$$

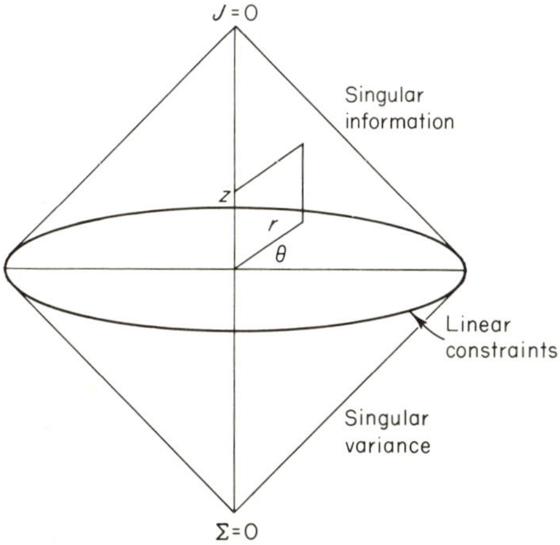

Fig. 1

The multinormal distribution with singular variance matrix is well known; in the following section, we introduce the multinormal distribution with singular information matrix.

3. THE MULTINORMAL DISTRIBUTION WITH SINGULAR INFORMATION MATRIX

Heuristic Introduction

The multinormal distribution with a singular variance matrix Σ is a distribution of a vector variate $\mathbf{x} \in R^n$ such that with probability one, its deviation $\mathbf{x} - \boldsymbol{\mu}$ from expectation $\boldsymbol{\mu}$ lies in a subspace. Now the dual of a subspace

$\mathscr{L}$, i.e., the set of linear functionals on it, is a quotient space (not a subspace), namely, the quotient space $R^{n*}/\mathscr{L}^{\perp}$ of the dual space R^{n*} of R^n over the annihilator $\mathscr{L}^{\perp} = \mathscr{S}$ of $\mathscr{L}$, $\mathscr{S} \subset R^{n*}$ being the set of all linear functionals which vanish on $\mathscr{L}$. A point $x^* + \mathscr{S}$ of the quotient space $R^{n*}/\mathscr{S}$ is a coset

$$\mathbf{x}^* + \mathscr{S} \stackrel{\text{def}}{=} \{\mathbf{x}^* + y^* | y^* \in \mathscr{S}\}.$$

The coset $x^* + \mathscr{S}$ clearly consists of all functionals $x^* \in R^{n*}$ which have the same values when restricted to $\mathscr{L}$. Any element of the coset determines it. Geometrically, the cosets may be represented by hyperplanes (Fig. 2).

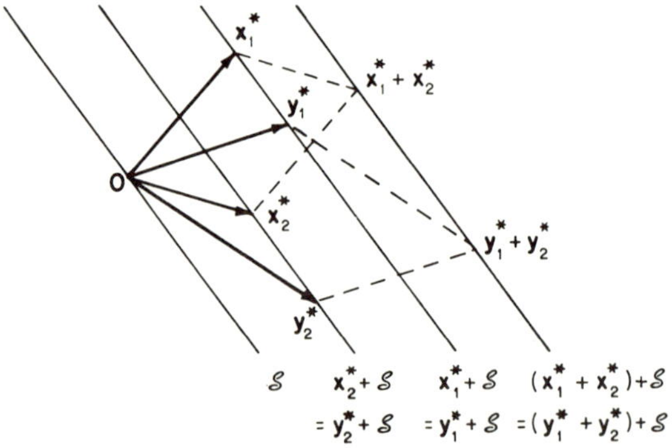

Fig. 2

The quotient space $R^{n*}/\mathscr{S}$ forms a vector space under addition and scalar multiplication (Fig. 3):

$$(\mathbf{x}_1{}^* + \mathscr{S}) + (\mathbf{x}_2{}^* + \mathscr{S}) = ((\mathbf{x}_1{}^* + \mathbf{x}_2{}^*) + \mathscr{S}),$$
$$\lambda(\mathbf{x}^* + \mathscr{S}) = (\lambda\mathbf{x}^* + \mathscr{S}).$$

The dual of a multinormal distribution with a singular variance matrix will then be a distribution with singular information matrix which will be a distribution on the quotient space $R^n/\mathscr{S}$. Notice that we now find it convenient to switch to the space, R^n, rather than its dual, R^{n*}.

If $\mathscr{S}$ is a subspace of R^n, then we have, from the definition of conditional probability, $P(\mathbf{x}|\mathbf{x} + \mathscr{S})$,

$$P(\mathbf{x}) = P(\mathbf{x} + \mathscr{S})P(\mathbf{x}|\mathbf{x} + \mathscr{S}).$$

The second factor is a distribution with singular variance matrix; the first factor will be an example of a distribution with singular information matrix that we are about to define.

The situation in normal regression theory suggests a method of definition. Suppose a vector variate $\mathbf{y} \in R^n$ is distributed as $N(X\boldsymbol{\beta}, \sigma^2 I_n)$ where X is an $N \times p$ matrix of rank $r < p$, and $\boldsymbol{\beta} \in R^p$.

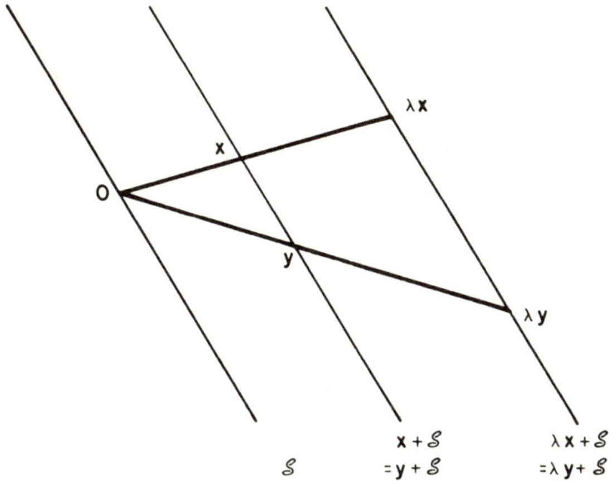

Fig. 3

Since all vectors in the coset $\boldsymbol{\beta} + \mathscr{K}(X)$ determine the same expectation of $\mathbf{y}$, where $\mathscr{K}(X)$ is the kernel or null space of X consisting of all vectors annihilated by X, our estimator can be considered as a coset $\mathbf{b} + \mathscr{R}(X)$ consisting of all solutions of the normal equations

$$X'X\mathbf{b} = X'\mathbf{y}.$$

Note that $\mathscr{K}(X'X) = \mathscr{K}(X)$.

When $r = p$, the matrix $J = (1/\sigma^2)X'X$ is the inverse of the variance matrix of $\mathbf{b}$. Hence when $r < p$ and J is singular, we can define that $\mathbf{b}$ or $\mathbf{b} + \mathscr{K}(X)$ has a singular information matrix $J = (1/\sigma^2)X'X$.

By dividing the normal equations by σ^2,

$$J\mathbf{b} = \frac{1}{\sigma^2} X'\mathbf{y},$$

we can see how to proceed with our definition in the general case, because $J\mathbf{b}$ is distributed as

$$N\left(\frac{1}{\sigma^2} X'X\boldsymbol{\beta}, \frac{1}{\sigma^2} X'X\right) = N(J\boldsymbol{\beta}, J).$$

Definition. *A vector variate* $\mathbf{x} \in R^n$ *is normally distributed with information matrix* J, *which may be singular, if* $J\mathbf{x}$ *is distributed as* $N(\boldsymbol{\mu}^*, J)$ *for some* $\boldsymbol{\mu}^* \in \mathscr{R}(J)$.

Since the distribution is only defined by the values of $J\mathbf{x}$, we define an equivalence relation, $\sim$, on R^n, namely, $\mathbf{x}_1 \sim \mathbf{x}_2$ if $J\mathbf{x}_1 = J\mathbf{x}_2$, i.e., if

$J(\mathbf{x}_1 - \mathbf{x}_2) = \mathbf{0}$, i.e., if $\mathbf{x}_1 - \mathbf{x}_2 \in \mathscr{K}(J)$. The equivalence classes are then the cosets $\mathbf{x} + \mathscr{K}(J) \in R^n/\mathscr{K}(J)$.

Just as a distribution with singular variance matrix may formally be regarded as either a distribution within R^n which has all its probability on a hyperplane or alternatively as a distribution defined on a hyperplane; similarly, we may regard the sample space of a distribution with singular information matrix as either R^n or $R^n/\mathscr{K}(J)$, according to convenience. If the sample space is taken to be R^n, then the only measurable sets are unions of cosets.

As a distribution on cosets, the expectation coset $\boldsymbol{\mu} + \mathscr{K}(J)$ is defined to be the set of all solutions $\boldsymbol{\mu}$ of $J\boldsymbol{\mu} = \boldsymbol{\mu}^*$.

Another way of considering the distribution in R^n is as an improper distribution, because in some directions it is uniform with infinite variance.

4. DERIVATION VIA THE DISTRIBUTION OF LINEAR FUNCTIONS

Suppose that $L'\mathbf{x}$ has variance matrix Σ_1, where L is an $n \times p$ constant matrix and Σ_1 is $p \times p$. If M were square and nonsingular, then $\mathbf{x}$ would have variance matrix $L'^{-1}\Sigma_1 L^{-1}$. If Σ_1 is nonsingular, then $\mathbf{x}$ would have information matrix $L\Sigma_1^{-1}L'$. This suggests the

Lemma. *If the distribution of* $\mathbf{x} \in R^n$ *is given by the fact that* $L'\mathbf{x}$ *has a nonsingular variance* Σ_1, *where* L *is an* $n \times p$ *constant matrix, then* $\mathbf{x}$ *has information matrix* $L\Sigma_1^{-1}L'$.

Proof. To prove that $\mathbf{x}$ has information matrix J, we must prove that $J\mathbf{x}$ has variance matrix J. Now

$$\begin{aligned} V(L\Sigma_1^{-1}L'\mathbf{x}) &= V((L\Sigma_1^{-1})(L'\mathbf{x})) \\ &= L\Sigma_1^{-1}\Sigma_1\Sigma_1^{-1}L' \\ &= L\Sigma_1^{-1}L', \end{aligned}$$

and the lemma follows.

5. APPLICATION TO THE ANALYSIS OF EXPERIMENTAL DESIGNS

The multinormal distributions with singular information matrices are useful in the analysis of experimental designs. The experiment often splits into r easily analyzable parts or groups, each of which supplies information about the vector $\boldsymbol{\beta}$ of treatment effects. The normal equations can be written separately for each of the ith parts of the experiment, and after eliminating the estimates of block effects one obtains *reduced* normal equations

$$J_i \mathbf{b}_i = \mathbf{h}_i, \qquad i = 1, \ldots, r,$$

of which the solutions $\mathbf{b}_i + \mathscr{K}(J_i)$ are estimates of $\boldsymbol{\beta}$ with singular information matrix J_i, quasi-sufficient for each ith part of the experiment. The combined quasi-sufficient estimator is then given by the solution of the equation

$$\left(\sum_{i=1}^{r} J_i\right)\mathbf{b} = \sum_{i=1}^{r} \mathbf{h}_i .$$

6. REPRESENTATION AS THE MARGINAL DISTRIBUTION OF A NONSINGULAR DISTRIBUTION

The distribution with singular information matrix J may be represented as the marginal distribution of a nonsingular distribution. Namely, if Σ is a nonsingular generalized inverse of J and $\boldsymbol{\mu}$ is any vector in the expectation coset of the singular distribution, then if $\mathbf{x}$ is a variate distributed as $N(\boldsymbol{\mu}, \Sigma)$, then $J\mathbf{x}$ is distributed as $N(J\boldsymbol{\mu}, J)$ because $J\Sigma J = J$. Hence the marginal distribution of $\mathbf{x} + \mathscr{K}(J)$ has information matrix J.

The singular information matrices appear in the general decomposition of a multinormal distribution.

7. DECOMPOSITION OF A MULTINORMAL DISTRIBUTION

Let the vectors $\mathbf{x}$ in the sample space R^n be called covariant. Then the (column) vectors $\mathbf{x}^*$ of coefficients of linear functions or contrasts belong to the dual space R^{n*} and are thus contravariant. The variance matrix Σ gives the natural inner product $\mathbf{x}^{*\prime}\Sigma\mathbf{y}^*$ in the space R^{n*} of contrasts. As its indices must both contract with indices of contravariant vectors $\mathbf{x}^*$, $\mathbf{y}^*$, it must be doubly covariant.

The information matrix Σ^{-1} supplies the natural inner product $\mathbf{x}'\Sigma^{-1}\mathbf{y}$ in the sample space R^n and is hence doubly contravariant.

The matrix of a linear transformation of sample space is covariant $\times$ contravariant and for a linear transformation of contrast space the matrix is contravariant $\times$ covariant.

Let $\Sigma = TT'$. Then the equation

$$I = E_1 + E_2$$

for the decomposition of the identity matrix I into symmetric idempotent matrices E_1, E_2 of ranks r and $n - r$ has four clearly distinguishable meanings which are shown up by the four different ways in which it can be transformed; namely,

$E_i \to TE_iT'$	cogrediently × cogrediently,
$E_i \to T^{-1\prime}E_iT^{-1}$	contragrediently × contragrediently,
$E_i \to TE_iT^{-1}$	cogrediently × contragrediently,
$E_i \to T^{-1\prime}E_iT'$	contragrediently × cogrediently.

We obtain a set of useful expressions for the four transforms of the decomposition equation if we let Q be an $n \times r$ matrix whose columns are covariant vectors spanning $\mathscr{S} \subset R^n$ and L be an $n \times (n-r)$ matrix of contravariant column vectors spanning $\mathscr{L} = \mathscr{S}^{\perp} \subset R^{n*}$; $L'Q = \mathbf{0}$:

cov. × cov.	$\Sigma = Q(Q'\Sigma^{-1}Q)^{-1}Q' + \Sigma L(L'\Sigma L)^{-1}L'\Sigma,$
contr. × contr.	$\Sigma^{-1} = \Sigma^{-1}Q(Q'\Sigma^{-1}Q)Q'\Sigma^{-1} + L(L'\Sigma L)^{-1}L',$
linear transformations of R^n, cov. × contr.	$I_n = Q(Q'\Sigma^{-1}Q)^{-1}Q'\Sigma^{-1} + \Sigma L(L'\Sigma L)^{-1}L',$
linear transformations of R^{n*}, contr. × cov.	$I_n = \Sigma^{-1}Q(Q'\Sigma^{-1}Q)^{-1}Q' + L(L'\Sigma L)^{-1}L'\Sigma.$

The first two equations are the decompositions of the variance and information matrices into singular variance and information components associated with the subspace $\mathscr{S}$ and its conjugate space

$$\Sigma\mathscr{L} = \{\mathbf{x} = \Sigma\mathbf{x}^* \in R^n \mid \mathbf{x}^* \in \mathscr{L}\} = \{\mathbf{x} \in R^n \mid \mathbf{y}'\Sigma^{-1}\mathbf{x} = 0 \quad \text{for all } \mathbf{y} \in \mathscr{S}\}.$$

Notice that a component, e.g., $Q(Q'\Sigma^{-1}Q)Q'$, of the variance matrix is uniquely determined by two of its properties:

1. the information matrix Σ^{-1} is a generalized inverse of it;
2. its range is $\mathscr{R}(Q) = \mathscr{S}$ or its kernel is $\mathscr{L}$.

The same is true for the components of the information matrix, the variance matrix Σ being their generalized inverse.

The third equation gives the idempotents projecting on $\mathscr{S}$ and $\Sigma\mathscr{L}$ and the fourth equation, the idempotents projecting on $\Sigma^{-1}\mathscr{S}$ and $\mathscr{L}$.

The structure of the matrices follows from some simple rules:

1. Matrix multiplication of indices of order n must always involve summation (or contraction) of a cogredient index with a contragredient one.

2. As the matrices depend only upon the ranges of Q and L, they must be invariant under transformations

$$Q \to QP', \qquad L \to LM'$$

where P and M are nonsingular matrices of respective orders r and $(n-r)$.

3. The covariance or contravariance of the first and last indices depends upon which of the four equations that one wants.

8. INVARIANT METRIC

From the Wishart density

$$\text{const. det}\,\Sigma^{-1/2n}\,\text{etr}(-\tfrac{1}{2}n\Sigma^{-1}S)$$

we have the log likelihood function

$$\mathscr{L} = \text{const.} - \tfrac{1}{2}n(\log\det\Sigma + \text{tr}\,\Sigma^{-1}S)$$

and its two differentials

$$d\mathscr{L} = -\tfrac{1}{2}n\,\text{tr}(\Sigma^{-1}d\Sigma - \Sigma^{-1}d\Sigma\Sigma^{-1}S),$$
$$d^2\mathscr{L} = -\tfrac{1}{2}n\,\text{tr}(-\Sigma^{-1}d\Sigma\Sigma^{-1}d\Sigma + 2\Sigma^{-1}d\Sigma\Sigma^{-1}d\Sigma\Sigma^{-1}S),$$

and the differential form in the information matrix as the expectation of $(-d^2\mathscr{L})$

$$E[-d^2\mathscr{L}] = \tfrac{1}{2}n\,\text{tr}(\Sigma^{-1}d\Sigma\Sigma^{-1}d\Sigma).$$

The differential form

$$(ds)^2 = \text{tr}(\Sigma^{-1}d\Sigma\Sigma^{-1}d\Sigma) = \text{tr}(J^{-1}dJJ^{-1}dJ),$$

where $J = \Sigma^{-1}$, is a useful metric on VIM.

Maass [1] has shown that such a metric differential form $(ds)^2 = \text{tr}(X^{-1}dXX^{-1}dX)$ on the space of $m \times m$ positive definite symmetric matrices X is invariant under congruence transformation

$$X \to LXL'$$

where L is an $m \times M$ nonsingular matrix.

If we write

$$X = HYH'$$

with H orthogonal and $Y = \text{diag}(y_i)$, then the metric differential form becomes

$$(ds)^2 = \sum_{i=1}^{m} \frac{(dy_i)^2}{y_i^2} + \sum_{i<j}^{m} \frac{(y_i - y_j)^2}{y_i y_j}(d\theta_{ij})^2$$

where the $d\theta_{ij}$ denote the differential forms of the skew symmetric matrix $H'dH$. Upon putting $z_i = \log y_i$ for $i = 1, \ldots, m$, we have

$$(ds)^2 = \sum_{i=1}^{m} (dz_i)^2 + \sum_{i<j}^{m} (\sinh(\tfrac{1}{2}(z_i - z_j)))^2\, d\theta_{ij}.$$

From the metric differential form, we obtain the Laplace–Beltrami operator as

$$\sum_{i=1}^{m} \left\{ y_i^2 \left(\frac{\partial^2}{\partial y_i^2} \right) + \left(\sum_{j=1,\, j \neq i}^{m} y_i^2 (y_i - y_j) - \tfrac{1}{2}(m-3) y_i \right) \left(\frac{\partial}{\partial y_i} \right) \right\} + \frac{1}{4} \sum_{i<j} y_i y_j (y_i - y_j)^{-1} \left(\frac{\partial^2}{\partial \theta_{ij}^2} \right).$$

9. GEODESIC DISTANCE BETWEEN TWO MATRICES

To find the distance from X_1 to X_2, we let L be a nonsingular matrix which by congruence transformation reduces X_1 to the identity matrix and X_2 to a diagonal matrix $Y = \mathrm{diag}(y_i)$ whose elements y_i are the latent roots of the determinantal equation

$$\det(X_2 - yX_1) = 0; \tag{9.1}$$

$LX_1L' = I_m$, $LX_2L' = Y$.

Then since the distance ds is invariant under congruence transformation, the distance between X_1 and X_2 is given by the following integral taken along a path which will minimize it. Hence we have

$$d(X_1, X_2) = \int_{X_1}^{X_2} ds = \int_{I_m}^{Y} ds = \int_{I_m}^{Y} \left(\sum_{i=1}^{m} \left(\frac{dy_i}{y_i} \right)^2 \right)^{1/2}$$
$$= \int_{I_m}^{Z} \left(\sum_{i=1}^{m} (dz_i)^2 \right)^{1/2} = \left(\sum_{i=1}^{m} z_i^2 \right)^{1/2}.$$

It is fairly clear that this is the minimum and it can be checked by the calculus of variations. Hence we have the

Theorem. *If X_1, X_2 are two positive definite symmetric matrices, the geodesic distance between them is*

$$\left(\sum_{i=1}^{m} (\log y_i)^2 \right)^{1/2}$$

where the y_i are the roots of the determinantal equation (9.1).

Since $\frac{1}{2}n$ times our metric differential form is a quadratic form in the information matrix for S, it follows that:

Theorem. *If $y_1, \ldots, y_m$ are the latent roots of the equation*

$$\det(S - y_i \Sigma) = 0,$$

then the statistic

$$d^2 = \tfrac{1}{2}n \sum_{i=1}^{m} (\log y_i)^2$$

is asymptotically distributed as χ^2 *on* $\tfrac{1}{2}m(m+1)$ *degrees of freedom.*

The statistic d^2 could be used as a test of a hypothesis which prescribes Σ.

10. ZONAL POLYNOMIALS

Definition. *A zonal polynomial is a symmetric homogeneous polynomial of degree k in the latent roots* $y_1, \ldots, y_k$ *of a matrix X which is an eigenfunction of the operator*

$$\Delta = \sum_{i=1}^{m} y_i^2\left(\frac{\partial^2}{\partial y_i^2}\right) + \sum_{\substack{i, j=1 \\ i \neq j}}^{m} y_i^2(y_i - y_j)^{-1}\left(\frac{\partial}{\partial y_i}\right) \tag{10.1}$$

derived from the Laplace–Beltrami operator in the latent roots y_i *omitting the multiple of the Euler operator,* $\sum y_i(\partial/\partial y_i)$.

Theorem. *For each monomial* $y_1^{k_1} \cdots y_m^{k_m}$ *with* $k_1 \geq k_2 \geq \cdots \geq k_m$, *there is exactly one zonal polynomial with this as the term of highest weight. The eigenvalue is* $\rho_\kappa + k(m-1)$ *where* $\rho_\kappa = \sum_{i=1}^{m} k_i(k_i - i)$, $\kappa = (k_1, \ldots, k_m)$, $k_1 + \cdots + k_m = k$.

Proof. One can verify that

$$\Delta y_1^{k_1} \cdots y_m^{k_m} = (\rho_\kappa + k(m-1))y_1^{k_1} \cdots y_m^{k_m} + \text{terms of lower weight.}$$

The zonal polynomial

$$C_\kappa(Y) = c_\kappa y_1^{k_1} \cdots y_1^{k_m} + \text{terms of lower weight}$$

satisfies the differential equation

$$\Delta C_k(Y) = (\rho_\kappa + k(m-1))C_\kappa(Y), \tag{10.2}$$

which yields a recurrence relation from which the coefficients of terms of lower weight can be determined from the first coefficient c_κ.

Let $C_\kappa^*(Y)$ be a renormalized zonal polynomial, so that it is unity at the identity matrix

$$C_\kappa^*(I_m) = 1.$$

Theorem. *If X and Y are arbitrary symmetric matrices, then*

$$\int_{O(m)} C_\kappa^*(XHYH')(dH) = C_\kappa^*(X)C_\kappa^*(Y). \tag{10.3}$$

Proof. Since

$$C_\kappa^*(XHYH') = C_\kappa^*(X^{1/2}HY(X^{1/2}H)') = C_\kappa^*(LYL')$$

where $L = X^{1/2}H$, and the operator Δ is invariant under congruence transformation

$$\Delta_{LYL'} = \Delta_Y$$

and can be taken under the integral sign, we see that the left-hand side of (10.3), when considered as a function of Y for fixed X, is an eigenfunction of Δ, namely,

$$\begin{aligned}\Delta_Y \int_{O(m)} C_\kappa^*(XHYH')(dH) &= \int_{O(m)} \Delta_{LYL'} C_\kappa^*(LYL')(dH)\\ &= (\rho_\kappa + k(m-1)) \int_{O(m)} C_\kappa^*(LYL')(dH)\\ &= (\rho_\kappa + k(m-1)) \int C_\kappa^*(XHYH')(dH).\end{aligned}$$

Hence the left-hand side of (10.3) must be a multiple of the zonal polynomial

$$\int_{O(m)} C_\kappa^*(XHYH')(dH) = \lambda C_\kappa^*(Y).$$

Putting $Y = I_m$, we have $C_\kappa^*(X) = \lambda$ and the theorem follows.

Now suppose both $X = \operatorname{diag}(x_i)$ and $Y = \operatorname{diag}(y_i)$ are diagonal matrices.

By equating coefficients of $x_1^{k_1} x_2^{k_2} \cdots x_m^{km}$ on both sides of Eq. (10.3), one obtains the following integral representation of the zonal polynomial

$$C_\kappa^*(Y) = \int_{O(m)} \Big(\sum_i h_{i1}^2 y_i\Big)^{k_1-k_2} \Big(\sum_{i_1<i_2} \det\begin{bmatrix} h_{i_1 1} & h_{i_1 2}\\ h_{i_2 1} & h_{i_2 2}\end{bmatrix}^2 y_i y_i\Big)^{k_2-k_3} \cdots \det Y^{k_m}(dH). \tag{10.4}$$

This integral over the orthogonal group $O(m)$ can be transformed to an integral over $\frac{1}{2}m(m-1)$-dimensional Euclidean space by representing the orthogonal matrix H as a Gramm–Schmidt orthonormalization of a triangular matrix with unities on the diagonal:

$$Z = (z_{ij}) = \begin{bmatrix} 1 & 0 & 0\\ z_{21} & 1 & \\ \vdots & & \ddots \\ z_{m1} & \cdots & 1\end{bmatrix} = [\mathbf{z}_1\, \mathbf{z}_2 \cdots \mathbf{z}_m].$$

The volume element (dH) becomes

$$(dH) = \frac{\Gamma_m(\frac{1}{2}m)}{\pi^{1/2\, m^2}} \frac{1}{D_1 D_2 \cdots D_{m-1}} (dZ)$$

where D_k is the first kth-order principal minor of $Z'Z$:

$$D_k = \det\begin{bmatrix} \mathbf{z}_1{}'\mathbf{z}_1 & \cdots & \mathbf{z}_1{}'\mathbf{z}_k \\ \vdots & & \vdots \\ \mathbf{z}_k{}'\mathbf{z}_1 & \cdots & \mathbf{z}_k{}'\mathbf{z}_k \end{bmatrix}$$

and the integral transforms to a form similar to one given by Bhanu Murti [2]

$$C_\kappa{}^*(Y) = \frac{\Gamma_m(\tfrac{1}{2}m)}{\pi^{1/2\,m^2}}$$

$$\cdot \int_{R^{1/2\,m(m-1)}} \frac{\left(\sum_i z_{i1}^2 y_i\right)^{k_1-k_2} \left(\sum_{i_1<i_2} \det\begin{bmatrix} z_{i_1 1} & z_{i_1 2} \\ z_{i_2 1} & z_{i_2 2}\end{bmatrix}^2 y_{i_1} y_{i_2}\right)^{k_2-k_3} \cdots \det Y^{k_n}}{D_1^{k_1-k_2+1} D_2^{k_2-k_3+1} \cdots D_{n-1}^{k_{n-1}-k_n+1}} (dZ). \tag{10.5}$$

REFERENCES

1. Maass, H. (1955). Die Bestimmung der dirichletreihen mit Grossencharakteren zu den Modulformen n-ten Grades. *J. Indian Math. Soc.* **19** 1–23.
2. Bhanu Murti, T. S. (1960). Plancherel's measure for the factor space **SL(n, R)/SO(n, R)**. *Dokl. Akad. Nauk. S.S.S.R.*, **133** 503–506.

Stopping Time in Sequential Samples from Multivariate Exponential Families[1]

R. A. WIJSMAN

UNIVERSITY OF ILLINOIS AT URBANA-CHAMPAIGN, ILLINOIS

A general theorem is presented that proves exponential boundedness of the stopping time N in invariant sequential probability ratio tests if certain conditions are satisfied. The theorem is applied to two examples, the first being a test about the characteristic roots of the covariance matrix in a bivariate normal population with zero mean, the second a test about the norm of the mean vector in a multivariate normal population with identity covariance matrix (sequential χ^2 test).

1. INTRODUCTION

Let $X, X_1, X_2, \ldots$ be iid (independent and identically distributed) random vectors with values in k-dimensional Euclidean space R^k and common distribution P. Later P will be allowed to be any distribution, but at first we shall assume that P is a member of a k-variate exponential family, i.e., it is of the form

$$P(dx) = c(\theta)e^{\theta' x}\mu(dx) \tag{1.1}$$

with μ some sigma-finite measure on R^k and the parameter θ taking values in a certain subset of R^k. For a particular choice of μ and the parameter space this will be called the *model.* Two (usually composite) hypotheses, H_1 and H_2, about θ are being made and it is desired to test sequentially H_1 against H_2. We shall assume that it is possible to reduce the composite hypotheses to simple ones by using the principle of invariance, so that it is possible then to employ a sequential probability ratio test (SPRT). The latter will be called an *invariant* SPRT.

We shall, for short, denote $(X_1, \ldots, X_n)$ by X^n and, similarly, $(x_1, \ldots, x_n)$ by x^n. Let L_n be the log probability ratio of X^n; then the invariant SPRT has

[1] Research supported by the National Science Foundation under grant GP-28154.

the following form, for some choice of constants l_1 and l_2: continue sampling as long as

$$l_1 < L_n < l_2; \tag{1.2}$$

stop sampling at the first positive integer n when (1.2) is violated and accept H_1 or H_2 according as $L_n \leq l_1$ or $\geq l_2$. Let N be the random sample size, i.e., the smallest positive integer n when (1.2) is violated. For any P, not necessarily in the model, we shall say that N is *exponentially bounded* under P if for any choice of $l_1 < l_2$ there are constants $c > 0$ and $0 < \rho < 1$ such that

$$P(N > n) < c\rho^n, \qquad n = 1, 2, \ldots. \tag{1.3}$$

It is desired to show N exponentially bounded under P for as large a class of P's as possible.

The problem stated above is quite general and is not restricted to the assumption that under the model X has distribution (1.1). The reason for having the title of this paper suggest that restriction is twofold. First, the requirement that the composite hypotheses can be reduced to simple ones by invariance is a strong one and in all known interesting parametric examples of this kind X has distribution (1.1) (in fact, the underlying distribution is usually multivariate normal). Second, in order to apply Theorem 2.2, L_n has to depend on X^n only through $\sum_1^n X_i$. This practically forces X to have distribution (1.1). Thus, although Theorem 2.2 is stated without the explicit assumption that X have distribution (1.1) under the model, there are strong reasons to believe that the theorem is only relevant to exponential models.

Example 1.1. Let $(x, y), (x_1, y_1), (x_2, y_2), \ldots$ be iid, (x, y) being bivariate normal under the model, with mean 0 and covariance matrix Σ. For $j = 1, 2$, the hypothesis H_j specifies the characteristic roots of Σ^{-1} to be σ_j, τ_j, with $\sigma_j \geq \tau_j > 0$ given and such that $(\sigma_1, \tau_1) \neq (\sigma_2, \tau_2)$. This example is treated in [3] as Example 2, Section 5, under the added assumption that $\sigma_j > \phi_j$ for both j. Here the vector X may be taken to have the three components x^2, xy, and y^2. In [3, Eq. (5.5)], an expression is given for L_n in terms of the Bessel function I_0 of imaginary argument.

Example 1.2. Let $Z, Z_1, Z_2, \ldots$ be iid, where under the model Z is k-variate normal with identity covariance matrix and unknown mean vector ζ. For $j = 1, 2$, H_j specifies $\|\zeta\| = \gamma_j$, where $\gamma_1 \neq \gamma_2$ are given and ≥ 0. This example is treated in [3] as Example 3, Section 6, under the added assumption that the γ_j are >0. Here X can be taken to be Z. In [3, Eq. (6.2)], an expression is given for L_n, involving a hypergeometric function ${}_0F_1$.

It was shown in [3] that both examples are special cases of the following problem. Given $Y, Y_1, Y_2, \ldots$ iid with values in R^k. Set

$$U_n = \sum_{i=1}^{n} \|Y_i\|, \qquad V_n = \left\| \sum_{i=1}^{n} Y_i \right\|, \tag{1.4}$$

then after division by a suitable constant, L_n may be written in the form

$$L_n = V_n + aU_n + bn \tag{1.5}$$

(in which a and b are given constants) plus a random variable that is uniformly bounded and may therefore be ignored for the purpose of the investigation of exponential boundedness of N. However, in the derivation of (1.5) it was essential to exclude the possibility $\sigma_j = \tau_j$ for $j = 1$ or 2 in Example 1.1, and $\gamma_j = 0$ for $j = 1$ or 2 in Example 1.2. If we relax these restrictions, then consultation of the asymptotic expansion of I_0 and of ${}_0F_1$, as given in [3], results in the following expression for L_n:

$$L_n = V_n + aU_n + bn + c\log(1 + V_n) \tag{1.6}$$

(again modulo a multiplicative constant and an additive bounded random variable). If for both $j = 1$ and 2, $\sigma_j > \tau_j$ in Example 1.1, or $\gamma_j > 0$ in Example 1.2, then in (1.6) $c = 0$ and (1.6) reduces to (1.5). The added logarithmic term in (1.6) could not be handled by the method of proof in [3].

In Theorem 2.2 we shall present a general proof of (1.3) valid under certain conditions. The theorem is applied to (1.6) as a special case. The proof given here of (1.3) in Examples 1.1 and 1.2 is very different from the proof in [3]. The new proof is both more general, in that it can cope with (1.6) rather than only with (1.5), and simpler conceptually. There is also hope that the method of proof can be generalized to cover hitherto unsolved problems.

2. THE MAIN THEOREM

Before being able to prove the main theorem, Theorem 2.2, we shall need a generalization of the well-known theorem that if $\{Y_n\}$ is a sequence of real-valued random variables, then, as $n \to \infty$, $Y_n \to Y$ a.e. implies $Y_n \to Y$ in probability and in law. The generalization consists of considering $Y_t \to Y$, where t takes values in a more general topological space.

Theorem 2.1. *Let T be a separable topological space such that the neighborhoods system of each $t \in T$ has a countable base. Let $\{Y_t,\ t \in T\}$ be a family of real-valued random variables on a probability space $(\Omega, \mathscr{A}, P)$ such that*

$Y_s(\omega) \to Y_t(\omega)$ as $s \to t$ for every $\omega \in \Omega - N_0$, where $PN_0 = 0$. Then given $\varepsilon > 0$, for each $t \in T$ there is a neighborhood V of t such that $s \in V$ implies

$$P\{|Y_s - Y_t| > \varepsilon\} < \varepsilon, \tag{2.1}$$

and for each $t \in T$ and continuity point y of the distribution function F_{Y_t} of Y_t there is a neighborhood V of t such that $s \in V$ implies

$$|F_{Y_s}(y) - F_{Y_t}(y)| < \varepsilon. \tag{2.2}$$

Proof. The proof that (2.2) follows from (2.1) is the same as in the case of a sequence $\{Y_n\}$, so we shall show only (2.1). Fix $t \in T$ and put $Z_s = |Y_s - Y_t|$. By removing N_0 from Ω, which does not affect (2.1), we may suppose $Z_s \to 0$ everywhere as $s \to t$. Let $V_1, V_2, \ldots$ be a base for the system of neighborhoods of t. Without loss of generality we may suppose $V_1 \supset V_2 \supset \cdots$. Define $A_n = \{\sup_{s \in V_n} Z_s > \varepsilon\}$, then $A_n \in \mathscr{A}$ because the sup over all $s \in V_n$ may be replaced by the sup over a dense countable subset since T is separable and Z_s continuous in s. Furthermore, $A_n \supset A_{n+1}$, $n = 1, 2, \ldots$. Take $\omega \in \Omega$ arbitrarily. Since $Z_s(\omega) \to 0$ as $s \to t$, there exists integer $n(\omega)$ such that $Z_s(\omega) < \varepsilon$ for all $s \in V_{n(\omega)}$. Hence $\omega \notin A_{n(\omega)}$ and therefore $\omega \notin \bigcap A_n$. Since ω was arbitrary, $\bigcap A_n = \varnothing$, so that $PA_n \to 0$ as $n \to \infty$. Take n such that $PA_n < \varepsilon$, then for any $s \in V_n$ we have $P(Z_s > \varepsilon) \leq PA_n < \varepsilon$, which is (2.1) and concludes the proof.

Theorem 2.2 will be stated in terms of a sequence of iid random vectors, which, for economy of notation, will be denoted $X_1, X_2, \ldots$, even though this may not be the sequence with which we started in Section 1. For instance, the X_i of Theorem 2.2 are really the Y_i of (1.4)–(1.6). It is true that in Example 1.2 the Y_i are identical to the X_i, but in Example 1.1 they are different.

Let $X, X_1, X_2, \ldots$ be iid with values in $\mathscr{X} = R^k$ and common distribution P, not necessarily of the form (1.1). We assume throughout that $P(X = 0) < 1$. Define

$$S_n = \sum_{i=1}^{n} X_i \tag{2.3}$$

and for any sequence $x_1, x_2, \ldots$ with values in $\mathscr{X}$ define

$$s_n = \sum_{i=1}^{n} x_i. \tag{2.4}$$

Let f_n be a real-valued Borel measurable function on $\mathscr{X}^n = \mathscr{X}$ crossed with itself n times, and let L_n be the random variable

$$L_n = f_n(X^n) \tag{2.5}$$

(the notation X^n and x^n was introduced in Section 1). With L_n defined in (2.5) let N be the smallest integer n for which (1.2) is violated. Using Stein's

[2] method as adapted by Sethuraman [1], in order to prove (1.3) it suffices to show that there exists $\varepsilon > 0$ and integer $r > 0$ such that for $n = 1, 2, \ldots$

$$P\{|L_{n+j} - L_{n+i}| > d \text{ for some } 0 \leq i < j \leq r \mid X^n\} > \varepsilon, \tag{2.6}$$

in which

$$d = l_2 - l_1 \tag{2.7}$$

is the distance between the stopping bounds in (1.2). If (2.6) is satisfied, then if stopping has not occurred by stage n, it will occur with probability at least ε by stage $n + r$.

Define

$$U = \{u \in R^k : \|u\| = 1\} \tag{2.8}$$

and provide U with the metric topology inherited from R^k. For any $x \in R^k$, $x \neq 0$, define

$$u(x) = x/\|x\| \tag{2.9}$$

so that $u(x) \in U$ and $x = \|x\| u(x)$.

Theorem 2.2. *N is exponentially bounded if the following conditions are satisfied.*

(i) *For any positive integers n, r, and any $x_{n+1}, \ldots, x_{n+r} \in \mathscr{X}$, $f_{n+r}(x^{n+r}) - f_n(x^n)$ depends on n and x^n only through s_n given by* (2.4), *and is a continuous function of s_n;*

(ii) *there exists a function $f: \mathscr{X} \times U \to R$ such that $f(\cdot, u)$ is Borel measurable for every $u \in U$, $f(x, \cdot)$ is continuous on U for every $x \in \mathscr{X}$, and for all positive integers n, r, for every $u \in U$ and for all $x_{n+1}, \ldots, x_{n+r} \in \mathscr{X}$ we have*

$$f_{n+r}(x^{n+r}) - f_n(x^n) \to \sum_{i=1}^{r} f(x_{n+i}, u) \tag{2.10}$$

as $\|s_n\| \to \infty$ and $u(s_n) \to u$;

(iii) *for every $u \in U$, $P\{f(X, u) = 0\} < 1$.*

Proof. We shall demonstrate the validity of (2.6). Let R^+ be the positive real half line, with the usual topology; then $R^+ \times U$, with the product topology, is homeomorphic to $\mathscr{X} - \{0\}$ if we let to $x \in \mathscr{X} - \{0\}$ correspond $(\|x\|, u(x)) \in R^+ \times U$, by (2.9). Extend R^+ to $R_\infty{}^+ = R^+ \cup \{+\infty\}$ with the usual topology. Define $T = R_\infty{}^+ \times U$, with the product topology. Then T can be described as $\mathscr{X} - \{0\}$ with a point at infinity added in each direction $u \in U$. Since both $R_\infty{}^+$ and U are separable and have the property that each point has a countable base for its system of neighborhoods, the same is true of T. Thus, T satisfies the conditions in the hypothesis of Theorem 2.1.

Using the compactness of U, the continuity of f as a function of u, and (iii) of the hypothesis of Theorem 2.2, it is essentially proved in [3, Lemma 5.2] that there exist $\delta_1 > 0$ and $\varepsilon_3 > 0$ such that for every $u \in U$

$$P\{f(X, u) > \delta_1\} > \varepsilon_3 \qquad \text{or} \qquad P\{f(X, u) < -\delta_1\} > \varepsilon_3 . \tag{2.11}$$

Choose an integer $p > d/\delta_1$ (d defined in (2.7)); then (2.11) implies that for every $u \in U$

$$P\left\{\left|\sum_{i=1}^{p} f(X_i, u)\right| > d\right\} > \varepsilon_3{}^p = 2\varepsilon_1 \qquad \text{(say)}. \tag{2.12}$$

According to (i) of the hypothesis, there is a function $g: \mathscr{X}^{p+1} \to R$ such that

$$f_{n+p}(x^{n+p}) - f_n(x^n) = g(s_n, x_{n+1}, \ldots, x_{n+p}) \tag{2.13}$$

where g is continuous in its first argument. Then (ii) of the hypothesis states that for fixed $(x_{n+1}, \ldots, x_{n+p})$ and $u \in U$,

$$g(s_n, x_{n+1}, \ldots, x_{n+p}) \to \sum_{i=1}^{p} f(x_{n+i}, u) \tag{2.14}$$

as $\|s_n\| \to \infty$, $u(s_n) \to u$. If $s_n \neq 0$, we can identify s_n with $t = (\|s_n\|, u(s_n)) \in R^+ \times U$. We shall with slight notational abuse replace s_n by t in (2.14) and also replace x_{n+i} by x_i; then (2.14) reads

$$g(t, x^p) \to \sum_{i=1}^{p} f(x_i, u) \qquad \text{as} \quad t \to (\infty, u). \tag{2.15}$$

Now define a random variable Y_t for each $t \in T$ as follows.

$$\begin{aligned} Y_t &= g(t, X^p) && \text{if} \quad t \in R^+ \times U \\ Y_t &= \sum_{i=1}^{p} f(X_i, u) && \text{if} \quad t = (\infty, u). \end{aligned} \tag{2.16}$$

Then (2.12) can be written

$$P\{|Y_t| > d\} > 2\varepsilon_1, \quad t = (\infty, u) \qquad \text{for every} \quad u \in U. \tag{2.17}$$

From the continuity of $g(\cdot, x^p)$ and from (2.15) it follows that the family $\{Y_t, t \in T\}$ defined in (2.16) satisfies $Y_s \to Y_t$ as $s \to t$, for every $t \in T$. Thus, Theorem 2.1 can be applied, and combining (2.2) (after replacing ε by $\varepsilon_1/2$) with (2.17), we conclude that for every $u \in U$ there exists a neighborhood $V(u)$ of $t = (\infty, u)$ such that

$$P\{|Y_s| > d\} > \varepsilon_1 \qquad \text{if} \quad s \in V(u). \tag{2.18}$$

We may take $V(u)$ of the form $V(u) = V_1(u) \times V_2(u)$, where $V_1(u)$ is a neighborhood of $\infty \in R_\infty{}^+$, i.e., $V_1(u) = \{y: y > B(u)\}$ for some $0 < B(u) < \infty$, and

$V_2(u)$ is an open neighborhood of $u \in U$. The neighborhoods $\{V_2(u),\ u \in U\}$ form an open covering of the compact U, so there is a finite subcovering, say, $V_2(u_1), \ldots, V_2(u_m)$. Let $B = \max(B(u_1), \ldots, B(u_m))$, then if $s = (y, u)$ with $y > B$ and $u \in V_2(u_j)$ for some $1 \le j \le m$, then $s \in V(u_j)$, which is one of the neighborhoods $V(u)$, so that (2.18) implies

$$P\{|Y_s| > d\} > \varepsilon_1 \quad \text{if} \quad s = (y, u) \quad \text{and} \quad B < y < \infty. \tag{2.19}$$

Using the definition (2.16) of Y_s for $s \in R^+ \times U$ and replacing X_i by X_{n+i} (permitted since the X_i are identically distributed) we rewrite (2.19) as

$$P\{|g(s, X_{n+1}, \ldots, X_{n+p})| > d\} > \varepsilon_1 \quad \text{if} \quad s = (y, u) \quad \text{and} \quad y > B. \tag{2.20}$$

Then replacing s by s_n and using (2.13), we obtain from (2.20)

$$P\{|f_{n+p}(x^n, X_{n+1}, \ldots, X_{n+p}) - f_n(x^n)| > d\} > \varepsilon_1 \quad \text{if} \quad \|s_n\| > B. \tag{2.21}$$

It is important to note that (2.21) holds for all $n = 1, 2, \ldots$. Now replacing in (2.21) x^n by X^n and using (2.3) and (2.5) gives

$$P\{|L_{n+p} - L_n| > d | X^n\} > \varepsilon_1 \quad \text{on} \quad \|S_n\| > B, \tag{2.22}$$

valid for $n = 1, 2, \ldots$.

Next we have to cope with the possibility $\|S_n\| \le B$. By assumption, $p(X = 0) < 1$. Therefore, there exists $u \in U$ and δ_2, $\varepsilon_4 > 0$ such that

$$P\{u'X > \delta_2\} > \varepsilon_4. \tag{2.23}$$

Choose an integer $q > 2B/\delta_2$; then (2.23) implies

$$P\left\{u' \sum_{i=1}^{q} X_i > 2B\right\} > \varepsilon_4{}^q = \varepsilon_2 \qquad \text{(say)}. \tag{2.24}$$

Since always $\|S_q\| \ge u' \sum_1^q X_i$, it follows from (2.24) that

$$P\{\|S_q\| > 2B\} > \varepsilon_2 \tag{2.25}$$

and replacing in (2.25) X_i by X_{n+i}, we get

$$P\left\{\left\|\sum_{i=1}^{q} X_{n+i}\right\| > 2B\right\} > \varepsilon_2. \tag{2.26}$$

Furthermore, if $\|S_n\| \le B$ and $\|\sum_1^q X_{n+i}\| > 2B$, then necessarily $\|S_{n+q}\| > B$. Hence by (2.26) we have

$$P\{\|S_{n+q}\| > B | X^n\} > \varepsilon_2 \quad \text{on} \quad \|S_n\| \le B. \tag{2.27}$$

Combining this with (2.22), after replacing in (2.22) n by $n + q$, we have

$$P\{|L_{n+q+p} - L_{n+q}| > d | X^n\} > \varepsilon_1 \varepsilon_2 \quad \text{on} \quad \|S_n\| \le B. \tag{2.28}$$

Putting $\varepsilon_1\varepsilon_2 = \varepsilon > 0$, we can combine (2.22) and (2.28) into a single statement:

$$P\{|L_{n+p} - L_n| > d \text{ or } |L_{n+q+p} - L_{n+q}| > d|\, X^n\} > \varepsilon \qquad \text{everywhere.} \quad (2.29)$$

Then (2.29) implies (2.6) if in (2.6) we take $r = p + q$, and then (i, j) in (2.6) can be chosen $(0, p)$ or (q, r). This concludes the proof of Theorem 2.2.

3. APPLICATION TO EXAMPLES 1.1 AND 1.2

We have to check the validity of conditions (i), (ii), and (iii) in the hypothesis of Theorem 2.2 if L_n is given by (1.6). Observing (1.4) (with Y_i replaced by X_i), (2.3), (2.4), and (2.5), we compute

$$f_{n+r}(x^{n+r}) - f_n(x^n) = \sum_{i=1}^{r} a\|x_{n+i}\| + br + \|s_{n+r}\|$$
$$- \|s_n\| + c\log(1 + \|s_{n+r}\|) - c\log(1 + \|s_n\|). \quad (3.1)$$

Since $s_{n+r} = s_n + \sum_1^r x_{n+i}$ it is seen that for fixed $(x_{n+1}, \ldots, x_{n+r})$ the right-hand side of (3.1) depends on n and x^n only through s_n, and continuously so. Thus, (i) of the hypothesis of Theorem 2.2 is verified. In order to verify (ii) of the hypothesis, define, for $x \in \mathscr{X}$, $u \in U$,

$$f(x, u) = a\|x\| + b + u'x; \quad (3.2)$$

then f is obviously Borel measurable in x and continuous in u. We still have to verify (2.10), that is, consulting (3.1) and (3.2), we have to verify

$$\|s_{n+r}\| - \|s_n\| + c\log(1 + \|s_{n+r}\|) - c\log(1 + \|s_n\|) \to u' \sum_{i=1}^{r} x_{n+i} \quad (3.3)$$

as $\|s_n\| \to \infty$, $u(s_n) \to u$. The right-hand side in (3.3) equals $u'(s_{n+r} - s_n)$ by (2.4). For simplicity of notation write x for s_n, $x + y$ for s_{n+r}. Then (3.3) is implied by

$$\|x + y\| - \|x\| \to u'y, \quad (3.4)$$

$$\log(1 + \|x + y\|) - \log(1 + \|x\| \to 0 \quad (3.5)$$

as $\|x\| \to \infty$, $u(x) \to u$, for any fixed $y \in R^k$. To prove (3.4), it is elementary to verify the double inequality

$$0 \le \|x + y\| - \|x\| - u(x)'y \le \|y\|^2/2\|x\|, \quad (3.6)$$

assuming $x \neq 0$. Then writing $\|x + y\| - \|x\| - u'y = \|x + y\| - \|x\| - u(x)'y + (u(x) - u)'y$ and employing the triangle inequality, we obtain

$$\big|\, \|x + y\| - \|x\| - u'y \,\big| \le (\|y\|^2/2\|x\|) + \|u(x) - u\|\, \|y\| \quad (3.7)$$

and (3.4) is an immediate consequence. The verification of (3.5) is trivial.

Condition (iii) in the hypothesis of Theorem 2.2, in conjunction with (3.2), reads

$$P\{a\|X\| + b + u'X = 0\} < 1 \qquad \text{for every} \quad u \in U. \tag{3.8}$$

Thus, for any P satisfying (3.8) the conclusion of Theorem 2.2 states that N is exponentially bounded in Examples 1.1 and 1.2.

REFERENCES

1. Sethuraman, J. (1970). Stopping time of a rank-order sequential probability ratio test based on Lehmann alternatives—II. *Ann. Math. Statist.* **41** 1322–1333.
2. Stein, C. (1946). A note on cumulative sums. *Ann. Math. Statist.* **17** 498–499.
3. Wijsman, R. A. (1972). Examples of exponentially bounded stopping time of invariant sequential probability ratio tests when the model may be false. *Proc. Sixth Berkeley Symp. Math. Statist. Prob.* **1** 109–128.

PART III

Characteristic Functions and Characterizations

An Isomorphism Method for the Study of I_0^n

ROGER CUPPENS[1]

UNIVERSITÉ PAUL SABATIER, TOULOUSE, FRANCE,
and
THE CATHOLIC UNIVERSITY OF AMERICA, WASHINGTON, D.C.

We denote by I_0^n the set of n-variate probability laws which have no indecomposable factor. Using a new method based on isomorphisms between some semigroups (for the convolution) of probability laws, we give a complete characterization of finite products of n-variate Poisson laws belonging to I_0^n and prove some other results. The simplest of these is the following theorem.

Let f be an n-variate characteristic function admitting the representation $\log f(t) = i(\alpha, t) + \int (e^{i(t,x)} - 1)\mu(dx)$, *where $\alpha \in R^n$ and μ is a bounded measure. If μ is concentrated on an independent (with respect to the rationals) set, then the probability of f belongs to I_0^n. From these results (which are new even in the case $n = 1$), we deduce the existence of n-variate probability laws belonging to I_0^n and having nonanalytic characteristic functions.*

1. INTRODUCTION

The theory of decompositions of probability laws has its origin in the well-known result conjectured by Lévy and proved by Cramér [2] in 1936: any factor of a normal law is a normal law. The most important problem of this theory, stated by Raikov [25] in 1938, is the problem of the characterization of the class (which we denote by I_0^n) of all the n-variate infinitely divisible laws having no indecomposable factor, which is, by a theorem due to Khintchine, identical with the class of all the n-variate probability laws having only infinitely divisible factors. We can precise the meaning of this problem: Lévy [15] has proved that if p is an n-variate infinitely divisible probability law, its characteristic function f admits a unique representation

$$\log f(t) = i(\alpha, t) - Q(t) + \int [\exp(i(t, x)) - 1 - i(t, x)(1 + |x|^2)^{-1}]\mu(dx) \qquad (t \in R^n) \tag{1}$$

[1] The preparation of this paper was supported by National Science Foundation grant GP-22585.

where log means the branch of logarithm defined by continuity from $\log f(0) = 0$, $\alpha \in R^n$, Q is a nonnegative quadratic form, and μ is a nonnegative measure satisfying the two conditions

$$\mu(\{0\}) = 0, \qquad \int |x|^2(1 + |x|^2)^{-1}\mu(dx) < +\infty.$$

The problem of the characterization of $I_0{}^n$ can be stated now: What necessary and sufficient condition on Q and μ implies that the probability of the characteristic function f defined by (1) belongs to $I_0{}^n$ (α has no importance here)?

This problem has not yet been solved, even in the simplest case $n = 1$. Nevertheless, many necessary or sufficient conditions have been given in this case. They can be found in the books by Linnik [18], Lukacs [21], and Ramachandran [26]. In the case $n > 1$, except for the cases of normal and Poisson laws, all the results are quite recent [1, 3–13, 19, 20, 22–24]. They can be classified in three groups, according to the methods which are used for their proofs:

1. results obtained by the projection method, for example, Cramér's theorem [2]: n-variate normal laws belong to $I_0{}^n$ (a law is normal if and only if all its projections are normal). A variant (using margins instead of projections) gives Teicher's theorem [28]: n-variate Poisson laws belong to $I_0{}^n$;
2. results obtained by induction on the dimension n, for example, the multivariate extension of Linnik's theorem [3, 22]: convolutions of an n-variate normal law and an n-variate Poisson law belong to $I_0{}^n$;
3. results obtained by the same method as in the univariate case. For example, we have the following result [8] (for the definitions, see Section 2).

Theorem A. *If $Q = 0$ and if μ is a bounded measure concentrated in a Borel convex set A satisfying $A \cap (2)A = \varnothing$, then the probability of the characteristic function f defined by* (1) *belongs to $I_0{}^n$. This condition is also necessary if $Q = 0$ and if the measure μ is absolutely continuous with respect to the Lebesgue measure of R^n.*

We give here a new method, which can be called the isomorphism method. This method gives many new results, even in the case $n = 1$: for example, we obtain a complete characterization of finite products of Poisson laws belonging to $I_0{}^n$ (this completes some old results by Lévy [16, 17]) and we deduce from Theorem A some extensions of the following result.

Theorem B. *If $Q = 0$ and if the measure μ is bounded and concentrated in a Borel independent set A, then the probability of the characteristic function f defined by* (1) *belongs to $I_0{}^n$.*

This result has a long history: it is due to Raikov [25] when $A \subset R$ is finite and positive, and to Lévy [17] when $A \subset R$ is finite; then it has been proved, with various additional assumptions, by Ostrovskiĭ [23], Čistyakov [1], and the author [9, 10]. Theorem B gives examples of probability laws belonging to I_0^n and having nonanalytic characteristic functions.

Most of the results presented here have been given previously in the literature [9, 12, 13].

2. DEFINITIONS AND NOTATIONS

If A and B are two sets of R^n, we denote by $A + B$ the vectorial sum of these two sets

$$A + B = \{z \in R^n \mid \exists(x, y) \in A \times B: z = x + y\}$$

and define inductively $(j)A$ by

$$(0)A = \{0\}, \qquad (1)A = A, \qquad (j)A = (j-1)A + A \qquad \text{if} \quad j > 1$$

and $(\infty)A$ by

$$(\infty)A = \bigcup_{j=0}^{\infty} (j)A.$$

$(\infty)A$ is the set of linear combinations of elements of A with nonnegative integer coefficients. More generally, we denote by $(Z)A$ (resp. $(Q^+)A$, $(Q)A$) the set of linear combinations of elements of A with integer (resp. nonnegative rational, rational) coefficients. If j is an integer, we denote by jA the set

$$jA = \{z \in R^n \mid \exists x \in A: z = jx\}.$$

Definition 1. *A set $A \subset R^n$ is independent if it is independent with respect to the rationals, that is, if*

$$\sum_{j=1}^{m} \lambda_j a_j = 0, \qquad \lambda_j \in Q, \quad a_j \in A,$$

implies

$$\lambda_1 = \cdots = \lambda_m = 0$$

for any m. More generally, we say that a set $A \subset R^n$ is a generalized independent set of type λ if

$$A = \bigcup_{j=1}^{\infty} \bigcup_{k=p_j}^{q_j} (k)B_j$$

where $B = \bigcup_{j=1}^{\infty} B_j$ is an independent set and p_j and q_j are integers satisfying $0 < p_j \leq q_j < \lambda p_j$ $(j = 1, 2, \ldots)$.

Definition 2. *If $A \subset (\infty)B$ where B is an independent set, we say that B is a basis of A. The sets $A_j \subset (\infty)B_j$ have an independent basis if $\bigcup_{j=1}^{\infty} B_j$ is an independent set.*

Definition 3. *An n-variate measure μ is concentrated in a Borel set A of R^n if*

$$\mu(A \cap B) = \mu(B)$$

for any Borel subset B of R^n.

If A is a Borel set of R^n, we denote by $\mathscr{P}(A)$ (resp. $\mathscr{M}(A)$) the semigroup (for the convolution) of all the n-variate probabilities (resp. measures) concentrated in $(\infty)A$.

Definition 4. *If μ is an n-variate bounded measure, a characteristic function f without zeros admits a Lévy–Khintchine representation with Poisson measure μ if*

$$\log f(t) = i(\alpha, t) + \int [\exp(i(t, x)) - 1]\mu(dx) \qquad (t \in R^n) \tag{2}$$

where $\alpha \in R^n$ and log *means the branch of logarithm defined by continuity from* $\log f(0) = 0$.

Remark 1. If the measure μ is nonnegative, then (2) can be reduced to (1) and the characteristic function f is infinitely divisible, but the measure μ in (2) can be negative on some subsets of R^n.

Remark 2. It is well known that if such a representation exists, this representation is unique.

3. ISOMORPHISM METHOD

Let $A \subset R^n$ and $A' \subset R^{n'}$ be two equipotent independent Borel sets and φ be a bijection of A on A'. We can extend φ to a bijection (denoted again by φ) of $(\infty)A$ on $(\infty)A'$ by the formula

$$\varphi\left(\sum_{j=1}^{m} h_j a_j\right) = \sum_{j=1}^{m} h_j \varphi(a_j)$$

for any $m \in N$, $a_j \in A$, $h_j \in N$.

If $p \in \mathscr{P}(A)$, we can define an n'-variate probability $\Phi(p)$ by

$$\Phi(p)(B') = p[\varphi^{-1}(B' \cap (\infty)A')]$$

for any Borel subset B' of $R^{n'}$. It is clear that $\Phi(p) \in \mathscr{P}(A')$ and Φ is an isomorphism of the semigroup $\mathscr{P}(A)$ on the semigroup $\mathscr{P}(A')$. If $p \in \mathscr{P}(A)$ and if

$$p = q * r,$$

there exists $m \in R^n$ such that $q' = q * \varepsilon_m$ and $r' = r * \varepsilon_{-m}$ belong to $\mathscr{P}(A)$, ε_m being the degenerate probability at the point m ("$\mathscr{P}(A)$ is factor closed"). Since Φ is an isomorphism, the set of the factors of p belonging to $\mathscr{P}(A)$ is transformed by Φ on the set of the factors of $\Phi(p)$ belonging to $\mathscr{P}(A')$. In particular, if p belongs to I_0^n, $\Phi(p)$ belongs to I_0^n, and conversely.

Finally, we make the following remarks.

1. If the characteristic function of a probability p admits a Lévy–Khintchine representation with a Poisson measure μ belonging to $\mathscr{M}(A)$, then p belongs to $\mathscr{P}(A)$ and $\Phi(p)$ has a characteristic function admitting a Lévy–Khintchine representation with Poisson measure $\Phi(\mu)$ defined by

$$\Phi(\mu)(B') = \mu[\varphi^{-1}(B' \cap (\infty)A')]$$

for any Borel subset B' of $R^{n'}$. $\Phi(\mu)$ belongs to $\mathscr{M}(A')$.

2. The same isomorphism can be defined for the set of probabilities concentrated in $(Z)A$, $(Q^+)A$, $(Q)A$.

3. When $n = 1$ and A is finite, this method is equivalent to Lévy's method of "uniformized generating functions" [17], but is easier to handle.

4. For any independent Borel set A of R^n and any real constants a, b ($a < b$), there exists an independent Borel set A' which is equipotent to A and contained in $[a, b]$. This is a direct consequence of the existence of an independent Borel set which is equipotent to R and contained in $[a, b]$ (an example is given by Kahane and Salem [14, p. 20]).

4. APPLICATIONS IN THE GENERAL CASE

We will use the following two results.

Lemma 1. [21, p. 282]. *If f is a univariate characteristic function admitting a Lévy–Khintchine representation with a Poisson measure concentrated in $A \subset [a, b]$, $0 < a < b < +\infty$, then any factor of f admits a Lévy–Khintchine representation with a Poisson measure which is concentrated in $(\infty)A \cap [a, b]$ and nonnegative on $[a, 2a[\cup \{b\}$.*

Lemma 2 [13]. *Let f be an n-variate characteristic function of the kind*

$$f(t_1, \ldots, t_n) = f_1(t_1, \ldots, t_m) f_2(t_{m+1}, \ldots, t_n)$$

where f_j is an entire characteristic function without zeros admitting a Lévy–Khintchine representation with a Poisson measure μ_j concentrated in a bounded convex set T_j ($j = 1, 2$). If a factor of f admits a Lévy–Khintchine representation with Poisson measure v, then v is concentrated in $T_1 \cup T_2$.

The main result is the following

Theorem 1. *Let f be a characteristic function admitting a Lévy–Khintchine* *representation with a Poisson measure concentrated in a generalized independent set A of the type* $\lambda(A = \bigcup_{j=1}^{\infty} \bigcup_{k=p_j}^{q_j} (k)B_j)$. *If g divides f, then g admits a Lévy–Khintchine representation with a Poisson measure ν which is concentrated in A and nonnegative on* $T = \bigcup_{j=1}^{\infty} \bigcup_{k_p=j}^{2p_j-1} (k)B_j$.

Proof. The proof presented here is a modification of the one given in [13]. We choose an independent set $B' \subset R$ which is equipotent to $B = \bigcup_{j=1}^{\infty} B_j$ and such that

$$\varphi(B_j) \subset \left]\frac{1}{p_j}, \frac{2}{2p_j - 1}\right[,$$

φ meaning the bijection of B on B'. Then $\Phi(\mu)$ is concentrated in $\varphi(A) \subset]1, 2\lambda[$. It follows from Lemma 1 and the isomorphism method that g admits a Lévy–Khintchine representation with a Poisson measure ν which is concentrated in $(\infty)B \cap \varphi^{-1}(]1, 2\lambda[)$ and nonnegative in $\varphi^{-1}(]1, 2[)$. Since $\varphi(T) \subset]1, 2[$, ν is nonnegative on T. It remains to prove that ν is concentrated in A, but this follows easily from the following two lemmas.

Lemma 3. *Let f be a characteristic function admitting a Lévy–Khintchine representation with a Poisson measure μ concentrated in* $A = \bigcup_{k=p}^{q} (k)B$ *where B is an independent set and p and q are two positive integers. If g divides f, its Poisson measure is concentrated in A.*

Proof. We choose an independent set $B' \subset]1, 1 + 1/q[$ which is equipotent to B and let φ be a bijection of B on B'. Then $\Phi(\mu)$ is concentrated in $\varphi(A) \subset]p, q + 1[$. It follows then from Lemma 1 that the Poisson measure of g is concentrated in $(\infty)B \cap \varphi^{-1}(]p, q + 1[) = A$.

Lemma 4. *Let f be a characteristic function admitting a Lévy–Khintchine representation with a Poisson measure μ concentrated in a generalized independent set A of type λ. If* $A = \bigcup_{j=1}^{\infty} A_j$ *where* A_j *are independent sets with an independent basis, then*

(a)

$$f(t) = e^{i(\alpha, t)} \prod_{j=1}^{\infty} f_j(t)$$

where $\alpha \in R^n$ *and* f_j *is a characteristic function admitting a Lévy–Khintchine representation with a Poisson measure concentrated in* A_j $(j = 1, 2, \ldots)$;

(b) *if g divides f,*

$$g(t) = e^{i(\beta, t)} \prod_{j=1}^{\infty} g_j(t)$$

where $\beta \in R^n$ *and* g_j *is a characteristic function dividing* f_j $(j = 1, 2, \ldots)$.

Proof. Let B_j be the basis of A_j and m be a positive integer. We consider an independent set $B_m' \subset R^2$ which is equipotent to $B = \bigcup_{j=1}^{\infty} B_j$ and has the following properties (φ_m means the bijection of B on B_m').

$$\varphi_m(B_m) \subset \{(x, y): y = 0, 1 < x < 1 + (1/q_m)\},$$

$$\varphi_m(B_j) \subset \left\{(x, y): x = 0, \frac{1}{p_j} < y < \frac{2}{2p_j - 1}\right\} \qquad \text{if} \quad j \neq m.$$

If f_m' is the characteristic function of $\Phi_m(p)$, p being the probability of f, then

$$f_m'(u, v) = f'_{m,1}(u) f'_{m,2}(v)$$

where $f'_{m,1}$ and $f'_{m,2}$ admit Lévy–Khintchine representations with Poisson measures concentrated in $T_1 = \{(x, y): y = 0, p_m < x < q_m + 1\}$ and $T_2 = \{(x, y): x = 0, 1 < y < 2\lambda\}$, respectively. It follows then that

$$f(t) = e^{i(\alpha, t)} f_m(t) f_m^*(t)$$

where f_m and f_m^* are characteristic functions admitting Lévy–Khintchine representations with Poisson measures concentrated in A_m and $A - A_m$, respectively. Since m is arbitrary, this implies by induction the first part of the lemma. The second part follows from the same idea and Lemma 2.

From this theorem, we deduce immediately the following corollary, which contains Theorem B of Section 1.

Corollary. *If f is a characteristic function admitting a Lévy–Khintchine representation with a Poisson measure concentrated in a generalized independent set of type* 2, *then the probability of f belongs to I_0^n.*

We can extend this corollary by the following result.

Theorem 2. *Let f be a characteristic function admitting a Lévy–Khintchine representation with a Poisson measure concentrated in $A \subset R^n$. If the projection of A on a subspace of R^n is a generalized independent set of type* 2, *then the probability of f belongs to I_0^n.*

For the proof, see [13]. This theorem contains a result by Livsič and Ostrovskiĭ [20]. It implies, as in the work of Livsič and Ostrovskiĭ [20], that I_0^n is dense (with respect to weak convergence) in the set of all infinitely divisible n-variate probabilities.

5. APPLICATIONS TO A FINITE INDEPENDENT SET

Let now $B = \{b_1, \ldots, b_m\}$ be a finite independent set of R^n. We can use the bijection of B on the canonical basis of R^m. All the results for probabilities concentrated in points of R^m with integer or rational coordinates have analogues for B. For example, we have the following extensions of Theorem 1.

Theorem 3. (a) *Let f be an entire characteristic function admitting a Lévy–Khintchine representation with a Poisson measure concentrated in* $(\infty)B$ (resp. $(Z)B$). *If g divides f, then g admits a Lévy–Khintchine representation with a Poisson measure concentrated in* $(\infty)B$ (resp. $(Z)B$).

(b) *Let f be a characteristic function admitting a Lévy–Khintchine representation with a Poisson measure concentrated in $A = \{x \in R^n : x = \sum_{j=1}^{m} k_j b_j, p_j \le k_j \le q_j\}$ where p_j and q_j are integers satisfying $-\infty < p_j \le q_j < +\infty$. If g divides f, then the Poisson measure of g is concentrated in A.*

Proof. If B is the canonical basis of R^m, for some $\beta \in R^m$, $\log g - i(\beta, \cdot)$ is an entire function which is periodic with respect to each variable. The first result follows from the fact that $\log g - i(\beta, \cdot)$ can be expanded in an m-variate Fourier series and the second result follows from the first and from the Plancherel–Pólya theorem.

We deduce also from Theorems 6.3 and 7.3 of [3]:

Theorem 4. *Let f be a characteristic function admitting the representation*

$$\log f(t) = \sum_{j=-\infty}^{\infty} \sum_{\varepsilon} \lambda_{j,\varepsilon}\left[\exp\left(i\left(t, \sum_{k=1}^{m} \varepsilon_k a_{j,k} b_k\right)\right) - 1\right] \qquad (t \in R^n)$$

where the following conditions are satisfied.

(a) $B = \{b_1, \ldots, b_m\}$ *is an independent set;*

(b) $\varepsilon_j = 0$ or 1 *and $\sum_\varepsilon$ indicates the summation on the $2^m - 1$ values of* $\varepsilon = (\varepsilon_1, \ldots, \varepsilon_m) \neq (0, \ldots, 0)$;

(c) *the $a_{j,k}$ are rationals such that $a_{j',k} a_{j,k}^{-1}$ is either negative or an integer greater than* 1 $(j' > j;\ k = 1, \ldots, m)$;

(d) *the $\lambda_{j,\varepsilon}$ are nonnegative constants satisfying*

$$\sum_{j=-\infty}^{+\infty} \sum_{\varepsilon} \lambda_{j,\varepsilon} < +\infty \qquad \text{and} \qquad \lambda_{j,\varepsilon} = O\left[\exp\left(-K \sum_{k=1}^{m} \varepsilon_k a_{j,k}^2\right)\right] \qquad (j \to +\infty)$$

for some $K > 0$. Then the probability of f belongs to I_0^n.

Theorem 5. *Let f be a characteristic function admitting the representation*

$$\log f(t) = \sum_{j=1}^{\infty} \sum_{\varepsilon} \lambda_{j,\varepsilon}\left[\exp\left(i\left(t, \sum_{k=1}^{m} \varepsilon_k a_{j,k} b_k\right)\right) - 1\right] \qquad (t \in R^n)$$

where the following conditions are satisfied.

(a) $B = \{b_1, \ldots, b_m\}$ *is an independent set;*

(b) $\varepsilon_j = 0$ or 1 *and $\sum_\varepsilon$ indicates the summation on the $2^m - 1$ values of* $\varepsilon = (\varepsilon_1, \ldots, \varepsilon_m) \neq (0, \ldots, 0)$

(c) *the $a_{j,k}$ are integers such that $a_{j',k} a_{j,k}^{-1}$ is either negative or an integer greater than* 1 $(j' > j;\ k = 1, \ldots, m)$;

(d) *the* $\lambda_{j,\varepsilon}$ *are nonnegative constants satisfying*

$$\lambda_{j,\varepsilon} = o\left[\exp\left(-2\sum_{k=1}^{m} \varepsilon_k |a_{j,k}| \log|a_{j,k}|\right)\right] \qquad (j \to +\infty).$$

Then the probability of f belongs to I_0^n.

From Theorem A of Section 1, we have also

Theorem 6. *Let f be a characteristic function admitting the representation*

$$\log f(t) = \sum_{(j_1,\ldots,j_m)\in A} \lambda_{j_1,\ldots,j_m}\left[\exp\left(i\left(t, \sum_{k=1}^{m} j_k b_k\right)\right) - 1\right] \qquad (t \in R^n)$$

where the following conditions are satisfied.

(a) $B = \{b_1, \ldots, b_m\}$ *is an independent set;*

(b) *A is a convex set of* R^m *such that* $A \cap (2)A = \varnothing$;

(c) $j_1, \ldots, j_m$ *are rationals;*

(d) *the* $\lambda_{j_1,\ldots,j_m}$ *are nonnegative constants satisfying*

$$\sum_{(j_1,\ldots,j_m)\in A} \lambda_{j_1,\ldots,j_m} < +\infty.$$

Then the probability of f belongs to I_0^n.

6. APPLICATIONS TO AN ENUMERABLE INDEPENDENT SET

In this case we can extend Theorems 5 and 6 by the following results.

Theorem 7. *Let f be a characteristic function admitting the representation*

$$\log f(t) = \sum_{j=-\infty}^{+\infty} \sum_{\varepsilon} \lambda_{j,\varepsilon}\left[\exp\left(i\left(t, \sum_{k=1}^{\infty} \varepsilon_k a_{j,k} b_k\right)\right) - 1\right] \qquad (t \in R^n)$$

where the following conditions are satisfied.

(a) $B = \{b_1, b_2, \ldots\}$ *is an enumerable independent set;*

(b) $\varepsilon_j = 0$ or 1 *and* $\sum_\varepsilon$ *indicates the summation on the sequences* $\varepsilon = (\varepsilon_1, \varepsilon_2, \ldots)$ *where all the* ε_j *are null except for a finite number;*

(c) *the* $a_{j,k}$ *are rationals such that* $a_{j+1,k}\, a_{j,k}^{-1}$ *is an integer greater than one;*

(d) *the* $\lambda_{j,\varepsilon}$ *are nonnegative constants satisfying*

$$\sum_{j=-\infty}^{+\infty} \sum_{\varepsilon} \lambda_{j,\varepsilon} < +\infty$$

and

$$\lambda_{j,\varepsilon} = O\left[\exp\left(-K\sum_{k=1}^{\infty} \varepsilon_k a_{j,k}^2\right)\right] \qquad (j \to +\infty)$$

for some $K > 0$. *Then the probability of f belongs to* I_0^n.

Theorem 8. *Let f be a characteristic function admitting the representation*

$$\log f(t) = \sum_{j=1}^{+\infty} \sum_{\varepsilon} \lambda_{j,\varepsilon}\left[\exp\left(i\left(t, \sum_{k=1}^{\infty} \varepsilon_k a_{j,k} b_k\right)\right) - 1\right] \qquad (t \in R^n)$$

where the following conditions are satisfied.

(a) $B = \{b_1, b_2, \ldots,\}$ *is an independent set;*

(b) $\varepsilon_j = 0$ or 1 *and* $\sum_\varepsilon$ *indicates the summation on the sequences* $\varepsilon = (\varepsilon_1, \varepsilon_2, \ldots)$ *where all the* ε_j *are null except for a finite number;*

(c) *the* $a_{j,k}$ *are integers such that* $a_{j+1,k}\, a_{j,k}^{-1}$ *is an integer greater than one;*

(d) *the* $\lambda_{j,\varepsilon}$ *are nonnegative constants satisfying*

$$\lambda_{j,\varepsilon} = o\left[\exp\left(-2\sum_{k=1}^{\infty} \varepsilon_k |a_{j,k}| \log|a_{j,k}|\right)\right] \qquad (j \to +\infty).$$

Then the probability of f belongs to I_0^n.

For the proofs, see [12]. If in Theorem 8 we put $a_{1,k} = 1$, $a_{2,k} = 2$, $\lambda_{j,\varepsilon} = 0$ for $j \geq 3$, we obtain the following result.

Corollary. *If the Poisson measure of a characteristic function f is concentrated in $A \cup (2)A$ where A is an enumerable independent set, then the probability of f belongs to I_0^n.*

We can also precise Theorem 1 using the following result.

Theorem 9. *Let f be a characteristic function admitting a Lévy–Khintchine representation with a Poisson measure concentrated in $A = \bigcup_{j=1}^{\infty} \bigcup_{k=p_j}^{q_j} kB_j$ where $\bigcup_{j=1}^{\infty} B_j$ is an enumerable independent set and p_j and q_j are nonnegative integers satisfying $0 < p_j \leq q_j < \lambda p_j$ for some $\lambda > 0$. If g divides f, the Poisson measure ν of g is concentrated in A and positive in*

$$\bigcup_{j=1}^{\infty}\left(\bigcup_{k=p_j}^{2p_j-1} kB_j \cup q_j B_j\right).$$

Proof. From Lemma 4, we can consider only a set $A = \bigcup_{k=p}^{q} kB$ where B is an enumerable independent set and p and q are positive integers. If $m \in (k)B - kB$, then $m = \alpha_1 b_1 + \cdots + \alpha_r b_r$ where α_j is a positive integer, $b_j \in B$ $(j = 1, \ldots, r; r \geq 2)$. If we apply Lemma 4 with $A = A_1 \cup \cdots \cup A_r \cup A'$ where $A_j = \bigcup_{k=p}^{q} k\{b_j\}$, $A' = A - \bigcup_{j=1}^{r} A_j$, we find that $\nu(\{m\}) = 0$. This implies that ν is concentrated in $\bigcup_{k=p}^{q} kB$.

If $m \in qB$, then $m = qb$ with $b \in B$. If we apply Lemma 4 and Lemma 1 with $A = C' \cup C''$ where $C' = \bigcup_{k=p}^{q} k\{b\}$ and $C'' = A - C'$, we find that $\nu(\{m\}) \geq 0$, so that the theorem is proved.

Corollary. *If the Poisson measure of a characteristic function f is concentrated in $\bigcup_{j=1}^{\infty} \bigcup_{k=p_j}^{2p_j} kB_j$ where $\bigcup_{j=1}^{\infty} B_j$ is an enumerable independent set, then the probability of f belongs to I_0^n.*

We do not know if these results remain valid for nonenumerable independent sets. Recently, some generalizations of Theorems 7 and 8 have been obtained by Rousseau [27].

7. FINITE PRODUCTS OF POISSON LAWS

If f is the characteristic function of a finite product p of Poisson laws, then f admits the representation

$$\log f(t) = \sum_{j=1}^{r} \lambda_j[\exp(i(t, a_j)) - 1] \qquad (t \in R^n)$$

where $a_j \in R^n$ and $\lambda_j > 0$. Then $a_j \in (Z)B$ where B is an independent set of m $(1 \leq m \leq r)$ points. If B' is the canonical basis of R^m and φ is a bijection of B on B', then the characteristic function f' of $\Phi(p)$ is of the kind

$$\log f'(t_1, \ldots, t_m) = P(e^{it_1}, \ldots, e^{it_m}) - P(1, \ldots, 1) \qquad ((t_1, \ldots, t_m) \in R^m) \quad (3)$$

where

$$P(x_1, \ldots, x_m) = \sum_{k_1=r_1}^{s_1} \cdots \sum_{k_m=r_m}^{s_m} \lambda_{k_1, \ldots, k_m} x_1^{k_1} \cdots x_m^{k_m} \quad (4)$$

and f belongs to I_0^n if and only if f' belongs to I_0^m.

If q divides p, then from Theorem 3 we deduce that the characteristic function g' of $\Phi(q)$ admits the representation

$$\log g'(t_1, \ldots, t_m) = Q(e^{it_1}, \ldots, e^{it_m}) - Q(1, \ldots, 1) \qquad ((t_1, \ldots, t_m) \in R^m) \quad (5)$$

where

$$Q(x_1, \ldots, x_m) = \sum_{k_1=r_1}^{s_1} \cdots \sum_{k_m=r_m}^{s_m} \mu_{k_1, \ldots, k_m} x_1^{k_1} \cdots x_m^{k_m} \quad (6)$$

and from the uniqueness of Lévy–Khintchine representation g' is not infinitely divisible if and only if one of the coefficients μ is negative.

Therefore the problem whether f belongs to I_0^n is now reduced to one of the following equivalent problems.

(a) Is the function g' defined by (5) a characteristic function?

(b) Are all the coefficients $a_{k_1, \ldots, k_m}$ of the development

$$\exp[Q(x_1, \ldots, x_m)] = \sum_{k_1=-\infty}^{+\infty} \cdots \sum_{k_m=-\infty}^{+\infty} a_{k_1, \ldots, k_m} x_1^{k_1}, \ldots, x_m^{k_m} \quad (7)$$

nonnegative? Here Q is the function defined by (6).

In the case $m = 1$, this problem has been solved by Lévy [17]. We recall now his results. Let Q be the function

$$Q(x) = \sum_{k=p}^{q} \mu_k x^k \qquad (-\infty < p \leq q < +\infty). \quad (8)$$

We can suppose (without loss of generality) that $\mu_0 = 0$. We denote by Q_1 and Q_2 the polynomials defined by

$$Q(x) = Q_1(x) + Q_2(x^{-1})$$

and by $l_1, \ldots, l_\alpha$ (resp. $l_1', \ldots, l'_{\alpha'}$) the degrees of nonnull terms of Q_1 (resp. Q_2). Let q_1 (resp. q_2) be the g.c.d. of $l_1, \ldots, l_\alpha$ (resp. $l_1', \ldots, l'_{\alpha'}$) and d_m be the g.c.d. of degrees of nonnull terms of Q satisfying $jm^{-1} > 1$ $(p < m < q)$. We introduce now the following conditions (we conserve Lévy's notation).

Condition A. If μ_m is negative, then d_m divides m.

Condition C. If, for a positive (resp. negative) m, μ_m is negative, then m belongs to $(\infty)L$ (resp. $(\infty)L'$) where $L = \{q_2, l_1, \ldots, l_\alpha\}$ (resp. $L' = \{q_1, l_1', \ldots, l'_{\alpha'}\}$).

Then Lévy's main results are the following.

1. If Q is the function defined by (8), in the development

$$\exp(Q(x)) = \sum_{k=-\infty}^{+\infty} a_k x^k$$

all the coefficients a_k are nonnegative if and only if the two following conditions are satisfied.

(a) Q satisfies Conditions A and C;
(b) the absolute values of the negative coefficients μ_k are small enough.

2. Let f be a characteristic function defined by

$$f(t) = \exp[P(e^{it}) - P(1)]$$

where

$$P(x) = \sum_{k=p}^{q} \lambda_k x^k, \qquad \lambda_k \geq 0.$$

Then f does not belong to I_0^1 if and only if it is possible to replace one of the λ_k by a negative constant so that Conditions A and C are still satisfied.

In the case of several variables, these results become

Theorem 10. *Let Q be the function defined by* (6). *In the development* (7) *all the coefficients $a_{k_1, \ldots, k_m}$ are nonnegative if and only if the two following conditions are satisfied.*

(a) *The function Q' defined by*

$$Q'(y) = Q(y^{p_1}, \ldots, y^{p_m})$$

satisfies Conditions A *and* C *for any integers $p_1, \ldots, p_m$;*
(b) *the absolute values of negative coefficients $\lambda_{k_1, \ldots, k_m}$ are small enough.*

Theorem 11. *The characteristic function f' defined by* (3) *does not belong to I_0^m if and only if it is possible to replace one of the $\lambda_{k_1,\ldots,k_m}$ by a negative constant so that the function P' defined by*

$$P'(y) = P(y^{p_1}, \ldots, y^{p_m})$$

satisfies Conditions A *and C for any integers $p_1, \ldots, p_m$.*

For the proof, see [9].

Example. Let f be the characteristic function defined by

$$\log f(t) = \lambda_1 \exp(i(pb_1, t)) + \lambda_2 \exp(i(pb_2, t)) + \lambda_3 \exp(i(2pb_1, t)) + \lambda_4 \exp(i(2pb_2, t)) + \lambda_5 \exp(i((2p-1)b_1 + b_2, t)) \qquad (t \in R^n)$$

where λ_j is positive ($j = 1, \ldots, 5$). If $\{b_1, b_2\}$ is an independent set and p an integer greater than 1, then f does not belong to I_0^n.

Proof. We can replace the coefficient of $\exp(i(p(b_1 + b_2), t))$ by a negative constant.

This example shows that the corollaries of Theorems 1, 8, and 9 are in a certain sense the best possible. For some other examples, see Cuppens [9].

8. α-DECOMPOSITIONS

Definition. *A probability with characteristic function g is an α-factor of a probability with characteristic function f without real zeros if there exist characteristic functions $h_1, \ldots, h_m$ and positive constants $\alpha, \beta_1, \ldots, \beta_m$ such that*

$$f(t) = g^{\alpha}(t) h_1^{\beta_1}(t) \ldots h_m^{\beta_m}(t)$$

for any $t \in R^n$.

Let A be an enumerable set of real numbers. We denote by $\mathscr{A}(A)$ the set of probabilities p which have the two properties:

(a) p is concentrated in $(Z)A$;

(b) the characteristic function of p has no real zeros. If q is an α-factor of $p \in \mathscr{A}(A)$, it follows from a result due to Ostrovskiĭ [24] that there exists a real constant c such that q is concentrated in $(Z)A + \{c\}$. This implies the existence of a probability $q' \in \mathscr{A}(A)$ which is equivalent to q and is an α-factor of p. ($\mathscr{A}(A)$ is closed for α-decompositions.)

Let now A and A' be two enumerable independent sets of real numbers and φ a bijection of A on A'. According to Section 3, φ implies an isomorphism Φ of the set of probabilities concentrated in $(Z)A$ on the set of probabilities concentrated in $(Z)A'$. If p is a probability such that $p \in \mathscr{A}(A)$ and $\Phi(p) \in \mathscr{A}(A')$, since Φ is continuous with respect to weak convergence, it follows easily that Φ maps an α-factor of p on an α-factor of $\Phi(p)$. In particular, p has no indecomposable α-factor if and only if $\Phi(p)$ has no indecomposable α-factor. We have then the following extension of Theorem B.

Theorem 12. *If f is a characteristic function admitting a Lévy–Khintchine representation with a Poisson measure concentrated in an enumerable independent set of real numbers, then the probability of f has no indecomposable α-factor.*

Proof. When A is bounded, this theorem is a particular case of a result due to Ostrovskiĭ [24]. The general case follows from the isomorphism between $\mathscr{A}(A)$ and $\mathscr{A}(A')$, A' being a bounded enumerable independent set of real numbers.

The other results admit similar extensions.

REFERENCES

1. Čistyakov, G. P. (1971). On belonging to the class I_0 of laws with non-analytic characteristic functions (in Russian). *Dokl. Akad. Nauk SSSR* **201** 280–283.
2. Cramér, H. (1936). Über eine Eigenschaft der normalen Verteilungsfunktion. *Math. Z.* **41** 405–414.
3. Cuppens, R. (1967). Décomposition des fonctions caractéristiques des vecteurs aléatoires. *Publ. Inst. Statist. Univ. Paris* **16** 61–153.
4. Cuppens, R. (1968). Sur un théorème de Paul Lévy. *Publ. Inst. Statist. Univ. Paris* **17** No. 3, 1–6.
5. Cuppens, R (1969). On the decomposition of infinitely divisible characteristic func ions of several variables. *Z. Wahrscheinlichkeitstheorie und Verw. Gebiete* **12** 59–72.
6. Cuppens, R. (1969). On the decomposition of infinitely divisible probability laws without normal factor. *Pacific J. Math.* **28** 61–76.
7. Cuppens, R. (1969). On finite products of Poisson-type characteristic functions of several variables. *Ann. Math. Statist.* **40** 434–444.
8. Cuppens, R. (1969). Décomposition des fonctions caractéristiques indéfiniment divisibles de plusieurs variables à spectre de Poisson continu. *Ann. Inst. H. Poincaré Sect. B* **5** 123–133.
9. Cuppens, R. (1970). Application de la théorie des fonctions caractéristiques de plusieurs variables à l'étude de certains problèmes de décomposition des fonctions caractéristiques d'une variable. *Z. Wahrscheinlichkeitstheorie und Verw. Gebiete* **15**, 144–156.
10. Cuppens, R. (1970). Quelques nouveaux résultats en arithmétique des lois de probabilité. *Colloq. Clermont-Ferrand Les Probabilités sur les Structures Algébriques, June* 30–*July* 5, 1969, pp. 97–111. CNRS, Paris.
11. Cuppens, R. (1971). Sur un problème de Shimizu. *J. Multivariate Anal.* **1** 276–287.
12. Cuppens, R. (1972). Isomorphismes et arithmétique des semigroupes de lois de probabilité. *Z. Wahrscheinlichkeitstheorie und Verw. Gebiete* **21** 147–153.
13. Cuppens, R. (1972). Independent sets and factorization of probability laws. *J. Multivariate Anal.* **2**, 239–248.
14. Kahane, J. P. and Salem, R. (1963). *Ensembles Parfaits et Séries Trigonométriques.* Actualités Sci. Indust. **No. 1301**. Hermann, Paris.
15. Lévy, P. (1937). *Théorie de l'Addition des Variables Aléatoires.* Gauthier-Villars, Paris.

16. Lévy, P. (1937). Sur les exponentielles de polynômes et sur l'arithmétique des produits de lois de Poisson. *Ann. Sci. École Norm. Sup.* **54** 231–292.
17. Lévy, P. (1938). L'arithmétique des lois de probabilité et les produits finis de lois de Poisson. Colloque de Genève, 1938. *Actualités Sci. Indust.* **No. 736** 25–59.
18. Linnik, Yu. V. (1960). *Decomposition of Probability Laws* (in Russian). Leningrad.
19. Livsič, L. Z. (1970). A sufficient condition that an infinitely divisible bivariate law has only infinitely divisible components (in Russian). *Teor. Funkčiĭ Funkcional. Anal. Priložen.* **12** 36–59.
20. Livsič, L. Z. and Ostrovskiĭ, I. V. (1971). On multidimensional infinitely divisible laws, all of whose components are infinitely divisible. *Dokl. Akad. Nauk SSSR* **198** 1273; *Soviet Math. Dokl.* **12** 978–979.
21. Lukacs, E. (1970). *Characteristic Functions*, 2nd ed. Griffin, London.
22. Ostrovskiĭ, I. V. (1965). A multidimensional analogue of Yu. V. Linnik's theorem on the decomposition of compositions of Gauss and Poisson laws (in Russian). *Teor. Verojatnost. i. Primenen.* **10** 742–745.
23. Ostrovskiĭ, I. V. (1966). Decomposition of infinitely divisible multivariate laws without Gaussian component (in Russian). *Vestnik Har'kov. Gos. Univ.* **No. 14** 51–72.
24. Ostrovskiĭ, I. V. (1970). On some classes of infinitely divisible laws (in Russian). *Izv. Akad. Nauk SSSR Ser. Mat.* **34** 923–944.
25. Raikov, D. A. (1938). On the decomposition of Gauss and Poisson law (in Russian). *Izv. Akad. Nauk SSSR Ser. Mat.* **1** 91–124.
26. Ramachandran, B. (1967). *Advanced Theory of Characteristic Functions.* Statistical Publ. Soc., Calcutta.
27. Rousseau, B. (1973). Arithmétique des lois de probabilité définies sur en espace de Hilbert. *C. R. Acad. Sci. Paris Sér. A-B* **276** A978–979.
28. Teicher, H. (1954). On the multivariate Poisson distribution. *Skand. Aktuarietidskr.* **37** 1–9.

A Characterization of the Multivariate Geometric Distribution

EUGENE LUKACS[1]

THE CATHOLIC UNIVERSITY OF AMERICA, WASHINGTON, D.C.[2]

1. INTRODUCTION

The multivariate negative binomial distribution was introduced by Bates and Neyman [1], who derived a number of properties of this distribution and discussed interesting applications (see also Neyman [4]). The characteristic function of the p-variate negative binomial distribution is

$$f(t_1, t_2, \ldots, t_p) = \left\{1 + \sum_{j=1}^{p} a_j[1 - \exp(it_j)]\right\}^{-\alpha}$$

where $\alpha > 0$ and $a_j > 0$ $(j = 1, 2, \ldots, p)$. For $\alpha = 1$ one obtains a distribution with characteristic function

$$f(t_1, t_2, \ldots, t_p) = \left\{1 + \sum_{j=1}^{p} a_j[1 - \exp(it_j)]\right\}^{-1} \qquad (a_j > 0) \tag{1.1}$$

which is called the p-variate geometric distribution.

In the present paper we derive a characterization theorem for the p-variate geometric distribution.

2. A REGRESSION PROPERTY

The characterization theorem will be based on a regression property which we discuss in this section.

Let Z be a random variable and $\mathbf{Y} = (Y_1, Y_2, \ldots, Y_p)$ be a p-dimensional random vector, both defined on a probability space $(\Omega, \mathfrak{A}, P)$. Suppose that the expectations $\mathscr{E}(Z)$ and $\mathscr{E}(\mathbf{Y}) = (\mathscr{E}(Y_1), \mathscr{E}(Y_2), \ldots, \mathscr{E}(Y_p))$ exist. The random variable Z is said to have constant regression on the vector $\mathbf{Y}$ if the relation

$$\mathscr{E}(Z \mid \mathbf{Y}) = \beta_0 \tag{2.1}$$

[1] Research supported by National Science Foundation grant NSF-GP-22585.

[2] Present address: Bowling Green State University, Bowling Green, Ohio.

holds almost everywhere. Here β_0 is a real constant. If $\beta_0 = 0$, then we say that Z has zero regression on $\mathbf{Y}$.

We shall need the following lemma.

Lemma 2.1. *Let Z be a random variable and let $\mathbf{Y}$ be a p-dimensional random vector and suppose that $\mathscr{E}(Z)$ and $\mathscr{E}(\mathbf{Y})$ exist. Then Z has constant regression on $\mathbf{Y}$ if, and only if, the relation*

$$\mathscr{E}(Ze^{i\mathbf{t}'\mathbf{Y}}) = \beta_0 \mathscr{E}(e^{i\mathbf{t}'\mathbf{Y}}) \tag{2.2}$$

holds for all real (nonrandom) vectors $\mathbf{t} = (t_1, t_2, \ldots, t_p)$.

Lemma 2.1 is an immediate generalization of a univariate result (see Lukacs and Laha [3, p. 103]) and is proved in exactly the same way. The proof of the lemma is therefore omitted.

Let $\mathbf{X} = (X_1, X_2, \ldots, X_q)$ be a q-dimensional random vector and let $\mathbf{Y} = (Y_1, Y_2, \ldots, Y_p)$ be a p-dimensional random vector such that $\mathscr{E}(\mathbf{X})$ and $\mathscr{E}(\mathbf{Y})$ exist. It is convenient to say that the random vector $\mathbf{X}$ has constant regression on the random vector $\mathbf{Y}$ if all components of $\mathbf{X}$ have constant regression on $\mathbf{Y}$, that is, if the q relations

$$\mathscr{E}(X_j \mid \mathbf{Y}) = \beta_j \qquad (j = 1, 2, \ldots, q) \tag{2.3}$$

are satisfied almost everywhere; here the β_j are some constants. We say that $\mathbf{X}$ has zero regression on $\mathbf{Y}$ if

$$\mathscr{E}(X_j \mid \mathbf{Y}) = 0 \qquad (j = 1, 2, \ldots, q) \tag{2.3a}$$

almost everywhere.

3. THE CHARACTERIZATION THEOREM

Let $\mathbf{X}_\alpha = (X_{1\alpha}, X_{2\alpha}, \ldots, X_{p\alpha})$ $(\alpha = 1, 2, \ldots, n)$ be a sample of size n taken from a p-variate population; that is, the $\mathbf{X}_\alpha$ are independently and identically distributed random vectors. In the following we denote components of these vectors by Latin subscripts j, k, etc., while Greek subscripts α, β, γ, ... will be used to distinguish different vectors of the sample. Accordingly, summations over Latin subscripts will run from 1 to p, while summations over Greek subscripts will run from 1 to n. We form the following functions.

$$S_j = \frac{1}{n}\sum_\alpha X_{j\alpha}^2 - \frac{2}{n(n-1)} \sum_{\substack{\alpha,\beta \\ \alpha \neq \beta}} X_{j\alpha} X_{j\beta} - \frac{1}{n}\sum_\alpha X_{j\alpha} \qquad (j = 1, 2, \ldots, n), \tag{3.1}$$

$$T_{jk} = T_{kj} = \frac{1}{n}\sum_\alpha X_{j\alpha} X_{k\alpha} - \frac{2}{n(n-1)} \sum_{\substack{\alpha,\beta \\ \alpha \neq \beta}} X_{j\alpha} X_{k\beta} \qquad (j, k = 1, \ldots, n; \quad j \neq k). \tag{3.2}$$

We note that the n functions S_j and the $\binom{n}{2}$ functions T_{jk} are univariate random variables which depend on the n vectors $\mathbf{X}_\alpha$ of the sample. It is convenient to introduce the $\binom{n+1}{2}$-dimensional random vector

$$\mathbf{Z} = (S_1, \ldots, S_n, T_{12}, \ldots, T_{1n}, T_{23}, \ldots, T_{n-1,n}), \tag{3.3}$$

whose components are the random variables S_j and T_{jk}. We also introduce the p-dimensional random vector

$$\mathbf{\Lambda} = \left(\sum_\alpha X_{1\alpha}, \sum_\alpha X_{2\alpha}, \ldots, \sum_\alpha X_{p\alpha}\right). \tag{3.4}$$

We are now able to formulate our result.

Theorem. *Let* $\mathbf{X}_\alpha$ $(\alpha = 1, 2, \ldots, n)$ *be a sample of size* n *taken from a* p*-variate population with distribution function* $F(x_1, x_2, \ldots, x_p)$ *and characteristic function* $f(t_1, t_2, \ldots, t_p)$. *Suppose that the components of* $\mathbf{X}_\alpha$ *are nonnegative random variables and that the population distribution function is nondegenerate and that its second moments exist. Let* $\mathbf{Z}$ *and* $\mathbf{\Lambda}$ *be the random vectors defined by* (3.3) *and* (3.4). *The population distribution function is a* p*-variate geometric distribution if, and only if,* $\mathbf{Z}$ *has zero regression on* $\mathbf{\Lambda}$.

4. DERIVATION OF THE DIFFERENTIAL EQUATIONS

We first show that the condition of the theorem is sufficient and assume that $\mathbf{Z}$ has zero regression on $\mathbf{\Lambda}$. It follows then from our definition that the $\binom{n+1}{2}$ relations

$$\mathscr{E}(S_j \mid \mathbf{\Lambda}) = 0 \qquad (j = 1, 2, \ldots, p),$$
$$\mathscr{E}(T_{jk} \mid \mathbf{\Lambda}) = 0 \qquad (j, k = 1, 2, \ldots, p; \quad j \neq k)$$

hold almost everywhere. We apply Lemma 2.1 and see that

$$\mathscr{E}(S_j e^{i\mathbf{t}'\mathbf{\Lambda}}) = 0, \qquad \mathscr{E}(T_{jk} e^{i\mathbf{t}'\mathbf{\Lambda}}) = 0 \qquad (j, k = 1, \ldots, p; \quad j \neq k). \tag{4.1}$$

We rewrite these equations, using (3.1) and (3.2), and get

$$\begin{aligned} &\frac{1}{n} \sum_\alpha \mathscr{E}(X_{j\alpha}^2 e^{i\mathbf{t}'\mathbf{\Lambda}}) - \frac{2}{n(n-1)} \sum_{\substack{\alpha, \beta \\ \alpha \neq \beta}} \mathscr{E}(X_{j\alpha} X_{j\beta} e^{i\mathbf{t}'\mathbf{\Lambda}}) - \frac{1}{n} \sum_\alpha \mathscr{E}(X_{j\alpha} e^{i\mathbf{t}'\mathbf{\Lambda}}) = 0, \\ &\frac{1}{n} \sum_\alpha \mathscr{E}(X_{j\alpha} X_{k\alpha} e^{i\mathbf{t}'\mathbf{\Lambda}}) - \frac{2}{n(n-1)} \sum_{\substack{\alpha, \beta \\ \alpha \neq \beta}} \mathscr{E}(X_{j\alpha} X_{k\beta} e^{i\mathbf{t}'\mathbf{\Lambda}}) = 0. \end{aligned} \tag{4.2}$$

Let

$$f = f(t_1, \ldots, t_p) = \mathscr{E}(e^{i\mathbf{t}'\mathbf{X}}) = \mathscr{E}\left[\exp\left(\sum_{j=1}^{p} it_j \mathbf{X}_j\right)\right]$$

be the common characteristic function of the n random vectors $\mathbf{X}_\alpha$; then

$$\frac{\partial f}{\partial t_j} = i\mathscr{E}(X_j e^{i\mathbf{t}'\mathbf{X}}), \qquad \frac{\partial^2 f}{\partial t_j^2} = -\mathscr{E}(X_j^2 e^{i\mathbf{t}'\mathbf{X}}), \qquad \frac{\partial^2 f}{\partial t_j \, \partial t_k} = -\mathscr{E}(X_j X_k e^{i\mathbf{t}'\mathbf{X}}).$$

Since the vectors $\mathbf{X}_\alpha$ are identically distributed with characteristic function f, we see that

$$\begin{aligned} \mathscr{E}(X_{j\alpha} e^{i\mathbf{t}'\mathbf{\Lambda}}) &= -i \frac{\partial f}{\partial t_j} f^{n-1}, \\ \mathscr{E}(X_{j\alpha}^2 e^{i\mathbf{t}'\mathbf{\Lambda}}) &= -\frac{\partial^2 f}{\partial t_j^2} f^{n-1}, \\ \mathscr{E}(X_{j\alpha} X_{j\beta} e^{i\mathbf{t}'\mathbf{\Lambda}}) &= -\left(\frac{\partial f}{\partial t_j}\right)^2 f^{n-2}, \end{aligned} \tag{4.3a}$$

and similarly

$$\begin{aligned} \mathscr{E}(X_{j\alpha} X_{k\beta} e^{i\mathbf{t}'\mathbf{\Lambda}}) &= -\frac{\partial f}{\partial t_j}\frac{\partial f}{\partial t_k} f^{n-2}, \\ \mathscr{E}(X_{j\alpha} X_{k\alpha} e^{i\mathbf{t}'\mathbf{\Lambda}}) &= -\frac{\partial^2 f}{\partial t_j \partial t_k} f^{n-1}. \end{aligned} \tag{4.3b}$$

We substitute (4.3a) into the first equation of (4.2) and (4.3b) into the second equation of (4.2) and obtain the following differential equations for the characteristic function f.

$$f^{n-1}\frac{\partial^2 f}{\partial t_j^2} - 2f^{n-2}\left(\frac{\partial f}{\partial t_j}\right)^2 - if^{n-1}\frac{\partial f}{\partial t_j} = 0 \qquad (j = 1, \ldots, p),$$

$$f^{n-1}\frac{\partial^2 f}{\partial t_j \, \partial t_k} - 2f^{n-2}\frac{\partial f}{\partial t_j}\frac{\partial f}{\partial t_k} = 0 \qquad (j, k = 1, \ldots, p; \quad j \neq k).$$

There exists a neighborhood N of the origin such that $f(\mathbf{t}) \neq 0$ if $\mathbf{t} \in N$. We restrict ourselves from now on to this neighborhood. It is then possible to divide the preceding equations by f^n, and one obtains

$$\frac{1}{f}\frac{\partial^2 f}{\partial t_j^2} - 2\left(\frac{1}{f}\frac{\partial f}{\partial t_j}\right)^2 - i\frac{1}{f}\frac{\partial f}{\partial t_j} = 0 \qquad (j = 1, \ldots, p), \tag{4.4a}$$

$$\frac{1}{f}\frac{\partial^2 f}{\partial t_j \, \partial t_k} - 2\left(\frac{1}{f}\frac{\partial f}{\partial t_j}\right)\left(\frac{1}{f}\frac{\partial f}{\partial t_k}\right) = 0 \qquad (j, k = 1, \ldots, p; \quad j \neq k). \tag{4.4b}$$

We introduce the function

$$\varphi = \varphi(t_1, \ldots, t_p) = \log f(t_1, \ldots, t_p),$$

which is defined for $\mathbf{t} \in N$. It is then easily seen that

$$\frac{1}{f}\frac{\partial f}{\partial t_j} = \frac{\partial \varphi}{\partial t_j} \qquad (j = 1, \ldots, p),$$

$$\frac{1}{f}\frac{\partial^2 f}{\partial t_j^2} = \frac{\partial^2 \varphi}{\partial t_j^2} + \left(\frac{\partial \varphi}{\partial t_j}\right)^2 \qquad (j = 1, \ldots, p),$$

$$\frac{1}{f}\frac{\partial^2 f}{\partial t_j\,\partial t_k} = \frac{\partial^2 \varphi}{\partial t_j\,\partial t_k} + \frac{\partial \varphi}{\partial t_j}\frac{\partial \varphi}{\partial t_k} \qquad (j, k = 1, \ldots, p;\quad j \neq k).$$

Using these equations one obtains from (4.4a) and (4.4b) the differential equations

$$\frac{\partial^2 \varphi}{\partial t_j^2} = i\frac{\partial \varphi}{\partial t_j} + \left(\frac{\partial \varphi}{\partial t_j}\right)^2 \qquad (j = 1, \ldots, p), \tag{4.5}$$

$$\frac{\partial^2 \varphi}{\partial t_j\,\partial t_k} = \frac{\partial \varphi}{\partial t_j}\frac{\partial \varphi}{\partial t_k} \qquad (j, k = 1, \ldots, p;\quad j \neq k). \tag{4.6}$$

These equations are valid in the neighborhood N of the origin.

5. COMPLETION OF THE PROOF OF SUFFICIENCY

We shall deal here with functions which depend either on all p variables $t_1, t_2, \ldots, t_p$ or only on a part of these variables. We use capital Latin letters for functions of these variables and write $A = A(t_1, \ldots, t_p)$ for functions which depend on all p variables. We indicate by subscripts those variables which do not occur among the arguments; for example, we write $A_{12 \ldots k} = A(t_{k+1}, \ldots, t_p)$ for a function which does not depend on the variables $t_1, \ldots, t_k$. It will be convenient to denote the set $(1, 2, \ldots, k-1)$ of subscripts by $(k-1)$, so that $A_{(k-1)k} = A_{(k)} = A_{12 \ldots k} = A(t_{k+1}, \ldots, t_p)$ and similarly $A_{(k-1)l} = A_{1 \ldots k-1\, l} = A(t_k, \ldots, t_{l-1}, t_{l+1}, \ldots, t_p)$. The symbol $A_{(0)l}$ will also occur and is defined by $A_{(0)l} = A_l$. Symbols with p subscripts, such as $A_{1 \ldots p} = A_{(p-1)p} = A_{(p)}$, denote constants. The same kind of notation is also used with other capital Latin letters. Superscripts, such as in $A^j_{(k-1)l}$ or $A^j_{(k)l}$, will be used to distinguish different functions which do not depend on the variables indicated by the subscripts.

We need two lemmas.

Lemma 5.1. *Suppose that* φ *satisfies Eq.* (4.5)*; then*

$$\varphi = -\log[-A^j_{(0)j}\exp(it_j) + A^0_{(0)j}] \qquad (j = 1, 2, \ldots, p).$$

It follows easily from Eq. (4.5) that

$$\frac{\partial}{\partial t_j}\log\left[\frac{\dfrac{\partial\varphi}{\partial t_j}+i}{\dfrac{\partial\varphi}{\partial t_j}}\right] = -i,$$

so that

$$\frac{\dfrac{\partial\varphi}{\partial t_j}+i}{\dfrac{\partial\varphi}{\partial t_j}} = B_j\exp(-it_j);$$

hence

$$\frac{\partial\varphi}{\partial t_j} = \frac{i}{B_j\exp(-it_j)-1} = \frac{i\exp(it_j)}{B_j-\exp(it_j)};$$

therefore

$$\frac{\partial\varphi}{\partial t_j} = -\frac{\partial}{\partial t_j}\log[B_j - \exp(it_j)]$$

and

$$\varphi = -\log[B_j - \exp(it_j)] - \log C_j.$$

We write

$$C_j = A^j_{(0)j} \qquad \text{and} \qquad C_j B_j = A^0_{(0)j}$$

and see that

$$\varphi = -\log[-A^j_{(0)j}\exp(it_j) + A^0_{(0)j}] \qquad (j = 1, \ldots, p). \tag{5.1}$$

This is the statement of the lemma.

Lemma 5.2. *Let* k *be a fixed positive integer* $(1 \le k < p)$ *and suppose that* φ *satisfies the differential equations*

$$\frac{\partial^2\varphi}{\partial t_k\,\partial t_l} = \frac{\partial\varphi}{\partial t_k}\frac{\partial\varphi}{\partial t} \tag{5.2}$$

for $l = k+1, k+2, \ldots, p$. *Assume further that*

$$\varphi = -\log\left[-\sum_{j=1}^{k-1} A^j_{(k-1)l}\exp(it_j) - A^l_{(k-1)l}\exp(it_l) + A^0_{(k-1)l}\right] \tag{5.3}$$

for $l = k, k+1, \ldots, p$. *Then*

$$\varphi = -\log\left[-\sum_{j=1}^{k} A^j_{(k)l}\exp(it_j) - A^l_{(k)l}\exp(it_l) + A^0_{(k)l}\right] \tag{5.4}$$

for $l = k+1, \ldots, p$. [*If* $k = 1$ *the sum* $\sum_{j=1}^{k-1}$ *is omitted in* (5.3).]

According to the assumptions of the lemma, (5.3) is valid for $l = k$ and also for l such that $k < l \le p$. Therefore

$$\begin{aligned} &-\sum_{j=1}^{k-1} A^j_{(k)}\exp(it_j) - A^k_{(k)}\exp(it_k) + A^0_{(k)} \\ &\quad = -\sum_{j=1}^{k-1} A^j_{(k-1)l}\exp(it_j) - A^l_{(k-1)l}\exp(it_l) + A^0_{(k-1)l} \qquad (l = 1, 2, \ldots, p). \end{aligned} \tag{5.5}$$

We differentiate (5.5) with respect to t_j and see that

$$-iA^j_{(k)}\exp(it_j) = -iA^j_{(k-1)l}\exp(it_j);$$

therefore, $A^j_{(k)} = A^j_{(k-1)l}$. The left-hand side of this equation is independent of $t_1, \ldots, t_k$; the right-hand side, of $t_1, \ldots, t_{k-1}, t_l$; we can therefore write

$$A^j_{(k)} = A^j_{(k-1)l} = A^j_{(k)l} \qquad (j = 1, 2, \ldots, k-1). \tag{5.6}$$

Equation (5.5) becomes then

$$A^k_{(k)}\exp(it_k) - A^l_{(k-1)l}\exp(it_l) - A^0_{(k)} + A^0_{(k-1)l} = 0. \tag{5.7}$$

We differentiate (5.7) first with respect to t_k, then with respect to t_l, and obtain

$$iA^k_{(k)}\exp(it_k) - \frac{\partial A^l_{(k-1)l}}{\partial t_k}\exp(it_l) + \frac{\partial A^0_{(k-1)l}}{\partial t_k} = 0, \tag{5.8a}$$

$$\frac{\partial A^k_{(k)}}{\partial t_l}\exp(it_k) - iA^l_{(k-1)l}\exp(it_l) - \frac{\partial A^0_{(k)}}{\partial t_l} = 0. \tag{5.8b}$$

We see from (5.8b) that

$$iA^l_{(k-1)l} = \left(\frac{\partial A^k_{(k)}}{\partial t_l}\exp(it_k) - \frac{\partial A^0_{(k)}}{\partial t_l}\right)\exp(-it_l).$$

We differentiate this equation with respect to t_k and see that

$$\frac{\partial A^l_{(k-1)l}}{\partial t_k} = \frac{\partial A^k_{(k)}}{\partial t_l} \exp(it_k - it_l)$$

or

$$\exp(-it_k) \frac{\partial A^l_{(k-1)l}}{\partial t_k} = \exp(-it_l) \frac{\partial A^k_{(k)}}{\partial t_l} = D_{(k)l} .$$

Therefore

$$\frac{\partial A^l_{(k-1)l}}{\partial t_k} = D_{(k)l} \exp(it_k)$$

and

$$A^l_{(k-1)l} = -iD_{(k)l} \exp(it_k) + A^l_{(k)l} . \tag{5.9a}$$

In the same way we obtain

$$A^k_{(k)} = -iD_{(k)l} \exp(it_l) + A^k_{(k)l} . \tag{5.9b}$$

From Eqs. (5.9a), (5.9b), and (5.8b) we conclude that

$$\frac{\partial A^0_{(k)}}{\partial t_l} = -iA^l_{(k)} \exp(it_l).$$

In the same way we see from (5.9a), (5.9b), and (5.8a) that

$$\frac{\partial A^0_{(k-1)l}}{\partial t_k} = -iA^k_{(k)} \exp(it_k),$$

so that

$$A^0_{(k-1)l} = -A^k_{(k)l} \exp(it_k) + A^0_{(k)l} .$$

Suppose that $l > k$; we see then from (5.3) and (5.6) that

$$\varphi = -\log\left[-\sum_{j=1}^{k} A^j_{(k)l} \exp(it_j) - A^l_{(k)l} \exp(it_l) + A^0_{(k)l} + iD_{(k)l} \exp(it_k + it_l)\right]. \tag{5.10}$$

In order to simplify the notation we write

$$M_{(k)l} = -\sum_{j=1}^{k} A^j_{(k)l} \exp(it_j) - A^l_{(k)l} \exp(it_l) + A^0_{(k)l} + iD_{(k)l} \exp(it_k + it_l),$$

so that $\varphi = -\log M_{(k)l}$. A simple computation shows that

$$\frac{\partial^2 \varphi}{\partial t_k\, \partial t_l} = -i \frac{D_{(k)l}}{M_{(k)l}} \exp(it_k + it_l) + \frac{\partial \varphi}{\partial t_k} \frac{\partial \varphi}{\partial t_l} .$$

It follows then from (5.2) that $D_{(k)l} = 0$, so that (5.10) reduces to (5.4) and the lemma is proved.

We proceed to prove the sufficiency of the condition of the theorem. It follows from the assumptions of the theorem that Eqs. (4.5) hold, so that the conditions of Lemma 5.1 are satisfied. Applying Lemma 5.1, one sees that

$$\varphi = -\log[-A^j_{(0)j}\exp(it_j) + A^0_{(0)j}] \qquad (j = 1, \ldots, p).$$

But this equation means that condition (5.3) is satisfied for $k = 1$. Since φ satisfies (4.6), we conclude from Lemma 5.2 that

$$\varphi = -\log[-A^1_{(1)l}\exp(it_1) - A^l_{(1)l}\exp(it_l) - A^0_{(1)l}]$$

for $l = 2, \ldots, p$. But this means that (5.3) is satisfied for $k = 2$. We proceed in this manner until $k = p - 1$; in this case $l - p$ and (5.4) becomes

$$\varphi = -\log\left[-\sum_{j=1}^{p-1} A^j_{(p)}\exp(it_j) - A^p_{(p)}\exp(it_p) + A^0_{(p)}\right]. \tag{5.11}$$

The $A^j_{(p)}$ $(j = 0, 1, \ldots, p)$ are, according to our convention, constants. We use the fact that $\varphi(\mathbf{0}) = 0$ and conclude that $A^0_{(p)} = 1 + \sum_{j=1}^{p} A^j_{(p)}$. Finally we change the notation for the coefficients by writing

$$a_j = A^j_{(p)} \qquad (j = 1, 2, \ldots, p)$$

and see that

$$\varphi(\mathbf{t}) = -\log\left\{1 - \sum_{j=1}^{p} a_j[\exp(it_j) - 1]\right\},$$

so that

$$f(\mathbf{t}) = \left\{1 + \sum_{j=1}^{p} a_j[1 - \exp(it_j)]\right\}^{-1}. \tag{5.12}$$

Formula (5.12) was derived under the restriction that $\mathbf{t} \in N$. It follows from the properties of multivariate analytic characteristic functions (see Cuppens [2]) that it is valid for all $\mathbf{t}$. From the Hermitian property of characteristic functions we conclude that the a_j are real, and from the nonnegativity of the random vectors $\mathbf{X}_\alpha$ that the a_j are positive. The proof of the sufficiency of the condition is therefore completed.

6. PROOF OF THE NECESSITY

To prove the necessity of the condition of the theorem we assume that the characteristic function $f(\mathbf{t})$ of the population has the form (5.12). It is then easily seen that $f(\mathbf{t})$ satisfies the differential equations (4.4a) and (4.4b). We use then (4.3a) and (4.3b) to show that (4.2), and therefore also (4.1), holds. The necessity of the condition of the theorem follows immediately from Lemma 2.1.

REFERENCES

1. Bates, G. E. and Neyman, J. (1952). Contribution to the theory of accident proneness I. *Univ. California Publ. Statist.* **1** 215–253.
2. Cuppens, R. (1967). Décomposition des fonctions caractéristiques des vecteurs aléatoires. *Publ. Inst. Statist. Univ. Paris* **16** 61–153.
3. Lukacs, E. and Laha, R. G. (1964). *Applications of Characteristic Functions.* Griffin, London.
4. Neyman, J. (1965). Certain chance mechanisms involving discrete distributions. *Proc. Symp. Classical and Contagious Distributions, Montreal, 1963*, pp. 4–14. Statist. Publ. Soc., Calcutta.

On Infinitely Decomposable Probability Distributions, and Helical Varieties in Hilbert Space

P. MASANI[1]

DEPARTMENT OF MATHEMATICS
UNIVERSITY OF PITTSBURGH
PITTSBURGH, PENNSYLVANIA

1. Introduction
2. The canonical association of helical varieties with infinitely decomposable distributions
3. A Hilbert space proof of the Lévy-Khinchine Theorem for $\mathbb{R}^q$
4. The operator-measure theoretic treatment
5. On probability and Hilbert spaces
6. Bibliographical and concluding remarks

1. INTRODUCTION

We claim that the theory of infinitely decomposable probability distributions is mirrored by the geometry of helical varieties in Hilbert space, or more accurately by the multi-time global kinematics of such helical motions.[2] In this paper we shall justify this claim to the extent of giving a probability-free, Hilbert space theoretic proof of the Levy–Khinchine formula for such distributions over $\mathbb{R}^q$. The paper will thus clarify, extend, and render explicit relations between infinitely decomposable distributions over $\mathbb{R}$ and helixes discerned earlier by I. J. Schoenberg [20] and M. G. Krein [10, 11].

In §2 we shall state modifications and extensions of results of R. Gangolly [3] and A. N. Kolmogorov [9] on positive definite kernels, and show how they lead us from infinitely decomposable distributions to helical varieties in

[1] Work supported by the National Science Foundation under Grant GP 31497X.

[2] It is hard to describe briefly the relationship involved. Several standard distributions $P(\cdot)$ yield differentiable helical varieties $x(\cdot)$ in Hilbert space, and the time derivatives of $x(\cdot)$, i.e., velocities, accelerations, etc., get related to the variances and moments of $P(\cdot)$. Hence *kinematical* rather than geometric analysis is involved. When the variety $x(\cdot)$ is a hypersurface rather than a curve, the kinematics is with respect to a *multidimensional time*. On the other hand, many distributions $P(\cdot)$ yield nowhere differentiable helical varieties $x(\cdot)$. Hence the kinematics has to involve *global* or *integral concepts* as well as local or differential concepts such as velocity.

Hilbert space. In §3 we shall deduce the Levy–Khinchine Thm. for infinitely decomposable distributions over $\mathbb{R}^q$ from Yaglom's Thm. on helical hypersurfaces [25], thereby providing a probability-free proof. In §4 we shall trace recent advances in helix theory [16], based on operator-measure theoretic considerations, which shed new light on infinitely decomposable distributions over $\mathbb{R}$. We shall allude to the situation for $\mathbb{R}^q$, and conjecture what may prevail for other locally compact abelian groups. In §5 we shall comment on the relationship between the theories of probability and Hilbert spaces suggested by these developments. Finally, in §6 we shall indicate why we deem this Hilbert space theoretic approach to the subject to be a fruitful one, worth pursuing.

2. THE CANONICAL ASSOCIATION OF HELICAL VARIETIES WITH INFINITELY DECOMPOSABLE DISTRIBUTIONS

Let P be a probability measure on the σ-algebra Bl(X) of Borel subsets of a locally compact abelian (l.c.a.) group X, and ϕ be the characteristic function of P. By definition, this means that

$$\forall\, t \in \Lambda, \qquad \phi(t) = \int_X t(x)P(dx) \tag{2.1}$$

where Λ is the character group of X, its members $t(\cdot)$ being "characters" of X, i.e., continuous homomorphisms on X onto the multiplicative group of complex numbers $z \ni |z| = 1$. We know from Bochner's Thm. that ϕ is such a characteristic function, iff it is complex-valued continuous and positive definite (CPD) on the group Λ, and $\phi(0) = 1$.

Readers not interested in l.c.a. groups may assume that[3] $X = \Lambda = \mathbb{R}^q$, and replace $t(x)$ by $e^{i(t,x)}$, where (t, x) is the inner product of $t, x \in \mathbb{R}^q$. This restriction suffices for purposes of multivariate analysis, but we shall not impose it at the outset in view of what has already been accomplished for arbitrary l.c.a. groups X by Parthasarathy *et al.* [19], and for Hilbert spaces X by Varadhan [23]. However, to avoid some of the complexities which accrue when X is an arbitrary l.c.a. group, we shall deal only with X subject to the first of the following assumptions:

2.2 Assumptions. (i) *X is an (additive) divisible, second-countable, Hausdorff, l.c.a. group with simply connected components, which has no nontrivial compact subgroups*$_\lambda$

[3] $\mathbb{R}$ and $\mathbb{C}$ denote the real number and complex number fields, and $\mathbb{N}$ the domain of integers. $\mathbb{R}_{0+}$, $\mathbb{N}_{0+}$ denote the subsets of non-negative elements of $\mathbb{R}$, $\mathbb{N}$.

(ii) Λ *is the (multiplicative) character group of* X;

(iii) P *is a c.a. probability measure on the* σ-*algebra* $\mathrm{Bl}(X)$ *of Borel subsets of* X.

With these assumptions on X, we may define infinite decomposability as follows:

2.3 Def. (a) *We say that* P *is* ***infinitely decomposable***, *iff* $\forall\, n \in \mathbb{N}_+$, $\exists$ *a probability measure* P_n *on* $\mathrm{Bl}(X)$ *such that* $P = P_n * \cdots * P_n$ *(n times),* $*$ *denoting convolution.*

(b) *We say that* $\phi \in \mathrm{CPD}(\Lambda, \mathbb{C})$[4] *is* ***infinitely factorable*** *(or* ***divisible****), iff* $\forall\, n \in \mathbb{N}_+$, $\exists$ *a* $\phi_n \in \mathrm{CPD}(\Lambda, \mathbb{C})$ *such that* $\phi = \phi_n{}^n$.

We know, of course, that P is infinitely decomposable, iff its characteristic function ϕ is infinitely factorable. It therefore suffices to deal exclusively with ϕ. This we shall do from here on. We start with the following result:

2.4 Thm. *Let* X, Λ *be as in* 2.2, $\phi \in \mathrm{CPD}(\Lambda, \mathbb{C})$ *and* $\phi(0) = 1$. *If* ϕ *is infinitely factorable, then* $\exists$ *a unique* $\psi \in C(\Lambda, \mathbb{C}) \ni$

$$\psi(0) = 0, \qquad \psi(-t) = \overline{\psi(t)} \quad \& \quad \phi(t) = e^{\psi(t)}, \qquad \forall\, t \in \Lambda.$$

Proof Remarks. For the classical case $X = \mathbb{R} = \Lambda$, see Chung [1b, pp. 222–224]. The assumption in 2.2(i) that X has no proper compact subgroups ensures that the measure P characterized by ϕ has no nondegenerate idempotent factors, and that ϕ vanishes nowhere on Λ [19, p. 78]. The rest then follows from the other assumptions in 2.2(i). To avoid a digression we defer the detailed proof to another paper.[5]

Following Chung [1b, p. 224] we shall call $\psi(\cdot)$ the *distinguished logarithm of the infinitely factorable* ϕ. In view of the reciprocity between ϕ and ψ affirmed in 2.4, we need only deal with ψ from here on. We accordingly make the following stipulation:

2.5 Hyp. $\psi(\cdot)$ *is the distinguished logarithm of an infinitely factorable function* ϕ *in* $\mathrm{CPD}(\Lambda, \mathbb{C})$ *such that* $\phi(0) = 1$, Λ *being as in* 2.2(ii) and (i).

2.6 Cor. $\forall\, r \in \mathbb{R}_{0+}, \qquad e^{r \cdot \psi(\cdot)} \in \mathrm{CPD}(\Lambda, \mathbb{C})$.

Gist of Proof. This is obvious for $r = n \in \mathbb{N}_{0+}$. But since ϕ is infinitely factorable, it is valid also for $r = 1/n$, and so far all non-negative rational numbers r. By a simple limiting argument we establish its validity for any $r \in \mathbb{R}_{0+}$. ∎

[4] $\mathrm{CPD}(\Lambda, \mathbb{C})$ denotes the class of continuous positive definite functions on Λ to $\mathbb{C}$. $C(\Lambda, Y)$ denotes the class of continuous functions on Λ to Y.

[5] We are grateful to Professors R. Gangolly, E. Hewitt, and W. Rudin for some relevant information on this question.

We now come to a crucial lemma:

2.7 Complexified Gangolly Lma. (cf. [3, Lma. 2.5]). *Let* (i) $\psi \in C(\Lambda, \mathbb{C})$ *be* $\ni$ $\psi(0) = 0$ & $\psi(-t) = \overline{\psi(t)}$,

(ii) $\forall\, s, t \in \Lambda$, $f_\psi(s, t) =_d \frac{1}{2}\{\psi(s-t) - \psi(s) - \psi(-t)\}$.

Then (a) *the following conditions are equivalent:*

$$\forall\, r \in \mathbb{R}_{0+}, \qquad e^{r\cdot\psi(\cdot)} \in \mathrm{CPD}(\Lambda, \mathbb{C}),$$

$$f_\psi(\cdot\cdot) \textit{ is a } \mathrm{CPD} \textit{ kernel on } \Lambda \times \Lambda \textit{ to } \mathbb{C}.^6$$

(b) *If, for* ψ_1, ψ_2 *satisfying* (i) *we have* $f_{\psi_1} = f_{\psi_2}$, *then*

$$\psi_2(t) = \psi_1(t) + i\delta(t), \qquad \forall\, t \in \Lambda,$$

where $\delta(\cdot)$ *is a "real character" of* Λ, *i.e., a continuous homomorphism on* Λ *to* $\mathbb{R}$.

Remarks. 1. The result 2.7 is a complexified adaptation of Gangolly's Lma. 2.5 in [3], wherein Λ is any topological space, and on $\Lambda \times \Lambda$ is given a *real-valued*, continuous symmetric kernel $r(\cdot\cdot)$ vanishing at some point $(0,0) \in \Lambda \times \Lambda$. If when Λ is a l.c.a. group, we let $-r(s, 0) = \psi(s)$, we get our 2.7(a) for the special case of a *real-valued* ψ. In this case 2.7(ii) can be inverted:

$$\psi(s) = \tfrac{1}{2} f_\psi(s, s) \qquad \forall\, s \in \Lambda,$$

and so $\psi(\cdot)$ is uniquely determined by the kernel $f_\psi(\cdot\cdot)$. Our 2.7(b) shows the extent of departure from uniqueness when ψ is complex-valued; to wit, the imaginary part of $\psi(\cdot)$ is undetermined up to a term $i\delta(t)$, where $\delta(t)$ is real-valued and $\delta(s + t) = \delta(s) + \delta(t)$.

2. Cor. 2.7 with $\Lambda = \mathbb{R}$ occurs in a piecemeal and submerged form in Krein's early papers [10, 11]. Cognate ideas occur in Gelfand and Vilenkin's book [4, Thm. 4, p. 279], and in some later literature. However, we owe to Gangolly [3] the clear-cut association of the kernel $f_\psi(\cdot\cdot)$ with $\psi(\cdot)$ for the first time. It is this association which plays a crucial role in our work, cf. footnote 9. We shall therefore call $f_\psi(\cdot\cdot)$ the *Gangolly kernel* of $\psi(\cdot)$, despite the designation "Lévy–Schoenberg kernel" used by Gangolly.

What about the Gangolly kernel $f_\psi(\cdot\cdot)$ of the distinguished logarithm ψ appearing in Hyp. 2.5? From 2.6 and 2.7 we see at once that

$$f_\psi(\cdot\cdot) \text{ is a CPD kernel on } \Lambda \times \Lambda \text{ to } \mathbb{C}. \tag{2.8}$$

[6] I.e. $\forall\, n \in \mathbb{N}_+$, $\forall\, t_1, \ldots, t_n \in \Lambda$ & $\forall\, c_1, \ldots, c_n \in \mathbb{C}$,

$$\sum_{j=1}^{n} \sum_{k=1}^{n} f_\psi(t_j, t_k) c_j \overline{c_k} \geq 0.$$

This suggests a transition to Hilbert spaces via appeal to a theorem of Kolmogorov:

2.9 Generalized Kolmogorov Thm.[7] *Let Λ be a separable Hausdorff space, and $\mathscr{H}$ be an infinite dimensional Hilbert space over $\mathbb{C}$. Then the following conditions are equivalent*:

$$f(\cdot\cdot) \text{ is a CPD kernel on } \Lambda \times \Lambda \text{ to } \mathbb{C},$$

$$\exists\, x(\cdot) \in C(\Lambda, \mathscr{H}) \ni \forall\, s, t \in \Lambda, \qquad (x(s), x(t)) = f(s, t).$$

By 2.2(i),(ii), Λ is separable; hence from (2.8) and 2.9 we conclude that

$$\exists\, x(\cdot) \in C(\Lambda, \mathscr{H}) \ni \forall\, s, t \in \Lambda, \qquad (x(s), x(t)) = f_\psi(s, t). \tag{2.10}$$

Since by 2.7 $f_\psi(0, 0) = 0$, so $x(0) = 0$. Even so the $x(\cdot)$ in (2.10) is not uniquely determined by $f_\psi(\cdot\cdot)$.

2.11 Def. *For any $x(\cdot) \in C(\Lambda, \mathscr{H})$ we define the* ***chordal covariance kernel*** *of $x(\cdot)$ to be the function γ_x on Λ^4 to $\mathbb{C}$ such that $\forall\, a, b, c, d \in \Lambda$,*

$$\gamma_x(a, b, c, d) = (x(b) - x(a), x(d) - x(c)).$$

For the $x(\cdot)$ associated with our distinguished logarithm ψ, we have $x(0) = 0$, and so there is a simple relation between the chordal covariance kernel $\gamma_x(\cdot\cdot\cdot)$ of $x(\cdot)$ and the Gangolly kernel $f_\psi(\cdot\cdot)$ of ψ, viz.

$$\gamma_x(0, b, 0, d) = (x(b), x(d)) = f_\psi(b, d). \tag{2.12}$$

More generally, we find from 2.11 and (2.12) that

$$\gamma_x(a, b, c, d) = f_\psi(b, d) + f_\psi(a, c) - f_\psi(b, c) - f_\psi(a, d). \tag{2.13}$$

Replacing $f_\psi(b, d)$, etc., à la 2.7(ii), we readily get

$$\gamma_x(a, b, c, d) = \psi(b - d) + \psi(a - c) - \psi(b - c) - \psi(a - d). \tag{2.14}$$

This yields at once the crucial property:

$$\forall\, t \in \Lambda, \qquad \gamma_x(a + t, b + t, c + t, d + t) = \gamma_x(a, b, c, d). \tag{2.15}$$

2.16 Def. *A function $x(\cdot) \in C(\Lambda, \mathscr{H})$ for which γ_x, defined in* 2.11, *has the translation-invariance property* (2.15) *is called a* ***helical variety in*** $\mathscr{H}$. *For $\Lambda = \mathbb{R}$, $x(\cdot)$ is called a* ***helix*** (*or* ***screw-curve***)*; for $\Lambda = \mathbb{R}^2$, $x(\cdot)$ is called a* ***helical surface***, *and so on.*

2.17 *Remark.* Helical varieties for which $\Lambda = \mathbb{R}^q$ & $\mathscr{H} = L_2(\Omega, \mathscr{B}, P)$ are of course familiar to probabilists: for $q = 1$ or $q > 1$, they are *random*

[7] In his fundamental paper, Kolmogorov [9] proved this result for $\Lambda = \mathbb{N}_+$ without any topology (or, as we may say, with the discrete topology). But it can be extended. For an elegant proof of the extension 2.9 see the forthcoming paper by G. D. Allen [1a].

processes or random fields with wide sense stationary increments, cf. [2, p. 551]. E.g. the Brownian movement SP is a helix. Indeed Wiener's paper [24] on the Brownian motion is the first in which a helix in an infinite dimensional space appears. In finite dimensional Euclidean spaces helixes have been known to differential geometers for a long time, and are discussed in many elementary textbooks. We should point out, however, that in infinite dimensional spaces many interesting helixes and helical varieties are nowhere differentiable, and hence differential geometry or kinematics has to give way to global geometry or kinematics.

We may sum up our findings up to this point in the following theorem:

2.18 Thm. *To the distinguished logarithm* $\psi(\cdot)$ *in Hyp.* 2.5 *corresponds a helical variety* $x(\cdot)$ *in the Hilbert space* $\mathscr{H}$ *such that* $x(0) = 0$ *and* $\forall\, s, t \in \Lambda$,

$$\gamma_x(0, s, 0, t) = (x(s), x(t)) = f_\psi(s, t),$$

$f_\psi(\cdot\cdot)$ *being the Gangolly kernel of* $\psi(\cdot)$.

Every helical variety in $\mathscr{H}$ is governed by a group of unitary operators. This result was first proved for $\Lambda = \mathbb{R}$ by Schoenberg and von Neumann [22] by an intricate argument involving the everywhere denseness in $\mathbb{R}$ of the rational numbers. We have a simple proof valid for any l.c.a. group Λ along the lines of similar proofs for orthogonally scattered and quasi-isometric measures of stationary type, cf. [13, p. 94; 15, p. 494]. This will appear elsewhere; here we shall only cnunciate the result:

2.19 Generalized Schoenberg–von Neumann Thm. *Let* (i) $x(\cdot) \in C(\Lambda, \mathscr{H})$ *be a helical variety*, (ii) $\mathscr{S}_x =_{\mathrm{d}} \mathfrak{S}\{x(b) - x(a);\ a, b \in \Lambda\}$ *be the closed subspace of* $\mathscr{H}$ *spanned by all chords* $x(b) - x(a)$ *of* $x(\cdot)$. *Then* $\exists$ *a unique strongly continuous group* $(U(t), t \in \Lambda)$ *of unitary operators on* $\mathscr{S}_x$ *onto* $\mathscr{S}_x \ni \forall\, a, b, t \in \Lambda$,

$$x(b + t) - x(a + t) = U(t)\{x(b) - x(a)\}.$$

We call $\mathscr{S}_x$ the *chordal subspace* of $x(\cdot)$, and $(U(t), t \in \Lambda)$ the *shift group* of the helical variety $x(\cdot)$. It provides the tool for the systematic study of such varieties from a global standpoint.

At this point our knowledge of helical varieties over arbitrary l.c.a. groups ceases, and we have to turn to the special case $\Lambda = \mathbb{R}^q$. This suffices of course for purposes of multivariate analysis.

3. A HILBERT SPACE PROOF OF THE LÉVY–KHINCHINE THEOREM FOR $\mathbb{R}^q$

Given a helical hypersurface $x(\cdot) \in C(\mathbb{R}^q, \mathscr{H})$, it is possible to prove the following Thm. 3.1 due to Yaglom on the representation of the chordal covariance kernel γ_x by considering the spectral resolution of its shift group and certain "average-vectors" associated with $x(\cdot)$. But to keep this paper oriented towards infinitely decomposable distributions we shall omit our proof of this result and deduce the Levy–Khinchine formula from it. In §4 we shall present our proof of Thm. 3.1 but only for $q = 1$, deferring the proof for $q > 1$ to another paper.

The following version is easily obtained from Thm. 6 in Yaglom's paper [25] in the special case discussed in Remark 3 [25, p. 289], our $\mu_x(\cdot)$ being related to Yaglom's $F(\cdot)$ by the equation $\mu_x(B) = \int_B \{|\lambda|^2/(1 + |\lambda|^2)\}F(d\lambda)$, $B \in \mathrm{Bl}(\mathbb{R}^q)$.

3.1 Yaglom's Thm. *Let* $x(\cdot) \in C(\mathbb{R}^q, \mathscr{H})$ *be a helical hypersurface. Then* $\exists$ *a unique non-negative definite, symmetric linear operator* A_x *on* $\mathbb{R}^q$ *to* $\mathbb{R}^q$, *and* $\exists$ *a unique bounded, non-negative c.a. measure* μ_x *on* $\mathrm{Bl}(\mathbb{R}^q) \ni \forall\, a, b, c, d \in \mathbb{R}^q$,

$$\gamma_x(a, b, c, d) = (A\{b - a\}, d - c)$$
$$+ \int_{\mathbb{R}^q - 0} \{e^{i(b,\lambda)} - e^{i(a,\lambda)}\}\{e^{-i(d,\lambda)} - e^{-i(c,\lambda)}\} \frac{1 + \lambda^2}{\lambda^2}\, \mu_x(d\lambda).$$

Our deduction of the Levy–Khinchine formula for infinitely factorable characteristic functions on $\mathbb{R}^q$ runs as follows.

Let ψ be as in Hyp. 2.5 where now $\Lambda = \mathbb{R}^q$, and let $x(\cdot) \in C(\mathbb{R}^q, \mathscr{H})$ be a helical hypersurface corresponding to ψ as in Thm. 2.18, so that

$$\gamma_x(0, s, 0, t) = f_\psi(s, t), \qquad \forall\, s, t \in \mathbb{R}^q. \tag{1}$$

Now taking the A_x and μ_x given us by Thm. 3.1, define the new function $\psi_x(\cdot)$ on $\mathbb{R}^q$ by

$$\psi_x(t) \underset{\mathrm{d}}{=} -\tfrac{1}{2}(A_x t, t) + \int_{\mathbb{R}^q - \{0\}} \left\{e^{i(t,\lambda)} - 1 - \frac{i(t,\lambda)}{1 + |\lambda|^2}\right\} \frac{1 + |\lambda|^2}{|\lambda|^2}\, \mu_x(d\lambda). \tag{2}$$

Then obviously

$$\psi_x(\cdot) \in C(\mathbb{R}^q, \mathbb{C}), \qquad \psi_x(0) = 0 \;\; \& \;\; \psi_x(-t) = \overline{\psi_x(t)}. \tag{3}$$

When we compute the Gangolly kernel $f_{\psi_x}(s, t)$ of the function ψ_x defined by (2) we easily get the expression on the RHS of Yaglom's equation in 3.1 with $0, s, 0, t$ instead of a, b, c, d. Thus, by (2) and 3.1

$$f_{\psi_x}(s, t) = \gamma_x(0, s, 0, t), \qquad \forall\, s, t \in \mathbb{R}^q. \tag{4}$$

It follows at once from (3), (4), (1), and Lma. 2.7(b) that

$$\forall\, t \in \mathbb{R}^q, \qquad \psi(t) = i\delta(t) + \psi_x(t), \tag{5}$$

where $\delta(\cdot)$ is a continuous homomorphism on $\mathbb{R}^q$ to $\mathbb{R}$.

But we know that if X, Y are topological vector spaces over $\mathbb{R}$, and $\delta(\cdot)$ is a continuous homomorphism on X to Y, then δ is a continuous linear operator on X to Y, cf. e.g. [6, p. 24]. Thus the $\delta(\cdot)$ in (5) is a continuous linear functional on $\mathbb{R}^q$ to $\mathbb{R}$. As is well known, this means that

$$\exists \text{ a unique } a \in \mathbb{R}^q \ni \forall\, t \in \mathbb{R}^q, \qquad \delta(t) = (a, t). \tag{6}$$

Substituting from (6) and (2) in (5), we get

$$\psi(t) = i(a, t) - \tfrac{1}{2}(A_x t, t) + \int_{\mathbb{R}^q - \{0\}} \left\{ e^{i(t,\lambda)} - 1 - \frac{i(t, \lambda)}{1 + |\lambda|^2} \right\} \frac{1 + |\lambda|^2}{|\lambda|^2}\, \mu_x(d\lambda). \tag{3.2}$$

This is the Levy–Khinchine formula for ψ.

We have thus given a "probability-free" proof of the following:

3.3 Lévy–Khinchine Thm. *Let $\psi(\cdot)$ be the distinguished logarithm of an infinitely factorable characteristic function on $\mathbb{R}^q$. Then $\exists$ a unique non-negative definite, symmetric, linear operator A on $\mathbb{R}^q$ to $\mathbb{R}^q$, and $\exists$ a unique bounded, non-negative, c.a. measure μ on* Bl($\mathbb{R}^q$) *such that* (3.2) *holds $\forall\, t \in \mathbb{R}^q$ with A, μ replacing A_x, μ_x.*

4. OPERATOR-MEASURE THEORETIC TREATMENT

In this section we shall first outline some recent advances in helix theory [16], and then point out their bearing on the theory of infinitely decomposable distributions over $\mathbb{R}$. We will also allude to the case $\mathbb{R}^q$, $q > 1$, and make some conjectures for other l.c.a. groups.

Let

$$\left.\begin{array}{ll} \text{(a)} & x(\cdot) \in C(\mathbb{R}, \mathscr{H}) \text{ be a helix} \\ \text{(b)} & (U_x(t)\colon t \in \mathbb{R}) \text{ be its shift group, cf. 2.19} \\ \text{(c)} & \forall\, a, b \in \mathbb{R} \ni a \le b, \\ & \qquad T_x(a, b] \underset{\mathrm{d}}{=} (1/\sqrt{2})\left\{ U_x(b) - U_x(a) - \int_a^b U_x(t)\, dt \right\}. \end{array}\right\} \tag{4.1}$$

Then [16, 2.11, 2.12] the Bochner integral

$$\alpha_x \underset{\mathrm{d}}{=} \sqrt{2} \int_0^\infty e^{-t}\{x(0) - x(t)\}\, dt \qquad \text{exists} \quad \& \quad \in \mathscr{S}_x. \tag{4.2}$$

We call α_x the *average vector* of the helix $x(\cdot)$. The operator-valued interval measure $T_x(\cdot)$ and the average vector α_x control the helix, as the following theorem shows:

4.3 Thm. [16, 2.19] $\forall\, a, b \in \mathbb{R} \ni a \le b$,

$$x(b) - x(a) = T_x(a, b](\alpha_x).$$

4.4 *Remarks.* Since by 4.3

$$x(t) = x(0) + T_x(0, t](\alpha_x), \qquad t \geq 0,$$
$$x(t) = x(0) - T_x(t, 0](\alpha_x), \qquad t < 0,$$

we might call $T_x(0, t]$ the *propagator* of the helix $x(\cdot)$. For instance for the Brownian motion SP $(x_t(\omega),\ t \in \mathbb{R}_{0+},\ \omega \in \Omega)$ with $x_0(\omega) = 0$, for which $\mathscr{H} = L_2(\Omega, \mathscr{B}, P)$, it follows from 4.3 that

$$x_t(\omega) = \{T_x(0, t]\alpha_x\}(\omega) \qquad \text{a.e. } (P),$$

where α_x, the average vector, is a fixed random variable over $(\Omega, \mathscr{B}, P)$. The entire Brownian movement is governed by the action of the propagator on this random variable.

Now let $E_x(\cdot)$ be the spectral measure of the shift group (4.1)(b), so that

$$\forall\, t \in \mathbb{R} \qquad U_x(t) = \int_{\mathbb{R}} e^{it\lambda} E_x(d\lambda). \tag{4.5}$$

It then easily follows from (4.1)(c) that

$$T_x(a, b] = (1/\sqrt{2}) \int_{\mathbb{R}} \{e^{ib\lambda} - e^{ia\lambda}\} \frac{\lambda + i}{\lambda} E_x(d\lambda), \tag{4.6}$$

the integrand at $\lambda = 0$ being defined (by continuity) to be $-(b - a)/\sqrt{2}$. Consequently by Thm. 4.3

$$x(b) - x(a) = (1/\sqrt{2}) \int_{\mathbb{R}} (e^{ib\lambda} - e^{ia\lambda}) \frac{\lambda + i}{\lambda} E_x(d\lambda)\alpha. \tag{4.7}$$

From this it is utterly trivial to deduce that

$$\gamma_x(a, b, c, d) = \frac{1}{2} \int_{\mathbb{R}} (e^{ib\lambda} - e^{ia\lambda})(e^{-id\lambda} - e^{-ic\lambda}) \frac{\lambda^2 + 1}{\lambda^2} |E_x(d\lambda)\alpha_x|^2. \tag{4.8}$$

This is the Yaglom Thm. 3.1 for $q = 1$.[8] This proof, apart from its simplicity, identifies the measure μ_x and operator A_x appearing in Yaglom's Thm. for $q = 1$; thus

$$\begin{aligned} \mu_x(B) &= |E_x(B)(\alpha_x/\sqrt{2})|^2, \qquad B \in \mathrm{Bl}(\mathbb{R}), \\ (A_x(b - a), (d - c)) &= |E_x\{0\}(\alpha_x/\sqrt{2})|^2 (b - a)(d - c), \end{aligned} \tag{4.9}$$

i.e., A_x is multiplication by the number $|E_x\{0\}(\alpha_x/\sqrt{2})|^2$.

[8] Due in this case actually to Kolmogorov [7]; our formulation involving $\int_{\mathbb{R}}$ and not $\int_{\mathbb{R}-0}$ is due to Doob [2].

These developments based on the concepts of average vector and propagator shed new light on infinitely decomposable distributions over $\mathbb{R}$, cf. [17]. We have shown in [15, 15.9] that $\forall\, s, t \in \mathbb{R}_{0+}$

$$T_x(0, t]^* T_x(0, s] = \tfrac{1}{2}\{R_x(s-t) - R_x(s) - R_x(-t)\}, \tag{4.10}$$[9]

where

$$R_x(t) \underset{\mathrm{d}}{=} U_x(t) - 1 - \int_0^t (t-s) U_x(s)\, ds. \tag{4.11}$$

From (4.2) and (4.10) we see at once that

$$\begin{aligned} \gamma_x(0, s, 0, t) &= (T_x(0, s]\alpha_x, T_x(0, t]\alpha_x) \\ &= \tfrac{1}{2}(\{R_x(s-t) - R_x(s) - R_x(-t)\}(\alpha_x), \alpha_x) \\ &= \tfrac{1}{2}\{\psi_x(s-t) - \psi_x(s) - \psi_x(-t)\} = f_{\psi_x}(s, t), \end{aligned} \tag{4.12}$$

where

$$\psi_x(t) \underset{\mathrm{d}}{=} (R_x(t)\alpha_x, \alpha_x) \tag{4.13}$$

clearly satisfies 2.7(i), and $f_{\psi_x}(\cdot\,\cdot)$ is its Gangolly kernel, cf. 2.7(ii).

Now let the helix $x(\cdot)$ just considered arise as in Thm. 2.18 from the distinguished logarithm ψ of an infinitely factorable characteristic function of $\mathbb{R}$. Then by 2.18 and (4.12)

$$f_\psi(s, t) = f_{\psi_x}(s, t).$$

Hence by Lma. 2.7(b)

$$\psi(t) = i\delta(t) + \psi_x(t),$$

i.e. cf. §3(6),

$$\psi(t) = iat + \psi_x(t), \qquad a \in \mathbb{R}.$$

To sum up, we have proved the following theorem announced in [17]:

4.14 Thm. *Let* (i) $\psi(\cdot)$ *be the distinguished logarithm of an infinitely factorable characteristic function on* $\mathbb{R}$;

(ii) $x(\cdot) \in C(\mathbb{R}, \mathscr{H})$ *be a helix associated with* ψ *as in Thm.* 2.18, α_x *be its average vector, and*

$$\forall\, t \in \mathbb{R}, \qquad R_x(t) \underset{\mathrm{d}}{=} U_x(t) - 1 - \int_0^t (t-s) U_x(s)\, ds,$$

[9] The striking similarity between the equation (4.10), which came from the study of the operator-valued measure $T(\cdot)$, and the complexified Gangolly formula 2.7(ii) is what finally convinced us of the feasibility of a Hilbert space approach to infinitely decomposable distributions via helix theory. Indeed, we may call $T_x(0, \cdot\,]^* T(0, \cdot\,]$ the "Gangolly kernel" of the operator-valued function $R_x(\cdot)$.

where $(U_x(t),\ t \in \mathbb{R})$ *is the shift group of* $x(\cdot)$. *Then* $\exists$ *a unique* $a \in \mathbb{R} \ni \forall\ t \in \mathbb{R}$

$$\psi(t) = iat + (R_x(t)\alpha_x, \alpha_x).$$

We might call the last equation the *Lévy–Khinchine formula for* $\psi(\cdot)$ *in the time-domain*. A routine computation combining (4.11) and (4.5) yields

$$R_x(t) = \int_{\mathbb{R}} \left(e^{it\lambda} - 1 - \frac{it\lambda}{1+\lambda^2}\right) \frac{1+\lambda^2}{\lambda^2} E_x(d\lambda). \tag{4.15}$$

From 4.14 and (4.15) we get at once the usual (spectral domain) Lévy–Khinchine formula for ψ, i.e. (3.2) for $q = 1$. As bonus we also learn that the canonical measure $\mu_x(\cdot)$ appearing therein is actually $|E_x(\cdot)(\alpha_x/\sqrt{2})|^2$. We thus get another Hilbert space proof of the Lévy–Khinchine Thm. 3.3, more direct than the one in §3 which appealed to Yaglom's Thm. and had the drawback of starting with the very expression which had to be derived.

The relationship between infinitely factorable characteristic functions ϕ on $\mathbb{R}$ and helixes $x(\cdot)$ in $\mathcal{H}$ is a reciprocal one. To every $x(\cdot)$ we can canonically attach a ϕ, as the following converse of Thm. 4.14 shows:

4.16 Thm. *Given a helix* $x(\cdot) \in C(\mathbb{R}, \mathcal{H})$, *build the operator-valued function* $R_x(\cdot)$ *from its shift group as in* (4.11), *and taking its average vector* α_x, *define the complex-valued function* $\psi_x(\cdot)$ *as in* (4.13). *Then* $e^{\psi_x(\cdot)}$ *is an infinitely factorable characteristic function on* $\mathbb{R}$.

Proof. Obviously from (4.11) and (4.13)

$$\psi_x(\cdot) \in C(\mathbb{R}, \mathbb{C}), \qquad \psi_x(0) = 0 \ \ \& \ \ \psi_x(-t) = \overline{\psi_x(t)}, \tag{1}$$

and by (4.12)

$$f_{\psi_x}(s, t) = \gamma_x(0, s, 0, t) = (x(s) - x(0), x(t) - x(0)). \tag{2}$$

By (2) and Kolmogorov's Thm. 2.9

$$f_{\psi_x}(\cdot\cdot) \textit{ is a } \text{CPD} \textit{ kernel on } \mathbb{R} \times \mathbb{R} \textit{ to } \mathbb{C}.$$

By (1), (3) and Gangolly's Lma. 2.7(a)

$$\forall\ r \in \mathbb{R}_{0+}, \qquad e^{r \cdot \psi_x(\cdot)} \in \text{CPD}(\mathbb{R}, \mathbb{C}).$$

This means, of course, that $e^{\psi_x(\cdot)}$ is infinitely factorable. ∎

For helical hypersurfaces $x(\cdot) \in C(\mathbb{R}^q, \mathcal{H})$ with $q \geq 2$, there is a corresponding time-domain theory involving the concepts of "average-vector" and "propagator," from which the spectral theory follows trivially. But the complication for $q \geq 2$ that spectral integrals $\int_{\mathbb{R}^q}$ have to be split into $\int_{\mathbb{R}^q - 0}$ and a term corresponding to $\int_{\{0\}}$ makes the time-domain analysis rather difficult. We shall defer the presentation of this analysis to another paper.

For helical varieties over l.c.a. groups the existence of a shift group (cf. Thm. 2.19) leads us to conjecture that average-vectors and propagators will again exist, at least for connected groups of a wide class. But it is far from clear if these can be expressed in a reasonable way in terms of $x(\cdot)$ and the shifts $U_x(t)$ as in (4.2) and (4.1)(c).

5. ON PROBABILITY AND HILBERT SPACES

What does our probability-free proof of the Lévy–Khinchine theorem signify in regard to the relationship between probability and Hilbert spaces?

Generally speaking, Hilbert spaces have entered only marginally into the calculus of probabilities. We encounter them when we are concerned with second order moments to the exclusion of the others. Doob has made this relationship between probability and Hilbert spaces explicit by giving contrasting wide sense and strict sense versions of the same concepts, e.g. stationarity, Markov process, martingale, etc. [2]. The wide sense concepts belong to the theory of Hilbert spaces. For instance, the wide sense version of stochastic independence in uncorrelatedness, which is just orthogonality in Hilbert spaces; the wide sense version of regression is linear regression, which is just orthogonal projection in Hilbert spaces. With the RMS error criterion linear theories usually become wide sense versions of nonlinear ones, and can therefore be handled in the framework of Hilbert spaces, e.g. linear prediction. Only in cases dominated by normality do wide sense and strict sense concepts merge. All this has led to the view that the theories of probability and Hilbert spaces are only marginally related.

But we have now established a theorem governing all infinitely decomposable probability distributions, quite irrespective of possession of moments, by Hilbert space considerations. Further analysis of the situation reveals that the theory of helical varieties in Hilbert spaces in fact completely mirrors that of infinitely decomposable distributions. It is reasonable to expect that the Central Limit Theorem and related limit laws are also amenable to a similar approach via Hilbert spaces. For instance, it is easily seen that to the normal distributions over $\mathbb{R}^q$ correspond linear helical varieties in $\mathscr{H}$: straight lines when $q = 1$, planes when $q = 2$, and so on. To a question of convergence of a sequence of infinitely divisible distributions to a normal distribution corresponds a question of the approach of a sequence of equivalence classes of nonlinear helical varieties to an equivalence class of linear ones.[10]

It thus seems necessary to supplement the current view of the marginal relationship between the two subjects by the recognition that the global geo-

[10] We have to consider equivalence classes, for as stressed after (2.10) the correspondence between the infinitely factorable ϕ and the helical varieties $x(\cdot)$ is not one–one.

metry of certain varieties in Hilbert space, or more accurately *the multi-time global kinematics of certain motions in Hilbert space may include full-fledged, momentwise-unrestricted, replicas of parts of probability theory.*

6. BIBLIOGRAPHICAL AND CONCLUDING REMARKS

The significant developments which culminated in the discovery of the Lévy–Khinchine formula are traced in [5]. Here we shall indicate that the Hilbert space approach to this subject described above also emerged from work on important questions of mathematical analysis and probability.

The very concept of helix came from the Brownian motion, and is exemplified by all processes with stationary increments, cf. [2]. It is these origins of the concept which led Kolmogorov and Krein to their investigations [7, 8, 10, 11]. Schoenberg and von Neumann were led to helixes because of their bearing on the embedding of metrics over $\mathbb{R}$ into Euclidean spaces [20, 21, 22]. This question is itself tied up with fundamental ones concerning positive definiteness, complete monotonicity, etc. The ideas used by the writer again came from research in other areas. The average vector and the propagator appear in the joint paper [18], which is devoted to a direct time-domain derivation of the Wold decomposition of a stationary continuous-time process. The measure $T(\cdot)$ also plays a crucial role in the papers [12, 14] devoted to the representation of isometric semi-groups and cognate representations in scattering theory.

Our investigations suggest new lines of attack on certain problems. We shall mention just one example. Let iH be the infinitesimal generator of the shift group of a helix $x(\cdot)$ in $\mathscr{H}$, and $\mathfrak{D}_H$ be its domain. Then we can show that

$$x(\cdot) \textit{ is differentiable at } 0 \textit{ and therefore throughout } \mathbb{R}, \textit{ iff } \alpha_x \in \mathfrak{D}_H. \quad (1)$$

Now call $w \in \mathscr{H}$ a *wandering vector* of the associated measure $T_x(\cdot)$, iff

$$a < b \leq c < d \Rightarrow T_x(a, b](w) \perp T_x(c, d](w),$$

and let W_T be the set of all such vectors. Then we can show that

$$W_T \cap \mathfrak{D}_H = \{0\}. \quad (2)$$

Now a Brownian motion helix x has orthogonal increments, and so by Thm. 4.3 $\alpha_x \in W_T$. It follows from (1) and (2) that the Brownian motion $x(\cdot)$ is nowhere differentiable on $\mathbb{R}$. This is of course in the topology of $\mathscr{H}$, i.e. $L_2(\Omega, \mathscr{B}, P)$. Can the nowhere differentiability of Brownian paths be deduced from this "wide sense" nowhere differentiability by some powerful lemmas? It would be interesting to find out.

We might add that when the action of our propagator measure $T(\cdot)$ is restricted to a wandering subspace W of $\mathscr{H}$, the resulting W-to-$\mathscr{H}$ operator-valued measure $T_0(\cdot)$ is *quasi-isometric*. The quasi-isometric measures yield precisely the integration theory required for the explicit rendition of many representation theorems and transform theorems of functional analysis [15]. When W is one-dimensional, the quasi-isometric measures reduce to $\mathscr{H}$-valued *orthogonally scattered* ones [13]. These of course are the ones used in stochastic integration [2, p. 426].

All this suggests that the approach followed in this paper is intrinsic, and has both a unifying- and research-potential.

REFERENCES

1a. Allen, G. D. *Proc. Amer. Math. Soc.* to be published.
1b. Chung, K. L. (1968). *A Course in Probability Theory.* Harcourt, New York.
2. Doob, J. L. (1953). *Stochastic Processes.* Wiley, New York.
3. Gangolly, R. (1967). Positive definite kernels on homogeneous spaces and certain stochastic processes related to Lévy's Brownian motion of several parameters. *Ann. Inst. H. Poincare Sect. B* **3** 121–225.
4. Gelfand, I. M. and Vilenkin, N. Ya. (1964). *Generalized Functions*, Vol. IV. Academic Press, New York.
5. Gnedenko, B. V. and Kolmogorov, A. N. (1954). *Limit Distributions for Sums of Independent Random Variables.* Addison-Wesley, Reading, Massachusetts.
6. Hille, E. and Phillips, R. S. (1957). *Functional Analysis and Semi-Groups* (Amer. Math. Soc. Colloq. Publ.), Vol. 31. Amer. Math. Soc., Providence, Rhode Island.
7. Kolmogorov, A. (1940). Kurven im Hilbertschen Raum die gegenuber einer einparametrigen Gruppe von Bewegungen invariant sind. *Dokl. Akad. Nauk SSSR* **26** 6–9.
8. Kolmogorov, A. (1940). Wienersche Spiralen und einige andere interessante Kurven im Hilbertschen Raum. *Dokl. Akad. Nauk SSSR* **26** 115–118.
9. Kolmogorov, A. (1941). Stationary sequences in Hilbert space. *Bull. Math. Univ. Moscow* **2** 1–40. (Engl. transl. by N. Artin.)
10. Krein, M. G. (1944). On the logarithm of an infinitely decomposable Hermite-positive function. *Dokl. Akad. Nauk SSSR* **45** 91–94.
11. Krein, M. G. (1944). On the problem of continuation of helical arcs in Hilbert space. *Dokl. Akad. Nauk SSSR* **45** 139–142.
12. Masani, P. (1962). Isometric flows on Hilbert space. *Bull. Amer. Math. Soc.* **68** 624–632.
13. Masani, P. (1968). Orthogonally scattered measures. *Advances in Math.* **2** 61–117.
14. Masani, P. (1968). On the representation theorem of scattering. *Bull. Amer. Math. Soc.* **74** 618–624.
15. Masani, P. (1970). Quasi-isometric measures and their applications. *Bull. Amer. Math. Soc.* **76** 427–528.
16. Masani, P. (1972). Helixes in Hilbert space, I. *Theor. Probability Appl.* (*USSR*) **17** 3–20. (English Edition (SIAM) **17** 1–19.)
17. Masani, P. (1972). Operator-measure theoretic approach to infinitely divisible probability distributions. *Notices Amer. Math. Soc.* **19** A-371.
18. Masani, P. and Robertson, J. (1962). The time-domain analysis of continuous parameter weakly stationary stochastic processes. *Pacific J. Math.* **12** 1361–1378.

19. Parthasarathy, K. R., Ranga Rao, R. and Varadhan, S. R. S. (1963). Probability distributions on locally compact abelian groups. *Illinois J. Math.* **7** 337–369.
20. Schoenberg, I. J. (1938). Metric spaces and positive definite functions. *Trans. Amer. Math. Soc.* **44** 522–536.
21. Schoenberg, I. J. (1938). Metric spaces and completely monotone function. *Ann. of Math.* (2), **39** 811–841.
22. Schoenberg, I. J. and von Neumann, J. (1941). Fourier integrals and metric geometry. *Trans. Amer. Math. Soc.* **50** 226–251.
23. Varadhan, S. R. S. (1962). Limit theorems for sums of independent random variables with values in a Hilbert space. *Sankhyā Ser. A* **24** 213–238.
24. Wiener, N. (1923). Differential space. *J. Math. and Phys.* **2** 131–174.
25. Yaglom, A. M. (1957). Some classes of random fields in *n*-dimensional space related to stationary random processes. *Theor. Probability Appl.* (*USSR*) English Edition (SIAM) **4** 289–320.

Limit Laws for Sequences of Normed Sums Satisfying Some Stability Conditions

K. URBANIK

INSTITUTE OF MATHEMATICS,
UNIVERSITY OF WROCŁAW,
WROCŁAW, POLAND

In this paper we discuss the limit laws arising from normed sums of independent random variables satisfying some stability conditions. These are, roughly speaking, sequences for which the limit properties of suitably normed sums are similar to those for sequences of identically distributed random variables. The result we obtain is an analogue of the Lévy–Khinchin representation of infinitely divisible laws. The present investigation arose from a study of self-decomposable probability measures.

Throughout this paper we denote by P the set of all probability measures on the real line. With the topology of weak convergence and multiplication defined by the convolution, P becomes a topological semigroup. We denote the convolution of two measures λ and μ by $\lambda * \mu$. Moreover, by δ_a we denote the probability measure concentrated at the point a. Further, for any real number a ($a \neq 0$) and any measure μ from P we denote by $a\mu$ the measure defined by the formula $a\mu(E) = \mu(a^{-1}E)$ for all Borel subsets E of the real line. The characteristic function $\hat{\mu}$ of a measure $\mu \in P$ is defined by the formula

$$\hat{\mu}(t) = \int_{-\infty}^{\infty} e^{itx}\mu(dx).$$

It is easy to check the equation $\widehat{a\mu}(t) = \hat{\mu}(at)$.

By a triangular array we shall understand a system

$$\begin{matrix} X_{11} & & & \\ X_{12}, & X_{22} & & \\ \vdots & \vdots & & \\ X_{1n}, & X_{2n}, & \ldots, & X_{nn} \\ \vdots & \vdots & & \vdots \quad \cdots \end{matrix}$$

of random variables such that $X_{1n}, X_{2n}, \ldots, X_{nn}$ are independent and for every $c > 0$

$$\lim_{n\to\infty} \max_{1\leq k\leq n} P(|x_{kn}| > cn) = 0.$$

Further, we say that a triangular array $\{X_{kn}\}$ is generated by a sequence $\{X_n\}$ of random variables if $X_{kn} = X_k$ $(k = 1, 2, \ldots, n; n = 1, 2, \ldots)$.

A probability measure is said to be a limit distribution of the triangular array $\{X_{kn}\}$ if it is the weak limit of the sequence of probability distributions of normed sums $(1/n)\sum_{k=1}^{n} X_{kn} - a_n$ for suitably chosen constants a_n. It is obvious that the limit distribution of $\{X_{kn}\}$ is defined uniquely up to a shift transformation. Moreover, this limit distribution is infinitely divisible (see [2], p. 309). Further, we call two triangular arrays equivalent if they have the same limit distribution. In particular, $\{X_{kn}\}$ is equivalent to an array generated by a sequence $\{X_n\}$ if and only if for suitably chosen a_n and b_n the sequences of normed sums

$$\frac{1}{n}\sum_{k=1}^{n} X_{kn} - a_n \qquad \text{and} \qquad \frac{1}{n}\sum_{k=1}^{n} X_k - b_n$$

have the same limit distribution.

We define classes S_m $(m = 0, 1, \ldots)$ of sequences $\{X_n\}$ of independent random variables recursively. Let S_0 be the class of all sequences $\{X_n\}$ generating convergent triangular arrays, i.e., the class of all sequences for which the sequence $(1/n)\sum_{k=1}^{n} X_k - a_n$ with suitably chosen constants a_n has a limit distribution. Further, $\{X_n\} \in S_m$ $(m \geq 1)$ whenever $\{X_n\} \in S_0$ and the following stability condition is fulfilled: for every positive number c the triangular array $X_{kn} = X_{[cn]+k}$ $(k = 1, 2, \ldots, n; n = 1, 2, \ldots)$ is equivalent to an array generated by a sequence from S_{m-1}. The square brackets here denote the integral part of the real number. It is clear that the classes S_m form a contracting sequence. Put $S_\infty = \bigcap_{m=1}^{\infty} S_m$. The sequences belonging to S_∞ will be called slowly varying. For instance, each sequence of independent identically distributed random variables generating a convergent triangular array belongs to all classes S_m and, consequently, is slowly varying. We get a less trivial example of slowly varying sequences from results of Koroljuk and Zolotarev [1] (see also [5]), namely, each sequence of independent random variables generating a convergent triangular array and such that there are no more than two different distribution laws among the laws of the random variables $X_1, X_2, \ldots$ is slowly varying.

Let L_m $(m = 0, 1, \ldots, \infty)$ be the set of all possible limit distributions of normed sums $(1/n)\sum_{k=1}^{n} X_k - a_n$ where $\{X_n\} \in S_m$ and a_n are constants. The problem of a description of probability measures from L_0 was solved by Lévy, who obtained an explicit representation of the characteristic function of those measures [2, p. 324]. Another characterization of the set L_0 was given [3]. The aim of the present paper is to give a characterization of all sets L_m ($m = 1, 2, \ldots, \infty$).

It is easy to see that the set L_m is invariant under the shift transformations $\mu \to \mu * \delta_c$ and under the transformations $\mu \to a\mu$. Moreover, it is closed under convolution. We characterize L_m by a decomposability property, and then we characterize the corresponding characteristic functions.

Proposition 1. *A probability measure μ belongs to L_m ($m = 0, 1, \ldots$) if and only if for every number $a \in (0, 1)$ there exists a measure $\mu_a \in L_{m-1}$ such that $\mu = a\mu * \mu_a$. L_{-1} denotes here the set P of all probability measures.*

Proof. We shall prove Proposition 1 by induction with respect to m. Owing to Lévy's results, it is true for $m = 0$ [2, p. 323]. Suppose that $m \geq 1$ and for indices less than m the statement is true. Consider a measure μ from L_m. Suppose that it is a limit distribution of normed sums $(1/n) \sum_{k=1}^{n} X_k - a_n$ where $\{X_n\} \in S_m$. Given $a \in (0, 1)$, we put

$$Y_n = \frac{1}{n} \sum_{k=1}^{[an]} X_k - \frac{[an]}{n} a_{[an]}, \qquad Z_n = \frac{1}{n} \sum_{k=[an]+1}^{n} X_k + \frac{[an]}{n} a_{[an]} - a_n .$$

It is evident that Y_n, Z_n are independent, and that μ and $a\mu$ are limit distributions of $\{Y_n + Z_n\}$ and $\{Y_n\}$, respectively. Further, taking into account that $a\mu$, being infinitely divisible, has a nonvanishing characteristic function, we infer that the sequence $\{Z_n\}$ also has a limit distribution, say, μ_a. Of course, $\mu = a\mu * \mu_a$. By the assumption the triangular array $X_{kn} = X_{[an]+k}$ ($k = 1, 2, \ldots, n$; $n = 1, 2, \ldots$) is equivalent to an array generated by a sequence from S_{m-1}. Since

$$Z_n = \frac{1}{n} \sum_{k=1}^{r_n} X_{kr_n} - c_n ,$$

where $r_n = n - [an]$ and c_n are constants, we infer that its limit distribution μ_a belongs to L_{m-1}, which completes the proof of the necessity of the condition.

Suppose now that μ is a probability distribution satisfying, for every $a \in (0, 1)$, the condition $\mu = a\mu * \mu_a$ where $\mu_a \in L_{m-1}$. We have to prove that $\mu \in L_m$. It is clear that $\mu \in L_0$ and consequently, being infinitely divisible, has a nonvanishing characteristic function. Setting $\nu_1 = \mu$, $\nu_n = n\mu_{(n-1)/n}$ ($n = 2, 3, \ldots$), we have the formula

$$\hat{\nu}_n(t) = \frac{\hat{\mu}_{(nt)}}{\hat{\mu}((n-1)t)}$$

whence the relation

$$\lim_{n\to\infty} \max_{1 \leq k \leq n} \left| \hat{\nu}_k\left(\frac{t}{n}\right) - 1 \right| = 0 \tag{1}$$

follows. Let $\{X_n\}$ be a sequence of independent random variables with probability distributions $\nu_1, \nu_2, \ldots$. By (1) we have the convergence

$$\lim_{n\to\infty} \max_{1\le k\le n} P(|X_k| > cn) = 0$$

for every positive number c. Thus $\{X_n\}$ generates a triangular array. Given a positive number c, we put $X_{kn} = X_{[cn]+k}$ ($k = 1, 2, \ldots, n$; $n = 1, 2, \ldots$). Let λ_n be the probability distribution of the sum $(1/n)\sum_{k=1}^{n} X_{kn}$. Then

$$\hat{\lambda}_n(t) = \frac{\hat{\mu}(n^{-1}([cn] + n)t)}{\hat{\mu}(n^{-1}[cn]t)},$$

which yields the relation

$$\hat{\lambda}_n(t) \to \frac{\hat{\mu}((1 + c)t)}{\hat{\mu}(ct)}.$$

Now taking into account the equation $\hat{\mu}((1 + c)t) = \hat{\mu}(ct)\hat{\mu}_{c/(1+c)}((1 + c)t)$, we infer that the probability measure $(1 + c)\mu_{c/(1+c)}$ is a limit distribution of the array $\{X_{kn}\}$. This probability measure, being an element of L_{m-1}, is, by the induction assumption, a limit distribution of an array generated by a sequence from S_{m-1}. Thus $\{X_{kn}\}$ is equivalent to an array generated by a sequence from S_{m-1} and, consequently, $\{X_n\} \in S_m$. Since μ is the probability distribution of the normed sums $(1/n)\sum_{k=1}^{n} X_k$ ($n = 1, 2, \ldots$), we have $\mu \in L_m$, which completes the proof.

As a direct consequence of Proposition 1 we get the following

Corollary. *A probability measure μ belongs to L_∞ if and only if every number $a \in (0, 1)$ there exists a probability measure $\mu_a \in L_\infty$ such that $\mu = a\mu * \mu_a$.*

Our next aim is to give a representation of the characteristic functions of probability measures from L_m. The following representation formula was established in [3]: a measure μ is self-decomposable, i.e., belongs to L_0, if and only if its characteristic function is of the form

$$\hat{\mu}(t) = \exp\left\{iqt + \int_{-\infty}^{\infty}\left(\int_0^{tu}\frac{\exp(iv) - 1}{v}\,dv - it\arctan u\right)\frac{Q(du)}{\log(1 + u^2)}\right\}, \tag{2}$$

where q is a real constant, Q is a finite Borel measure on the real line, and the integrand is defined as its limiting value $-\frac{1}{4}t^2$ when $u = 0$. Moreover, the measure μ determines the couple q, Q uniquely. We shall call Q the spectral measure for μ. It is evident that the convolution of measures there corresponds to the sum of spectral measures of factors. Further, it is easy to check that if Q is the spectral measure for μ, then

$$Q_a = \int_E \frac{\log(1 + u^2)}{\log(1 + (u/a)^2}\,Q(a^{-1}\,du) \tag{3}$$

is the spectral measure for a.

By L_m^+ ($m = 0, 1, \ldots, \infty$) we shall denote the subset of L_m consisting of probability measures with vanishing constant q in (2) and whose spectral measure is concentrated on the open right half-line $(0, \infty)$. If $\upsilon \in L_m^+$, then, of course, $(-1)\upsilon \in L_m$ and its spectral measure is concentrated on the half-line $(-\infty, 0)$. Let us introduce the operation $\mu \to \mu^+$, which associates with every measure μ from L_0 having the spectral measure Q a probability measure μ^+ from L_0^+ with the spectral measure $Q^+(E) = Q(E \cap (0, \infty))$. It is easy to verify the equation $(a\mu)^+ = a\mu^+ * \delta_b$ for $a > 0$ where

$$b = \int_0^\infty (\arctan u - \arctan au) \frac{Q(du)}{\log(1 + u^2)}.$$

Further, it is clear that $(\mu * \lambda)^+ = \mu^+ * \lambda^+$. Thus the relation $\mu = a\mu * \mu_a$ yields the following one $\mu^+ = a\mu^+ * (\mu_a)^+ * \delta_b$. Hence, by a simple induction, we get the relation $\mu^+ \in L_m$ whenever $\mu * L_m$. Suppose that, $\mu \in L_m$ and $\hat{u}$ is given by (2). Denoting by μ^0 the Gaussian probability measure with the characteristic function

$$\hat{\mu}^0(t) = \exp(iqt - \tfrac{1}{4}Q(\{0\})t^2),$$

we have the equation

$$\mu = \mu^0 * \mu^+ * (-1)((-1)\mu)^+.$$

It is evident that Gaussian measures belong to L_∞. Consequently, formula (4) reduces the investigation of L_m to that of L_m^+. Moreover, we have the following criterion: $\lambda \in L_m^+$ if and only if for every $a \in (0, 1)$, $\lambda = a\lambda * y_a * \delta_c$ where c is a constant and $\lambda_a \in L_{m-1}^+$ ($m = 1, 2, \ldots, \infty$).

With every probability measure λ from L_0^+ we associate a function F_λ defined on the real line by means of the formula

$$F_\lambda(x) = \int_{e^x}^\infty \frac{Q(du)}{\log(1 + u^2)}. \tag{5}$$

Here Q denotes the spectral measure for λ. It is clear that this correspondence is one-to-one and

$$Q(E) = -\int_E \log(1 + x^2)\, dF_\lambda(\log x) \tag{6}$$

for any Borel subset E of $(0, \infty)$. Moreover

$$F_{\lambda * \nu} = F_\lambda + F_\nu \tag{7}$$

and, by (3), for $a > 0$

$$F_{a\lambda}(x) = F_\lambda(x - \log a). \tag{8}$$

First we characterize L_m^+ in terms of the function F_λ.

Proposition 2. *Let $m = 0, 1, \ldots.$ A function F is associated with a probability measure λ from L_m^+, i.e., $F = F_\lambda$, if and only if*

$$F(x) = \int_{e^x}^{\infty} \frac{(\log y - x)^m H(dy)}{\log^{m+1}(1 + y^{2/(m+1)})}, \tag{9}$$

where H is a finite Borel measure on $(0, \infty)$.

Proof. We shall prove Proposition 2 by induction with respect to m. For $m = 0$ it is a direct consequence of formula (5). Suppose now that $m > 0$ and that for indices less than m the statement is true. Let λ be a probability measure from L_m^+. Since it belongs to L_{m-1}^+, we have, by the induction assumption,

$$F_\lambda(x) = \int_{e^x}^{\infty} \frac{(\log y - x)^{m-1} H_0(dy)}{\log^m(1 + y^{2/m})}, \tag{10}$$

where H_0 is a finite Borel measure on $(0, \infty)$. Moreover, for any $a \in (0, 1)$ we have the equation $\lambda = a\lambda * \lambda_a * \delta_c$ where $\lambda_a \in L_{m-1}^+$. Consequently, by the induction assumption,

$$F_{\lambda a}(x) = \int_{e^x}^{\infty} \frac{(\log y - x)^{m-1} G_a(dy)}{\log^m(1 + y^{2/m})}$$

where G_a are finite Borel measures on $(0, \infty)$. On the other hand, by (7), (8), and (10),

$$F_a(x) = \int_{e^x}^{\infty} \frac{(\log y - x)^{m-1}}{\log^m(1 + y^{2/m})} \left(H_0(dy) - \frac{\log^m(1 + y^{2/m})}{\log^m(1 + (y/a)^{2/m})} H_0(a^{-1}\, dy) \right).$$

Hence we get the equation

$$G_a(E) = H_0(E) - \int_E \frac{\log^m(1 + y^{2/m}) H_0(a^{-1}\, dy)}{\log^m(1 + (y/a)^{2/m})}$$

for any Borel subset E of $(0, \infty)$. Consequently,

$$\int_E \frac{H_0(dy)}{\log^m(1 + y^{2/m})} - \int_E \frac{H_0(a^{-1}\, dy)}{\log^m(1 + (y/a)^{2/m})} \geq 0. \tag{11}$$

Put

$$g(x) = \int_{e^x}^{\infty} \frac{H_0(dy)}{\log^m(1 + \lambda^{2/m})}. \tag{12}$$

By (11) for any $a \in (0, 1)$ and $u < v$ we have the inequality

$$g(u) - g(v) - g(u - \log a) + g(v - \log a) \geq 0,$$

which for $u = v + \log a$ yields

$$g(v) \leq \tfrac{1}{2}(g(v - \log a) + g(v + \log a).$$

Thus the function g is convex and, consequently, can be represented in the form $g(x) = \int_x^\infty h(u)\,du$, where h is a nonnegative monotone nonincreasing function. Further, by (12) we have the formula

$$H_0(E) = \int_E \log^m(1 + y^{2/m})h(\log y)\frac{dy}{y}.$$

Consequently,

$$\int_0^\infty \log^m(1 + y^{2/m})h(\log y)\frac{dy}{y} < \infty. \tag{13}$$

Moreover, by (10),

$$F_\lambda(x) = \int_{e^x}^\infty (\log y - x)^{m-1}h(\log y)\frac{dy}{y}. \tag{14}$$

By (13) the limit inferior of the function $h(\log y)\log^{m+1}(1 + y^{2/(m+1)})$ at 0 and ∞ is equal to 0. Consequently, integration of (14) by parts gives

$$F_\lambda(x) = -\frac{1}{m}\int_{e^x}^\infty (\log y - x)^m\,dh(\log y). \tag{15}$$

Moreover, by (13),

$$-\int_0^\infty \log^{m+1}(1 + y^{2/(m+1)})\,dh(\log y)$$

$$= 2\int_0^\infty \frac{y^{2/(m+1)}}{1 + y^{2/(m+1)}}\log^m(1 + y^{2/(m+1)})h(\log y)\frac{dy}{y} < \infty.$$

Thus the measure H defined by means of the formula

$$H(E) = -\frac{1}{m}\int_E \log^{m+1}(1 + y^{2/(m+1)})\,dh(\log y)$$

is finite on $(0, \infty)$. Setting it into (15), we get desired representation (9).

Suppose now that the function F is given by formula (9). We note that F can be written in the form

$$F(x) = \int_{e^x}^\infty \frac{(\log y - x)^{m-1}G(dy)}{\log^m(1 + y^{2/m})}$$

where the measure

$$G(E) = m\int_E \log^m(1 + y^{2/m})\int_y^\infty \log^{-m-1}(1 + x^{2/(m+1)})H(dx)\frac{dy}{y}$$

is also finite on $(0, \infty)$. Consequently, by the induction assumption, $F = F_\lambda$ for a measure λ from L_{m-1}^+. In order to prove that $\lambda \in L_m{}^+$, consider for any

$a \in (0, 1)$ the decomposition $\lambda = a\lambda * \lambda_a$ with $\lambda_a \in L_{m-2}$. Here L_{-1} denotes the set of all probability measures on the real line. By virtue of (8) we have the formula

$$F_\lambda(x) - F_{a\lambda}(x) = \int_{e^x}^{\infty} \frac{(\log y - x)^{m-1} H_a(dy)}{\log^m(1 + y^{2/m})}$$

where the measure

$$H_a(E) = m \int_E \log^m(1 + y^{2/m}) \int_y^{y/a} \log^{-m-1}(1 + x^{2/(m+1)}) H(dx) \frac{dy}{y}$$

is finite on $(0, \infty)$. Hence, by (7) and the induction assumption, we infer that $\lambda_a = \lambda_a^+ * \delta_c$ where $\lambda_a^+ \in L_{m-1}^+$ and c is a constant. Thus $\lambda_a \in L_{m-1}$, which, by virtue of Proposition 1, shows that $\lambda \in L_m^+$. This completes the proof of Proposition 2.

As an immediate consequence of formula (6) and Proposition 2 we get the following characterization of L_m^+ in terms of the spectral measures.

Proposition 3. *Let $m = 1, 2, \ldots$. A measure Q defined on $(0, \infty)$ is the spectral measure for a probability distribution from L_m^+ if and only if it is of the form*

$$Q(E) = \int_E \log(1 + x^2) \int_x^{\infty} \left(\log \frac{y}{x}\right)^{m-1} \log^{-m-1}(1 + y^{2/(m+1)}) N(dy) \frac{dx}{x} \tag{16}$$

where N is a finite Borel measure on $(0, \infty)$.

Setting this representation of the spectral measure into (2), we obtain the following result.

Proposition 4. *Let $m = 1, 2, \ldots$. A function φ is the characteristic function of a probability measure from L_m^+ if and only if it is of the form*

$$\varphi(t) = \exp \int_0^{\infty} k_m(t, y) N(dy) \tag{17}$$

where N is an arbitrary finite Borel measure on $(0, \infty)$ and the kernel k_m is defined by the formula

$$k_m(t, y) = \frac{1}{(m-1)!} \int_0^y \left(\int_0^{tx} \frac{e^{iu} - 1}{u} \, du - it \arctan x \right) \times \left(\log \frac{y}{x}\right)^{m-1} \frac{dx}{x} \log^{-m-1}(1 + y^{2/(m+1)}).$$

We note that the function φ determines the measure N in (17). Indeed, by a simple calculation (17) can be transformed into formula (2), whence, by the uniqueness of the spectral measure, Eq. (16) follows. Now it is obvious that the spectral measure, and consequently the function φ, determines the measure N.

Integration by parts shows that

$$k_m(t, y) = \left(\frac{it}{(m+1)!}\int_0^y e^{itx}\left(\log\frac{y}{x}\right)^{m+1} dx - it \arctan y\right) \times \log^{-m-1}(1 + y^{2/(m+1)}) + itr_m(y) \quad (18)$$

where r_m is a bounded continuous function on the right half-line tending to 0 as $y \to 0$. Moreover, it is easy to check that

$$\lim_{y\to 0} k_m(t, y) = -\frac{t^2}{2^{m+2}}. \quad (19)$$

We turn now to probability measures from L_m $(m = 1, 2, \ldots)$. Given $\mu \in L_m$, we denote by N_+ and N_- the Borel measures corresponding by Proposition 4 to μ^+ and $((-1)\mu)^+$, respectively. Further, by d^2 we denote the variance of the Gaussian component μ_0 of μ. Put for any Borel subset E of the real line

$$M(E) = N_+(E \cap (0, \infty)) + N_-((-E) \cap (0, \infty)) + 2^{m+1}\, d^2\, \delta_0(E)$$

where $-E = \{-x : x \in E\}$. From decomposition formula (4) and Proposition 4 we get a characterization of L_m in terms of the measures M. Using relations (18) and (19), we finally obtain the following theorem.

Theorem 1. *Let $m = 1, 2, \ldots$. A function φ is the characteristic function of a probability measure from L_m if and only if*

$$\phi(t) = \exp\left\{ict + it\left(\frac{1}{(m+1)!}\int_0^y \exp(itx)\left(\log\frac{y}{x}\right)^{m+1} dx - \arctan y\right) \times \frac{M(dy)}{\log^{m+1}(1 + |y|^{2/(m+1)})}\right\}$$

where c is a constant, M is a finite Borel measure on the real line, and the integrand is defined as its limiting value $it/2^{m+2}$ when $y = 0$. Moreover, the function φ determines c and M uniquely.

We proceed now to a characterization of the set L_∞. We begin with that of $L_\infty{}^+$.

Proposition 5. *A function F is associated with a probability measure λ from L_∞^+, i.e., $F = F_\lambda$, if and only if*

$$F(x) = \int_0^2 \sin[(\pi/2)y]e^{-xy}N(dy), \tag{20}$$

where N is a finite Borel measure on $(0, 2)$.

Proof. Suppose that $\lambda \in L_\infty^+$. Then, by Proposition 2, for every m we have the formula

$$F_\lambda(x) = \int_{e^x}^\infty \frac{(\log y - x)^m H_m(dy)}{\log^{m+1}(1 + y^{2/(m+1)})}$$

for some finite measures H_m on $(0, \infty)$. Hence it follows that F_λ is infinitely differentiable and

$$F_{(\lambda)}^m(x) = (-1)^m m! \int_{e^x}^\infty \frac{H_m(dy)}{\log^{m+1}(1 + y^{2/(m+1)})} \qquad (m = 0, 1, \ldots).$$

Thus the function F_λ is completely monotonic on the real line, and consequently, by Bernstein's theorem [4, p. 155], has a representation

$$F_\lambda(x) = \int_0^\infty e^{-xy}R(dy) \tag{21}$$

where R is a finite Borel measure on the right half-line. By (6) for the spectral measure Q corresponding to λ we have the equation

$$Q((0, \infty)) = \int_0^\infty \left(\int_{-\infty}^\infty \log(1 + x^2)e^{-xy}\, dx\right) yR(dy).$$

Since the integrand is equal to $\pi/(\sin(\pi/2)y)$ in the interval $(0, 2)$ and is infinite outside it, we infer that the measure R is concentrated on $(0, 2)$ and

$$\int_0^2 \frac{R(dy)}{\sin(\pi/2)y}$$

is finite. Setting the measure

$$N(E) = \int_E \frac{R(dy)}{\sin(\pi/2)y}$$

into (21), we get representation (20). The necessity of the condition is thus proved.

Suppose now that the function F is given by formula (20). Note first that

$$\int_0^\infty \log^{m+1}(1 + x^{2/(m+1)}) \frac{dx}{x^{1+y}} \leq \frac{b}{y^{m+2}} + \frac{c}{2 - y}$$

for $y \in (0, 2)$, where b and c are constants. Hence it follows that the measure H_m defined on $(0, \infty)$ by means of the formula

$$H_m(E) = \frac{1}{m!}\int_0^2 y^{m+1} \sin\frac{\pi}{2} y \int_E \log^{m+1}(1 + x^{2/(m+1)}) \frac{dx}{x^{1+y}} N(dy)$$

is finite. Consequently, by Proposition 2, the function

$$G_m(x) = \int_{e^x}^{\infty} \frac{(\log y - x)^m H_m(dy)}{\log^{m+1}(1 + y^{2/(m+1)})}$$

is associated with a probability measure, say, λ_m, from L_m^+. By a simple calculation we get the formula

$$G_m^{(m)}(x) = (-1)^m \int_0^2 y^m \sin\frac{\pi}{2} y \exp(-xy) N(dy)$$

which, by (20), yields the equation $G_m^{(m)} = F^{(m)}$. Since both functions F and G_m approach 0 at infinity, the last equation implies $G_m = F$. Thus $F = F_{\lambda_m}$ $(m = 1, 2, \ldots)$. Since the correspondence between measures λ and functions F_λ is one-to-one, we have the equations $\lambda_1 = \lambda_2 = \cdots$, which show that F is associated with a measure belonging to all sets L_m^+ and, consequently, to L_∞^+. The sufficiency of the condition is thus proved.

Proposition 5 and formula (6) yield the following characterization of L_∞^+ in terms of the spectral measures.

Proposition 6. *A measure Q defined on $(0, \infty)$ is the spectral measure for a probability distribution from L_∞^+ if and only if it is of the form*

$$Q(E) = \int_E \log(1 + x^2) \int_0^2 x^{-y-1} y \sin[(\pi/2)y] N(dy)\, dx$$

where N is a finite Borel measure on $(0, 2)$.

Setting this expression into (2) we get the following result.

Proposition 7. *A function φ is the characteristic function of a probability measure from L_∞^+ if and only if*

$$\varphi(t) = \exp \int_0^2 k_\infty(t, y) N(dy) \tag{22}$$

where N is an arbitrary finite Borel measure on $(0, 2)$ and the kernel k_∞ is defined by the formula

$$k_\infty(t, y) = y \sin\frac{\pi}{2} y \int_0^\infty \left(\int_0^{tx} \frac{e^{iu} - 1}{u}\, du - it \arctan x \right) \frac{dx}{x^{1+y}}.$$

In the same way as in Proposition 4 we conclude that the function φ determines the measure N uniquely. Moreover, integrating by parts, we have

$$k_\infty(t, y) = \sin\frac{\pi}{2}y\int_0^\infty\left(e^{itx} - 1 - \frac{itx}{1+x^2}\right)\frac{dx}{x^{1+y}}.$$

This integral occurs in the investigation of stable laws [2, p. 329]. For $y \in (0, 1) \cup (1, 2)$ we have the formula

$$k_\infty(t, y) = -it\frac{\pi}{2}\tan\frac{\pi}{2}y - \frac{\Gamma(2-y)\sin(\pi/2)y}{y(1-y)}|t|^y\left(\cos\frac{\pi}{2}y - i\frac{t}{|t|}\sin\frac{\pi}{2}y\right),$$

which by the continuity of the kernel yields

$$k_\infty(t, 1) = it(1 - C) - it\log|t| - \frac{\pi}{2}|t|$$

where C is Euler's constant. Further, by a simple calculation we get the formula

$$k_\infty(t, y) = -\frac{\Gamma(2-y)\sin(\pi/2)y}{y(1-y)}\left[|t|^y\left(\cos\frac{\pi}{2}y - i\frac{t}{|t|}\sin\frac{\pi}{2}y\right) + ity\right] + itr(y) \tag{23}$$

where r is a bounded continuous function on $(0, 2)$ tending to $-\pi/2$ as $y \to 2$. For $y = 1$ we take the limiting value of the kernel. It is also easy to verify the formula

$$\lim_{y\to 2} k_\infty(t, y) = -\frac{\pi t^2}{4}. \tag{24}$$

We note that the function $y^{-1}\Gamma(2-y)\sin(\pi/2)y$ is bounded and has a positive greatest lower bound in the interval $(0, 2)$. Consequently, we may replace the measure N in representation formula (22) by the measure I defined as follows.

$$I(E) = \int_E y^{-1}\Gamma(2-y)\sin\frac{\pi}{2}yN(dy).$$

Taking into account (23), we get an equivalent form of representation (22)

$$\varphi(t) = \exp\left\{ibt - \int_0^2\left[|t|^y\left(\cos\frac{\pi}{2}y - i\frac{t}{|t|}\sin\frac{\pi}{2}y\right) + ity\right]\frac{I(dy)}{1-y}\right\} \tag{25}$$

where I is an arbitrary finite Borel measure on $(0, 2)$ and b a suitably chosen constant.

Consider a probability measure μ from L_∞. Let I_+ and I_- be the measures corresponding in representation (25) to μ^+ and $((-1)\mu)^+$, respectively. Suppose that the characteristic function of the Gaussian component μ^0 of μ is given by the formula

$$\widehat{\mu^0}(t) = \exp(iat - d^2t^2).$$

For any subset E of $(-2, 0) \cup (0, 2]$ we put

$$M(E) = I_+(E \cap (0, 2)) + I_-((-E) \cap (0, 2)) + d^2\, \delta_2(E).$$

From decomposition formula (4) and representation (25) we get a characterization of L_∞ in terms of the measures M. Namely, we have the following theorem.

Theorem 2. *A function φ is the characteristic function of a probability measure from L_∞ if and only if*

$$\varphi(t) = \exp\left\{ict - \int_{-2}^{2} \left[|t|^{|y|}\left(\cos\frac{\pi}{2}y - i\frac{t}{|t|}\sin\frac{\pi}{2}y\right) + ity\right]\frac{M(dy)}{1-|y|}\right\},$$

where c is a real constant, M is a finite Borel measure on $(-2, 0) \cup (0, 2]$, and the integrand is defined as its limiting values $(\pi/2)|t| + it\log|t| - it$ and $(\pi/2)|t| - it\log|t| + it$ when $y = 1$ and $y = -1$, respectively.

Corollary 1. *The set L_∞ is the smallest set containing all stable probability measures and closed under convolutions and passages to the limit.*

Corollary 2. *Each sequence of independent random variables with stable probability distributions (not necessarily with the same exponent) generating a convergent triangular array is slowly varying.*

REFERENCES

1. Koroljuk, V. S. and Zolotarev, V. M. (1961). On a hypothesis proposed by B. V. Gnedenko. *Teor. Verojatnost. i Primenen.* **6** 469–474.
2. Loève, M. (1950). *Probability Theory.* New York.
3. Urbanik, K. (1968). A representation of self-decomposable distributions. *Bull. Acad. Polon. Sci. Sér. Sci. Math. Astronom. Phys.* **16** 209–214.
4. Widder, D. V. (1971). *An Introduction to Transform Theory.* New York and London.
5. Zinger, A. A. (1965). On a class of limit distributions for normed sums of independent random variables. *Teor. Verojatnost. i Primenen.* **10** 672–692.

PART IV

Design and Analysis of Experiments

The Analysis of Time Series Collected in an Experimental Design[1]

DAVID R. BRILLINGER

DEPARTMENT OF STATISTICS
THE UNIVERSITY OF CALIFORNIA
BERKELEY, CALIFORNIA

1. INTRODUCTION

We consider the frequency analysis of stretches of time series resulting from runs made in accordance with an experimental design. The analyses proposed may be viewed as extensions of the familiar ANOVA procedures employed when each run of a design leads to the measurement of a real-valued variate. Our approach consists of the application of a complex version of the computations of ANOVA to the components of the discrete Fourier transforms of the observed stretches of series. Initially the time series examined are taken to be sections of ordinary stationary time series. In Section 5, however, we consider a case where the observed series form sections of stationary point processes. The analyses suggested are investigated in the case that the series are generated by linear models. The computations suggested are illustrated, in the stationary series case, by an analysis of a sample of ten temperature series classified as of European or North American origin. The point process computations are illustrated by an analysis of the times of earthquakes at two California locations.

We begin with a consideration of series collected in accordance with a balanced one-way classification. For example, the experiment might consist of a single factor varying through I levels with J replicates being carried out at each level. The time series segments recorded could then be denoted

$$x_{ij}(t); \qquad t = 0, 1, \ldots, T-1 \tag{1.1}$$

where $t = 0, \pm 1, \ldots$ indexes time, $j = 1, \ldots, J$ indexes replicates, and $i = 1, \ldots, I$ indexes factor levels. A variety of models might be considered for such data. These include:

[1] This paper was prepared with the support of the NSF grant GP-31411.

The Fixed Effects Model. *The series are given by*

$$x_{ij}(t) = \mu_{ij} + \gamma(t) + \delta_i(t) + \varepsilon_{ij}(t) \tag{1.2}$$

where the μ_{ij}, $\gamma(t)$, *and* $\delta_i(t)$ $(t = 0, \pm 1, \ldots; i = 1, \ldots, I; j = 1, \ldots, J)$ *are constants and where the* $\{\varepsilon_{ij}(t), t = 0, \pm 1, \ldots\}$ $(i = 1, \ldots, I; j = 1, \ldots, J)$ *are independent realizations of a stationary series, satisfying Assumption I below, with mean* 0 *and power spectrum* $f_{\varepsilon\varepsilon}(\lambda)$, $0 \le \lambda \le \pi$.

The Random Effects Model. *The series are given by* (1.2) *where the* μ_{ij} *are constants, where the* $\{\gamma(t), t = 0, \pm 1, \ldots\}$, $\{\delta_i(t), t = 0, \pm 1, \ldots\}$, $\{\varepsilon_{ij}(t), t = 0, \pm 1, \ldots\}$ $(i = 1, \ldots, I; j = 1, \ldots, J)$ *are independent realizations of stationary series, satisfying Assumption I below, with mean* 0 *and power spectra* $f_{\gamma\gamma}(\lambda)$, $f_{\delta\delta}(\lambda)$, $f_{\varepsilon\varepsilon}(\lambda)$, *respectively.*

The Fixed Effects Model including Transients. *The series are given by*

$$x_{ij}(t) = \mu_{ij} + \omega_{ij}(t) + \gamma(t) + \delta_i(t) + \varepsilon_{ij}(t) \tag{1.3}$$

where the μ_{ij}, $\gamma(t)$, $\delta_i(t)$, $\varepsilon_{ij}(t)$ *are as in the Fixed Effects Model above and the* $\omega_{ij}(t)$, $t = 0, 1, 2, \ldots$, *are constants satisfying*

$$\sum_{t=0}^{\infty} |t|\,|\omega_{ij}(t)| < \infty. \tag{1.4}$$

The details of the analyses are illustrated only for the case of a balanced one-way classification. The procedures may clearly be generalized to handle more complex designs, however. An important role in our analysis will be played by the discrete Fourier transform of a stretch of series. We therefore turn to a discussion of its statistical properties.

2. THE FINITE FOURIER TRANSFORM

Suppose the values $\mathbf{x}(t)$, $t = 0, \ldots, T-1$, from an r vector-valued series are available. The finite Fourier transform of this stretch of values is defined by

$$\mathbf{d}_x^{(T)}(\lambda) = \sum_{t=0}^{T-1} \exp\{-i\lambda t\}\mathbf{x}(t) \tag{2.1}$$

$-\infty < \lambda < \infty$, where $i = \sqrt{-1}$. Suppose that $\mathbf{x}(t)$, $t = 0, \pm 1, \ldots$, is a stationary series with spectral density matrix $\mathbf{f}_{xx}(\lambda)$. In the case where values of the series, well-separated in time, are only weakly statistically dependent, $\mathbf{d}_x^{(T)}(\lambda)$ is asymptotically complex normal with variance $2\pi T\mathbf{f}_{xx}(\lambda)$. (This is demonstrated under strong mixing in Rosenblatt [7] and Hannan and Thomson [5]. It is demonstrated under the absolute summability of certain cumulants in Leonov and Shiryaev [6] and Brillinger [1, 3].) An assumption that will be enough for what we conclude in this paper is

Assumption I. *The r vector-valued series* $\mathbf{x}(t)$, $t = 0, \pm 1, \ldots$ *is stationary; all moments exist and satisfy*

$$\sum_{u_1, \ldots, u_{k-1}} |\text{cum}\{x_{a_1}(t + u_1), \ldots, x_{a_{k-1}}(t + u_{k-1}), x_{a_k}(t)\}| < \infty$$

for $a_1, \ldots, a_k = 1, \ldots, r$; $k = 2, 3, \ldots$.

The appearance of $\mathbf{f}_{xx}(\lambda)$ in the limiting distribution above suggests basing an estimate of it on $\mathbf{d}_x^{(T)}(\lambda)$. However, such an estimate would be based, essentially, on a single observation. There are two different procedures available for effectively increasing the number of observations; one can evaluate the finite Fourier transform at a number of frequencies in the neighborhood of λ, or one can evaluate the finite Fourier transform for a number of disjoint stretches of the series. We begin with a discussion of the first procedure.

Let $s(T)$, $T = 1, 2, \ldots$ denote a sequence of integers with $2\pi s(T)/T \to \lambda$. Suppose $\lambda \not\equiv 0 \pmod{\pi}$. Then, under Assumption I for example, one can show that the K values

$$\mathbf{X}_k = (2\pi T)^{-1/2} \mathbf{d}_x^{(T)}(2\pi[s(T) + k]/T); \qquad k = 1, \ldots, K \tag{2.2}$$

are asymptotically independent r vector-valued complex normal variates with mean $\mathbf{0}$ and covariance matrix $\mathbf{f}_{xx}(\lambda)$, as $T \to \infty$. (Throughout the paper we will use the notation of a capital letter denoting the Fourier transform of a series with a given lowercase letter.) This result suggests taking

$$\mathbf{f}_{xx}^{(T)}(\lambda) = K^{-1} \sum_{k=1}^{K} \mathbf{X}_k \mathbf{X}_k^* \tag{2.3}$$

as an estimate of $\mathbf{f}_{xx}(\lambda)$. It also suggests approximating the distribution of (2.3) by a complex Wishart with K degrees of freedom. In the case where T is highly composite, the values (2.2) may be computed rapidly with the use of a fast Fourier transform algorithm (see Cooley and Tukey [4]).

The second procedure involves the splitting of the period of observation into K segments, of length $V = T/K$, and then the computation of the K values

$$\mathbf{X}_k = (2\pi V)^{-1/2} \sum_{t=(k-1)V}^{kV-1} \mathbf{x}(t) \exp\{-i\lambda t\}; \qquad k = 1, \ldots, K \tag{2.4}$$

If $\lambda \not\equiv 0 \pmod{\pi}$, the values (2.4) are also asymptotically independent r vector-valued complex normal variates with mean 0 and covariance matrix $\mathbf{f}_{xx}(\lambda)$, as $T \to \infty$ under Assumption I (see Brillinger [3]). We are led to form the statistic (2.3) as an estimate of $\mathbf{f}_{xx}(\lambda)$ and the estimate will again be complex Wishart with K degrees of freedom.

Results of a similar character may be developed in the case that one has collected data from a stationary point process, provided one modifies the definitions (2.2), (2.4) suitably. Specifically, suppose the segment $\mathbf{n}(t)$, $0 < t \leq T$, of a stationary r vector-valued point process $\mathbf{n}(t)$, $-\infty < t < \infty$, with spectral density matrix $\mathbf{f}_{nn}(\lambda)$ is available. In place of (2.1) form

$$\mathbf{d}_n^{(T)}(\lambda) = \int_0^T \exp\{-i\lambda t\}\, d\mathbf{n}(t)$$

An estimate of $\mathbf{f}_{nn}(\lambda)$ may then be formed, in the manner of (2.3) using

$$\mathbf{N}_k = (2\pi T)^{-1/2}\mathbf{d}_n^{(T)}(2\pi[s(T) + k]/T); \qquad k = 1, \ldots, K$$

or

$$\mathbf{N}_k = (2\pi V)^{-1/2}\int_{(k-1)V}^{kV} \exp\{-i\lambda t\}\, d\mathbf{n}(t); \qquad k = 1, \ldots, K \tag{2.5}$$

We set down

Assumption II. *The r vector-valued point process $\mathbf{n}(t)$, $-\infty < t < \infty$, is stationary; all moments exist and satisfy*

$$\int_{u_1} \cdots \int_{u_{k-1}} |\mathrm{cum}\{dn_{a_1}(t + u_1), \ldots, dn_{a_{k-1}}(t + u_{k-1}), dn_{a_k}(t)\}/dt| < \infty$$

for $a_1, \ldots, a_k = 1, \ldots, r$; $k = 2, 3, \ldots$.

Under this assumption the $\mathbf{N}_k$ may be shown to be asymptotically independent r vector-valued complex normal variates with mean $\mathbf{0}$ and covariance matrix $\mathbf{f}_{nn}(\lambda)$ when $\lambda \not\equiv 0 \pmod{\pi}$ (see Brillinger [2]). The statistic (2.5) is the one employed for the computations described later in the paper.

It will be convenient for us to write Fourier transforms, such as $\mathbf{X}_k$ or $\mathbf{N}_k$, as equal to normal variates plus a remainder term tending to $\mathbf{0}$ almost surely as $T \to \infty$. Following a theorem of Skorokhod [12] (see also Wichura [14]), we may do this provided the $\mathbf{X}_k$, $k = 1, \ldots, K$ (or $\mathbf{N}_k$) tend in distribution to normal variates. Specifically we may then write

$$\mathbf{X}_k = \zeta_k + o_{\text{a.s.}}(1); \qquad k = 1, \ldots, K \tag{2.6}$$

where the ζ_k are independent complex normal variates with mean $\mathbf{0}$ and covariance matrix $\mathbf{f}_{xx}(\lambda)$. The representation (2.6) allows us to write (2.3) as

$$\mathbf{f}_{xx}^{(T)}(\lambda) = K^{-1}\sum_{k=1}^{K} \zeta_k \zeta_k^* + o_{\text{a.s.}}(1)$$

giving the Wishart limit directly.

3. THE FIXED EFFECTS MODEL

Consider time series $x_{ij}(t)$ ($t = 0, \ldots, T-1$; $i = 1, \ldots, I$; $j = 1, \ldots, J$), collected in accordance with a one-way classification, in a situation where the Fixed Effects Model of Section 1 seems appropriate. In this case

$$Ex_{ij}(t) = \mu_{ij} + \gamma(t) + \delta_i(t)$$

and so, up to an additive constant, series from the same class have the same mean function. In the case that $\delta_i(t) = \bar{\delta}(t)$ ($t = 0, \ldots, T-1$; $i = 1, \ldots, I$) up to an additive constant, all the series involved have the same mean function. Also, the series are all second-order stationary with power spectra $f_{\varepsilon\varepsilon}(\lambda)$ and cross-spectra 0. We turn to an analysis of such series.

Suppose X_{ijk} denotes either

$$(2\pi T)^{-1/2} d_{x_{ij}}^{(T)}(2\pi[s(T) + k]/T) \qquad \text{or} \qquad (2\pi V)^{-1/2} \sum_{(k-1)V}^{kV-1} x_{ij}(t) \exp\{-i\lambda t\}$$

$k = 1, \ldots, K$, in the manner of the previous section. Carrying out this step across the model (1.2) gives

$$X_{ijk} = \Gamma_k + \Delta_{ik} + E_{ijk}$$

$i = 1, \ldots, I$; $j = 1, \ldots, J$; $k = 1, \ldots, K$. As the E_{ijk} are asymptotically independent complex normal variates with mean 0 and covariance matrix $f_{\varepsilon\varepsilon}(\lambda)$, we may write them as $E_{ijk} = \zeta_{ijk} + o_{\text{a.s.}}(1)$ where the ζ_{ijk} are independent complex normal variates with mean 0 and variance $f_{\varepsilon\varepsilon}(\lambda)$. The model thus gives

$$X_{ijk} = \Gamma_k + \Delta_{ik} + \zeta_{ijk} + o_{\text{a.s.}}(1) \tag{3.1}$$

Expression (3.1) is seen to take, approximately, the form of the model of a balanced two-way hierarchical classification with fixed effects. In order to proceed with an analysis, we are led to compute the following overall and class mean series

$$\bar{x}_{i.}(t) = J^{-1} \sum_{j=1}^{J} x_{ij}(t)$$

$$\bar{x}_{..}(t) = I^{-1} J^{-1} \sum_{i=1}^{I} \sum_{j=1}^{J} x_{ij}(t)$$

These have Fourier transforms $\bar{X}_{i.k}$, $\bar{X}_{..k}$, respectively. In order to estimate the error spectrum $f_{\varepsilon\varepsilon}(\lambda)$, expression (3.1) leads us to compute the following within-class sum of squares (using the notation (2.3)),

$$\sum_{i,j,k} |X_{ijk} - \bar{X}_{i.k}|^2 = K \sum_{i,j} f_{x_{ij}-\bar{x}_{i.},\, x_{ij}-\bar{x}_{i.}}^{(T)}(\lambda) \tag{3.2}$$

In order to examine the hypothesis $\delta_i(t) = \bar{\delta}(t)$ $(t = 0, \ldots, T-1; i = 1, \ldots, I)$, we are led to compute the following between class sum of squares

$$J \sum_{i,k} |\bar{X}_{i.k} - \bar{X}_{..k}|^2 = JK \sum_i f^{(T)}_{\bar{x}_{i.}-\bar{x}_{..},\, \bar{x}_{i.}-\bar{x}_{..}}(\lambda) \tag{3.3}$$

and in order to examine the hypothesis $\gamma(t) + \bar{\delta}(t) = 0$, $t = 0, \ldots, T-1$, we are led to compute the following grand mean sum of squares

$$IJ \sum_k |\bar{X}_{..k}|^2 = IJK f^{(T)}_{\bar{x}_{..}\bar{x}_{..}}(\lambda) \tag{3.4}$$

The total of these last three expressions is

$$\sum_{i,j,k} |X_{ijk}|^2 = K \sum_{i,j} f^{(T)}_{x_{ij}x_{ij}}(\lambda) \tag{3.5}$$

If desired, the expressions (3.2), (3.3), (3.4), (3.5) could be collected together into an ANOPOW Table.

Using the representation (3.1), we see that expressions (3.2), (3.3), (3.4), (3.5) may be written, respectively,

$$\sum_{i,j,k} |\zeta_{ijk} - \bar{\zeta}_{i.k}|^2 + o_{\text{a.s.}}(1) \tag{3.6}$$

$$J \sum_{i,k} |\Delta_{ik} - \bar{\Delta}_{.k} + \bar{\zeta}_{i.k} - \bar{\zeta}_{..k}|^2 + o_{\text{a.s.}}(1) \tag{3.7}$$

$$IJ \sum_k |\Gamma_k + \bar{\Delta}_{.k} + \bar{\zeta}_{..k}|^2 + o_{\text{a.s.}}(1) \tag{3.8}$$

$$\sum_{i,j,k} |\zeta_{ijk} + \Gamma_k + \Delta_{ik}|^2 + o_{\text{a.s.}}(1) \tag{3.9}$$

Using the complex extension of the Fisher–Cochran theorem given in the Appendix, we see that expressions (3.6), (3.7), (3.8) have the forms

$$f_{\varepsilon\varepsilon}(\lambda)\chi^2_{2I(J-1)K}/2 + o_{\text{a.s.}}(1)$$

$$f_{\varepsilon\varepsilon}(\lambda)\chi^2_{2(I-1)K}\left(JK \sum_i f^{(T)}_{\delta_i-\bar{\delta},\, \delta_i-\bar{\delta}}(\lambda)/f_{\varepsilon\varepsilon}(\lambda)\right)\Big/2 + o_{\text{a.s.}}(1)$$

$$f_{\varepsilon\varepsilon}(\lambda)\chi^2_{2K}(IJK f^{(T)}_{\gamma+\bar{\delta},\, \gamma+\bar{\delta}}(\lambda)/f_{\varepsilon\varepsilon}(\lambda))/2 + o_{\text{a.s.}}(1)$$

respectively, with the chi-squared variates appearing being statistically independent.

The F ratio

$$\left\{J \sum_i f^{(T)}_{\bar{x}_{i.}-\bar{x}_{..},\, \bar{x}_{i.}-\bar{x}_{..}}(\lambda)/(I-1)\right\}\Big/\left\{\sum_{i,j} f^{(T)}_{x_{ij}-\bar{x}_{i.},\, x_{ij}-\bar{x}_{i.}}(\lambda)/[I(J-1)]\right\}$$

useful in examining the hypothesis $\delta_i(t) = \bar{\delta}(t)$ $(t = 0, \ldots, T-1; i = 1, \ldots, I)$ may therefore be written

$$F_{2(I-1)K;\, 2I(J-1)K}\left(JK \sum_i f^{(T)}_{\delta_i-\bar{\delta},\, \delta_i-\bar{\delta}}(\lambda)/f_{\varepsilon\varepsilon}(\lambda)\right) + o_{\text{a.s.}}(1)$$

while the F ratio

$$IJf^{(T)}_{\bar{x}..\bar{x}..}(\lambda) \Big/ \left\{\sum_{i,j} f^{(T)}_{x_{ij}-\bar{x}_{i.},\, x_{ij}-\bar{x}_{i.}}(\lambda)/[I(J-1)]\right\}$$

useful in examining the hypothesis $\gamma(t) + \delta(t) = 0$, $t = 0, \ldots, T-1$, has the form

$$F_{2K;\, 2I(J-1)K}(IJKf^{(T)}_{\gamma+\delta,\, \gamma+\delta}(\lambda)/f_{\varepsilon\varepsilon}(\lambda)) + o_{\text{a.s.}}(1)$$

These last two results suggest appropriate null points against which the sample statistics may be put in a formal significance test of the hypotheses.

We note that Shumway [9, 10] considers the model

$$Y_j(t) = s(t) + n_j(t)$$

$j = 1, \ldots, N$, where $s(t)$ is a fixed unknown signal and $n_j(t)$ a random noise series. He suggests the consideration of F ratios computed in the frequency domain. Shumway and Saikia [11] develop an empirical Bayes procedure for the model

$$Y_{jk}(t) = m_0(t) + a_j(t) + e_{jk}(t)$$

We turn next to an investigation of the Fixed Effects Model including Transients. Here the Fixed Effects Model is modified by the addition of $\omega_{ij}(t)$, $t = 0, 1, 2, \ldots$, which is to be thought of as a brief superimposed transient series. In the case where condition (1.4) is satisfied and $2\pi s(T)/T = \lambda + O(T^{-1})$,

$$d^{(T)}_{\omega_{ij}}(2\pi[s(T) + k]/T) = d^{(T)}_{\omega_{ij}}(\lambda) + O(1)$$

and so the model gives

$$\begin{aligned} X_{ijk} &= \Omega_{ijk} + \Gamma_k + \Delta_{ik} + E_{ijk} \\ &= \Omega_{ij} + \Gamma_k + \Delta_{ik} + \zeta_{ijk} + o_{\text{a.s.}}(1) \end{aligned} \tag{3.10}$$

where the ζ_{ijk} are IJK independent complex normal variates with mean 0 and variance $f_{\varepsilon\varepsilon}(\lambda)$. Ignoring the o term, we see that (3.10) has the form of an ANOVA model.

Generally we would not be interested in the ω_{ij} themselves, as the series $\omega_{ij}(t)$ is just a transient. These values may be removed from the setup through a covariance analysis. Specifically we simply define the covariate $Z_{ijk} = 1$ for all i, j, k and rewrite (3.10) in the form

$$X_{ijk} = \Omega_{ij}Z_{ijk} + \Gamma_k + \Delta_{ik} + \zeta_{ijk} + o_{\text{a.s.}}(1)$$

The expressions (3.2), (3.3), (3.4) are replaced by

$$\sum_{i,j,k} |X_{ijk} - \bar{X}_{ij.} - \bar{X}_{i.k} + \bar{X}_{i..}|^2$$

$$J\sum_{i,k} |\bar{X}_{i.k} - \bar{X}_{..k} - \bar{X}_{i..} + \bar{X}_{...}|^2$$

$$IJ\sum_{k} |\bar{X}_{..k} - \bar{X}_{...}|^2$$

respectively. Using the theorem of the Appendix, we see that these may be written

$$f_{\varepsilon\varepsilon}(\lambda)\chi^2_{2I(J-1)(K-1)}/2 + o_{\text{a.s.}}(1)$$
$$f_{\varepsilon\varepsilon}(\lambda)\chi^2_{2(I-1)(K-1)}(J\sum |\Delta_{ik} - \bar{\Delta}_{.k} - \bar{\Delta}_{i.} + \bar{\Delta}_{..}|^2/f_{\varepsilon\varepsilon}(\lambda))/2 + o_{\text{a.s.}}(1)$$
$$f_{\varepsilon\varepsilon}(\lambda)\chi^2_{2(K-1)}(IJ\sum |\Gamma_k + \bar{\Delta}_{.k} - \bar{\Gamma}_. - \bar{\Delta}_{..}|^2/f_{\varepsilon\varepsilon}(\lambda))/2 + o_{\text{a.s.}}(1)$$

with the chi-squared variates appearing being statistically independent. The effect on the analysis of including the transient series in the model is essentially the loss of two degrees of freedom in the chi-squared variates appearing. This seems a very small price to pay in return for the substantial gain in robustness of the procedure against transients.

4. THE RANDOM EFFECTS MODEL

Consider a two-way array of time series $x_{ij}(t)$ ($t = 0, \pm 1, \ldots; i = 1, \ldots, I; j = 1, \ldots, J$) satisfying the Random Effects Model presented in Section 1. The series are assumed to have the form

$$x_{ij}(t) = \mu_{ij} + \gamma(t) + \delta_i(t) + \varepsilon_{ij}(t) \tag{4.1}$$

where the series $\gamma(t)$, $\delta_i(t)$ are now random rather than fixed. $\delta_i(t)$ provides a component series common to all series in the ith class, while $\gamma(t)$ provides a component series common to all series. Under the random effects model the series are jointly stationary with

$$\begin{aligned} Ex_{ij}(t) &= \mu_{ij}, \\ f_{x_{ij}x_{ij}}(\lambda) &= f_{\gamma\gamma}(\lambda) + f_{\delta\delta}(\lambda) + f_{\varepsilon\varepsilon}(\lambda) \\ f_{x_{ij}x_{ij'}}(\lambda) &= f_{\gamma\gamma}(\lambda) + f_{\delta\delta}(\lambda), \qquad j \neq j' \\ f_{x_{ij}x_{i'j'}}(\lambda) &= f_{\gamma\gamma}(\lambda), \qquad i \neq i' \end{aligned} \tag{4.2}$$

Series in the same class are seen to have coherency

$$[f_{\gamma\gamma}(\lambda) + f_{\delta\delta}(\lambda)]/[f_{\gamma\gamma}(\lambda) + f_{\delta\delta}(\lambda) + f_{\varepsilon\varepsilon}(\lambda)]$$

while series in different classes have coherency

$$f_{\gamma\gamma}(\lambda)/[f_{\gamma\gamma}(\lambda) + f_{\delta\delta}(\lambda) + f_{\varepsilon\varepsilon}(\lambda)]$$

The relative magnitude of the spectra $f_{\gamma\gamma}(\lambda)$, $f_{\delta\delta}(\lambda)$, $f_{\varepsilon\varepsilon}(\lambda)$ is seen to determine the extent and character of the dependence of the series. In the case that $f_{\delta\delta}(\lambda) \equiv 0$, series in the same class are no more linearly dependent than any pair of series. In the case that $f_{\gamma\gamma}(\lambda) \equiv 0$, series in different classes are not linearly dependent. It will clearly be of interest to estimate these three parameters.

As an example to illustrate the model of this section, we took $I = 2$ classes corresponding climatic regions in North America and Europe. We then took $J = 5$ replicates, points at random in each of the two regions, and as series we took monthly mean temperatures for the period 1918–1960, recorded at the nearest station to the selected points and listed in World Weather Records [15]. The regions considered were those having humid temperate climates as given by Trewartha [13, Plate II]. For North America this comprised, approximately, the area east of the Rocky Mountains and south of the Great Lakes. For Europe the region comprised much of Western Europe. The stations actually selected are shown in Table I. There were approximately 25 stations that could have been selected in each region.

TABLE I

United States	*Western Europe*
New Haven	Copenhagen
Cape Hatteras	De Bilt
Cincinnati	Paris
Nashville	Odessa
St. Louis	Valentia, Eire

Suppose that one is interested in estimating the strength of the weather component common to the stations of the same continent and also of estimating the strength of the weather component common to the North American and European continents. This problem might be approached through the model (4.1) taking $i = 1, 2$ to index continents and $j = 1, \ldots, 5$ to index stations sampled within continents. We turn to the problem of developing estimates of the spectra of the series of the model (4.1).

Suppose we take the Fourier transform of the model (4.1). We may write

$$X_{ijk} = \Gamma_k + \Delta_{ik} + E_{ijk} \tag{4.3}$$

for $i = 1, \ldots, I$; $j = 1, \ldots, J$; $k = 1, \ldots, K$. Following the discussion of Section 2 we can write

$$\Gamma_k = \eta_k + o_{\text{a.s.}}(1) \qquad \Delta_{ik} = \theta_{ik} + o_{\text{a.s.}}(1) \qquad E_{ijk} = \zeta_{ijk} + o_{\text{a.s.}}(1) \tag{4.4}$$

where the η_k are independent complex normal variates with mean 0 and variance $f_{\gamma\gamma}(\lambda)$, the θ_{ik} are independent complex normal variates with mean 0 and variance $f_{\delta\delta}(\lambda)$, and the ζ_{ijk} are independent complex normal variates with mean 0 and variance $f_{\varepsilon\varepsilon}(\lambda)$.

The model is now seen to have the approximate form of that of a balanced two-way hierarchical classification with random effects. We now follow the usual procedure in random effect models of considering the statistical properties, under the random effect model, of the sums of squares suggested by the corresponding fixed effects model. (There is clearly nothing to prevent our considering any of the other random effects estimates that have been proposed, however.) Here these sums of squares are given by (3.2), (3.3), (3.4). Using the representations (4.4), we see that we can write them as

$$\sum_{i,j,k} |X_{ijk} - \bar{X}_{i.k}|^2 = \sum_{i,j,k} |\zeta_{ijk} - \bar{\zeta}_{i.k}|^2 + o_{\text{a.s.}}(1)$$

$$J \sum_{i,k} |\bar{X}_{i.k} - \bar{X}_{..k}|^2 = J \sum_{i,k} |\theta_{ik} - \bar{\theta}_{.k} + \bar{\zeta}_{i.k} - \bar{\zeta}_{..k}|^2 + o_{\text{a.s.}}(1)$$

$$IJ \sum_{k} |\bar{X}_{..k}|^2 = IJ \sum_{k} |\eta_k + \bar{\theta}_{.k} + \bar{\zeta}_{..k}|^2 + o_{\text{a.s.}}(1)$$

For fixed θ, η, following the discussion of Section 3, the first terms on the right here will be distributed independently as

$$f_{\varepsilon\varepsilon}(\lambda)\chi^2_{2I(J-1)K}/2$$

$$f_{\varepsilon\varepsilon}(\lambda)\chi^2_{2(I-1)K}\left(J \sum_{i,k} |\theta_{ik} - \bar{\theta}_{.k}|^2 / f_{\varepsilon\varepsilon}(\lambda)\right) \Big/ 2$$

$$f_{\varepsilon\varepsilon}(\lambda)\chi^2_{2K}\left(JK \sum_{k} |\eta_k + \bar{\theta}_{.k}|^2 / f_{\varepsilon\varepsilon}(\lambda)\right) \Big/ 2$$

respectively. The noncentrality parameters

$$J \sum_{i,k} |\theta_{ik} - \bar{\theta}_{.k}|^2; \qquad JK \sum_{k} |\eta_k + \bar{\theta}_{.k}|^2$$

will be independent

$$Jf_{\delta\delta}(\lambda)\chi^2_{2(I-1)K}; \qquad [JKf_{\gamma\gamma}(\lambda) + Jf_{\delta\delta}(\lambda)]\chi^2_{2K}$$

respectively. We may now use the lemma of the Appendix to see that

$$\sum_{i,j,k} |X_{ijk} - \bar{X}_{i.k}|^2 = f_{\varepsilon\varepsilon}(\lambda)\chi^2_{2I(J-1)K}/2 + o_{\text{a.s.}}(1)$$

$$J \sum_{i,k} |\bar{X}_{i.k} - \bar{X}_{..k}|^2 = [f_{\varepsilon\varepsilon}(\lambda) + Jf_{\delta\delta}(\lambda)]\chi^2_{2(I-1)K}/2 + o_{\text{a.s.}}(1)$$

$$IJ \sum_{k} |\bar{X}_{..k}|^2 = [f_{\varepsilon\varepsilon}(\lambda) + Jf_{\delta\delta}(\lambda) + JKf_{\gamma\gamma}(\lambda)]\chi^2_{2K}/2 + o_{\text{a.s.}}(1)$$

with the chi-squared variates appearing being independent.

In order to examine the hypothesis $f_{\delta\delta}(\lambda) = 0$, we are led to compute

$$\frac{J \sum |\bar{X}_{i.k} - \bar{X}_{..k}|^2/[2(I-1)K]}{\sum |X_{ijk} - \bar{X}_{i.k}|^2/[2I(J-1)K]} = \frac{J \sum f^{(T)}_{\bar{x}_{i.} - \bar{x}_{..},\, \bar{x}_{i.} - \bar{x}_{..}}(\lambda)/(I-1)}{\sum f^{(T)}_{x_{ij} - \bar{x}_{i.},\, x_{ij} - \bar{x}_{i.}}(\lambda)/[I(J-1)]}$$

$$= \frac{f_{\varepsilon\varepsilon}(\lambda) + Jf_{\delta\delta}(\lambda)}{f_{\varepsilon\varepsilon}(\lambda)} F_{2(I-1)K;\, 2I(J-1)K} + o_{\text{a.s.}}(1) \tag{4.5}$$

In order to examine the hypothesis $f_{\gamma\gamma}(\lambda) = 0$, we are likewise led to compute the ratio

$$\frac{IJ \sum |\bar{X}_{..k}|^2/[2K]}{\sum |X_{ijk} - \bar{X}_{i.k}|^2/[2I(J-1)K]} = \frac{IJ f^{(T)}_{\bar{x}..\bar{x}..}(\lambda)}{\sum f^{(T)}_{x_{ij}-\bar{x}_{i.}, x_{ij}-\bar{x}_{i.}}(\lambda)/[I(J-1)]}$$
$$= \frac{f_{\varepsilon\varepsilon}(\lambda) + Jf_{\delta\delta}(\lambda) + JKf_{\gamma\gamma}(\lambda)}{f_{\varepsilon\varepsilon}(\lambda) + Jf_{\delta\delta}(\lambda)} F_{2K;\,2I(J-1)K} + o_{a.s.}(1) \quad (4.6)$$

Returning to the empirical example introduced at the beginning of this section, Figure 4.1 presents, from the top down, the curves

$$\log_{10} IJf^{(T)}_{\bar{x}..\bar{x}..}(\lambda)$$
$$\log_{10} J \sum_i f^{(T)}_{\bar{x}_{i.}-\bar{x}.., \bar{x}_{i.}-\bar{x}..}(\lambda)$$
$$\log_{10} \sum_{i,j} f^{(T)}_{x_{ij}-\bar{x}_{i.}, x_{ij}-\bar{x}_{i.}}(\lambda)$$

for $0 \le \lambda/2\pi \le 0.5$. The estimates were computed using the first procedure discussed in Section 2 with $K = 10$. The bandwidth of the estimates is 0.03. Their asymptotic standard errors are 0.14, 0.14, 0.05, respectively. The pronounced peaks in each of the curves occur at the seasonal frequency of one cycle per year. The top two curves are not much different, except in the region

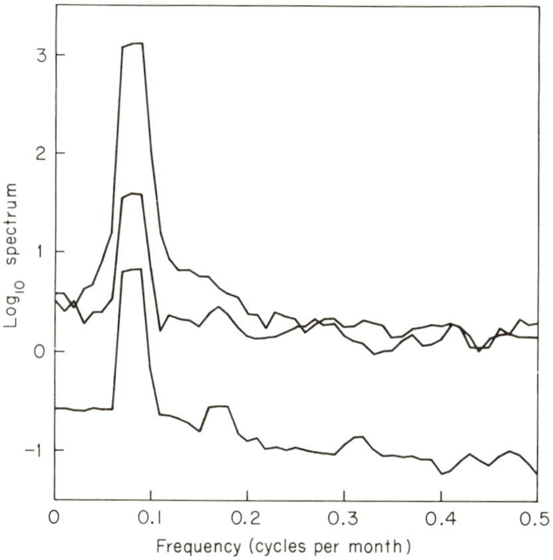

Fig. 4.1

of the seasonal frequency. The bottom two curves do seem to differ across the whole frequency domain.

Figure 4.2 presents $\log_{10}$ (4.5) and $\log_{10}$ (4.6). The second curve is the one fluctuating more or less around the level 1.0. The 95% points of the asymptotic null distributions here are $\log_{10} F_{20;20;.95} = 0.33$ and $\log_{10} F_{20;160;.95} = 0.27$. The first curve is above 0.33 only in the neighborhood of the seasonal frequency. This suggests that any component series common to the two

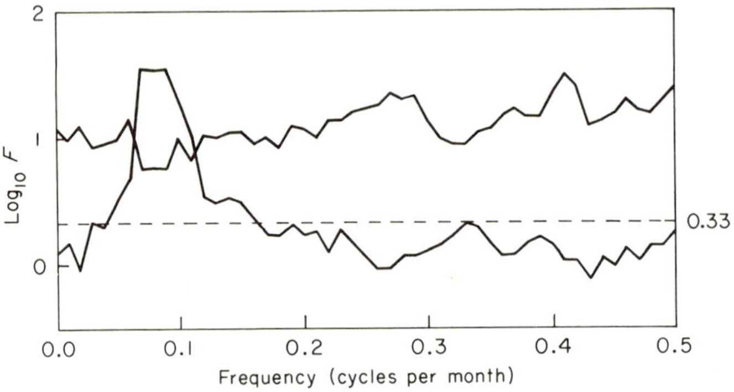

Fig. 4.2

continents is only a seasonal effect. The second curve is above 0.27 across the whole frequency domain. This suggests that there is a substantial within-continent components series.

In order to further investigate the suggestion that any component series common to the two continents is simply a seasonal effect, we recomputed the first curve using the series seasonally adjusted by removing monthly means. This gives us Figure 4.3. There is no longer any suggestion of a component common to the two continents.

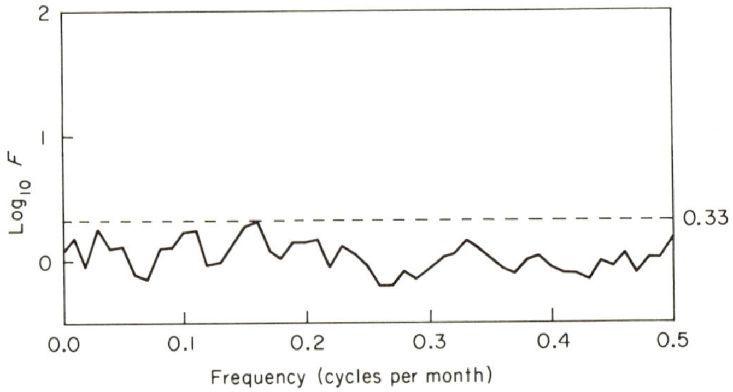

Fig. 4.3

The analysis of this section may clearly be modified to handle the case of a random effects model including transients, or to handle random effects data collected in more complicated experimental designs.

5. THE POINT PROCESS CASE

We next consider a situation in which the series observed $n_i(t)$, $0 < t \leq T$, $i = 1, \ldots, I$, are the components of a stationary point process $\mathbf{n}(t)$, $-\infty < t < \infty$. We discuss the simplest situation of data collected in a one-way classification with one replicate in each class. A model we might envisage is

$$n_i(t) = m_i(t) + p_i(t) \tag{5.1}$$

where the $p_i(t)$ are I independent realizations of a stationary point process with power spectrum $f_{pp}(\lambda)$, while independently the $m_i(t)$ are the components of an I vector-valued stationary point process $\mathbf{m}(t)$, $-\infty < t < \infty$, whose components are symmetrically dependent. Specifically, its spectral density matrix, $\mathbf{f}_{mm}(\lambda)$, has diagonal elements $f_{mm}(\lambda)$ and off-diagonal elements $R(\lambda)f_{mm}(\lambda)$ for some coherency coefficient $R(\lambda)$. In this situation $\mathbf{n}(t)$ is a stationary point process with spectral density matrix having diagonal elements $f_{mm}(\lambda) + f_{pp}(\lambda)$ and off-diagonal elements $R(\lambda)f_{mm}(\lambda)$. The coherency of any two of its components is therefore

$$R(\lambda)f_{mm}(\lambda)/[f_{mm}(\lambda) + f_{pp}(\lambda)]$$

In practice it might be of interest to examine the hypothesis $R(\lambda) = 0$.

As an example of a situation in which the series $m_i(t)$ are related in this manner, consider I cluster processes having the same series of cluster centers. Here

$$m_i(t) = \sum_j m_i'(t - \tau_j, j)$$

where the τ_j, $j = 0, \pm 1, \ldots$ denote the times of events of the primary process of cluster centers, while $m_i'(t, j)$, $i = 1, \ldots, I$ denote independent realizations of a subsidiary process. One can show that

$$f_{mm}(\lambda) = f_{m'm'}(\lambda) \left| E\left\{\sum_k \exp(i\lambda\sigma_k)\right\} \right|^2 + (2\pi)^{-1} f_{m'} \operatorname{var}\left\{\sum_k \exp(i\lambda\sigma_k)\right\}$$

and

$$R(\lambda) = f_{m'm'}(\lambda) \left| E\left\{\sum_k \exp(i\lambda\sigma_k)\right\} \right|^2 \bigg/ f_{mm}(\lambda)$$

where the σ_k denote the times of events of a realization of the subsidiary process while $f_{m'}$, $f_{m'm'}(\lambda)$ respectively denote the intensity and spectral density of the primary process.

The Fourier transform of (5.1) takes the form

$$N_{ik} = M_{ik} + P_{ik} \tag{5.2}$$

$k = 1, \ldots, K$; $i = 1, \ldots, I$. If the series $m(t)$ satisfies Assumption II then we may write $M_{ik} = \eta_{ik} + o_{\text{a.s.}}(1)$ where $\boldsymbol{\eta}_k = [\eta_{ik}; i = 1, \ldots, I], k = 1, \ldots, K$, are independent complex normal variates with mean $\mathbf{0}$ and covariance matrix $\mathbf{f}_{mm}(\lambda)$. Because of the symmetric dependence of the $\boldsymbol{\eta}_k$, we may write them in the form

$$\eta_{ik} = \theta_k + \zeta_{ik} \tag{5.3}$$

$i = 1, \ldots, I$, where θ_k is a complex normal variate with mean 0 and variance $|R(\lambda)|^2 f_{mm}(\lambda)$ and the ζ_{ik}, $i = 1, \ldots, I$, are independent complex normal variates with mean 0 and variance $[1 - |R(\lambda)|^2] f_{mm}(\lambda)$. If the series $p(t)$ satisfies Assumption II, then we may write $P_{ik} = \Pi_{ik} + o_{\text{a.s.}}(1)$ where the Π_{ik} are independent complex normal variate with mean 0 and variance $f_{pp}(\lambda)$.

Using these representations, the model (5.2) may be written

$$N_{ik} = \theta_k + \zeta_{ik} + \Pi_{ik} + o_{\text{a.s.}}(1) \tag{5.4}$$

where the θ_k, ζ_{ik}, Π_{ik} are all independent with variances $|R(\lambda)|^2 f_{mm}(\lambda)$, $[1 - |R(\lambda)|^2] f_{mm}(\lambda)$, $f_{pp}(\lambda)$, respectively. Suppose we are interested in the hypothesis $R(\lambda) = 0$. The expression (5.4) suggests the computation of the statistics

$$I \sum_k |\bar{N}_{.k}|^2 = IK f_{\bar{n}.\bar{n}.}^{(T)}(\lambda); \qquad \sum_{i,k} |N_{ik} - \bar{N}_{.k}|^2 = K \sum_i f_{n_i - \bar{n}., n_i - \bar{n}.}^{(T)}(\lambda)$$

in order to investigate this hypothesis. These sums of squares may be written

$$I \sum_k |\theta_k + \bar{\zeta}_{.k} + \bar{\Pi}_{.k}|^2 + o_{\text{a.s.}}(1); \quad \sum_{i,k} |\zeta_{ik} + \Pi_{ik} - \bar{\zeta}_{.k} - \bar{\Pi}_{.k}|^2 + o_{\text{a.s.}}(1)$$

respectively. Using the theorem and lemma of the Appendix, these may be written

$$(I |R(\lambda)|^2 f_{mm}(\lambda) + [1 - |R(\lambda)|^2] f_{mm}(\lambda) + f_{pp}(\lambda)) \chi_{2K}^2/2 + o_{\text{a.s.}}(1)$$
$$([1 - |R(\lambda)|^2] f_{mm}(\lambda) + f_{pp}(\lambda)) \chi_{2(I-1)K}^2/2 + o_{\text{a.s.}}(1)$$

respectively, with the chi-squared variates independent. The ratio

$$I f_{\bar{n}.\bar{n}.}^{(T)}(\lambda) \Big/ \left\{ \sum_i f_{n_i - \bar{n}., n_i - \bar{n}.}^{(T)}(\lambda)/(I - 1) \right\} \tag{5.5}$$

may therefore be written

$$\frac{I |R(\lambda)|^2 f_{mm}(\lambda) + [1 - |R(\lambda)|^2] f_{mm}(\lambda) + f_{pp}(\lambda)}{[1 - |R(\lambda)|^2] f_{mm}(\lambda) + f_{pp}(\lambda)} F_{2K; 2(I-1)K} + o_{\text{a.s.}}(1)$$

The practical problem that led us to propose the computations of this section involved two series of time of California earthquakes. The series covered the period 1945–1968. One series referred to events along a segment of the San Andreas Fault. The other referred to events along a neighboring segment of the Calaveras Fault. It was of interest to ask whether the earthquakes were occurring independently or whether certain of them had a common source of excitation.

Figure 5.1 presents a plot of the ratio (5.5). The spectra involved were computed using the last procedure described in Section 2 with $K = 10$. The 95% point of the asymptotic null distribution is $F_{20;\,20;\,0.95} = 2.12$. There is not much suggestion of a relation between the series.

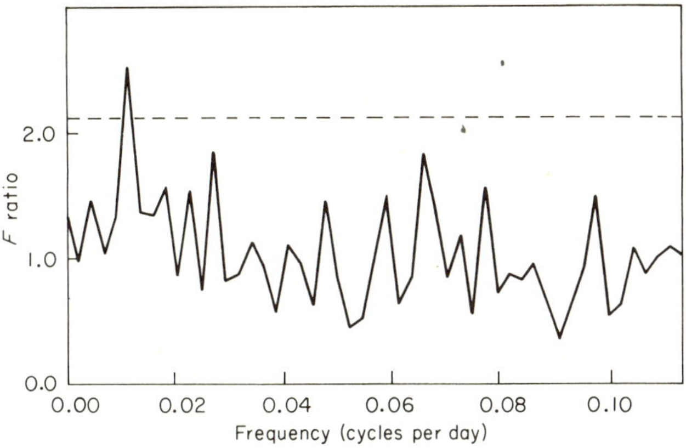

Fig. 5.1

ACKNOWLEDGMENT

I would like to thank Professor B. P. Bolt of the University of California Seismic Station for providing me with these data.

APPENDIX

Theorem. *Let* $\mathbf{Y}$ *be distributed as* $N_n^C(\boldsymbol{\mu}, \sigma^2\mathbf{I})$ *and let*

$$\mathbf{Y}^*\mathbf{Y} = \mathbf{Y}^*\mathbf{A}_1\mathbf{Y} + \cdots + \mathbf{Y}^*\mathbf{A}_K\mathbf{Y} \tag{A.1}$$

where $\mathbf{A}_k$ *is Hermitian of rank* n_k. *A necessary and sufficient condition for the forms* $\mathbf{Y}^*\mathbf{A}_k\mathbf{Y}$ *to be distributed independently with* $\mathbf{Y}^*\mathbf{A}_k\mathbf{Y}$ *distributed as* $\sigma^2\chi^2_{2n_k}(\boldsymbol{\mu}^*\boldsymbol{\mu}/\sigma^2)/2$ *is that* $n_1 + \cdots + n_K = n$.

Proof. We may write $\mathbf{Y} = \operatorname{Re}\mathbf{Y} + i \operatorname{Im}\mathbf{Y}$ with

$$\mathbf{X} = \begin{bmatrix} \operatorname{Re}\mathbf{Y} \\ \operatorname{Im}\mathbf{Y} \end{bmatrix} \quad \text{distributed as} \quad N_{2n}\left(\begin{bmatrix} \operatorname{Re}\boldsymbol{\mu} \\ \operatorname{Im}\boldsymbol{\mu} \end{bmatrix}, \frac{\sigma^2}{2}\mathbf{I}\right)$$

The identity (A.1) may be written

$$\mathbf{X}^\tau\mathbf{X} = \sum_1^K \mathbf{X}^\tau \begin{bmatrix} \operatorname{Re}\mathbf{A}_k & -\operatorname{Im}\mathbf{A}_k \\ \operatorname{Im}\mathbf{A}_k & \operatorname{Re}\mathbf{A}_k \end{bmatrix} \mathbf{X}$$

involving quadratic forms in real normal variables. The real analogue of the theorem (see Searle [8, p. 61]) now applies to give the indicated result.

Lemma. *Let λ be distributed as $a^2\chi_\nu^2$. Let the conditional distribution of u, given λ, be $b^2\chi_\nu^2(\lambda/b^2)$. Then the unconditional distribution of u is $(a^2 + b^2)\chi_\nu^2$.*

Proof. The characteristic function of λ/b^2 is

$$(1 - 2a^2t/b^2)^{-\nu/2}. \tag{A.2}$$

The conditional characteristic function of u is

$$(1 - 2b^2t)^{-\nu/2} \exp\{-[1 - (1 - 2b^2t)^{-1}]\lambda/b^2\}$$

Using (A.2), the unconditional characteristic function is

$$(1 - 2b^2t)^{-\nu/2}(1 + 2a^2[1 - (1 - 2b^2t)^{-1}]/b^2)^{-\nu/2} = (1 - 2[a^2 + b^2]t)^{-\nu/2}$$

which is the characteristic function of $(a^2 + b^2)\chi_\nu^2$.

REFERENCES

1. Brillinger, D. R. (1969). A search for a relationship between monthly sunspot numbers and certain climatic series. *Bull. I.S.I.* **43** 293–307.
2. Brillinger, D. R. (1972). The spectral analysis of stationary interval functions. *Proc. Sixth Berkeley Symp. Math. Statist. Prob.* (L. M. Le Cam, J. Neyman and E. L. Scott, eds.) **1** 483–513.
3. Brillinger, D. R. (1973). *The Frequency Analysis of Vector-valued Time Series.* Holt, New York.
4. Cooley, J. W. and Tukey, J. W. (1965). An algorithm for the machine calculation of complex Fourier series. *Math. Comp.* **19** 297–301.
5. Hannan, E. J. and Thomson, P. J. (1971). Spectral inference over narrow bands. *J. Appl. Probability* **8** 157–169.
6. Leonov, V. P. and Shiryaev, A. N. (1960). Some problems in the spectral theory of higher-order moments, II. *Theor. Probability Appl.* **5** 460–464.
7. Rosenblatt, M. (1956). A central limit theorem and a strong mixing condition. *Proc. Nat. Acad. Sci. U.S.A.* **42** 43–47.
8. Searle, S. R. (1971). *Linear Models.* Wiley, New York.
9. Shumway, R. H. (1970). Applied regression and analysis of variance for stationary time series. *J. Amer. Statist. Assoc.* **65** 1527–1546.
10. Shumway, R. H. (1971). On detecting a signal in N stationarily correlated noise series. *Technometrics* **13** 499–519.
11. Shumway, R. H. and Saikia, A. (1972). An empirical Bayes approach to stochastic signal estimation. Unpublished.
12. Skorokhod, A. V. (1956). Limit theorems for stochastic processes. *Theor. Probability Appl.* **1** 261–290.
13. Trewartha, G. T. (1954). *An Introduction to Climate.* McGraw-Hill, New York.
14. Wichura, M. (1970). On the construction of almost uniformly convergent random variables with given weakly convergent image laws. *Ann. Math. Statist.* **41** 284–291.
15. World Weather Records, Smithsonian Miscellaneous Collections, Vol. 79 (1927), Vol. 90 (1934) and Vol. 105 (1947). Smithsonian Inst., Washington.
16. World Weather Records, 1941–1950 (1959) and 1951–1960 (1965). U.S. Weather Bureau, Washington, D.C.

Max-Min Designs in the Analysis of Variance[1]

R. H. FARRELL

CORNELL UNIVERSITY

In a restricted context it is shown that M-optimality and E-optimality of analysis of variance designs are closely related. A simple matrix inequality is introduced that allows one to prove in a few lines the E-optimality of many designs having a high degree of symmetry.

1. INTRODUCTION AND MAX-MIN DESIGNS

The notations of this paper continue the usage started by Farrell [1]. Definitions of optimality may be found in the work of Kiefer [3], and from that paper one may obtain references to earlier work by A. Wald and S. Ehrenfeld.

We assume throughout that once a design C is decided and the random variable X is observed the experimenter will use an analysis of variance test. The problem considered here is that of comparison of designs in this limited context. X is an $H \times 1$ random vector and throughout we assume that the components of X are independent normally distributed random variables with common unknown variance σ^2. Our model takes the form

$$EX = (A, B_1, \ldots, B_h)\begin{pmatrix} \varphi_0 \\ \varphi_1 \\ \vdots \\ \varphi_h \end{pmatrix}$$

$A, B_1, \ldots, B_h$ are $H \times R$, $H \times S_1, \ldots, H \times S_h$ matrices, respectively, with 0, 1 entries and $\varphi_0, \varphi_1 \ldots, \varphi_n$ are $R \times 1$, $S_1 \times 1, \ldots, S_h \times 1$ vectors of unknown parameters. We suppose always that $h \geq 1$ and that e_k represents the $k \times 1$ vector with all entries equal to one. We assume $Ae_R = e_H$ and if $1 \leq i \leq h$, $B_i e_{S_i} = e_H$. We are then considering the problem of testing whether all contrasts of φ_0 are zero and optimality of designs will be considered

[1] Research supported in part under NSF grant GP24438.

relative to this problem. The number of degrees of freedom is a variable of the problem. In the discussion below we assume the degrees of freedom is fixed at a representative value $R-1$, $H-(R+S_1+\cdots+S_h-h)$ and we write

$$\beta\left(\frac{\varphi_0{}^T D\varphi_0}{\sigma^2}; \alpha, R-1, H-(R+S_1+\cdots+S_h-h)\right)$$

for the power function of the size α analysis of variance test. If $B=(B_1, \ldots, B_h)$, then $D = A^TA - (A^TB)(B^TB)^+(B^TA)$, as is shown by Farrell [1]. Here a superscript T means transpose and a superscript $+$ means generalized inverse.

The alternative will be represented by a set $\mathscr{A} \subset \mathbb{R}_R$ so that $(\varphi_0, \varphi_1, \ldots, \varphi_h, \sigma)$ is in the alternative if and only if $\varphi_0/\sigma \in \mathscr{A}$. We let $\mathscr{D}$ be the set of matrices D resulting from possible designs $C=(A, B_1, \ldots, B_h)$. We will say an alternative $\mathscr{A}$ which is topologically a closed set is *definite* if $\tau \in \mathscr{A}$ implies $\tau \neq 0$, and $\tau^T e_R = 0$.

In our more restricted problem M-optimality seeks a design C^* with associated matrix D^* satisfying

$$\inf_{\tau\in\mathscr{A}} \beta(\tau^T D^*\tau; \alpha, R-1, H-R-S+h) = \sup_{D\in\mathscr{D}} \inf_{\tau\in\mathscr{A}} \beta(\tau^T D\tau; \alpha, R-1, H-R-S+h).$$

Here we have set $S=S_1+\cdots+S_h$. Since the power function is a strictly increasing function, this is equivalent to the problem

$$\inf_{\tau\in\mathscr{A}} \tau^T D^*\tau = \sup_{D\in\mathscr{D}} \inf_{\tau\in\mathscr{A}} \tau^T D\tau = \sup_{D\in\mathscr{D}} \inf_{\lambda\in\bar{\mathscr{A}}} \operatorname{tr} D \int \tau^T\tau\lambda(d\tau)$$

where $\bar{\mathscr{A}}$ is the set of all probability measures on the Borel subsets of $\mathscr{A}$. Let $M(\lambda) = \int \tau^T\tau\lambda(d\tau)$. If the alternative $\mathscr{A}$ is definite, a standard game theory argument shows that the minimax problem has a value. That is, for some ε,

$$\varepsilon = \inf_{\lambda\in\bar{\mathscr{A}}} \sup_{D\in\mathscr{D}} \operatorname{tr} DM(\lambda) = \sup_{D\in\mathscr{D}} \inf_{\lambda\in\bar{\mathscr{A}}} \operatorname{tr} DM(\lambda) = \sup_{D\in\mathscr{D}} \inf_{\tau\in\mathscr{A}} \operatorname{tr} D\tau\tau^T.$$

A max-min design C^* with associated matrix D^* is one satisfying $\varepsilon = \inf_{\lambda\in\bar{\mathscr{A}}} \operatorname{tr} D^*M(\lambda)$. That is, max-min and D-optimality are the same. Since $\mathscr{D}$ is a finite set, such D^* exist.

Theorem 1. *Suppose the alternative $\mathscr{A}$ is definite. Then there exists $D^* \in \mathscr{D}$ and a number $\varepsilon \geq 0$ such that*

$$\varepsilon = \inf_{\lambda\in\bar{\mathscr{A}}} \operatorname{tr} D^*M(\lambda) = \sup_{D\in\mathscr{D}} \inf_{\tau\in\mathscr{A}} \operatorname{tr} D\tau^T\tau.$$

Suppose $C_1, \ldots, C_k$ are max-min designs and $(p_1, \ldots, p_k)$ is a probability vector. Suppose for some design C with associated matrix D that $D \geq \sum_{i=1}^{k} p_i D_i$. Then C is max-min.

Proof.

$$\varepsilon \geq \inf_{\lambda \in \bar{\mathscr{A}}} \operatorname{tr} DM(\lambda) \geq \inf_{\lambda \in \bar{\mathscr{A}}} \sum p_i \operatorname{tr} D_i M(\lambda) \geq \sum p_i \inf_{\lambda \in \bar{\mathscr{A}}} \operatorname{tr} D_i M(\lambda) = \varepsilon. \quad \|$$

Theorem 2. *Suppose for some $\delta > 0$ that $\mathscr{A} = \{\tau \mid \tau^T e_R = 0, \ \|\tau\| \geq \delta\}$. Then M-optimality (in the restricted problem) and E-optimality are equivalent.*

Proof. Write $\mu(D)$ for the minimum nonzero eigenvalue of D. Then if $\tau \in \mathscr{A}$, $\tau^T D\tau \geq \delta^2 \mu(D)$ (D has a single zero eigenvalue corresponding to the eigenvector e_R) and $\inf_{\tau \in \mathscr{A}} \tau^T D\tau = \delta^2 \mu(D)$. Thus $\varepsilon = \delta^2 \sup_{D \in \mathscr{D}} \mu(D)$. Therefore if D^* is M-optimal, then $\mu(D^*) = \sup_{D \in \mathscr{D}} \mu(D)$, and conversely if $\mu(D^*) = \sup_{D \in \mathscr{D}} \mu(D)$, then $\varepsilon = \inf_{\tau \in \mathscr{A}} \tau^T D^* \tau$. $\|$

Theorem 3 (Two-way classification). *Suppose $H \leq RS_1$, $h = 1$. Suppose $\mathscr{A}$ is definite, and $\mathscr{A}$ is invariant under all permutations of coordinates. Then a balanced incomplete block design, if it exists, is max-min (relative to the alternative $\mathscr{A}$).*

Proof. Let C be a balanced incomplete block design with associated matrix D. Then the diagonal elements of D are all equal to, say, d_1, and the off-diagonal elements are all equal to, say, d_2. As is well known, $\operatorname{tr} D = H - S_1 = Rd_1$. We use here the hypothesis that $H \leq RS_1$.

Suppose $C_1 = (A_1, B_1)$ is a max-min design with associated matrix D_1. To each permutation σ of $\{1, 2, \ldots, R\}$ let P_σ be the permutation matrix such that $(\tau_1, \ldots, \tau_R)P_\sigma^T = (\tau_{\sigma(1)}, \ldots, \tau_{\sigma(R)})$ for all $\tau \in \mathbb{R}_R$. Since $\mathscr{A}$ is invariant, $P_\sigma D_1 P_\sigma^T$ is again max-min, so that

$$\varepsilon \geq \inf_{\lambda \in \bar{\mathscr{A}}} \operatorname{tr}\left(\frac{1}{R!} \sum_\sigma P_\sigma D_1 P_\sigma^T\right) M(\lambda) \geq \inf_{\lambda \in \bar{\mathscr{A}}} \operatorname{tr} D_1 M(\lambda) \geq \inf_{\lambda \in \bar{\mathscr{A}}} \operatorname{tr} DM(\lambda).$$

However, because of the special form of D and $(1/R!) \sum_\sigma P_\sigma D_1 P_\sigma^T$, we find that $\operatorname{tr} DM(\lambda) = \mu(D) \int \|\tau\|^2 \lambda(d\tau)$ and

$$\operatorname{tr}\left(\frac{1}{R!} \sum_\sigma P_\sigma D_1 P_\sigma^T\right) M(\lambda) = \mu\left(\frac{1}{R!} \sum_\sigma P_\sigma D_1 P_\sigma^T\right) \int \|\tau\|^2 \lambda(d\tau).$$

Therefore

$$\operatorname{tr} D_1 = (R-1)\mu\left(\frac{1}{R!} \sum_\sigma P_\sigma D_1 P_\sigma^T\right) \geq (R-1)\mu(D) = \operatorname{tr} D.$$

It was shown by Farrell [1] that if ${A_1}^T B_1$ is not a matrix of 0's and 1's, then there is a design $C_2 = (A_2, B_2)$ such that ${A_2}^T B_2$ is a matrix of 0's and 1's and tr $D_2 >$ tr D_1. Since tr $D_2 = H - S_1$, this is a contradiction. Therefore, $D = (1/R!) \sum_\sigma P_\sigma D_1 {P_\sigma}^T$ and by Theorem 1, D is max-min. ||

2. A MATRIX INEQUALITY AND REGULAR DESIGNS

Theorem 4. *Let C be a design with associated $R \times R$ matrix D satisfying $De_R = 0$. Then* tr $D \le H(R-1)/R$.

Proof. Recall that $D = A^T A - (A^T B)(B^T B)^+ (B^T A) \le A^T A$. In reading this proof it will help to note that if the matrix $(A^T B)(B^T B)^+(B^T A)$ is of rank 1, then equality holds.

Let $\tau_1, \ldots, \tau_{R-1}$ be an orthonormal set of eigenvectors of D for the nonzero eigenvalues of D. And let $\tau_1', \ldots, \tau'_{R-1}$ be an orthonormal set of eigenvectors for the nonzero eigenvalues of $(1/R!) \sum_\sigma P_\sigma D {P_\sigma}^T$. Then

$$\begin{aligned}
\operatorname{tr} D &= \sum_{i=1}^{R-1} {\tau_i}^T D \tau_i = \operatorname{tr} D \sum_{i=1}^{R-1} \tau_i {\tau_i}^T = \operatorname{tr} D(Id - e_R {e_R}^T/R) \\
&= \operatorname{tr} D{P_\sigma}^T (Id - e_R {e_R}^T/R) P_\sigma = (1/R!) \sum_\sigma \operatorname{tr}(P_\sigma D {P_\sigma}^T)(Id - e_R {e_R}^T/R) \\
&= (\tau_1' + \cdots + \tau'_{R-1})^T (1/R!) \sum_\sigma (P_\sigma D {P_\sigma}^T)(\tau_1' + \cdots + \tau'_{R-1}) \\
&= (R-1) \sup_{\|\tau\|=1} \tau^T (1/R!) \sum_\sigma (P_\sigma D {P_\sigma}^T) \tau \\
&\le (R-1) \sup_{\|\tau\|=1} \tau^T (1/R!) \sum_\sigma P_\sigma A^T A {P_\sigma}^T \tau \\
&= (R-1) H/R. \quad \|
\end{aligned}$$

A design $C = (A, B_1, \ldots, B_h)$ will be said to be *regular* if $A^T A = (H/R)/Id$ (under our hypothesis that $Ae_R = e_H$ it follows that $A^T A$ is always a diagonal matrix) and $A^T B_i = K_i e_R e_{S_i}^T$, $i = 1, 2, \ldots, h$, where $K_1, \ldots, K_h$ are numbers. A design C will be said to be *balanced* if the associated matrix D has all its nonzero eigenvalues equal.

Theorem 5. *Suppose that $\mathscr{A}$ is definite and invariant under permutations of coordinates. Then a regular design, if it exists, is a balanced design and is max-min (relative to $\mathscr{A}$).*

Proof. Let $B = (B_1, \ldots, B_h)$. Then by hypothesis

$$A^T B = (K_1 e_R e_{S_1}^T, \ldots, K_h e_R e_{S_h}^T) = e_R f^T \qquad \text{where} \quad f^T = (K_1 e_{S_1}^T, \ldots, K_h e_{S_h}^T).$$

Therefore $A^T B$ is a rank 1 matrix and $(A^T B)(B^T B)^+(B^T A) = K(e_R {e_R}^T/R)$ is of rank 0 or 1. Therefore $D = (H/R)Id - K(e_R {e_R}^T/R)$. The eigenvalue 0 has the eigenvector e_R. This implies $K = H/R$ and $D = (H/R)(Id - e_R e_R^T/R)$. Then $\operatorname{tr} D = (H/R)(R-1)$ is maximal and this clearly implies that the minimum nonzero eigenvalue $\mu(D)$ is maximal. Repeating the proof of Theorem 3 shows D must be max-min. ||

The variables φ_0 may be interaction terms as well as main effects. This theory applies equally well. Designs exhibiting the regularity called for in Theorem 5 include all designs constructed using orthogonal arrays. See the work of Farrell *et al.* [2] for further discussion of and references about orthogonal arrays. In particular Latin Square designs are a special case of orthogonal arrays of strength 2, so Theorem 5 implies the E-optimality of all Latin Square designs (recall we are considering only analysis of variance tests!) and thus includes the results of Ehrenfeld. Also included are the class of designs introduced by Farrell [1, p. 1988].

Kiefer [3], has given an example of a two-way elimination ($h = 2$) problem in which the E-optimal design does not maximize the trace. Consequently the proof of Theorem 5 does not hold, whereas the proof of Theorem 2 does hold. This suggests that in some problems the set of alternatives for which a design is max-min may be larger than in other problems.

REFERENCES

1. Farrell, R. H. (1968). On the admissibility at ∞, within the class of randomized designs, of balanced designs. *Ann. Math. Statist.* **39** 1978–1994.
2. Farrell, R. H., Kiefer, J., and Walbran, A. (1967). Optimum multivariate designs. *Proc. Fifth Berkeley Symp. Math. Statist. Prob.* **1**.
3. Kiefer, J. (1958). On the nonrandomized and randomized nonoptimality of symmetrical designs. *Ann. Math. Statist.* **29** 675–699.

Analysis of Covariance Structures

K. G. JÖRESKOG
DEPARTMENT OF STATISTICS
UNIVERSITY OF UPPSALA
UPPSALA, SWEDEN

1. INTRODUCTION

Analysis of covariance structures [2, 3, 5, 23, 34, 42] is the common term for a number of different techniques for analyzing multivariate data where the variance covariance matrix is constrained to be of some particular form. The model considered in this paper is the same as in a previous paper [23]. Some additional results are given and some new applications are indicated.

The method to be described may be used to analyze data according to a model involving structures of a very general form on means, variances, and covariances of multivariate observations. With this method, a great deal of generality and flexibility is achieved in that the method is capable of handling most standard statistical models as well as many nonstandard and complicated ones.

When the variance-covariance matrix of the observed variables is unconstrained, the method may be used to estimate location parameters and to test linear hypotheses about these. For example, the method may be used to handle such standard problems as multivariate regression, ANOVA, and MANOVA. It can also be used for generalized MANOVA in the sense of Potthoff and Roy [35], Khatri [29], and Grizzle and Allen [14] (see also Rao [36–39], Gleser and Olkin [11, 12], and Geisser [10]. A unique feature is that the method can be used also when the variance-covariance matrix is constrained to be of a certain form. In this case one can estimate the covariance structure as well as location parameters and, in large samples, one can test various hypotheses about the structure of the variance-covariance matrix. This is useful in many areas and problems, particularly in the behavioral sciences. For example, one can handle such problems as analysis of multitrait–multimethod data, analysis of simplexes and circumplexes, analysis of multitest–multioccasion data and growth data in general, estimation of variance and covariance components, path analysis, and linear structural equations [21–25]. Various other models involving correlated errors can also be handled.

2. GENERAL RESULTS

2.1. The General Model

The general model considers a data matrix $\mathbf{X}(N \times p)$ of N observations on p variates and assumes that the rows of $\mathbf{X}$ are independently distributed, each having a multivariate normal distribution with the same variance-covariance matrix $\mathbf{\Sigma}$. It is assumed that

$$E(\mathbf{X}) = \mathbf{A\Xi P}, \tag{1}$$

where $\mathbf{A}(N \times g) = (a_{\alpha s})$ and $\mathbf{P}(h \times p) = (p_{ti})$ are known matrices of ranks g and h, respectively, $g \leq N$, $h \leq p$ and $\mathbf{\Xi}(g \times h) = (\xi_{st})$ is a matrix of parameters; and that $\mathbf{\Sigma}$ has the form

$$\mathbf{\Sigma} = \mathbf{B}(\mathbf{\Lambda\Phi\Lambda}' + \mathbf{\Psi}^2)\mathbf{B}' + \mathbf{\Theta}^2, \tag{2}$$

where the matrices $\mathbf{B}(p \times q) = (\beta_{ik})$, $\mathbf{\Lambda}(q \times r) = (\lambda_{km})$, the symmetric matrix $\mathbf{\Phi}(r \times r) = (\varphi_{mn})$, and the diagonal matrices $\mathbf{\Psi}(q \times q) = (\delta_{kl}\psi_k)$ and $\mathbf{\Theta}(p \times p) = (\delta_{ij}\theta_i)$ are parameter matrices. (Throughout this paper δ_{ij} denotes the Kronecker delta, which is one if $i = j$ and zero otherwise.) If the rank conditions on A and P do not hold, the model may be reparameterized in terms of a smaller matrix $\mathbf{\Xi}$ and new matrices A and P satisfying the rank conditions (see, e.g., Gleser and Olkin [12]).

Thus the general model is one where means, variances, and covariances are structured in terms of other sets of parameters that are to be estimated. In any application of this model, p, N, and $\mathbf{X}$ will be given by the data, and g, h, q, r, $\mathbf{A}$, and $\mathbf{P}$ will be given by the particular application. In any such application we shall allow for any one of the parameters in $\mathbf{\Xi}$, $\mathbf{B}$, $\mathbf{\Lambda}$, $\mathbf{\Phi}$, $\mathbf{\Psi}$, and $\mathbf{\Theta}$ to be known a priori and for one or more subsets of the remaining parameters to have identical but unknown values. Thus parameters are of three kinds: (i) *fixed parameters* that have been assigned given values; (ii) *constrained parameters* that are unknown but equal to one or more other parameters; and (iii) *free parameters* that are unknown and not constrained to be equal to any other parameter.

The general method attempts to estimate the free and constrained parameters of any such model by the maximum likelihood method and provides a test of goodness of fit of the whole model against the general alternative that $\mathbf{P}$ is square and $\mathbf{\Xi}$ and $\mathbf{\Sigma}$ are unconstrained. A test of a specified model (hypothesis) H_0 against any more general alternative model H_1 may be obtained, in large samples, by computing the maximum likelihood solution under the two models and then setting up the likelihood ratio test. In the special case when both $\boldsymbol{\Xi}$ and $\boldsymbol{\Sigma}$ are unconstrained, one may test a sequence of linear hypotheses of the form

$$\mathbf{C\Xi D} = \mathbf{0}, \tag{3}$$

where $\mathbf{C}(s \times g)$ and $\mathbf{D}(h \times t)$ are given matrices of ranks s and t, respectively.

2.2. Identification of Parameters

Before an attempt is made to estimate a model of this kind, the identification problem must be examined. The identification problem depends on the specification of fixed, free, and constrained parameters.

It should be noted that if $\mathbf{B}$ is replaced by $\mathbf{BT}_1^{-1}$, $\boldsymbol{\Lambda}$ by $\mathbf{T}_1\boldsymbol{\Lambda}\mathbf{T}_2^{-1}$, $\boldsymbol{\Phi}$ by $\mathbf{T}_2\boldsymbol{\Phi}\mathbf{T}_2'$, and $\boldsymbol{\Psi}^2$ by $\mathbf{T}_1\boldsymbol{\Psi}^2\mathbf{T}_1'$ while $\boldsymbol{\Theta}$ is left unchanged, then $\boldsymbol{\Sigma}$ is unaffected. This holds for all nonsingular matrices $\mathbf{T}_1(q \times q)$ and $\mathbf{T}_2(r \times r)$ such that $\mathbf{T}_1\boldsymbol{\Psi}^2\mathbf{T}_1'$ is diagonal. Hence in order to obtain a unique set of parameters and a corresponding unique set of estimates, some restrictions must be imposed. In most cases these restrictions are given in a natural way by the particular application of the model (see Section 3 of the paper). In other cases they can be chosen in any convenient way by specifying certain parameters to be fixed to certain values. In what follows it is assumed that all such indeterminacies have been eliminated by the specification of fixed and constrained parameters. To make sure that all indeterminacies have been eliminated, one should verify that the only transformations $\mathbf{T}_1$ and $\mathbf{T}_2$ that preserve the specifications about fixed and constrained parameters are identity matrices.

2.3. Matrices U, V, and W

The information provided by the matrices $\mathbf{X}$ and $\mathbf{A}$ is most conveniently summarized in three matrices $\mathbf{U}$, $\mathbf{V}$, and $\mathbf{W}$ of sums of squares and cross products, defined as follows.

$$\mathbf{U}(g \times g) = (1/N)\mathbf{A}'\mathbf{A}; \tag{4}$$

$$\mathbf{V}(g \times p) = (1/N)\mathbf{A}'\mathbf{X}; \tag{5}$$

$$\mathbf{W}(p \times p) = (1/N)\mathbf{X}'\mathbf{X}. \tag{6}$$

2.4. Special Case

We now consider the estimation of the free and constrained parameters of the general model and distinguish between two different cases as follows.

Special Case—Both $\boldsymbol{\Xi}$ and $\boldsymbol{\Sigma}$ unconstrained;
General Case—Otherwise.

The logarithm of the likelihood is

$$\log L = -\tfrac{1}{2}pN\log(2\pi) - \tfrac{1}{2}N\log|\boldsymbol{\Sigma}| - \frac{1}{2}\sum_{\alpha=1}^{N}\sum_{i=1}^{p}\sum_{j=1}^{p}(x_{\alpha i} - \mu_{\alpha i})\sigma^{ij}(x_{\alpha j} - \mu_{\alpha j}),$$

where $\mu_{\alpha i}$ and σ^{ij} are elements of $E(\mathbf{X})$ and $\boldsymbol{\Sigma}^{-1}$, respectively. Writing

$$\mathbf{T}(\boldsymbol{\Xi}) = (1/N)(\mathbf{X} - \mathbf{A}\boldsymbol{\Xi}\mathbf{P})'(\mathbf{X} - \mathbf{A}\boldsymbol{\Xi}\mathbf{P}) = \mathbf{W} - \mathbf{P}'\boldsymbol{\Xi}'\mathbf{V} - \mathbf{V}'\boldsymbol{\Xi}\mathbf{P} + \mathbf{P}'\boldsymbol{\Xi}'\mathbf{U}\boldsymbol{\Xi}\mathbf{P}, \quad (7)$$

$\log L$ may be written, omitting the constant term,

$$\log L = -\tfrac{1}{2}N[\log|\boldsymbol{\Sigma}| + \operatorname{tr}(\mathbf{T}\boldsymbol{\Sigma}^{-1})].$$

Maximizing $\log L$ is therefore equivalent to minimizing

$$F = \tfrac{1}{2}[\log|\boldsymbol{\Sigma}| + \operatorname{tr}(\mathbf{T}\boldsymbol{\Sigma}^{-1})]. \quad (8)$$

The function F is regarded as a function of $\boldsymbol{\Xi}$, $\mathbf{B}$, $\boldsymbol{\Lambda}$, $\boldsymbol{\Phi}$, $\boldsymbol{\Psi}$, and $\boldsymbol{\Theta}$, remembering that $\mathbf{T}$ is a function of $\boldsymbol{\Xi}$ by (7) and $\boldsymbol{\Sigma}$ is a function of $\mathbf{B}$, $\boldsymbol{\Lambda}$, $\boldsymbol{\Phi}$, $\boldsymbol{\Psi}$, and $\boldsymbol{\Theta}$ by (2).

Consider first the special case. This may be defined in terms of fixed, free, and constrained parameters by specifying $r = q = p$ and $\mathbf{B} = \mathbf{I}$, $\boldsymbol{\Lambda} = \mathbf{I}$, $\boldsymbol{\Psi} = \mathbf{0}$, $\boldsymbol{\Theta} = \mathbf{0}$. Then $\boldsymbol{\Sigma}$ is identical to $\boldsymbol{\Phi}$, which is unconstrained except for symmetry and positive definiteness. The mean vectors are supposed to satisfy (1) with all the elements of $\boldsymbol{\Xi}$ free.

In this case the maximum likelihood estimate of $\boldsymbol{\Xi}$ is

$$\hat{\boldsymbol{\Xi}} = \mathbf{U}^{-1}\mathbf{V}\mathbf{S}^{-1}\mathbf{P}'(\mathbf{P}\mathbf{S}^{-1}\mathbf{P}')^{-1} \quad (9)$$

and that of $\boldsymbol{\Sigma}$ is

$$\hat{\boldsymbol{\Sigma}} = \mathbf{S} + \hat{\mathbf{Q}}'\mathbf{U}\hat{\mathbf{Q}}, \quad (10)$$

where $\mathbf{S} = \mathbf{W} - \mathbf{V}'\mathbf{U}^{-1}\mathbf{V}$ and

$$\hat{\mathbf{Q}} = \mathbf{U}^{-1}\mathbf{V} - \hat{\boldsymbol{\Xi}}\mathbf{P}. \quad (11)$$

It should be noted that if P is square and nonsingular, formulas (9), (10), and (11) reduce to the ordinary formulas for MANOVA, i.e.,

$$\hat{\boldsymbol{\Xi}} = \mathbf{U}^{-1}\mathbf{V}\mathbf{P}^{-1}, \quad (12)$$

$$\hat{\boldsymbol{\Sigma}} = \mathbf{S}, \quad (13)$$

$$\hat{\mathbf{Q}} = \mathbf{0}. \quad (14)$$

To test the hypothesis $\mathbf{C}\boldsymbol{\Xi}\mathbf{D} = \mathbf{0}$ against $\mathbf{C}\boldsymbol{\Xi}\mathbf{D} \neq \mathbf{0}$ one uses

$$\mathbf{S}_e = \mathbf{D}'(\mathbf{P}\mathbf{S}^{-1}\mathbf{P}')^{-1}\mathbf{D}, \quad (15)$$

$$\mathbf{S}_h = (\mathbf{C}\hat{\boldsymbol{\Xi}}\mathbf{D})'(\mathbf{C}\hat{\mathbf{R}}\mathbf{C}')^{-1}(\mathbf{C}\hat{\boldsymbol{\Xi}}\mathbf{D}), \quad (16)$$

where $\hat{\mathbf{R}} = \mathbf{U}^{-1} + \hat{\mathbf{Q}}\mathbf{S}^{-1}\hat{\mathbf{Q}}'$. Let the eigenvalues of $\mathbf{S}_h\mathbf{S}_e^{-1}$ be $\lambda_1 \geq \lambda_2 \geq \cdots \geq \lambda_t$. One can then use any one of the three test statistics:

$$\text{largest root} = \lambda;$$

$$\text{sum of roots} = \sum_{i=1}^{t} \lambda_i;$$

$$\text{likelihood ratio} = \left[\prod_{i=1}^{t}(1 + \lambda_i)\right]^{-1}$$

The largest root test, due to Roy [40], can be used with Heck's [18] tables. The sum of roots test is due to Lawley [31] and Hotelling [19]. The likelihood ratio test is an extension of Wilks' [44] λ-test and can be used with correction tables provided by Schatzoff [41]. When N is large, $-[N - g - (p - h) - \frac{1}{2}(t - s + 1)]$ times the likelihood ratio is approximately distributed as χ^2 with st degrees of freedom.

2.5. General Case

The maximum likelihood estimates may be obtained numerically by minimizing the function F in (8) with respect to the free parameters in $\boldsymbol{\Xi}$, $\mathbf{B}$, $\boldsymbol{\Lambda}$, $\boldsymbol{\Phi}$, $\boldsymbol{\Psi}$, $\boldsymbol{\Theta}$.

However, it is better to apply the minimization method not directly to F but instead to

$$\begin{aligned} f(\mathbf{B}, \boldsymbol{\Lambda}, \boldsymbol{\Phi}, \boldsymbol{\Psi}, \boldsymbol{\Theta}) &= \min_{\boldsymbol{\Xi}} F(\boldsymbol{\Xi}, \mathbf{B}, \boldsymbol{\Lambda}, \boldsymbol{\Phi}, \boldsymbol{\Psi}, \boldsymbol{\Theta}) \\ &= F(\hat{\boldsymbol{\Xi}}_{\boldsymbol{\Sigma}}, \mathbf{B}, \boldsymbol{\Lambda}, \boldsymbol{\Phi}, \boldsymbol{\Psi}, \boldsymbol{\Theta}), \end{aligned}$$

where $\hat{\boldsymbol{\Xi}}_{\boldsymbol{\Sigma}}$ minimizes F for given $\boldsymbol{\Sigma}$. If $\boldsymbol{\Xi}$ is unconstrained,

$$\hat{\boldsymbol{\Xi}}_{\boldsymbol{\Sigma}} = \mathbf{U}^{-1}\mathbf{V}\boldsymbol{\Sigma}^{-1}\mathbf{P}'(\mathbf{P}\boldsymbol{\Sigma}^{-1}\mathbf{P}')^{-1}, \tag{17}$$

but this formula cannot be used if $\boldsymbol{\Xi}$ contains fixed and/or constrained elements. Nevertheless, $\hat{\boldsymbol{\Xi}}_{\boldsymbol{\Sigma}}$ can easily be evaluated since, for given $\boldsymbol{\Sigma}$, F is quadratic in $\boldsymbol{\Xi}$. The minimization of f takes into account the specification of fixed, free, and constrained parameters as described in Section 2.7. During the minimization, f is regarded as a function of the independent parameters $\boldsymbol{\theta}' = (\theta_1, \theta_2, \ldots, \theta_m)$, say.

The derivatives of f are

$$\partial f/\partial \mathbf{B} = \boldsymbol{\Omega}\mathbf{B}(\boldsymbol{\Lambda}\boldsymbol{\Phi}\boldsymbol{\Lambda}' + \boldsymbol{\Psi}^2), \tag{18}$$

$$\partial f/\partial \boldsymbol{\Lambda} = \mathbf{B}'\boldsymbol{\Omega}\mathbf{B}\boldsymbol{\Lambda}\boldsymbol{\Phi}, \tag{19}$$

$$\partial f/\partial \boldsymbol{\Phi} = \tfrac{1}{2}\boldsymbol{\Lambda}'\mathbf{B}'\boldsymbol{\Omega}\mathbf{B}\boldsymbol{\Lambda}, \tag{20}$$

$$\partial f/\partial \boldsymbol{\Psi} = \mathbf{B}'\boldsymbol{\Omega}\mathbf{B}\boldsymbol{\Psi}, \tag{21}$$

$$\partial f/\partial \boldsymbol{\Theta} = \boldsymbol{\Omega}\boldsymbol{\Theta}, \tag{22}$$

where

$$\boldsymbol{\Omega} = \boldsymbol{\Sigma}^{-1}[\boldsymbol{\Sigma} - \mathbf{T}(\hat{\boldsymbol{\Xi}}_{\boldsymbol{\Sigma}})]\boldsymbol{\Sigma}^{-1}. \tag{23}$$

In (20) the symmetry of $\boldsymbol{\Phi}$ has not been taken into account and in (21) and (22) the diagonality of $\boldsymbol{\Phi}$ and $\boldsymbol{\Theta}$ has not been taken into account. The symmetry of $\boldsymbol{\Phi}$ is handled by equality constraints on the off-diagonal elements and the diagonality of $\boldsymbol{\Psi}$ and $\boldsymbol{\Theta}$ is handled by fixed zero off-diagonal elements.

2.6. The Information Matrix

The elements of $T(\hat{\boldsymbol{\Xi}}_{\boldsymbol{\Sigma}})$ have an asymptotic multinormal distribution with mean $\boldsymbol{\Sigma}$. Let ε_{ij} be a typical element of $\mathbf{T}(\hat{\boldsymbol{\Xi}}_{\boldsymbol{\Sigma}}) - \boldsymbol{\Sigma}$. Then

$$NE(\varepsilon_{gh}\varepsilon_{ij}) = \sigma_{gi}\sigma_{hj} + \sigma_{gj}\sigma_{hi}. \tag{24}$$

We shall prove a general theorem concerning the asymptotic means of the second-order derivatives of any function of the type (8) and show how this theorem can be applied to compute all the elements of the information matrix. The inverse of the information matrix provides an asymptotic variance-covariance matrix of the maximum likelihood estimates.

We first prove the following

Lemma. *Let* $\mathbf{T} = \mathbf{T}(\hat{\boldsymbol{\Xi}}_{\boldsymbol{\Sigma}})$. *Then the asymptotic distribution of the elements of* $\boldsymbol{\Omega} = \boldsymbol{\Sigma}^{-1}(\boldsymbol{\Sigma} - \mathbf{T})\boldsymbol{\Sigma}^{-1}$ *is multivariate normal with means zero and variances and covariances given by*

$$NE(\omega_{\alpha\beta}\omega_{\mu\nu}) = \sigma^{\alpha\mu}\sigma^{\beta\nu} + \sigma^{\alpha\nu}\sigma^{\beta\mu}.$$

Proof. The proof follows immediately by multiplying

$$\omega_{\alpha\beta} = \sum_g \sum_h \sigma^{\alpha g}(\sigma_{gh} - t_{gh})\sigma^{h\beta} \qquad \text{and} \qquad \omega_{\mu\nu} = \sum_i \sum_j \sigma^{\mu i}(\sigma_{ij} - t_{ij})\sigma^{j\nu}$$

and using (24).

We can now prove the following general theorem.

Theorem. *Let the elements of* $\boldsymbol{\Sigma}$ *be functions of two parameter matrices* $\mathbf{M} = (\mu_{gh})$ *and* $\mathbf{N} = (\nu_{ij})$ *and let* $F(\mathbf{M}, \mathbf{N}) = \frac{1}{2}[\log|\boldsymbol{\Sigma}| + \operatorname{tr}(\mathbf{T}\boldsymbol{\Sigma}^{-1})]$, *where* $T = T(\hat{\boldsymbol{\Xi}}_{\boldsymbol{\Sigma}})$. *Then if* $\partial F/\partial\mathbf{M} = \mathbf{A\Omega B}$ *and* $\partial F/\partial\mathbf{N} = \mathbf{C\Omega D}$, *where* $\mathbf{A}, \mathbf{B}, \mathbf{C}, \mathbf{D}$ *are independent of* $\mathbf{T}$ *and* $\boldsymbol{\Omega} = \boldsymbol{\Sigma}^{-1}(\boldsymbol{\Sigma} - \mathbf{T})\boldsymbol{\Sigma}^{-1}$, *we have asymptotically*

$$E(\partial^2 F/\partial\mu_{gh}\,\partial\nu_{ij}) = (\mathbf{A}\boldsymbol{\Sigma}^{-1}\mathbf{C}')_{gi}(\mathbf{B}'\boldsymbol{\Sigma}^{-1}\mathbf{D})_{hj} + (\mathbf{A}\boldsymbol{\Sigma}^{-1}\mathbf{D})_{gj}(\mathbf{B}'\boldsymbol{\Sigma}^{-1}\mathbf{C}')_{hi}. \tag{25}$$

Proof. Writing $\partial F/\partial\mu_{gh} = a_{g\alpha}\omega_{\alpha\beta}b_{\beta h}$ and $\partial F/\partial\nu_{ij} = c_{i\mu}\omega_{\mu\nu}d_{\nu j}$, where it is assumed that every repeated subscript is to be summed over, we have (cf. Kendall and Stuart [28, Eq. 18.57])

$$
\begin{aligned}
E(\partial^2 F/\partial \mu_{gh}\, \partial \nu_{ij}) &= NE(\partial F/\partial \mu_{gh}\, \partial F/\partial \nu_{ij}) \\
&= NE(a_{g\alpha}\, \omega_{\alpha\beta}\, b_{\beta h}\, c_{i\mu}\, \omega_{\mu\nu}\, d_{\nu j}) \\
&= N a_{g\alpha}\, b_{\beta h}\, c_{i\mu}\, d_{\nu j}\, E(\omega_{\alpha\beta}\, \omega_{\mu\nu}) \\
&= a_{g\alpha}\, b_{\beta h}\, c_{i\mu} d_{\nu j} (\sigma^{\alpha\mu}\sigma^{\beta\nu} + \sigma^{\alpha\nu}\sigma^{\beta\mu}) \\
&= (a_{g\alpha}\sigma^{\alpha\mu} c_{i\mu})(b_{\beta h}\sigma^{\beta\nu} d_{\nu j}) + (a_{g\alpha}\sigma^{\alpha\nu} d_{\nu j})(b_{\beta h}\sigma^{\beta\mu} c_{i\mu}) \\
&= (\mathbf{A\Sigma}^{-1}\mathbf{C}')_{gi}(\mathbf{B}'\mathbf{\Sigma}^{-1}\mathbf{D})_{hj} + (\mathbf{A\Sigma}^{-1}\mathbf{D})_{gj}(\mathbf{B}'\mathbf{\Sigma}^{-1}\mathbf{C}')_{hi}.
\end{aligned}
$$

It should be noted that the theorem is quite general in that both $\mathbf{M}$ and $\mathbf{N}$ may be row or column vectors or scalars and $\mathbf{M}$ and $\mathbf{N}$ may be identical, in which case, of course, $\mathbf{A} \equiv \mathbf{C}$ and $\mathbf{B} \equiv \mathbf{D}$.

From (18)–(22) it is seen that the derivatives are all of the form required by the theorem, so that this makes it possible to compute the whole information matrix.

2.7. The Handling of Fixed and Constrained Parameters

Let $\boldsymbol{\mu}' = (\mu_1, \mu_2, \ldots, \mu_k)$ be a vector of *all* elements in all five parameter matrices $\mathbf{B}$, $\mathbf{\Lambda}$, $\mathbf{\Phi}$, $\mathbf{\Psi}$, and $\mathbf{\Theta}$, and consider f as a function of $\boldsymbol{\mu}$. This function is continuous and has continuous derivatives of first and second order except where $\mathbf{\Sigma}$ is singular. The totality of these derivatives is represented by a gradient vector $\partial f/\partial \boldsymbol{\mu}$ and a symmetric Hessian matrix $\partial^2 f/\partial \boldsymbol{\mu}\, \partial \boldsymbol{\mu}'$. Some $k - l$ of the μ's are fixed. Let the remaining l μ's form a vector $\mathbf{v}' = (\nu_1, \nu_2, \ldots, \nu_l)$. Derivatives $\partial f/\partial \mathbf{v}$ and $\partial^2 f/\partial \mathbf{v}\, \partial \mathbf{v}'$ are obtained from $\partial f/\partial \boldsymbol{\mu}$ and $\partial^2 f/\partial \boldsymbol{\mu}\, \partial \boldsymbol{\mu}'$ by elimination of the rows and columns corresponding to the fixed $\boldsymbol{\mu}$'s. Among $\nu_1, \nu_2, \ldots, \nu_l$ there are some distinct parameters $\theta_1, \theta_2, \ldots, \theta_m$, assumed to be identifiable. Let $k_{ig} = 1$ if $\nu_i = \theta_g$ and $k_{ig} = 0$ otherwise and let $\mathbf{K} = (k_{ig})$, $i = 1, 2, \ldots, l$, $g = 1, 2, \ldots, m$. Then we have

$$\partial f/\partial \boldsymbol{\theta} = \mathbf{K}'\, \partial f/\partial \mathbf{v}, \qquad \partial^2 f/\partial \boldsymbol{\theta}\ \partial \boldsymbol{\theta}' = \mathbf{K}'\, \partial^2 f/\partial \mathbf{v}\, \partial \mathbf{v}'\mathbf{K} \tag{26}$$

and from (26),

$$E(\partial^2 f/\partial \boldsymbol{\theta}\ \partial \boldsymbol{\theta}') = \mathbf{K}' E(\partial^2 f/\partial \mathbf{v}\ \partial \mathbf{v}')\mathbf{K} \tag{27}$$

The elements of the information matrix on the right-hand side of (27) are obtained as described in the previous section.

2.8. Basic Minimization Algorithm

The function $f(\boldsymbol{\theta})$ may be minimized numerically by Fisher's scoring method or the method of Fletcher and Powell [9] (see also Gruvaeus and Jöreskog [15]).

The minimization starts at an arbitrary starting point $\boldsymbol{\theta}^{(1)}$ and generates successively new points $\boldsymbol{\theta}^{(2)}, \boldsymbol{\theta}^{(3)}, \ldots,$ such that $f(\boldsymbol{\theta}^{(s+1)}) < f(\boldsymbol{\theta}^{(s)})$ until convergence is obtained. Let $\mathbf{g}^{(s)}$ be the gradient vector $\partial f/\partial\boldsymbol{\theta}$ at $\boldsymbol{\theta} = \boldsymbol{\theta}^{(s)}$ and let $\mathbf{E}^{(s)}$ be the information matrix $E(\partial^2 f/\partial\boldsymbol{\theta}\,\partial\boldsymbol{\theta}')$ evaluated at $\boldsymbol{\theta} = \boldsymbol{\theta}^{(s)}$. Then Fisher's scoring method computes a correction vector by solving the equation system

$$\mathbf{E}^{(s)}\boldsymbol{\delta}^{(s)} = \mathbf{g}^{(s)} \tag{28}$$

and then computes the new point as

$$\boldsymbol{\theta}^{(s+1)} = \boldsymbol{\theta}^{(s)} - \boldsymbol{\delta}^{(s)}. \tag{29}$$

This requires the computation of the inverse of $\mathbf{E}^{(s)}$ in each iteration and this is often quite time consuming. An alternative is to use the method of Fletcher and Powell, which evaluates only the inverse of $\mathbf{E}^{(1)}$ and in subsequent iterations $\mathbf{E}$ is improved, using information built up about the function, so that ultimately $\mathbf{E}$ converges to an approximation of $\partial^2 f/\partial\boldsymbol{\theta}\,\partial\boldsymbol{\theta}'$ at the minimum.

A computer program based on the Fletcher and Powell method has been written by Jöreskog *et al.* [27].

2.9. Tests of Hypotheses

Let H_0 be any specific hypothesis concerning the parametric structure of the general model and let H_1 be an alternative hypothesis. One can then test H_0 against H_1 by means of the likelihood ratio technique. Let F_0 be the minimum of F under H_0 and let F_1 be the minimum of F under H_1. Then $F_1 \leq F_0$ and minus two times the logarithm of the likelihood ratio becomes $\frac{1}{2}N(F_0 - F_1)$. Under H_0 this is distributed, in large samples, as a χ^2 distribution with degrees of freedom equal to the difference in the number of independent parameters estimated under H_1 and H_0.

In general this requires the computation of the solution under both H_0 and H_1. However, for most of the useful alternatives H_1, the solution is known and the value of F_1 can be computed from some simple sample statistics. One such general alternative is when $\mathbf{P}$ is square (i.e., $h = p$) and nonsingular, and $\boldsymbol{\Xi}$ and $\boldsymbol{\Sigma}$ unconstrained. Then, under H_1, the maximum likelihood estimates of $\boldsymbol{\Xi}$ and $\boldsymbol{\Sigma}$ are given by (12) and (13), respectively, and the test statistic becomes

$$u = N(2F_0 - \log|\hat{\boldsymbol{\Sigma}}| - p) \tag{30}$$

with degrees of freedom

$$d = gp + \tfrac{1}{2}p(p+1) - m \tag{31}$$

where m is the number of independent parameters estimated under H_0.

3. APPLICATIONS

3.1. Test Theory Models

Most measurements employed in the behavioral sciences contain sizeable errors of measurements and any adequate theory or model must take this fact into account. Of particular importance is the study of congeneric measurements, i.e., those measurements that are assumed to measure the same thing.

Classical test theory [33] assumes that a test score x is the sum of a true score τ and an error score e, where e and τ are uncorrelated. A set of test scores $x_1, \ldots, x_p$ with true scores $\tau_1, \ldots, \tau_p$ is said to be congeneric if every pair of true scores τ_i and τ_j has unit correlation. Such a set of test scores can be represented as

$$\mathbf{x} = \boldsymbol{\mu} + \boldsymbol{\beta}\tau + \mathbf{e},$$

where $\mathbf{x}' = (x_1, \ldots, x_p)$, $\boldsymbol{\beta}' = (\beta_1, \ldots, \beta_p)$ is a vector of regression coefficients, $\mathbf{e}' = (e_1, \ldots, e_p)$ is the vector of error scores, $\boldsymbol{\mu}$ is the mean vector of $\mathbf{x}$, and τ is a true score, for convenience scaled to zero mean and unit variance. The elements of $\mathbf{x}$, $\mathbf{e}$, and τ are regarded as random variables for a population of examinees. Let $\theta_1^2, \ldots, \theta_p^2$ be the variances of $e_1, \ldots, e_p$, respectively, i.e., the error variances. The corresponding true score variances are $\beta_1^2, \ldots, \beta_p^2$. One important problem is that of estimating these quantities. The variance-covariance matrix of $\mathbf{x}$ is

$$\boldsymbol{\Sigma} = \boldsymbol{\beta\beta}' + \boldsymbol{\Theta}^2, \tag{32}$$

where $\boldsymbol{\Theta} = \text{diag}(\theta_1, \ldots, \theta_p)$. This is a special case of (2) obtained by specifying $q = r = 1$, $\mathbf{B} = \boldsymbol{\beta}$, $\boldsymbol{\Lambda} = \boldsymbol{\Phi} = 1$, and $\boldsymbol{\Psi} = 0$.

Parallel tests and tau-equivalent tests, in the sense of Lord and Novick [33], are special cases of congeneric tests. Parallel tests have equal true score variances and equal error variances, i.e.,

$$\beta_1^2 = \cdots = \beta_p^2, \qquad \theta_1^2 = \cdots = \theta_p^2.$$

Tau-equivalent tests have equal true score variances but possibly different error variances. These two models are obtained from (2) by specification of equality of the corresponding set of parameters.

Recently Kristof [30] developed a model for tests which differ only in length. This model assumes that there is a "length" parameter β_i associated with each test score x_i in such a way that the true score variance is proportional to β_i^4 and the error variance proportional to β_i^2. It can be shown that the covariance structure for this model is of the form

$$\boldsymbol{\Sigma} = \mathbf{D}_\beta(\boldsymbol{\beta\beta}' + \psi^2\mathbf{I})\mathbf{D}_\beta$$

where $\mathbf{D}_\beta = \text{diag}(\beta_1, \beta_2, \ldots, \beta_p)$ and $\boldsymbol{\beta}' = (\beta_1, \beta_2, \ldots, \beta_p)$. This is a special case of (1), obtained by specifying $q = p$, $r = 1$, $\mathbf{B} = \mathbf{D}_\beta$, $\boldsymbol{\Lambda} = \boldsymbol{\beta}$, $\boldsymbol{\Phi} = 1$, $\boldsymbol{\Psi}^2 = \psi^2\mathbf{I}$, and $\boldsymbol{\Theta} = \mathbf{0}$. It should be noted that this model specifies equality constraints between the diagonal elements of $\mathbf{B}$ and the elements of the column vector $\boldsymbol{\Lambda}$ and also the equality of all the diagonal elements of $\boldsymbol{\Psi}$. The model has $p + 1$ independent parameters and is less restrictive than the parallel model but more restrictive than the congeneric model.

3.2. A Statistical Model for Several Sets of Congeneric Test Scores

The previous model generalizes immediately to several sets of congeneric test scores. If there are q sets of such tests, with $m_1, m_2, \ldots, m_q$ tests, respectively, we write $\mathbf{x}' = (\mathbf{x}_1', \mathbf{x}_2', \ldots, \mathbf{x}_q')$, where $\mathbf{x}_g'$, $g = 1, 2, \ldots, q$, is the vector of observed scores for the gth set. Associated with the vector x_g there is a true score τ_g and vectors $\boldsymbol{\mu}_g$ and $\boldsymbol{\beta}_g$ defined as in the previous section, so that

$$\mathbf{x}_g = \boldsymbol{\mu}_g + \boldsymbol{\beta}_g \tau_g + \mathbf{e}_g .$$

As before we may, without loss of generality, assume that τ_g is scaled to zero mean and unit variance. If the different true scores $\tau_1, \tau_2, \ldots, \tau_q$ are all mutually uncorrelated, then each set of tests can be analyzed separately as in the previous section. However, in most cases these true scores correlate with each other and an overall analysis of the entire set of tests must be made. Let $p = m_1 + m_2 + \cdots + m_q$ be the total number of tests. Then $\mathbf{x}$ is of order p. Let $\boldsymbol{\mu}$ be the mean vector of $\mathbf{x}$, and let $\mathbf{e}$ be the vector of error scores. Furthermore, let

$$\boldsymbol{\tau}' = (\tau_1, \tau_2, \ldots, \tau_q)$$

and let $\mathbf{B}$ be the matrix of order $p \times q$, partitioned as

$$\mathbf{B} = \begin{bmatrix} \boldsymbol{\beta}_1 & 0 & \ldots & 0 \\ 0 & \boldsymbol{\beta}_2 & \cdots & 0 \\ \cdot & \cdot & \cdots & \cdot \\ \cdot & \cdot & \cdots & \cdot \\ 0 & 0 & \cdots & \boldsymbol{\beta}_q \end{bmatrix}. \tag{33}$$

Then $\mathbf{x}$ is represented as

$$\mathbf{x} = \boldsymbol{\mu} + \mathbf{B}\boldsymbol{\tau} + \mathbf{e}. \tag{34}$$

Let $\boldsymbol{\Gamma}$ be the correlation matrix of $\boldsymbol{\tau}$. Then the variance-covariance matrix $\boldsymbol{\Sigma}$ of $\mathbf{x}$ is

$$\boldsymbol{\Sigma} = \mathbf{B}\boldsymbol{\Gamma}\mathbf{B}' + \boldsymbol{\Theta}^2 \tag{35}$$

where $\boldsymbol{\Theta}^2$ is a diagonal matrix of order p containing the error variances. This is a special case of (2) obtained by specifying $r = q$, $\mathbf{B}$ to be of the form (33), $\boldsymbol{\Lambda} = \mathbf{I}$, $\boldsymbol{\Phi} = \boldsymbol{\Gamma}$, $\boldsymbol{\Psi} = \mathbf{0}$, and $\boldsymbol{\Theta}$ diagonal as in (35).

3.3. Analysis of Multitrait–Multimethod Data

A particular instance when sets of congeneric tests are employed is in multitrait–multimethod studies, where each of a number of traits is measured with a number of different methods or measuring instruments (see, e.g., Campbell and Fiske [7]). One objective is to find the best method of measuring a given trait. In particular, one would like to get estimates of the trait, method, and error variance involved in each measure. A second objective is to study the internal relationships between the measures employed, in particular between the traits and between the methods.

Data from multitrait–multimethod studies are usually summarized in a correlation matrix giving correlations for all pairs of trait–method combinations. If there are m methods and n traits, this correlation matrix is of order $mn \times mn$. In analyzing such a correlation matrix, it seems natural to start out with the hypothesis that all methods are equivalent in measuring each trait, in the sense that scores obtained for a given trait with the different methods are congeneric. This hypothesis implies that all variation and covariation in the multitrait–multimethod matrix is due to trait factors only and may be tested by using a factor matrix $\mathbf{B}$ of order $mn \times n$ with one column for each trait. If the measurements are arranged with methods within traits, $\mathbf{B}$ is of the form (33). If, on the other hand, measurements are arranged with traits within methods, $\mathbf{B}$ has the form

$$\mathbf{B} = \begin{bmatrix} \mathbf{\Delta}_1 \\ \mathbf{\Delta}_2 \\ \vdots \\ \mathbf{\Delta}_m \end{bmatrix} \tag{36}$$

where each $\mathbf{\Delta}_i$ is a diagonal matrix of order $n \times n$. In both cases, the model is given by (34), where $\mathbf{\Gamma}$ is the correlation matrix for the trait factors and $\mathbf{\Theta}^2$ is the diagonal matrix of error variances. If this model fits the data, the interrelationships between the trait factors may be analyzed further by a factoring of $\mathbf{\Gamma}$ as described in Section 3.4. However, if the hypothesis of equivalent methods does not fit the data, this is an indication that method factors are present. It then seems best to postulate the existence of one method factor for each method. This leads to a factor matrix $\mathbf{B}$ of order $mn \times (m + n)$ of the form (with traits within methods)

$$\mathbf{B} = \begin{bmatrix} \mathbf{\Delta}_1 & \boldsymbol{\beta}_1 & \mathbf{0} & \cdots & \mathbf{0} \\ \mathbf{\Delta}_2 & 0 & \boldsymbol{\beta}_2 & \cdots & \mathbf{0} \\ \vdots & \vdots & & \vdots & \vdots \\ \mathbf{\Delta}_m & \mathbf{0} & \mathbf{0} & \cdots & \boldsymbol{\beta}_m \end{bmatrix} \tag{37}$$

where the Δ's are as before and each $\boldsymbol{\beta}_i$ is a column vector of order n. The correlation matrix $\boldsymbol{\Gamma}$ of the factors is defined to be

$$\boldsymbol{\Gamma} = \begin{pmatrix} \boldsymbol{\Gamma}_1 & \mathbf{0} \\ \mathbf{0} & \boldsymbol{\Gamma}_2 \end{pmatrix} \tag{38}$$

where $\boldsymbol{\Gamma}_1$ is the correlation matrix for the trait factors and $\boldsymbol{\Gamma}_2$ the correlation matrix for the method factors. In (38) it is thus assumed that trait factors and method factors are uncorrelated. This is our way of defining each method factor to be independent of the particular traits that the method is used to measure. In other words, method factors are sources of variation and covariation in the data that remain after all trait factors have been eliminated. Substituting (37) and (38) into (34) gives the variance-covariance matrix $\boldsymbol{\Sigma}$ under this model. An analysis of data under this model yields estimates of $\mathbf{B}$, $\boldsymbol{\Gamma}$, and $\boldsymbol{\Theta}$. If the two factor loadings in each row of $\mathbf{B}$ and the corresponding element of $\boldsymbol{\Theta}$ are squared, one obtains a partition of the total variance of each measurement into components due to traits, methods, and error, respectively. If the fit of the model is good and there are many traits and/or methods, one may analyze the interrelationships in $\boldsymbol{\Gamma}_1$ and $\boldsymbol{\Gamma}_2$ further in a way similar to that of Section 3.4.

In analyzing data in accordance with the above model it sometimes happens that one or more correlations in $\hat{\boldsymbol{\Gamma}}_2$ are close to unity or else that $\hat{\boldsymbol{\Gamma}}$ is not Gramian. This means that two or more factors are collinear and have to be combined into one factor.

3.4. Factor Analysis Models

Factor analysis is a widely used technique, especially among psychologists and other behavioral scientists. The basic idea is that for a given set of response variates $x_1, \ldots, x_p$ one wants to find a set of underlying or latent factors $f_1, \ldots, f_k$, fewer in number than the observed variates, that will account for the intercorrelations of the response variates, in the sense that when the factors are partialed out from the observed variates there no longer remains any correlation between these. This leads to the model

$$\mathbf{x} = \boldsymbol{\mu} + \boldsymbol{\Lambda}\mathbf{f} + \mathbf{z} \tag{39}$$

where $E(\mathbf{x}) = \boldsymbol{\mu}$, $E(\mathbf{f}) = \mathbf{0}$, and $E(\mathbf{z}) = \mathbf{0}$, $\mathbf{z}$ being uncorrelated with $\mathbf{f}$. Let $\boldsymbol{\Phi} = E(\mathbf{ff}')$, which may be taken as a correlation matrix, and $\boldsymbol{\Psi}^2 = E(\mathbf{zz}')$, which is diagonal. Then the variance covariance matrix $\boldsymbol{\Sigma}$ of $\mathbf{x}$ becomes

$$\boldsymbol{\Sigma} = \boldsymbol{\Lambda\Phi\Lambda}' + \boldsymbol{\Psi}^2. \tag{40}$$

If $(p - k)^2 < p + k$, this relationship can be tested statistically, unlike (39), which involves hypothetical variates and cannot be verified directly. Equation

(40) may be obtained from the general model (2) by specifying $\mathbf{B} = \mathbf{I}$ and $\boldsymbol{\Theta} = \mathbf{0}$

When $k > 1$, there is an indeterminacy in (40) arising from the fact that a nonsingular linear transformation of $\mathbf{f}$ changes $\boldsymbol{\Lambda}$ and in general also $\boldsymbol{\Phi}$ but leaves $\boldsymbol{\Sigma}$ unchanged. The usual way to eliminate this indeterminacy in exploratory factor analysis (see, e.g., Lawley and Maxwell [32], Jöreskog [20], Jöreskog and Lawley [26]) is to choose $\boldsymbol{\Phi} = \mathbf{I}$ and $\boldsymbol{\Lambda}'\boldsymbol{\Psi}^{-1}\boldsymbol{\Lambda}$ to be diagonal and to estimate the parameters in $\boldsymbol{\Lambda}$ and $\boldsymbol{\Psi}$ subject to these conditions. This leads to an arbitrary set of factors which may then be subjected to a rotation or a linear transformation to another set of factors which can be given a more meaningful interpretation.

In terms of the general model (2), the indeterminacy in (40) may be eliminated by assigning zero values, or any other values, to k^2 elements in $\boldsymbol{\Lambda}$ and/or $\boldsymbol{\Phi}$, in such a way that these assigned values will be destroyed by all nonsingular transformations of the factors except the identity transformation. There may be an advantage in eliminating the indeterminacy this way, in that, if the fixed parameters are chosen in a reasonable way, the resulting solution will be directly interpretable and the subsequent rotation of factors may be avoided.

Specification of parameters a priori may also be used in a confirmatory factor analysis, where the experimenter has already obtained a certain amount of knowledge about the variates measured and is in a position to formulate a hypothesis that specifies the factors on which the variates depend. Such an hypothesis may be specified by assigning values to some parameters in $\boldsymbol{\Lambda}$, $\boldsymbol{\Phi}$, and $\boldsymbol{\Psi}$; see, for example, Jöreskog and Lawley [26] and Jöreskog [21]. If the number of fixed parameters in $\boldsymbol{\Lambda}$ and $\boldsymbol{\Phi}$ exceeds k^2, the hypothesis represents a restriction of the common factor space and a solution obtained under such an hypothesis cannot be obtained by a rotation of an arbitrary solution such as is obtained in an exploratory analysis.

Model (32) is formally equivalent to a factor analytic model with one common factor and model (35) is equivalent to a factor analytic model with q correlated nonoverlapping factors. In the latter case the factors are the true scores $\tau' = (\tau_1, \ldots, \tau_q)$ of the tests. These true scores may themselves satisfy a factor analytic model, i.e.,

$$\boldsymbol{\tau} = \boldsymbol{\Lambda}\mathbf{f} + \mathbf{s}$$

where $\mathbf{f}$ is a vector of order k of common true score factors, $\mathbf{s}$ is a vector of order q of specific true score factors, and $\boldsymbol{\Lambda}$ is a matrix of order $q \times k$ of factor loadings. Let $\boldsymbol{\Phi}$ be the variance-covariance matrix of $\mathbf{f}$ and let $\boldsymbol{\Psi}^2$ be a diagonal matrix whose diagonal elements are the variances of the specific true score factors $\mathbf{s}$. Then $\boldsymbol{\Gamma}$, the variance-covariance matrix of $\boldsymbol{\tau}$, becomes

$$\boldsymbol{\Gamma} = \boldsymbol{\Lambda}\boldsymbol{\Phi}\boldsymbol{\Lambda}' + \boldsymbol{\Psi}^2. \tag{41}$$

Substituting (41) into (35) gives $\boldsymbol{\Sigma}$ as

$$\boldsymbol{\Sigma} = \mathbf{B}(\boldsymbol{\Lambda}\boldsymbol{\Phi}\boldsymbol{\Lambda}' + \boldsymbol{\Psi}^2)\mathbf{B}' + \boldsymbol{\Theta}^2. \tag{42}$$

Model (42) is a special case of (2) by specifying zero values in $\mathbf{B}$ as in (33). To define $\boldsymbol{\Lambda}$ and $\boldsymbol{\Phi}$ uniquely it is necessary to impose k^2 independent conditions on these matrices to eliminate the indeterminacy due to rotation. Model (42) is a special case of the second order factor analytic model.

3.5. Estimation of Variance and Covariance Components

Several authors [4, 5, 43] have considered a covariance structure analysis as an approach by which to study differences in test performances when the tests have been constructed by assigning items or subtests according to objective features of content or format to subclasses of a factorial or hierarchical classification.

Bock [4] suggested that the scores of N subjects on a set of tests classified in 2^n factorial design may be viewed as data from an $N \times 2^n$ experimental design, where the subjects represent a random mode of classification and the tests represent n fixed modes of classification. Bock pointed out that conventional mixed-model analysis of variance gives useful information about the psychometric properties of the tests. In particular, the presence of nonzero variance components for the random mode of classification provides information about the number of dimensions in which the tests are able to discriminate among subjects. The relative sizes of these components measure the power of the tests to discriminate among subjects along the respective dimensions.

Consider an experimental design that has one random way of classification $v = 1, 2, \ldots, N$, one fixed way of classification $i = 1, 2, 3$, and another fixed way of classification $j = 1, 2, 3$ for $i = 1, 2$ and $j = 1, 2$ for $i = 3$. One model that may be considered is

$$x_{vij} = a_v + b_{vi} + c_{vj} + e_{vij} \tag{43}$$

where a_v, b_{vi}, c_{vj}, and e_{vij} are uncorrelated random variables with means μ_a, μ_{b_i}, μ_{c_j}, and 0 and variances σ_a^2, $\sigma_{b_i}^2$, $\sigma_{c_j}^2$, and $\sigma_{e_{ij}}^2$, respectively. Writing $\mathbf{x}_v' = (x_{v11}, x_{v12}, x_{v13}, x_{v21}, x_{v22}, x_{v23}, x_{v31}, x_{v32})$, $\mathbf{u}_v' = (a_v, b_{v1}, b_{v2}, b_{v3}, c_{v1}, c_{v2}, c_{v3})$ and

$$\mathbf{B} = \begin{bmatrix} 1 & 1 & 0 & 0 & 1 & 0 & 0 \\ 1 & 1 & 0 & 0 & 0 & 1 & 0 \\ 1 & 1 & 0 & 0 & 0 & 0 & 1 \\ 1 & 0 & 1 & 0 & 1 & 0 & 0 \\ 1 & 0 & 1 & 0 & 0 & 1 & 0 \\ 1 & 0 & 1 & 0 & 0 & 0 & 1 \\ 1 & 0 & 0 & 1 & 1 & 0 & 0 \\ 1 & 0 & 0 & 1 & 0 & 1 & 0 \end{bmatrix},$$

we may write (43) as

$$\mathbf{x}_v = \mathbf{B}\mathbf{u}_v + \mathbf{e}_v$$

where $\mathbf{e}_v$ is a random error vector of the same form as $\mathbf{x}_v$. The mean of $\mathbf{x}_v$ is $\mathbf{B}\boldsymbol{\mu}$, where $\boldsymbol{\mu}' = (\mu_a, \mu_{b_1}, \mu_{b_2}, \mu_{b_3}, \mu_{c_1}, \mu_{c_2}, \mu_{c_3})$ and the variance-covariance matrix of $\mathbf{x}_v$ is

$$\boldsymbol{\Sigma} = \mathbf{B}\boldsymbol{\Phi}\mathbf{B}' + \boldsymbol{\Psi}^2 \tag{44}$$

where $\boldsymbol{\Phi}$ is a diagonal matrix whose diagonal elements are $\sigma_a{}^2$, $\sigma_{b_1}^2$, $\sigma_{b_2}^2$, $\sigma_{b_3}^2$, $\sigma_{c_1}^2$, $\sigma_{c_2}^2$, and $\sigma_{c_3}^2$, and $\boldsymbol{\Psi}^2$ is a diagonal matrix whose elements are the $\sigma_{e_{ij}}^2$. In terms of the general model (1) and (2), this model may be represented by choosing $p = 8$, $g = 1$, $h = 7$, $q = 8$, $r = 7$, $\boldsymbol{\Xi} = \boldsymbol{\mu}'$, $\mathbf{P} = \mathbf{B}'$, $\mathbf{B} = \mathbf{I}$, $\boldsymbol{\Lambda} = \mathbf{B}$, and $\boldsymbol{\Theta} = \mathbf{0}$. Matrices $\boldsymbol{\Phi}$ and $\boldsymbol{\Psi}^2$ are as defined in (17), and the matrix $\mathbf{A}$ in (1) is a column vector of order N of unities. The general method of analysis yields maximum likelihood estimates of the fixed effects and of the variance components $\sigma_a{}^2$, $\sigma_{b_i}^2$, $\sigma_{c_j}^2$, and $\sigma_{e_{ij}}^2$. In conventional mixed-model analysis of variance one usually makes the assumptions that $\sigma_{b_i}^2 = \sigma_b{}^2$ for all $i = 1, 2, 3$; $\sigma_{c_j}^2 = \sigma_c{}^2$ for all $j = 1, 2, 3$; and $\sigma_{e_{ij}}^2 = \sigma_e{}^2$ for all i and j.

In general, if $\mathbf{B}$ is of order $p \times r$ and of rank k, one may choose k independent linear functions of the u's, each one linearly dependent on the rows of $\mathbf{B}$, and estimate the mean vector and variance-covariance matrix of these functions. It is customary to choose linear combinations that are mutually uncorrelated, but this is not necessary in the analysis by our method. Let $\mathbf{L}$ be the matrix of coefficients of the chosen linear functions and let $\mathbf{K}$ be any matrix such that $\mathbf{B} = \mathbf{K}\mathbf{L}$. For example, $\mathbf{K}$ may be obtained from

$$\mathbf{K} = \mathbf{B}\mathbf{L}'(\mathbf{L}\mathbf{L}')^{-1}.$$

The model may then be reparameterized to full rank by defining $\mathbf{u}^* = \mathbf{L}\mathbf{u}$. We then have $\mathbf{x} = \mathbf{B}\mathbf{u} + \mathbf{e} = \mathbf{K}\mathbf{L}\mathbf{u} + \mathbf{e} = \mathbf{K}\mathbf{u}^* + \mathbf{e}$. The mean vector of $\mathbf{x}$ is $KE(\mathbf{u}^*)$ and the variance-covariance matrix of $\mathbf{x}$ is represented as

$$\boldsymbol{\Sigma} = \mathbf{K}\boldsymbol{\Phi}^*\mathbf{K}' + \boldsymbol{\Psi}^2$$

where $\boldsymbol{\Phi}^*$ is the variance-covariance matrix of $\mathbf{u}^*$ and $\boldsymbol{\Psi}^2$ is as before. The general method of analysis yields estimates of $E(\mathbf{u}^*)$,$\boldsymbol{\Psi}^2$, and $\boldsymbol{\Phi}^*$. The last matrix may be taken to be diagonal if desired.

In most applications in the behavioral sciences it may not be realistic to assume that the latent random variables would be uncorrelated. Bock *et al.* [6] and Wiley, *et al.* [43] gave examples of the inadequacy of the specification of uncorrelated latent variables, i.e., the inadequacy of $\boldsymbol{\Phi}$ being specified as a diagonal matrix, which is the case considered by Bock [4] and by Bock and Bargmann [5]. In our method of covariance structure analysis, the assumption that $\boldsymbol{\Phi}$ be diagonal is not necessary. If the model provides information enough, so that all the variances and covariances of the latent variables are identified, these may also be estimated, and the assumption of zero covariances may be examined empirically.

Wiley *et al.* [43] suggested a general class of components of covariance models. This class of models is a special case of (2), namely, when $\mathbf{B}$ is diagonal, $\mathbf{\Lambda}$ is known a priori, $\mathbf{\Phi}$ is symmetric and positive definite, and $\mathbf{\Psi}$ or $\mathbf{\Theta}$ are either zero or diagonal. The covariance matrix $\mathbf{\Sigma}$ will then be of the form

$$\mathbf{\Sigma} = \mathbf{\Delta A \Phi A' \Delta} + \mathbf{\Theta}^2 \qquad \text{or} \qquad \mathbf{\Sigma} = \mathbf{\Delta}(\mathbf{A \Phi A'} + \mathbf{\Psi}^2)\mathbf{\Delta}. \tag{45a,b}$$

The matrix $\mathbf{A}(p \times k)$ is known and gives the coefficients of the linear functions connecting the manifest and latent variables, $\mathbf{\Delta}$ is a $p \times p$ diagonal matrix of unknown scale factors, $\mathbf{\Phi}$ is the $k \times k$ symmetric and positive definite covariance matrix of the latent variables, and $\mathbf{\Theta}^2$ or $\mathbf{\Psi}^2\, \mathbf{\Delta}^2$ are $p \times p$ diagonal matrices of error variances.

3.6. Simplex Models

Guttman [16] introduced the notion of a simplex structure for a set of tests that involve the same kind of ability and are ordered according to increasing or decreasing complexity. In statistical terminology such simplex structures are equivalent to the correlation structures arising in first-order Markov processes (see, e.g., Anderson [1], or Jöreskog [24]).

Let $\boldsymbol{\eta}' = (\eta_1, \eta_2, \ldots, \eta_p)$ be a set of random variables generated by a first-order autoregressive series

$$\eta_i = \beta_i \eta_{i-1} + \zeta_i, \qquad i = 2, 3, \ldots, p,$$

where $\boldsymbol{\zeta}' = (\zeta_1 \equiv \eta_1, \zeta_2, \ldots, \zeta_p)$ are mutually uncorrelated and uncorrelated with the η's. Let $\mathbf{T}(p \times p)$ be a lower triangular matrix whose nonzero elements are all unity and let $\kappa_i = \beta_2 \beta_3 \cdots \beta_i$, $i = 2, 3, \ldots, p$. Then

$$\boldsymbol{\eta} = \mathbf{D}_\kappa \mathbf{T} \mathbf{D}_\kappa^{-1} \boldsymbol{\zeta} \tag{46}$$

where $\mathbf{D}_\kappa = \text{diag}(1, \kappa_2, \ldots, \kappa_p)$.

The η_i is not directly observed but

$$x_i = \mu_i + \eta_i + \varepsilon_i$$

is observed, where $\mu_i = E(x_i)$ and ε_i is an error of measurement. Then the variance-covariance matrix of $\mathbf{x}' = (x_1, x_2, \ldots, x_p)$ is

$$\mathbf{\Sigma} = \mathbf{D}_\kappa \mathbf{T} \mathbf{D}_\kappa^{-1} \mathbf{D}_\varphi \mathbf{D}_\kappa^{-1} \mathbf{T}' \mathbf{D}_\kappa + \mathbf{\Theta}^2 = \mathbf{D}_\kappa \mathbf{T} \mathbf{D}_{\varphi *} \mathbf{T}' \mathbf{D}_\kappa + \mathbf{\Theta}^2 \tag{47}$$

where $\mathbf{D}_\varphi = \text{diag}(\varphi_1, \varphi_2, \ldots, \varphi_p)$ is a diagonal matrix containing the variances of $\zeta_1, \zeta_2, \ldots, \zeta_p$ and $\mathbf{\Theta}^2$ is a diagonal matrix of the variances of the measurement errors $\varepsilon_1, \varepsilon_2, \ldots, \varepsilon_p$ and where $\mathbf{D}_{\varphi *} = \mathbf{D}_\kappa^{-1} \mathbf{D}_\varphi \mathbf{D}_\kappa^{-1}$. There is a one-to-one correspondence between the parameters $(\beta_2, \beta_3, \ldots, \beta_p, \varphi_1, \varphi_2, \ldots, \varphi_p)$ and $(\kappa_2, \kappa_3, \ldots, \kappa_p, \varphi_1^*, \varphi_2^*, \ldots, \varphi_p^*)$. The model (47) is a special case of (2) with $\mathbf{B} = \mathbf{D}_\kappa$, $\mathbf{\Lambda} = \mathbf{T}$, $\mathbf{\Phi} = \mathbf{D}_{\varphi^*}$, $\mathbf{\Psi} = \mathbf{0}$, and $\mathbf{\Theta} = \mathbf{\Theta}$. The model has $3p - 3$ independent parameters. The parameters φ_1 and θ_1 are not identified; only their sum $\sigma_{11} = \varphi_1 + \theta_1^2$ is. Similarly, only the sum $\mathrm{Var}(\eta_p) + \theta_p^2$ is identified.

Models of this kind for the structure of the variance-covariance matrix may be used in connection with various structures on the mean values. If one variate is measured on p occasions $t_1, \ldots, t_p$ and there are g independent groups of observations with n_s observations in the sth group, $n_1 + \cdots + n_g = N$, one may wish to consider polynomial growth curves like

$$E(x_t) = \mu_t = \xi_{s0} + \xi_{s1}\, t + \cdots + \xi_{sh}\, t^h \tag{48}$$

for the sth group ($s = 1, \ldots, g$). Such a model may be represented in the form of (2) by letting $\mathbf{A}$, $\mathbf{\Xi}$, and $\mathbf{P}$ be as follows. The matrix $\mathbf{A}$ is of order $N \times g$ and has n_1 rows $(1, 0, \ldots, 0)$, n_2 rows $(0, 1, 0, \ldots, 0)$, $\ldots$, and n_g rows $(0, \ldots, 0, 1)$. Further,

$$\mathbf{\Xi} = \begin{bmatrix} \xi_{10} & \xi_{11} & \cdots & \xi_{1h} \\ \xi_{20} & \xi_{21} & \cdots & \xi_{2h} \\ \cdot & \cdot & \cdots & \cdot \\ \xi_{g0} & \xi_{g1} & \cdots & \xi_{gh} \end{bmatrix}, \tag{49}$$

$$\mathbf{P} = \begin{bmatrix} 1 & 1 & \cdots & 1 \\ t_1 & t_2 & \cdots & t_p \\ t_1^2 & t_2^2 & \cdots & t_p^2 \\ \cdot & \cdot & \cdots & \cdot \\ t_1^h & t_2^h & \cdots & t_p^h \end{bmatrix}. \tag{50}$$

If this model is used together with (47), for example, one can estimate the parameters in $\mathbf{\Xi}$, the β's, φ's, and the θ's simultaneously. If desired, one can also estimate these parameters under the condition that some of the coefficients in the polynomial growth curves have assigned values or that one or more of them are the same for certain groups.

The above models may be extended to the case when several variates are measured at different occasions. To consider the simplest case of only two variates, let

$$E(x_{it}) = \xi_{s0}^{(i)} + \xi_{s1}^{(i)} t + \cdots + \xi_{sh}^{(i)} t^h \qquad (s = 1, \ldots, g; \quad i = 1, 2). \tag{51}$$

If the variates x_{it} are ordered so that

$$(x_{1t_1}, \ldots, x_{1t_p}, x_{2t_1}, \ldots, x_{2t_p}) \tag{52}$$

corresponds to a row of the data matrix **X**, then **A** is as before,

$$\boldsymbol{\Xi} = \begin{bmatrix} \xi_{10}^{(1)} & \xi_{11}^{(1)} & \cdots & \xi_{1h}^{(1)} & \xi_{10}^{(2)} & \xi_{11}^{(2)} & \cdots & \xi_{1h}^{(2)} \\ \xi_{20}^{(1)} & \xi_{21}^{(1)} & \cdots & \xi_{2h}^{(1)} & \xi_{20}^{(2)} & \xi_{21}^{(2)} & \cdots & \xi_{2h}^{(2)} \\ \cdot & \cdot & \cdots & \cdot & \cdot & \cdot & \cdots & \cdot \\ \xi_{g0}^{(1)} & \xi_{g1}^{(1)} & \cdots & \xi_{gh}^{(1)} & \xi_{g0}^{(2)} & \xi_{g1}^{(2)} & \cdots & \xi_{gh}^{(2)} \end{bmatrix}, \tag{53}$$

$$\mathbf{P} = \begin{bmatrix} \mathbf{P}^* & \mathbf{0} \\ \mathbf{0} & \mathbf{P}^* \end{bmatrix},$$

where $\mathbf{P}^*$ is the same as $\mathbf{P}$ in (50), and $\mathbf{0}$ is a zero matrix. The variance-covariance matrix $\boldsymbol{\Sigma}$ of the random vector in (52) may be assumed, for example, to be of the form,

$$\boldsymbol{\Sigma} = \begin{pmatrix} \mathbf{D}_1 & \mathbf{0} \\ \mathbf{0} & \mathbf{D}_2 \end{pmatrix} \begin{pmatrix} \mathbf{T} & \mathbf{0} \\ \mathbf{0} & \mathbf{T} \end{pmatrix} \begin{pmatrix} \mathbf{D}_3 & \mathbf{D}_4 \\ \mathbf{D}_4 & \mathbf{D}_5 \end{pmatrix} \begin{pmatrix} \mathbf{T}' & \mathbf{0} \\ \mathbf{0} & \mathbf{T}' \end{pmatrix} \begin{pmatrix} \mathbf{D}_1' & \mathbf{0} \\ \mathbf{0} & \mathbf{D}_2 \end{pmatrix} + \begin{pmatrix} \boldsymbol{\Theta}_1^2 & \mathbf{0} \\ \mathbf{0} & \boldsymbol{\Theta}_2^2 \end{pmatrix} \tag{54}$$

where all the $\mathbf{D}_i$ are diagonal matrices. Such a covariance structure results from $x_{it} = \mu_{it} + \eta_{it} + \varepsilon_{it}$ and $\eta_{it} = \beta_{it}\eta_{i,t-1} + \xi_{it}$ if the increments ξ_{it} are uncorrelated between occasions but correlated within occasions. In this case one can estimate the elements of the $\mathbf{D}_i$, $i = 1, 2, \ldots, 5$, and $\boldsymbol{\Theta}_i$, $i = 1, 2$. One can also test various hypotheses about the growth curves, for example, that the curves are the same for several groups or for the two variables.

3.7. Circumplex Models

Simplex models, as considered in the previous section, are models for tests that may be conceived of as having a linear ordering. The circumplex is another model considered by Guttman [16] and this yields a circular ordering instead of a linear one. The circular order has no beginning and no end but there is still a law of neighboring that holds.

The circumplex model suggested by Guttman is a circular moving average process (see Anderson [1]). Let $\zeta_1, \zeta_2, \ldots, \zeta_p$ be uncorrelated random latent variables. Then the perfect circumplex of order m with p variables is defined by

$$x_i = \zeta_i + \zeta_{i+1} + \cdots + \zeta_{i+m-1}$$

where $x_{p+i} = x_i$. In matrix form we may write this as $\mathbf{x} = \mathbf{C}\boldsymbol{\zeta}$, where $\mathbf{C}$ is a matrix of order $p \times p$ with zeros and ones. In the case of $p = 6$ and $m = 3$

$$\mathbf{C} = \begin{bmatrix} 1 & 1 & 1 & 0 & 0 & 0 \\ 0 & 1 & 1 & 1 & 0 & 0 \\ 0 & 0 & 1 & 1 & 1 & 0 \\ 0 & 0 & 0 & 1 & 1 & 1 \\ 1 & 0 & 0 & 0 & 1 & 1 \\ 1 & 1 & 0 & 0 & 0 & 1 \end{bmatrix}.$$

Let $\varphi_1, \varphi_2, \ldots, \varphi_p$ be the variances of $\zeta_1, \zeta_2, \ldots, \zeta_p$, respectively. Then the variance-covariance matrix of $\mathbf{x}$ is

$$\boldsymbol{\Sigma} = \mathbf{C}\mathbf{D}_\varphi \mathbf{C}' \tag{55}$$

where $\mathbf{D}_\varphi = \text{diag}(\varphi_1, \varphi_2, \ldots, \varphi_p)$.

The perfect circumplex is a too restrictive model. It cannot account for random measurement error in the test scores, and since the model is not scale free, it cannot be used to analyze correlations. However, these difficulties are easily remedied by considering the following quasi-circumplex

$$\mathbf{x} = \mathbf{D}_\alpha \mathbf{C}\boldsymbol{\zeta} + \mathbf{e}$$

with dispersion matrix

$$\boldsymbol{\Sigma} = \mathbf{D}_\alpha \mathbf{C}\mathbf{D}_\varphi \mathbf{C}'\mathbf{D}_\alpha + \boldsymbol{\Theta}^2 \tag{56}$$

where $\mathbf{e}$ is the vector of error scores with variances in the diagonal matrix $\boldsymbol{\Theta}^2$ and $\mathbf{D}_\alpha$ is a diagonal matrix of scale factors. One element in $\mathbf{D}_\alpha$ or $\mathbf{D}_\varphi$ must be fixed at unity.

3.8. Path Analysis Models

Path analysis, due to Wright [45], is a technique sometimes used to assess the direct causal contribution of one variable to another in a nonexperimental situation. The problem in general is that of estimating the parameters of a set of linear structural equations, representing the cause and effect relationships hypothesized by the investigator. Recently, several models have been studied which involve hypothetical constructs, i.e., latent variables which, while not directly observed, have operational implications for relationships among observable variables (see, e.g., Hauser and Goldberger [17]). In some models, the observed variables appear only as effects (indicators) of the hypothetical constructs, while in others, the observed variables appear as causes (components) or as both causes and effects of latent variables. We give one simple example of each kind of model to indicate how many such models may be handled within the framework of covariance structure analysis.

In presenting a path analysis model it is convenient to use a path diagram, where observed variables are enclosed in squares and hypothetical variables in circles. Other unobserved variables, such as residuals and measurement errors, are not enclosed. A one-way arrow indicates a direct causal influence of one variable on another, whereas a two-headed arrow indicates correlation between variables not dependent on other variables in the system.

As a first example consider the model discussed by Costner [8] and shown in Fig. 1. Note that the errors δ_3 and ε_3 are assumed to be correlated, as might be the case, for example, if x_3 and y_3 were scores from the same

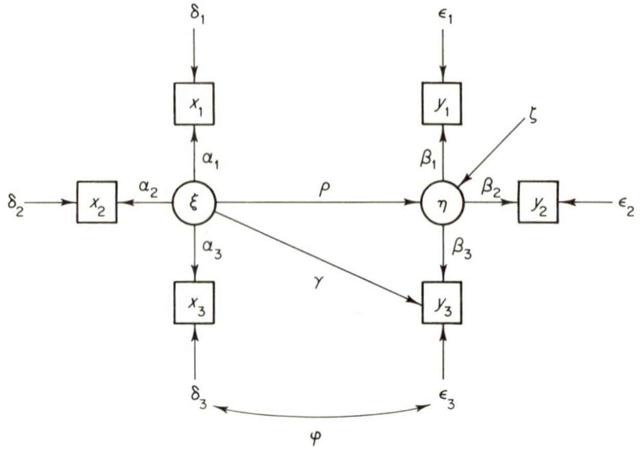

Fig. 1

measuring instrument used at two different occasions. In algebraic form the model may be written, ignoring the means of the observed variables,

$$\begin{pmatrix} x_1 \\ x_2 \\ x_3 \\ y_1 \\ y_2 \\ y_3 \end{pmatrix} = \begin{bmatrix} \alpha_1 & 0 & 1 & 0 & 0 & 0 & 0 & 0 \\ \alpha_2 & 0 & 0 & 1 & 0 & 0 & 0 & 0 \\ \alpha_3 & 0 & 0 & 0 & 1 & 0 & 0 & 0 \\ \gamma & \beta_1 & 0 & 0 & 0 & 1 & 0 & 0 \\ 0 & \beta_2 & 0 & 0 & 0 & 0 & 1 & 0 \\ 0 & \beta_3 & 0 & 0 & 0 & 0 & 0 & 1 \end{bmatrix} \begin{pmatrix} \xi \\ \eta \\ \delta_1 \\ \delta_2 \\ \delta_3 \\ \varepsilon_1 \\ \varepsilon_2 \\ \varepsilon_3 \end{pmatrix}. \qquad (57)$$

Let **Λ** be the matrix in (57) and **Φ** the variance-covariance matrix of the vector on the right-hand side. Then **Φ** is of the form

$$\mathbf{\Phi} = \begin{bmatrix} 1 & & & & & & & \\ \rho & 1 & & & & & & \\ 0 & 0 & \theta_1^2 & & & & & \\ 0 & 0 & 0 & \theta_2^2 & & & & \\ 0 & 0 & 0 & 0 & \theta_3^2 & & & \\ 0 & 0 & 0 & 0 & 0 & \theta_4^2 & & \\ 0 & 0 & 0 & 0 & 0 & 0 & \theta_5^2 & \\ 0 & 0 & 0 & 0 & \varphi & 0 & 0 & \theta_6^2 \end{bmatrix}$$

where ρ is the correlation between the latent variables ξ and η, φ the covariance between δ_3 and ε_3, and $\theta_1^2, \theta_2^2, \ldots, \theta_6^2$ the variances of the errors δ_1, δ_2, δ_3, ε_1, ε_2, ε_3. The variance-covariance matrix of the observed variables is

$$\mathbf{\Sigma} = \mathbf{\Lambda\Phi\Lambda}'. \qquad (58)$$

Note that in this example $\mathbf{\Lambda}$ has more columns than rows and includes also the error part of the model. This representation is necessary, since the covariance matrix of the errors is not diagonal.

This model has 15 parameters to be estimated and the covariance matrix in (58) has 6 degrees of freedom. The investigator may be interested in testing the specific hypothesis $\gamma = 0$, i.e., that ξ affects y_3 only via η. This may be done in large samples, assuming that the rest of the model holds, by a χ^2 with 1 degree of freedom.

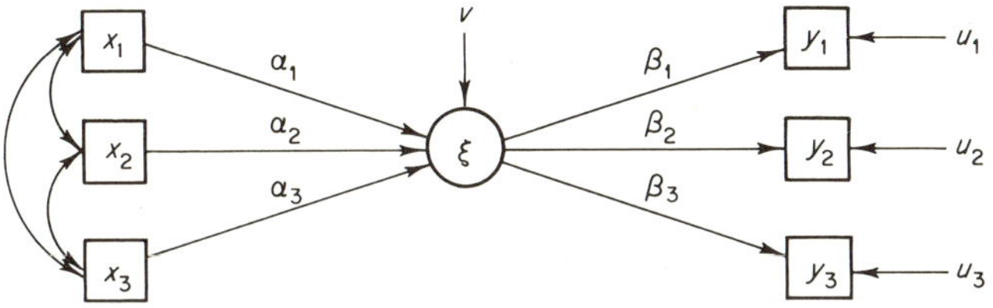

Fig. 2

As a second example consider the model discussed by Hauser and Goldberger [17] and shown in Fig. 2. This model involves a single hypothetical variable ξ which appears as both cause and effect variable. The equations are

$$\xi = \boldsymbol{\alpha}'\mathbf{x} + v, \qquad \mathbf{y} = \boldsymbol{\beta}\xi + \mathbf{u}.$$

The case where the residuals u_1, u_2, and u_3 are mutually correlated and $v = 0$ was considered by Hauser and Goldberger [17]. The case shown in the figure, where u_1, u_2, and u_3 are mutually uncorrelated, will be considered in detail in a forthcoming paper by Goldberger and Jöreskog [13]. In this case the structure of the variance-covariance matrix of the observed variables is

$$\textstyle\sum_{yy} = \boldsymbol{\beta\alpha}' \sum_{xx} \boldsymbol{\alpha\beta}' + \boldsymbol{\beta\beta}' + \boldsymbol{\Theta}^2, \qquad \sum_{yx} = \boldsymbol{\beta\alpha}' \sum_{xx}, \qquad \sum_{xx} \ \text{unconstrained.}$$

The residual v may be scaled to unit variance, as assumed here. Alternatively, the latent variable ξ may be scaled to unit variance or one of the α' fixed at some nonzero value. It is readily verified that this model may be represented in terms of (2) by specifying

$$\mathbf{B} = \begin{bmatrix} \beta_1 & 0 & 0 & 0 \\ \beta_2 & 0 & 0 & 0 \\ \beta_3 & 0 & 0 & 0 \\ 0 & 1 & 0 & 0 \\ 0 & 0 & 1 & 0 \\ 0 & 0 & 0 & 1 \end{bmatrix}, \qquad \mathbf{\Lambda} = \begin{bmatrix} \alpha_1 & \alpha_2 & \alpha_3 \\ 1 & 0 & 0 \\ 0 & 1 & 0 \\ 0 & 0 & 1 \end{bmatrix}, \qquad \mathbf{\Phi} = \mathbf{\Sigma}_{xx},$$

$$\mathbf{\Psi} = \operatorname{diag}(1, 0, 0, 0), \qquad \mathbf{\Theta} = \operatorname{diag}(\sigma_{u_1}, \sigma_{u_2}, \sigma_{u_3}, 0, 0, 0).$$

The model has 15 independent parameters and 6 degrees of freedom.

REFERENCES

1. Anderson, T. W. (1960). Some stochastic process models for intelligence test scores. *Mathematical Methods in the Social Sciences, 1959* (K. J. Arrow, S. Karlin and P. Suppes, eds.), 205–220. Stanford Univ. Press, Stanford, California.
2. Anderson, T. W. (1969). Statistical inference for covariance matrices with linear structure. *Multivariate Analysis II* (P. R. Krishnaiah, ed.), 55–66. Academic Press, New York.
3. Anderson, T. W. (1970). Estimation of covariance matrices which are linear combinations or whose inverses are linear combinations of given matrices. *Essays in Probability and Statistics* (R. C. Bose *et al.*, eds.), 1–24. Univ. of North Carolina Press, Chapel Hill, North Carolina.
4. Bock, R. D. (1960). Components of variance analysis as a structural and discriminal analysis for psychological tests. *British J. Math. Statist. Psychology* **13** 151–163.
5. Bock, R. D. and Bargmann, R. E. (1966). Analysis of covariance structures. *Psychometrika* **31** 507–534.
6. Bock, R. D., Dicken, D. and van Pelt, J. (1969). Methodological implications of content-acquiescence correlation in the MMPI. *Psychological Bull.* **71** 127–139.
7. Campbell, D. T. and Fiske, D. W. (1959). Convergent and discriminant validation by the multitrait-multimethod matrix. *Psychological Bull.* **56** 81–105.
8. Costner, H. L. (1969). Theory, deduction and rules of correspondence. *Amer. J. Sociology* **75** 245–263.
9. Fletcher, R. and Powell, M. J. D. (1963). A rapidly convergent descent method for minimization. *Comput. J.* **6** 163–168.
10. Geisser, S. (1970). Bayesian analysis of growth curves. *Sankhyā Ser. A* **32** 53–64.
11. Gleser, L. and Olkin, I. (1966). A k-sample regression model with covariance. *Multivariate Analysis II* (P. R. Krishnaiah, ed.), 59–72. Academic Press, New York.
12. Gleser, L. J. and Olkin, I. (1970). Linear models in multivariate analysis. *Essays in Probability and Statistics* (R. C. Bose *et al.*, eds.), 267–292. Univ. of North Carolina Press, Chapel Hill, North Carolina.
13. Goldberger, A. S. and Jöreskog, K. G. (1973). Estimation of a model with multiple causes and multiple indicators of a single latent variable. To be published.
14. Grizzle, J. E. and Allen, D. M. (1969). Analysis of growth and dose response curves. *Biometrics* **25** 357–381.
15. Gruvaeus, G. T. and Jöreskog, K. G. (1970). A computer program for minimizing a function of several variables. Res. Bull. 70–14. Educational Testing Service, Princeton, New Jersey.
16. Guttman, L. (1954). A new approach to factor analysis: the Radex. *Mathematical Thinking in the Social Sciences* (P. F. Lazarsfeld, ed.), 258–348. Columbia Univ. Press, New York.
17. Hauser, R. M. and Goldberger, A. S. (1971). The treatment of unobservable variables in path analysis. *Sociological Methodology 1971* (H. L. Costner, ed.), 81–117. Jossey-Bass, London.
18. Heck, D. L. (1960). Charts of some upper percentage points of the distribution of the largest characteristic root. *Ann. Math. Statist.* **31** 625–642.
19. Hotelling, H. (1951). A generalized T test and measure of multivariate dispersion. *Proc. Second Berkeley Symp. Math. Statist. Prob.* 23–41.
20. Jöreskog, K. G. (1967). Some contribution to maximum likelihood factor analysis. *Psychometrika* **32** 443–482.
21. Jöreskog, K. G. (1969). A general approach to confirmatory maximum likelihood factor analysis. *Psychometrika* **34** 183–202.

22. Jöreskog, K. G. (1970). Factoring the multitest-multioccasion correlation matrix. *Current Problems and Techniques in Multivariate Psychology.* (*Proc. Conf. Honoring Professor Paul Horst*), *Univ. of Washington, Seattle*, 68–100.
23. Jöreskog, K. G. (1970). A general method for analysis of covariance structures. *Biometrika* **57** 239–251.
24. Jöreskog, K. G. (1970). Estimation and testing of simplex models. *British J. Math. Statist. Psychology* **23** 121–145.
25. Jöreskog, K. G. (1971). Statistical analysis of sets of congeneric tests. *Psychometrika* **36** 109–133.
26. Jöreskog, K. G. and Lawley, D. N. (1968). New methods in maximum likelihood factor analysis. *British J. Math. Statist. Psychology* **21** 85–96.
27. Jöreskog, K. G., van Thillo, M. and Gruvaeus, G. T. (1971). ACOVSM—A general computer program for analysis of covariance structures including generalized MANOVA. Res. Bull. 71-01. Educational Testing Service, Princeton, New Jersey.
28. Kendall, M. G. and Stuart, A. (1961). *The Advanced Theory of Statistics*, Vol. 2, *Inference and Relationship*. Griffin, London.
29. Khatri, C. G. (1966). A note on a MANOVA model applied to problems in growth curves. *Ann. Inst. Statist. Math.* **18** 75–86.
30. Kristof, W. (1971). On the theory of a set of tests which differ only in length. *Psychometrika* **36** 207–225.
31. Lawley, D. N. (1938). A generalization of Fisher's z-test. *Biometrika* **30** 180–187.
32. Lawley, D. N. and Maxwell, A. E. (1971). *Factor Analysis as a Statistical Method*, 2nd ed. Butterworth, London.
33. Lord, F. M. and Novick, M. R. (1968). *Statistical Theories of Mental Test Scores* (with contributions by A. Birnbaum). Addison-Wesley, Reading, Massachusetts.
34. Mukherjee, B. N. (1970). Likelihood ratio tests of statistical hypotheses associated with patterned covariance matrices in psychology. *British J. Math. Statist. Psychology* **23** 89–120.
35. Potthoff, R. F. and Roy, S. N. (1964). A generalized multivariate analysis of variance model useful especially for growth curve problems. *Biometrika* **51** 313–326.
36. Rao, C. R. (1959). Some problems involving linear hypothesis in multivariate analysis. *Biometrika* **46** 49–58.
37. Rao, C. R. (1965). The theory of least squares when the parameters are stochastic and its application to the analysis of growth curves. *Biometrika* **52** 447–458.
38. Rao, C. R. (1966). Covariance adjustment and related problems in multivariate analysis. *Multivariate Analysis* (P. R. Krishnaiah, ed.), 87–103. Academic Press, New York.
39. Rao, C. R. (1967). Least squares theory using an estimated dispersion matrix and its application to measurement of signals. *Proc. Fifth Berkeley Symp. Math. Statist. Prob.* 355–372.
40. Roy, S. N. (1953). On a heuristic method of test construction and its uses in multivariate analysis. *Ann. Math. Statist.* **24** 220–238.
41. Schatzoff, M. (1966). Exact distributions of Wilks's likelihood ratio criterion. *Biometrika* **53** 347–358.
42. Srivastava, J. N. (1966). On testing hypothesis regarding a class of covariance structures. *Psychometrika* **31** 147–164.
43. Wiley, D. E., Schmidt, W. H. and Bramble, W. J. (1973). Studies of a class of covariance structure models. *J. Amer. Statist. Assoc.* **68** (to be published).
44. Wilks, S. S. (1932). Certain generalizations in the analysis of variance. *Biometrika* **24** 471–494.
45. Wright, S. (1918). On the nature of size factors. *Genetics* **3** 367–374.

Optimum Designs for Fitting Biased Multiresponse Surfaces[1]

J. KIEFER
CORNELL UNIVERSITY

1. INTRODUCTION

Let $\tilde{F}$ be a $k \times m$ matrix of real-valued functions on a set $\mathscr{X} \cup \mathscr{Z}$ (not necessarily disjoint), and suppose $\tilde{F}$ is continuous on the compact set $\mathscr{X}$. Let $\tilde{G}$ be an $s \times m$ matrix of functions on $\mathscr{Z}$. If N multiresponse row m-vectors $\tilde{Y}_1(x_1), \ldots, \tilde{Y}_N(x_N)$ are observed corresponding to (not necessarily distinct) levels $x_1, \ldots, x_N$ of the "controlled variable" taking on values in $\mathscr{X}$, we suppose the $\tilde{Y}_i$ uncorrelated with

$$E\,\tilde{Y}_i(x) = \theta'\tilde{F}(x), \qquad \text{Cov}\ \tilde{Y}_i(x) = \sigma^2 Q_x; \tag{1.1}$$

here θ is the unknown column k-vector of parameters and Q_x is a known positive definite $m \times m$ matrix, varying continuously on $\mathscr{X}$; σ^2 may be unknown. We suppose the law (1.1) to be meaningful for a "virtual" observation at x in $\mathscr{Z} - \mathscr{X}$, *but the experiment* $(x_1, x_2, \ldots, x_N)$ *is restricted to values in* $\mathscr{X}$. (The definition of Q_x for x in $\mathscr{Z} - \mathscr{X}$ will often be arbitrary, since its variation may be subsumed into that of v, defined below.)

The problem, to be made precise shortly, is to choose $X_N = (x_1, x_2, \ldots, x_N)$ in $\mathscr{X}$ in some "good" fashion for the purpose of estimating the response $\theta'\tilde{F}(x)$ for x in $\mathscr{Z}$. However, for economy of form we use *not* a linear combination of the rows of $\tilde{F}$, but, instead, a linear combination $t'\tilde{G}(x)$ of the rows of $\tilde{G}$.

If $\tilde{F} = \tilde{G}$ we have the classical problem of minimum variance curve-fitting. When $\mathscr{Z} = \mathscr{X}$ and $m = 1$, this was treated by various authors (see especially [15, 16, 12, 13, 3b, 4, 9]) under various optimality criteria. The multiresponse generalization $m > 1$ in this case was given in detail in [5]. For $\mathscr{Z} \neq \mathscr{X}$, we have extrapolation or interpolation problems, as treated in [17, 8, 18, 9, etc.] for $m = 1$.

[1] Prepared under NSF Grant GP24438.

When $m = 1$, $\mathscr{X} = \mathscr{Z}$, and $\tilde{G}$ is taken to be the first s rows $\tilde{F}_1$ of $\tilde{F}$ and $\theta' = (\theta_1', \theta_2')$ is partitioned correspondingly, we have the model of Box and Draper [1] for estimating $\theta'\tilde{F}$ by $t_1'\tilde{G} = t_1'\tilde{F}_1$. Further work on this model, with different recommendations (discussed in Section 2 below), was carried out by Karson, Manson, and Hader [10]. Recently Hader, Manson, and Cote considered this model when $\tilde{G}$ is not so restricted. Related work by Fedorov and Malyutov [6] treated particular $(\mathscr{X}, \tilde{F})$ in the case $m = 1$, $\mathscr{Z} = \mathscr{X}$, with the restricted form of $\tilde{G}$, but with the maximum of the mean squared error function replacing its integral as considered below.

A common formulation of a "biased estimation" problem, especially in fractional factorial settings, is to ask for estimates of $L\theta$ where $L = [L_1 \mid L_2]$ and L_1 is $r \times s$, and where the estimates are of the form $L_1 t_1$ in a notation consistent with that of the previous paragraph; θ_2 may not be estimable. This is fitted into the previous formulation by letting $\mathscr{Z} = \{1, 2, \ldots, r\}$ and $[\tilde{G}(1), \ldots, \tilde{G}(r)] = L_1'$, $[\tilde{F}(1), \ldots, \tilde{F}(r)] = L'$, if $m = 1$.

If $m > 1$ this last formulation has an obvious extension, since $\mathscr{Z} \cap \mathscr{X} = \varnothing$. Thus, for example, for each i the m columns of $\tilde{F}(i)$ and $\tilde{G}(i)$ can be taken to be the same as the single column of the previous paragraph, or $m - 1$ of them can be taken as 0. Phrased another way, we can avoid this artificial representation by noting that if $\mathscr{X} \cap \mathscr{Z} = \varnothing$, there is no need for $\tilde{F}$ and $\tilde{G}$ on $\mathscr{Z}$ to have m columns.

Thus, our theory includes the simultaneous fitting of several response surfaces $\tilde{G}_i$ on what can even be different $\mathscr{Z}_i$'s.

If the observable components of $\tilde{Y}_i(x)$ (and the number of such components) vary with x, this is easily subsumed into our model by assigning the value zero to the corresponding elements of $\tilde{F}(x)$ (or of $F(x)$, below).

The present note is intended to unify and extend these considerations to arbitrary m, $\mathscr{Z}$, $\tilde{G}$ (Section 2). At the same time, a modification of the previous approaches is related to the author's general theory of approximate design optimization [14] (Section 3). We show by example that the recommendations of [1] and [10] can lead to procedures whose risk can be much improved upon (Section 4).

Before proceeding further, we simplify notation by a reduction. Write $[\tilde{Y}_i(x), \tilde{F}(x), \tilde{G}(x)] = [Y_i(x), F(x), G(x)]Q_x^{-1/2}$ where $Q_x^{1/2}$ is the symmetric positive definite square root of Q_x. Then

$$EY_i(x) = \theta' F(x), \qquad \text{Cov } Y_i(x) = \sigma^2 I. \tag{1.2}$$

We hereafter treat the problem in terms of Y_i, F, G; the results are then easily transformed back into the original formulation.

To minimize confusion in the sequel, we shall use the dummy variable x when referring to the choice of experimental points, and z when referring to the estimated response on $\mathscr{Z}$.

Suppose that the estimator $t'G$ is used to estimate $\theta'F$ on $\mathscr{X}$. Write Y for the Nm-vector $(Y_1, Y_2, \ldots, Y_N)$. *We restrict consideration to estimators which are linear in* Y. Thus, $T = CY'$, where C is $s \times Nm$. We denote by $\bar{F}(X_N)$ the $k \times Nm$ matrix $[F(x_1) \mid F(x_2) \mid \cdots \mid F(x_N)]$, and also write $D = D(X_N) = \bar{F}(X_N)C'$ and $CC' = H$. Hence, we have

$$Et = D'\theta, \qquad \operatorname{Cov}(t) = \sigma^2 H. \tag{1.3}$$

Now suppose we are given an $m \times m$ matrix-valued *loss measure* v on $\mathscr{X}$, nonnegative definite in value, and such that F and G are square-integrable relative to v. We define $\Gamma_{FF}(k \times k)$, $\Gamma_{GG}(s \times s)$, and $\Gamma_{FG}(k \times s)$ or Γ_{GF} by

$$\Gamma_{A_1A_2} = \int_{\mathscr{X}} A_1(z)\, dv(z)\, A_2(z)'. \tag{1.4}$$

We assume the (quadratic) expected loss incurred if the design X_N and estimator t are used and θ is the true parameter value is then

$$\begin{aligned} R(\theta; X_N, t) &= E\int_{\mathscr{X}} [t'G(z) - \theta'F(z)]v(dz)[t'G(z) - \theta'F(z)]' \\ &= \sigma^2 \operatorname{tr} H\Gamma_{GG} + \theta'[D\Gamma_{GG}D' - D\Gamma_{GF} - \Gamma_{FG}D' + \Gamma_{FF}]\theta \\ &= \sigma^2(V + B) \qquad \text{(say)}, \end{aligned} \tag{1.5}$$

the last being a notation which specializes to that adopted by previous authors [1, 10] when certain restrictions on D are introduced, as described in the next section. As will be seen in the example of Section 4, the vector function (B, V) is often convenient to consider without combining as in (1.5).

We conclude this section by recording a simple quadratic minimization result, essentially the Gauss–Markov theorem. With the usual ordering of nonnegative definite matrices, we write $P_1 \prec P_2$ to mean $P_2 - P_1$ is nonnegative definite. Suppose the row space of $R(k \times h)$ is contained in that of $L(n \times h)$, so that there are matrices $U(k \times n)$ satisfying $UL = R$. Then, if

$$U_0 = R(L'L)^- L', \qquad U_0 U_0' = R(L'L)^- R', \tag{1.6}$$

we have $U_0 L = R$ and

$$U_0 U_0' \prec UU' \qquad \text{if} \quad UL = R. \tag{1.7}$$

Here $(\ )^-$ denotes generalized inverse.

2. FORMULATION OF BOX AND DRAPER

Without further restrictive criteria, the unknown relative magnitudes of $\theta'\theta$ and σ^2 make the problem of "choosing X_N and t so as to make (1.5) small" too vague or unfruitful. In the approach of [1] and [10] this difficulty is overcome by first considering B alone for minimization. Before discussing this more precisely, we must recall for the reader the distinction between the *exact* and *approximate* design theory developments; failure to make clear which development is being used has created some confusion in the previous literature, in the author's opinion. The *exact theory* follows the previous lines, wherein a design X_N in the present setting is an N-tuple of points (not necessarily distinct) in $\mathscr{X}$; or, equivalently, it is a discrete probability measure ξ on $\mathscr{X}$ restricted to the family of probability measures $\{\xi$: values of ξ are integral multiples of $N^{-1}\}$, where the ξ corresponding to X_N is given by $N\xi(x) =$ [number of $x_i = x$]. In the *approximate theory* ξ is permitted to be any member of the family of *all* probability measures on $\mathscr{X}$ relative to a specified σ-field which contains at least all the finite subsets of $\mathscr{X}$. The use of the approximate theory makes tractable various unwieldy minimization problems of the exact theory, but then necessitates implementation in terms of exact designs which may only be approximately optimum. See [11, 15, 4] for discussion. We write $M(\xi) = \int_{\mathscr{X}} F(x)F(x)'\xi(dx)$ for the information matrix (per observation) under ξ.

We continue with our description of the work of [1] and [10] in the case $m = 1$, $G =$ first s elements of F. Motivated by illustrative examples which indicated that minimization of B alone yielded a risk close to the minimum of R (assuming knowledge of $\sigma^{-2}\theta_2{}^2$), Box and Draper (hereafter BD) assumed t to be the least squares estimator of θ_1 under the assumption $\theta_2 = 0$ and then found, *in what we must in general regard as the approximate theory for reasons described in the next paragraph*, the design which minimized B. It is not at all evident that "minimization of B" makes mathematical sense in view of the dependence of B on the vector θ, but in the present setting it turns out that a single achievable matrix of the quadratic form in θ of (1.5) (second line) yields, for each possible value of θ, the smallest value of B; that is, the matrix of the quadratic form B is smaller than all other possible ones in the sense of the ordering $\prec$ defined at the end of Section 1. In Section 4 the reader will find an illustration of what we regard as a fairly common occurrence, in the failures of the BD motivation (almost minimizing R) to be an accurate premise.

The original BD considerations always contemplated, I believe, an F consisting of monomials on a convex Euclidean $\mathscr{X}$ with $\mathscr{X} = \mathscr{Z}$ and $\nu =$ Lebesgue measure, and with a large enough N (for a given k and s), that the solution could be implemented as an *exact design*. But it is easily seen that lack of convexity or smallness of N can make this solution sensible in the approximate theory alone.

The work of Karson, Manson, and Hader (hereafter KMH) in the same setting assumes it is satisfactory to begin by "minimizing B" (made precise two paragraphs above). They do *not* restrict t to be the least squares estimator of θ_1, but instead find, for any X_N for which $\theta_1 + \Gamma_{FF}^{-1}\Gamma_{FG}\theta_2$ is estimable, a t

which minimizes B; *their striking observation is that the minimizing function B is the same for all such X_N*. In illustrative examples they then choose an exact X_N to minimize V for this t, subject to some simplifying symmetry restrictions on X_N; this must do as well as BD and can often do quite better, as we will see in Section 4. (In subsequent papers these and other authors have used other criteria than simple minimization of V; these will not be discussed here.) We observe that N is quite small in the illustrative examples, and this is simply a reflection of the difficulty of solving such extremum problems in the exact theory.

The fact that G consists of the first s rows of F produces a solution in the KMH formulation in which the minimizing function B does not depend on θ_1, but the corresponding result for general G does not have this feature and hence is often unrealistic in practical terms, as we will describe in the second paragraph below. For the moment, let us nevertheless give the extension of the KMH solution to our general setting, since it does seem sensible in some cases and will enable us to describe the resulting problem of minimizing V, using approximate theory algorithms, in this simplest approach. The matrix of the quadratic form (B) in θ in (1.5) can obviously be rewritten as

$$\Gamma_{FF} - \Gamma_{FG}\Gamma_{GG}^{-}\Gamma_{GF} + (D - \Gamma_{FG}\Gamma_{GG}^{-})\Gamma_{GG}(D - \Gamma_{FG}\Gamma_{GG}^{-})'; \tag{2.1}$$

here $\Gamma_{GG}^{-}\Gamma_{GF}$ is well-defined and (2.1) holds even if Γ_{GG} is singular; note that

$$\begin{pmatrix} \Gamma_{FF} & \Gamma_{FG} \\ \Gamma_{GF} & \Gamma_{GG} \end{pmatrix}$$

is nonnegative definite. If X_N is such that $D(X_N)$ (that is, C) can be chosen to satisfy

$$\bar{F}(X_N)C' = D(X_N) = \Gamma_{FG}\Gamma_{GG}^{-}, \tag{2.2}$$

then any such choice clearly minimizes the *matrix* (2.1) in the sense described earlier and defined at the end of Section 1, and yields

$$B = \sigma^{-2}\theta'[\Gamma_{FF} - \Gamma_{FG}\Gamma_{GG}^{-}\Gamma_{GF}]\theta \tag{2.3}$$

for the minimum of the matrix. Of course, there is a C satisfying (2.2) if and only if $\theta'\Gamma_{FG}\Gamma_{GG}^{-}$ is estimable (has a linear unbiased estimator) when design X_N is used. For such an X_N, it follows from (1.6) that H achieves its (matrix) minimum among C satisfying (2.2) with the choice

$$C = \Gamma_{GG}^{-}\Gamma_{GF}(\bar{F}(X_N)\bar{F}(X_N)')^{-}\bar{F}(X_N), \tag{2.4}$$

which yields

$$H = \Gamma_{GG}^{-}\Gamma_{GF}(\bar{F}(X_N)\bar{F}(X_N)')^{-}\Gamma_{FG}\Gamma_{GG}^{-}. \tag{2.5}$$

As is usual in optimum design considerations, there will not generally be a design uniformly best in our matrix sense of making H smallest. However, if we continue along the KMH line, we must by (1.5) only minimize tr $\Gamma_{GG} H$. (The BD analogue will be clear; we shall omit it.)

Write $A = \Gamma_{FG} \Gamma_{GG}^{-} \Gamma_{GF}$. The minimization of tr $A[\bar{F}(X_N)\bar{F}(X_N)']^{-}$ is generally difficult in the exact theory. In the approximate theory, Fedorov [4] characterizes the solution, and obtains iterative methods for solving this problem; the optimum ξ^* (say) is characterized, if $M(\xi^*)$ is nonsingular, by

$$\max_{x \in \mathscr{X}} \operatorname{tr} F(x)' M^{-1}(\xi^*) A M^{-1}(\xi^*) F(x) = \operatorname{tr} A M^{-1}(\xi^*) \tag{2.6}$$

or by the fact that the left side of (2.6) cannot be decreased by altering ξ^*. (For $m = 1$, (2.6) is originally due to Elfving [3a] and Karlin and Studden [9]; Chernoff's work [2] is related.) There is a corresponding result for $M(\xi^*)$ singular. We omit further discussion, except to note that this approximate theory development often yields solutions without much difficulty, which can then be implemented to supply almost optimum solutions for even fairly small N.

We mentioned, two paragraphs above, that if $G = F_1$ = first s rows of F on $\mathscr{X}$, then the minimizing B does not depend on θ_1. In fact, abbreviating $\Gamma_{F_i F_j}$ by Γ_{ij} in this case, a trivial calculation shows the minimum to be

$$B = \sigma^{-2} \theta_2' (\Gamma_{22} - \Gamma_{21} \Gamma_{11}^{-} \Gamma_{12}) \theta_2. \tag{2.7}$$

In other cases, (2.3) need not simplify in this way. One can still use the generalized KMH line of development as described in the previous two paragraphs, but this may not even approximately reflect the experimenter's goals. The reason for this is that minimizing B may now so restrict C (and thus affect the choice of X_N) that it makes $V + B$ obtained by the KMH route much larger than the minimum of $V + B$ for a wide spectrum of parameter values. While (Section 4) the BD or KMH prescription may not be satisfactory even in the situation $G = F_1$ originally contemplated by BD, the motivation for considering B separately is now completely absent. For, if G is not of the form F_1, then F_1 and θ_1 no longer have the special meaning they had and which resulted in both BD and KMH using (possibly different) estimators which were unbiased when $\theta_2 = 0$. In the case of general G, only the (possibly zero) part of the row space of G contained in the row space of F has this meaning.

This motivates our return now to consideration of $V + B$, and of trying to make it small from the outset.

3. MINIMIZING V + B

We know we must make further assumptions before it can make sense to minimize (1.5). One possibility is to minimize the maximum of (1.5) subject to an assumption $\sigma^{-1}\theta \in S$ where S is a specified set in R^k. Somewhat simpler, but in the same direction, is the minimization of an integral of (1.5) with respect to a specified measure ψ on $\sigma^{-1}\theta$. This has obvious meaning to Bayesians, but it may appeal to others as a possible compromise. For, in order to inspect the risk functions of the *admissible* (X_N, t) (after the reduction to (1.5)), it suffices to consider the closure of such integral minimizers (more generally, without the invariance reduction to $\sigma^{-1}\theta$). Although our restriction to linear estimators could then well be questioned, nevertheless, as an illustration of the type of mathematics which arises, we now consider such a development.

Let us suppose, then, that $\int \sigma^{-2}\theta\theta'\psi(d\theta/\sigma) = \Phi$ is a specified nonnegative definite $k \times k$ matrix and that, in accordance with the discussion of the previous paragraph (and omitting the dependence of $\bar{F}$ on X_N), we seek to choose (X_N, t) to minimize

$$\bar{R} = \int \sigma^{-2}R\psi(d\theta/\sigma) = \operatorname{tr}\{\Gamma_{GG}CC' + \Phi[D\Gamma_{GG}D' - D\Gamma_{GF} - \Gamma_{FG}D' + \Gamma_{FF}]\}$$

$$= \operatorname{tr}\{\Gamma_{GG}^{1/2}C[I_{Nm} + \bar{F}'\Phi\bar{F}]C'\Gamma_{GG}^{1/2} - C\bar{F}'\Phi\Gamma_{FG} - \Gamma_{GF}\Phi\bar{F}C' + \Phi\Gamma_{FF}\}. \tag{3.1}$$

The minimum with respect to C of the *matrix* whose trace is taken in (3.1) is easily seen, by an argument like that used in conjunction with (2.1), to be attained when

$$C = \Gamma_{GG}^{-}\Gamma_{GF}\Phi\bar{F}[I_{Nm} + \bar{F}'\Phi\bar{F}]^{-1}. \tag{3.2}$$

The C of (3.2) yields

$$\bar{R} = \operatorname{tr}\{\Phi\Gamma_{FF} - \Phi\Gamma_{FG}\Gamma_{GG}^{-1}\Gamma_{GF}\Phi\bar{F}[I_{Nm} + \bar{F}'\Phi\bar{F}]^{-1}\bar{F}'\}. \tag{3.3}$$

(As before, matrix minimization with respect to $\bar{F}$ is unachievable, so we consider the trace.)

Since

$$\Phi^{1/2}\bar{F}[I_{Nm} + \bar{F}'\Phi\bar{F}]^{-1}\bar{F}'\Phi^{1/2} = I_k - [I_k + \Phi^{1/2}\bar{F}\bar{F}'\Phi^{1/2}]^{-1}, \tag{3.4}$$

we see that minimization of $\bar{R}$ with respect to X_N is equivalent to minimization of

$$\bar{\bar{R}} = \operatorname{tr}\Gamma_{FG}\Gamma_{GG}^{-1}\Gamma_{GF}\Phi^{1/2}[I_k + \Phi^{1/2}\bar{F}\bar{F}'\Phi^{1/2}]^{-1}\Phi^{1/2}$$

$$= \operatorname{tr}\Gamma_{FG}\Gamma_{GG}^{-1}\Gamma_{GF}[\Phi^{-1} + \bar{F}\bar{F}']^{-}, \tag{3.5}$$

where the last bracketed expression has the well-defined meaning obtained from the previous line in the obvious way.

So far we have not invoked the approximate theory. In a recent work on general approximate theory optimality criteria, the author [14] has shown that the analogue of (2.6) for a minimizing criterion of the form tr $A[\bar{A} + M(\xi)]^{-1}$, where A and $\bar{A}$ are symmetric nonnegative definite, is that the optimum ξ^* satisfy

$$\max_{x \in \mathscr{X}} \operatorname{tr} F(x)'[\bar{A} + M(\xi^*)]^{-1} A[\bar{A} + M(\xi^*)]^{-1} F(x)$$
$$= \operatorname{tr} M(\xi^*)[\bar{A} + M(\xi^*)]^{-1} A[\bar{A} + M(\xi^*)]^{-1}, \quad (3.6)$$

with an appropriate analogue if $M(\xi^*)$ is singular. This criterion differs from that of (2.6) or from the analogue for D-optimality, in that the left side of (3.6) need not be a minimum as was the case for (2.6); this is a consequence of the fact that tr $A[\bar{A} + M]^{-1}$ is not homogeneous in M. However, (3.6) leads to iterative procedures for finding an optimum design.

We note that one is often led to recommend what in the limit *appears* as a singular matrix replacing Φ^{-1} in the last line of (3.5). (It is not justified simply to substitute such a Φ^{-1} in (3.5).) For example, in the special case we have discussed wherein $G = F_1$, the treatment of θ_1 and θ_2 in comparable terms in the ψ of (3.1) is of doubtful applicability except in the rarest Bayesian models. More likely is the original BD spirit, implemented here by assuming no restriction on θ_1 and letting $\int \sigma^{-2}\theta_2\theta_2'\psi_2(d\theta_2/\sigma) = \Phi_2$ (specified). This case can be developed by requiring the estimator $t'F_1$ to have bounded risk as a function of θ_1, which means being unbiased when $\theta_2 = 0$, and then going through a line of reasoning analogous to that of (3.1)–(3.6). Alternatively, one can let

$$\Phi = \begin{pmatrix} \lambda I & 0 \\ 0 & \Phi_2 \end{pmatrix}$$

and treat the result of letting $\lambda \to +\infty$ in our development above. Illustrations of this will be given elsewhere.

4. AN ILLUSTRATIVE EXAMPLE

This is not a practical example, but rather one chosen to be arithmetically trivial (not necessitating the use of (2.6) or (3.6)), at the same time that it exhibits various general features which occur in more meaningful settings. Suppose $m = 1$, $k = 2$, and that $\mathscr{X} = \mathscr{Z}$ consists of the two points $\{-1, a\}$ where $a > 0$. The problem is one of "linear regression" with the estimated regression *homogeneous*. This may be expressed as $F(x)' = (x, 1)$ with $s = 1$ and $G(x) = F_1(x) = x$. For arithmetical simplicity suppose $v(-1) = a/(1 + a)$ and $v(a) = 1/(1 + a)$, so that $\Gamma_{FF} = \begin{pmatrix} a & 0 \\ 0 & 1 \end{pmatrix} = \Gamma$ (say) is diagonal. (In practice this might arise as an approximation to an example where there is a uniform measure v on a cluster of r_{-1} points near -1 and r_{-a} near a, in proportion $r_{-1}/r_a \sim a$.)

In the calculations which follow, σ^2 should be thought of as N^{-1} Var(Y_1).

We abbreviate $\xi(-1) = \alpha$, $\xi(a) = 1 - \alpha$. Thus,

$$M(\xi) = \begin{pmatrix} \alpha + (1-\alpha)a^2 & -\alpha + (1-\alpha)a \\ -\alpha + (1-\alpha)a & 1 \end{pmatrix}.$$

The BD prescription is $m_{11}^{-1}m_{12} = \gamma_{11}^{-1}\gamma_{12} = 0$, or $\alpha = a/(1+a)$. (As we have remarked, this may not be even approximately implementable for small N.) This yields (from (2.2) or (2.7))

$$B_{\text{BD}} = \sigma^{-2}\theta_2{}^2, \qquad V_{\text{BD}} = \gamma_{11}(M^{-1})_{11} = 1. \tag{4.1}$$

On the other hand, the KMH prescription for t yields a bias term equal to that of (4.1) for every design ξ, and corresponding value $V = a/\alpha(1-\alpha)(1+a)^2$ if the value α is used for $\xi(-1)$. This is minimized by $\alpha = \frac{1}{2}$ and yields

$$B_{\text{KMH}} = \sigma^{-2}\theta_2{}^2, \qquad V_{\text{KMH}} = 4a/(1+a)^2. \tag{4.2}$$

(We note that this design is also not approximately implementable when N is small and odd.)

As $a \to 0$ or $+\infty$ we see that $V_{\text{KMH}}/V_{\text{BD}} \to 0$. The corresponding ratio $R_{\text{KMH}}/R_{\text{BD}}$ of course depends on $\sigma^{-2}\theta_2{}^2$, but if this is suspected to be small (which is reasonable where one adopts the biased curve-fitting rationale of this model, especially if N is small), the KMH procedure will often yield great improvement.

Suppose we consider instead the approach of Section 3. In accordance with the last paragraph of that section, we consider only linear estimators $tG(x)$ which are unbiased estimators of $(\theta_1 + c\theta_2)x$ for some c. From (1.3) and (1.5) we obtain $B = (1 + c^2a)\theta_2{}^2/\sigma^2$. For fixed α, the variance of the Gauss–Markov estimator of $\theta_1 + c\theta_2$ is proportional to $(1, c)M^{-1}(\xi)(1, c)'$. Thus, if we have a value of $\int \theta_2{}^2\sigma^{-2}\psi_2(d\theta_2/\sigma) = \Phi_2$ (say) to be considered, the last paragraph of Section 3 tells us to minimize

$$a(1, c)M^{-1}(\xi)(1, c)' + \Phi_2(1 + c^2a) \tag{4.3}$$

with respect to c and α. Rather than to do this for general Φ_2, let us see what happens when Φ_2 is very small; that is, let us minimize V alone. (Recall this does not mean making $V = 0$, since t is restricted to unbiased estimators of $\theta_1 + c\theta_2$.) The minimum with respect to c (for fixed α) is easily seen to occur at $c = [(1-\alpha)a - \alpha]/[\alpha + (1-\alpha)a^2]$, for which $V = a/[\alpha + (1-\alpha)a^2]$. This is minimized by $\alpha = 1$ or 0 if $a < 1$ or > 1 (or by any α if $a = 1$). The corresponding V and B are

$$\begin{aligned} B &= \sigma^{-2}\theta_2{}^2(a+1), & V &= a, & &\text{if } a \le 1; \\ B &= \sigma^{-2}\theta_2{}^2(a^{-1}+1), & V &= a^{-1}, & &\text{if } a \ge 1. \end{aligned} \tag{4.4}$$

Comparing (4.4) with (4.2), we see that for a close to 0 or very large, there is a kind of "subadmissibility" advantage of (4.4) over (4.2): the B's are almost the same, and the V of (4.4) is about $\frac{1}{4}$ that of (4.2). When $\sigma^{-2}\theta_2{}^2 = N\theta_2{}^2/\mathrm{Var}(Y_1)$ is not suspected to be of larger order of magnitude than $\min(a, a^{-1})$, the resulting saving in $B + V$ can be large; and, even when $\sigma^{-2}\theta_2{}^2$ is large, the maximum possible relative increase in risk is small. The factor of decrease $\frac{1}{4}$ in the above example can be made arbitrarily close to 0 in suitable examples.

The moral of the above example is that a slight alteration in the design and estimator of KMH can often achieve a risk function R^* (say) such that R^*/R_{KMH} is much less than 1 over a large set of parameter values, while R^*/R_{KMH} is never more than very slightly above 1. Thus, the investigation of (and choice among) the risks R of many (X_M, t) seems preferable to reliance upon any simple single prescription.

The lack of reasonably close implementability of the approximate theory optimum for N small (and, in the above example, odd) reinforces these last comments; it may be preferable to depart from minimization of B in order to improve V over what it is for an inadequate discrete implementation of our KMH approximate theory optimum.

5. OTHER COMMENTS

Nonlinear models can be fitted into our scheme by the usual method of linearizing after a first stage of experimentation yields sufficiently accurate estimates.

Variable costs of observations (cost depending on x) can be included and then reduced to the present constant cost model in the manner used just above (1.2) to eliminate covariance variation. See, e.g., [2, 11, 15, 4] regarding the above.

The problem of curve-fitting is not really solved by these considerations. We have not treated such a difficult but more meaningful problem as that of deciding, for a given F, which rows of it to call G. A sensible decision-theoretic framework is then to let the loss function reflect both errors in estimation and also a penalty depending on the choice of G, for example, a function of the number of rows s (or coefficients θ_i estimated to be nonzero). There are not at present satisfactory prescriptions for this model even in the absence of the choice of design.

Analytical solutions of the problem described by (2.6) or (3.6) are often aided by invariance considerations or a conclusion about the support ξ^* must have in order that it assign all measure to the set of x where the maximum on the left side is attained. For example, if $\mathscr{X}$ is the q-cube, $m = 1$, and F consists of all monomials of degree ≤ 2, and if G is symmetric with respect to the cube (e.g., consists of monomials of degree ≤ 1, or these and the $x_i{}^2$, $1 \leq i \leq q$) along with $\mathscr{X}$ and v, then one concludes that there is a symmetric optimum design; the fact that the function maximized in (2.6) or (3.6) is a symmetric quartic then allows one to conclude, as in the D-optimality argument of [3b], that ξ^* is supported by a subset of the obvious 3^q array of points. This reduces the problem to a minimization in q variables, the weights on faces of various dimensions. Further reduction may then be possible, as in [3b].

Added in proof: Recent additional work on designs for biased curve fitting has been obtained in work of S. Stigler, C. Atwood, L. Jordan, and P. Huber, each using a different approach.

REFERENCES

1. Box, G. E. P. and Draper, N. R. (1959). A basis for the selection of a response surface design. *J. Amer. Statist. Assoc.* **54** 622.
2. Chernoff, H. (1953). Locally optimum designs for estimating parameters. *Ann. Math. Statist.* **24** 586.

3a. Elfving, G. (1959). Design of linear experiments. *Cramér Festschrift Volume*, 58. Wiley, New York.

3b. Farrell, R., Kiefer, J. and Walbran, A. (1965). Optimum multivariate designs. *Proc. Fifth Berkeley Symp. Math. Statist. Prob.* **1** 113.

4. Fedorov, V. V. (1972). *Theory of Optimal Experiments.* Academic Press, New York.
5. Fedorov, V. V. (1971). The design of experiments in the multiresponse case. *Theor. Probability Appl.* **16** 323.
6. Fedorov, V. V. and Malyutov, M. B. (1971). Preprint 18. LSM, Moscow State Univ., Moscow.
7. Hader, R. J., Manson, A. R. and Cote, R. (1971). Abstract. *Ann. Math. Statist.* **42** 2190.
8. Hoel, P. and Levine, A. (1964). Optimal spacing and weighting in polynomial prediction. *Ann. Math. Statist.* **35** 1553.
9. Karlin, S. and Studden, W. (1966). Optimal experimental designs. *Ann. Math. Statist.* **37** 783.
10. Karson, M. J., Manson, A. R. and Hader, R. J. (1969). Minimum bias estimation and experimental designs for response surfaces. *Technometrics* **11** 461.
11. Kiefer, J. (1959). Optimum experimental designs. *J. Roy. Statist. Soc. Ser. B.* **21** 272.
12. Kiefer, J. (1961). Optimal designs in regression problems, II. *Ann. Math. Statist.* **32** 298.
13. Kiefer, J. (1960). Optimum experimental designs with applications to systematic and rotatable designs. *Proc. Fourth Berkeley Symp. Math. Statist. Prob.* **1** 381.
14. Kiefer, J. (1972). Equivalence theory for general convex design criteria. To be published. (Abstract in *Bull. IMS* **1** 1972.)
15. Kiefer, J. and Wolfowitz, J. (1959). Optimum designs in regression problems. *Ann. Math. Statist.* **30** 271.
16. Kiefer, J. and Wolfowitz, J. (1960). The equivalence of two extremum problems. *Canad. J. Math.* **12** 363.
17. Kiefer, J. and Wolfowitz, J. (1964). Optimum extrapolation designs I and II. *Ann. Inst. Statist. Math.* **16** 79, 295.
18. Kiefer, J. and Wolfowitz, J. (1965). On a theorem of Hoel and Levine on extrapolation designs. *Ann. Math. Statist.* **36** 1627.

Asymptotic Properties of Some Sequential Nonparametric Estimators in Some Multivariate Linear Models

PRANAB KUMAR SEN[1]

DEPARTMENT OF BIOSTATISTICS
UNIVERSITY OF NORTH CAROLINA
CHAPEL HILL, NORTH CAROLINA

MALAY GHOSH

RESEARCH TRAINING SCHOOL
INDIAN STATISTICAL INSTITUTE
CALCUTTA, INDIA

1. INTRODUCTION

For a general multivariate linear model (which includes the one-sample and two-sample location models as special cases), robust sequential point as well as interval estimators based on suitable rank order statistics are proposed and studied. In a nonsequential setup, parallel procedures were considered by Sen and Puri [14]. Also, the sequential point estimation problem based on sample means (in the univariate case) has been studied earlier by Blum *et al.* [3], and later, in a more general setup, by Mogyorodi [10], among others. Finally, the sequential interval estimation procedure, based on the principles of Chow and Robbins [4], extends the univariate theory developed by Sen and Ghosh [13] and Ghosh and Sen [5–7] to the general multivariate case.

In Section 2, along with our basic model, we briefly sketch the problems. Preliminary notions and basic assumptions are then considered in Section 3. Section 4 is devoted to the study of the asymptotic properties of sequential point estimators based on robust rank order statistics. The problem of robust sequential interval estimation is then treated in Section 5. The last section is devoted to a comparison with the corresponding parametric procedures, and presents the allied asymptotic relative efficiency (ARE) results.

[1] Work by the author was sponsored by the Aerospace Research Laboratories, Air Force Systems Command, U.S. Air Force, Contract F33615-71-C-1927. Reproduction in whole or in part is permitted for any purpose of the U.S. Government.

2. THE PROBLEMS

Consider a sequence $\{\mathbf{X}_i = (X_{i1}, \ldots, X_{ip})', i \geq 1\}$ of $p(\geq 1)$-variate stochastic vectors, defined on a probability space $(\Omega, \mathscr{A}, P)$, where $\mathbf{X}_i$ has an absolutely continuous cumulative distribution function (cdf) $F_i(\mathbf{x})$, $\mathbf{x} \in R^p$, the p-dimensional Euclidean space. It is assumed that

$$F_i(\mathbf{x}) = F(\mathbf{x} - \boldsymbol{\alpha} - \boldsymbol{\beta} c_i), \qquad i \geq 1, \tag{2.1}$$

where $\boldsymbol{\alpha} = (\alpha_1, \ldots, \alpha_p)'$ and $\boldsymbol{\beta} = (\beta_1, \ldots, \beta_p)'$ are unknown parameters (vectors), and $\{c_i, i \geq 1\}$ is a sequence of known (scalar) constants.

Robust point as well as interval estimators of $(\boldsymbol{\alpha}, \boldsymbol{\beta})$ based on suitable rank order statistics when the sample size is large but nonrandom were studied in detail by Sen and Puri [14]. We are primarily concerned here with the following two sequential extensions of this theory.

Let $\{N_\nu, \nu \geq 1\}$ be a sequence of nonnegative integer-valued random variables, such that

$$\nu^{-1} N_\nu \to \lambda, \qquad \text{in probability, as} \qquad \nu \to \infty, \tag{2.2}$$

where λ is a positive random variable having an arbitrary distribution

$$H(u) = P\{\lambda \leq u\}, \qquad 0 < u < \infty, \tag{2.3}$$

and defined on the same probability space $(\Omega, \mathscr{A}, P)$. Consider then an estimator $(\hat{\boldsymbol{\alpha}}_{N_\nu}, \hat{\boldsymbol{\beta}}_{N_\nu})$ of $(\boldsymbol{\alpha}, \boldsymbol{\beta})$ based on $\mathbf{X}_1, \ldots, \mathbf{X}_{N_\nu}$ through a general class of rank order statistics, to be precisely defined in Section 3. Our first problem is to derive (along the lines of Blum *et al.* [3] and Mogyorodi [10]) the asymptotic normality of $N_\nu^{1/2}[(\hat{\boldsymbol{\alpha}}_{N_\nu} - \boldsymbol{\alpha}), (\hat{\boldsymbol{\beta}}_{N_\nu} - \boldsymbol{\beta})]$ (as $\nu \to \infty$). This enables us to study various asymptotic properties of $(\hat{\boldsymbol{\alpha}}_{N_\nu}, \hat{\boldsymbol{\beta}}_{N_\nu})$.

In the second problem, our sample size N_ν remains a random variable, but so determined by a "stopping rule" that we have a simultaneous confidence interval for $(\boldsymbol{\alpha}, \boldsymbol{\beta})$, with the property that the confidence coefficient is asymptotically equal to a predetermined $1 - \varepsilon$: $0 < \varepsilon < 1$, and the length of the interval for each component of $\boldsymbol{\alpha}$ (or $\boldsymbol{\beta}$) is bounded above by $2d$ (or by a known multiple of $2d$), where $d > 0$ is a predetermined (small) number. The theory is an extension of the corresponding univariate theory developed by Sen and Ghosh [13] and Ghosh and Sen [5–7]. It is also a sequential extension of the theory developed by Sen and Puri [14], and a nonparametric analogue of the theory developed by Gleser [8] and Albert [2].

3. PRELIMINARY NOTIONS AND BASIC ASSUMPTIONS

Let $\mathscr{F}_p$ be the class of all p-variate absolutely continuous cdf's with finite Fisher information matrix, and let $\mathscr{F}_p{}^0$ be the subclass of $\mathscr{F}_p$ for which the distribution is diagonally symmetric about $\mathbf{0}$.

Assumption I. *If we are only interested in* $\boldsymbol{\beta}$, *we assume that* $F \in \mathscr{F}_p$; *otherwise, we assume that* $F \in \mathscr{F}_p^{\,0}$, *where* F *is defined in* (2.1). *For every* $v \geq 1$, *let*

$$\bar{c}_v = v^{-1}\sum_{i=1}^{v} c_i, \qquad C_v^{\,2} = \sum_{i=1}^{v}(c_i - \bar{c}_v)^2. \tag{3.1}$$

We have then the following problems: (a) *estimation of* $\boldsymbol{\alpha}$ *assuming* $\boldsymbol{\beta} = \mathbf{0}$; (b) *estimation of* $\boldsymbol{\beta}$ *treating* $\boldsymbol{\alpha}$ *as a nuisance parameter;* (c) *simultaneous estimation of* $(\boldsymbol{\alpha}, \boldsymbol{\beta})$. *For* (a) *no assumptions are needed on* $\{c_i, i \geq 1\}$; *for* (b) *and* (c) *our assumptions are, respectively,* II *and* III, *and* II, III′, *and* IV, *where*

Assumption II. *As* $v \to \infty$,

$$\max_{1 \leq i \leq v} v(c_i - \bar{c}_v)^2/C_v^{\,2} = 0(1). \tag{3.2}$$

Assumption III.

$$\lim_{v \to \infty} C_v^{\,2} = \infty. \tag{3.3}$$

Assumption III′.

$$\lim_{v \to \infty} v^{-1}C_v^{\,2} = C^2 \qquad (0 < C < \infty). \tag{3.4}$$

Assumption IV.

$$\lim_{v \to \infty} \bar{c}_v = \bar{c} \qquad \text{(finite)}. \tag{3.5}$$

It is easy to verify that all these assumptions hold true for the multivariate one-sample (where $c_i = 0 \;\forall\, i$) and two-sample (where c_i is either 0 or 1) models.

For every $v \geq 1$, let $R_{vi}^{(j)}$ (or $R_{vi}^{(j)+}$) be the rank of X_{ij} (or $|X_{1j}|$) among $X_{1j}, \ldots, X_{vj}$ (or $|X_{1j}|, \ldots, |X_{vj}|$) for $1 \leq i \leq v$, $1 \leq j \leq p$. To estimate $\boldsymbol{\beta}$, consider the following linear (regression) rank statistics

$$S_{vj} = \sum_{i=1}^{v}(c_i - \bar{c}_v)a_v^{(j)}(R_{vi}^{(j)}), \qquad j = 1, 2, \ldots, p, \tag{3.6}$$

$$\mathbf{S}_v = (S_{v1}, \ldots, S_{vp})', \tag{3.7}$$

where the *rank scores* $a_v^{(j)}(i)$, $1 \leq i \leq v$, $j = 1, \ldots, p$, are defined by

$$a_v^{(j)}(i) = E\phi_j(U_{vi}) \qquad [\text{or } \phi_j(i/(v+1))], \qquad i = 1, \ldots, v; \tag{3.8}$$

$\phi_j(u)$ is nondecreasing and absolutely continuous inside [0, 1], $U_{v1} \leq \cdots < U_{vv}$ are the ordered random variables in a sample of size v from the rectangular [0, 1] distribution. Regarding the *score functions* $\phi_1, \ldots, \phi_p$, one assumes as in the work of Ghosh and Sen [6] that for every $j\,(= 1, \ldots, p)$,

$$|\phi_j(u)| \leq K[-\log(u(1-u))], \qquad |\phi_j'(u)| \leq K[u(1-u)]^{-1}, \qquad 0 < u < 1, \tag{3.9}$$

where $0 < K < \infty$. This implies the existence of a t_0 (> 0) such that

$$M_j(t) = \int_{-\infty}^{\infty} \exp(t\phi_j(u))\, du < \infty \qquad \text{for all} \quad t: |t| \leq t_0, \tag{3.10}$$

for all $j = 1, \ldots, p$.

For estimating $\boldsymbol{\alpha}$, we need an alignment procedure and the following type of one-sample rank order statistics

$$T_{\nu j} = \sum_{i=1}^{\nu} c(X_{ij}) a_\nu^{(j)*}(R_{\nu i}^{(j)+}), \qquad j = 1, \ldots, p, \tag{3.11}$$

$$\mathbf{T}_\nu = (T_{\nu 1}, \ldots, T_{\nu p})', \tag{3.12}$$

where $c(u) = 1, \frac{1}{2}$, or 0, according as $u >, =,$ or < 0,

$$a_\nu^{(j)*}(i) = E\phi_j^*(U_{\nu i}) \qquad [\text{or } \phi_j^*(i/(\nu+1))], \qquad 1 \leq i \leq \nu, \tag{3.13}$$

$$\phi_j^*(u) = \phi_j\left(\frac{1+u}{2}\right) \qquad \text{and assume that} \qquad \phi_j(u) + \phi_j(1-u) = 0. \tag{3.14}$$

Some well-known cases of $\mathbf{S}_\nu$ and $\mathbf{T}_\nu$ are the normal scores and the Wilcoxon scores statistics, which relate respectively to $\phi_j(u)$ as the inverse of the standard normal cdf and $\phi_j(u) = 2u - 1$, $0 < u < 1$. Let us also define for later use

$$\boldsymbol{\Gamma} = ((\gamma_{jl})), \quad \gamma_{jl} = \int_{-\infty}^{\infty} \int_{-\infty}^{\infty} \phi_j(F_{[j]}(x))\phi_l(F_{[l]}(y))\, dF_{[jl]}(x, y) - \mu_j \mu_l, \tag{3.15}$$

for $j, l = 1, \ldots, p$, where $F_{[j]}$ is the jth marginal cdf and $F_{[jl]}$ is the bivariate (j, l)th joint cdf in the joint cdf F, and

$$\mu_j = \int_0^1 \phi_j(u)\, du, \; \mu_j^* = \frac{1}{2}\int_0^1 \phi_j^*(u)\, du \; j = 1, \ldots, p. \tag{3.16}$$

Assumption V. *For every j ($= 1, \ldots, p$), the density function $f_{[j]} = F'_{[j]}$ and its first derivative $f'_{[j]}$ exist and are bounded for almost all x* (a.a.x), *and*

$$\lim_{x \to \pm\infty} |\phi_j'(F_{[j]}(x)) f_{[j]}(x)| \qquad \textit{is finite.} \tag{3.17}$$

Let us then denote by

$$B_j = B(F_{[j]}, \phi_j) = \int_{-\infty}^{\infty} (d/dx)\phi_j(F_{[j]}(x)) dF_{[j]}(x), \qquad j = 1, \ldots, p; \tag{3.18}$$

$$\mathbf{T} = ((\tau_{jl})), \qquad \tau_{jl} = \gamma_{jl}/[B_j B_l], \qquad j, l = 1, \ldots, p. \tag{3.19}$$

Note that $B_j > 0 \; \forall \, j$, and $\mathbf{T}$ is positive semidefinite.

4. ASYMPTOTIC PROPERTIES OF ROBUST SEQUENTIAL POINT ESTIMATORS OF (α, β)

We find it more convenient to consider separately the following three problems:

I. estimation of $\boldsymbol{\alpha}$ assuming that $\boldsymbol{\beta} = \mathbf{0}$ (one-sample model);
II. estimation of $\boldsymbol{\beta}$ treating $\boldsymbol{\alpha}$ as a nuisance parameter;
III. joint estimation of $(\boldsymbol{\alpha}, \boldsymbol{\beta})$.

In the first problem, assume that $F \in \mathscr{F}_p{}^0$, and denote by $R_{vi}^{(j)+}(a_j)$ the rank of $|X_{ij} - a|$ among $|X_{1j} - a|, \ldots, |X_{vj} - a|$, $1 \leq i \leq v$, $1 \leq j \leq p$; the resulting rank statistics, defined in (3.11), are denoted by $T_{vj}(a), j = 1, \ldots, p$. Note that

$$T_{vj}(a) \qquad \text{is} \downarrow \text{in } a \qquad \text{for all} \quad j = 1, \ldots, p. \tag{4.1}$$

Define for each positive integer v,

$$\hat{\alpha}_{vj}^{(1)} = \sup\{a: T_{vj}(a) > v\mu_j^*\}, \qquad \hat{\alpha}_{vj}^{(2)} = \inf\{a: T_{vj}(a) < v\mu_j^*\}; \tag{4.2}$$

$$\hat{\alpha}_{vj} = \tfrac{1}{2}(\hat{\alpha}_{vj}^{(1)} + \hat{\alpha}_{vj}^{(2)}), \qquad j = 1, \ldots, p; \tag{4.3}$$

$$\hat{\boldsymbol{\alpha}}_v = (\hat{\alpha}_{v1}, \ldots, \hat{\alpha}_{vp})'. \tag{4.4}$$

We intend to study various asymptotic properties of $\hat{\boldsymbol{\alpha}}_{N_v}$, and toward this goal, we have the following.

Theorem 4.1. *When* $F \in \mathscr{F}_p{}^0$ *and* $\boldsymbol{\beta} = \mathbf{0}$, *under* (2.1), (2.2), (2.3), (3.9), (3.13), (3.14), *and* (3.17), *as* $v \to \infty$

$$\mathscr{L}(N_v^{1/2}[\hat{\boldsymbol{\alpha}}_{N_v} - \boldsymbol{\alpha}]) \to \mathscr{N}_p(\mathbf{0}, \mathbf{T}), \tag{4.5}$$

where $\mathbf{T}$ *is defined by* (3.19).

Proof. We use a recent powerful result of Mogyorodi [10, Theorem 2], according to which we are only to show that for nonstochastic v,

$$\mathscr{L}(v^{1/2}[\hat{\boldsymbol{\alpha}}_v - \boldsymbol{\alpha}]) \to \mathscr{N}_p(\mathbf{0}, \mathbf{T}) \qquad \text{as} \quad v \to \infty, \tag{4.6}$$

and for every $\varepsilon > 0$ and $\eta > 0$, there exists a $\delta > 0$ and an $n_0 = n_0(\varepsilon, \eta)$ such that for $n \geq n_0$,

$$P\left\{\max_{k: |n-k| < \delta n} \left| \sqrt{n} \|\hat{\boldsymbol{\alpha}}_k - \hat{\boldsymbol{\alpha}}_n\| > \varepsilon \right.\right\} < \eta, \tag{4.7}$$

where $\|\mathbf{x}\| = \max_{1 \leq j \leq p} |x_j|$, $\mathbf{x} = (x_1, \ldots, x_p)'$. Now, (4.6) has already been proved by Puri and Sen [11, Theorem 6.2.3, p. 226]. On the other hand, the left-hand side of (4.7) is bounded above by

$$\sum_{j=1}^{p} P\left\{\max_{k: |k-n| < \delta n} \sqrt{n}\, |\hat{\alpha}_{k,j} - \hat{\alpha}_{n,j}| > \varepsilon\right\}, \tag{4.8}$$

and hence, by the same technique of Sen and Ghosh [13, Lemma 5.3], it can be shown that (4.8) can be bounded by η (>0) but a proper choice of δ (>0) and n. For brevity, the proof is therefore omitted.

Since λ, defined by (2.2), is a positive random variable, for every $0 < \varepsilon < 1$, there exists a λ_ε (>0) such that $P\{\lambda \geq \lambda_\varepsilon\} \geq 1 - \varepsilon$, and hence, $N_\nu \to \infty$, in probability, as $\nu \to \infty$. Consequently, by (4.5)

$$\hat{\boldsymbol{\alpha}}_{N_\nu} \to \boldsymbol{\alpha}, \qquad \text{in probability, as} \quad \nu \to \infty. \tag{4.9}$$

Consider now the problem of estimating $\boldsymbol{\beta}$ treating $\boldsymbol{\alpha}$ as a nuisance parameter. Assume that $F \in \mathscr{F}_p$ and that II and III hold. Let $R_{\nu i}^{(j)}(b)$ be the rank of $X_{ij} - bc_i$ among $X_{1j} - bc_1, \ldots, X_{nj} - bc_n$ ($1 \leq i \leq \nu$; $1 \leq j \leq p$), b real; the resulting rank statistics defined by (3.6) are then denoted by

$$S_{\nu j}(b) \qquad (1 \leq j \leq p;\ \nu \geq 1). \tag{4.10}$$

It follows from Sen [12, Sect. 6] that

$$S_{\nu j}(b) \qquad \text{is} \downarrow \text{in } b \text{ for all} \quad j = 1, \ldots, p. \tag{4.11}$$

Define for each $\nu \geq 1$,

$$\hat{\beta}_{\nu j}^{(1)} = \sup\{b\colon S_{\nu j}(b) > 0\}, \qquad \hat{\beta}_{\nu j}^{(2)} = \inf\{b\colon S_{\nu j}(b) < 0\}, \qquad 1 \leq j \leq p; \tag{4.12}$$

$$\hat{\beta}_{\nu j} = \tfrac{1}{2}(\hat{\beta}_{\nu j}^{(1)} + \hat{\beta}_{\nu j}^{(2)}), \qquad 1 \leq j \leq p; \tag{4.13}$$

$$\hat{\boldsymbol{\beta}}_\nu = (\hat{\beta}_{\nu 1}, \ldots, \hat{\beta}_{\nu p})'. \tag{4.14}$$

Then, parallel to Theorem 4.1, we have the following.

Theorem 4.2. *For* $F \in \mathscr{F}_p$, *when* (2.1), (2.2), (2.3), (3.2), (3.3), (3.9), *and* (3.17) *hold, as* $\nu \to \infty$,

$$\mathscr{L}(C_{N_\nu}[\hat{\boldsymbol{\beta}}_{N_\nu} - \boldsymbol{\beta}]) \to \mathscr{N}_p(\mathbf{0}, \mathbf{T}), \tag{4.15}$$

where $\mathbf{T}$ *is defined by* (3.19).

Proof. As in the proof of Theorem 4.1, we require only to show that for nonstochastic ν,

$$\mathscr{L}(C_\nu[\hat{\boldsymbol{\beta}}_\nu - \boldsymbol{\beta}]) \to \mathscr{N}_p(0, \mathbf{T}) \qquad \text{as} \quad \nu \to \infty, \tag{4.16}$$

and for every $\varepsilon > 0$ and $\eta > 0$, there exist a $\delta > 0$ and an $n_0 = n_0(\varepsilon, \eta)$, such that for $n \geq n_0$,

$$P\left\{\max_{k:|n-k|<\delta n} C_n \|\hat{\boldsymbol{\beta}}_k - \hat{\boldsymbol{\beta}}_n\| > \varepsilon\right\} < \eta. \tag{4.17}$$

Now, (4.16) has already been proved by Sen and Puri [14, Theorem 5.1], while (4.17), by virtue of an inequality similar to that in (4.7)–(4.8), follows from Ghosh and Sen [6, Lemma 4.4]. Hence, the details are omitted.

By (4.16), (3.3), and the discussion preceding (4.9), as $\nu \to \infty$,

$$\hat{\boldsymbol{\beta}}_{N_\nu} \to \boldsymbol{\beta}, \qquad \text{in probability.} \tag{4.18}$$

Finally, consider the joint estimation of $(\boldsymbol{\alpha}, \boldsymbol{\beta})$. Assume that $F \in \mathscr{F}_p{}^0$ and Assumptions II, III′, IV, and V hold. Define the estimators $\hat{\boldsymbol{\beta}}_\nu$ as in (4.12)–(4.14), and then for estimating $\boldsymbol{\alpha}$, consider the following aligned rank statistics.

Let $\tilde{R}_{\nu i}^{(j)+}(a)$ be the rank of $|X_{ij} - a - \hat{\beta}_{\nu j} c_i|$ among $|X_{1j} - a - \hat{\beta}_{\nu j} c_1|, \ldots, |X_{\nu j} - a - \hat{\beta}_{\nu j} c_\nu|$, $(1 \le i \le \nu,\ 1 \le j \le p)$. The resulting one-sample rank order statistics defined by (3.11) are denoted by $\tilde{T}_{\nu j}(a)$ $(1 \le j \le p,\ \nu \ge 1)$. Define

$$\tilde{\alpha}_{\nu j}^{(1)} = \sup\{a\colon \tilde{T}_{\nu j}(a) > 0\}, \qquad \tilde{\alpha}_{\nu j}^{(2)} = \inf\{a\colon \tilde{T}_{\nu j}(a) < 0\}, \qquad \nu \ge 1, \quad 1 \le j \le p; \tag{4.19}$$

$$\tilde{\alpha}_{\nu j} = \tfrac{1}{2}(\tilde{\alpha}_{\nu j}^{(1)} + \tilde{\alpha}_{\nu j}^{(2)}), \qquad 1 \le j \le p, \quad \nu \ge 1; \tag{4.20}$$

$$\tilde{\boldsymbol{\alpha}} = (\tilde{\alpha}_{\nu 1}, \ldots, \tilde{\alpha}_{\nu p})', \qquad \nu \ge 1. \tag{4.21}$$

For notational simplicity, let $\boldsymbol{\theta} = (\boldsymbol{\alpha}', \boldsymbol{\beta}')'$ and $\hat{\boldsymbol{\theta}}_\nu = (\tilde{\boldsymbol{\alpha}}_\nu', \hat{\boldsymbol{\beta}}_\nu')'$. Then we have the following theorem.

Theorem 4.3 *Under* (2.1)–(2.3), (3.9), (3.13), (3.14), *and Assumptions* I, II, III′, IV, *and* V, *as* $\nu \to \infty$,

$$\mathscr{L}(N_\nu^{1/2}(\tilde{\boldsymbol{\theta}}_{N_\nu} - \boldsymbol{\theta})) \to N_{2p}(\mathbf{0}, \boldsymbol{\Delta} \otimes \mathbf{T}), \tag{4.22}$$

where $\mathbf{T}$ *is defined by* (3.19) *and*

$$\boldsymbol{\Delta} = \begin{pmatrix} 1 + \bar{c}^2/C^2 & -\bar{c}/C^2 \\ -\bar{c}/C^2 & 1/C^2 \end{pmatrix}. \tag{4.23}$$

Proof. First note that by the same technique as in the proofs of results of Sen and Puri [14, Sect. 7] (who considered the particular case of $\bar{c}_\nu = 0$ for all $\nu \ge 1$), one gets

$$\mathscr{L}(\nu^{1/2}(\tilde{\boldsymbol{\theta}}_\nu - \tilde{\boldsymbol{\theta}})) \to N_{2p}(\mathbf{0}, \boldsymbol{\Delta}_2 \otimes \mathbf{T}). \tag{4.24}$$

Hence, as in Theorems 4.1 and 4.2, one needs to show that for every $\varepsilon > 0$ and $\eta > 0$, there exists a $\delta > 0$ and an $n_0 = n_0(\varepsilon, \eta)$ such that for $n \ge n_0$,

$$P\left\{ \max_{k\colon |k-n| < \delta n} \|n^{1/2}(\tilde{\boldsymbol{\theta}}_k - \tilde{\boldsymbol{\theta}}_n)\| > \varepsilon \right\} < 2\eta. \tag{4.25}$$

Now, the left-hand side of (4.25) is bounded above by

$$P\left\{ \max_{k\colon |k-n| < \delta n} \|n^{1/2}(\tilde{\boldsymbol{\alpha}}_k - \tilde{\boldsymbol{\alpha}}_n)\| > \varepsilon \right\} + P\left\{ \max_{k\colon |k-n| < \delta n} \|n^{1/2}(\hat{\boldsymbol{\beta}}_k - \hat{\boldsymbol{\beta}}_n)\| > \varepsilon \right\}. \tag{4.26}$$

By virtue of (4.17) and the Bonferroni inequality, it suffices to show now that

$$\sum_{j=1}^{p} P\left\{ \max_{k:|k-n|<\delta n} \|n^{1/2}(\tilde{\boldsymbol{\alpha}}_k - \tilde{\boldsymbol{\alpha}}_n)\| > \varepsilon \right\} < \eta \tag{4.27}$$

For simplicity, instead of proving (4.27) we shall consider the following. Let $\tilde{\tilde{R}}_{vj}^{(j)+}$ be the rank of $|X_{ij} - a - \hat{\beta}_{vj}(c_i - \bar{c}_v)|$ among $|X_{1j} - a - \hat{\beta}_{vj}(c_1 - \bar{c}_v)|, \ldots, |X_{vj} - a - \hat{\beta}_{vj}(c_v - \bar{c}_v)|$, $1 \le i \le v$, $1 \le j \le p$. The resulting one-sample rank order statistics defined by (3.11) will now be denoted by $\tilde{\tilde{T}}_{vj}(a)$ $(1 \le j \le p;\ v \ge 1)$. Define

$$\hat{\delta}_{vj}^{(1)} = \sup\{a\colon \tilde{\tilde{T}}_{vj}(a) > 0\}, \qquad \hat{\delta}_{vj}^{(2)} = \inf\{a\colon \tilde{\tilde{T}}_{vj}(a) < 0\}, \qquad v \ge 1, \quad 1 \le j \le p; \tag{4.28}$$

$$\hat{\delta}_{vj} = \tfrac{1}{2}(\hat{\delta}_{vj}^{(1)} + \hat{\delta}_{vj}^{(2)}), \qquad 1 \le j \le p, \quad v \ge 1; \tag{4.29}$$

$$\hat{\delta}_v = (\hat{\delta}_{v1}, \ldots, \hat{\delta}_{vp})', \qquad v \ge 1. \tag{4.30}$$

It follows from the results of Adichie [1] that

$$\hat{\delta}_{vj} = \tilde{\alpha}_{vj} + \hat{\beta}_{vj}\bar{c}_v \qquad (1 \le j \le p, \quad v \ge 1), \tag{4.31}$$

i.e.,

$$\hat{\boldsymbol{\delta}}_v = \tilde{\boldsymbol{\alpha}}_v + \hat{\boldsymbol{\beta}}_v \bar{c}_v \qquad (v \ge 1). \tag{4.32}$$

In view of (4.25)–(4.27) and (4.31)–(4.32) it now suffices to show that for every $\varepsilon > 0$ and $\eta > 0$, there exists a $\delta > 0$ and an $n_0 = n_0(\varepsilon, \eta)$ such that for $n \ge n_0$,

$$\sum_{j=1}^{p} P\left\{ \max_{k:|k-n|<\delta n} \|n^{1/2}(\hat{\boldsymbol{\delta}}_k - \hat{\boldsymbol{\delta}}_n)\| > \varepsilon \right\} < \eta. \tag{4.33}$$

To prove (4.33) we prove the following two lemmas. Since $\tilde{\boldsymbol{\alpha}}_v$ and $\hat{\boldsymbol{\beta}}_v$ (and hence $\hat{\boldsymbol{\delta}}_v$) are translation invariant for every v (see Sen and Puri [14]), we may, for proving these lemmas, assume that $\boldsymbol{\alpha} = \boldsymbol{\beta} = \mathbf{0}$.

Lemma 4.4. *Under the assumptions of Theorem* 4.3, *for every* $s > 0$, *there exist positive constants* $c_s^{(1)}$ *and* $c_s^{(2)}$ *and a positive integer* v_s *such that for* $\boldsymbol{\alpha} = \boldsymbol{\beta} = \mathbf{0}$, *and all* $v \ge v_s$,

$$P\left\{ \sup_{|a| \le K_0(\log v)^k v^{-1/2}} v^{-1/2} |\tilde{\tilde{T}}_{vj}(a) - \tilde{T}_{vj}(a)| > c_s^{(1)} v^{-\delta} (\log v)^{k+1} \right\} \le c_s^{(2)} v^{-s}, \tag{4.34}$$

where K_0 *is a positive constant*, k *any positive integer*, *and* δ *fixed* $(0 < \delta < \frac{1}{4})$.

Before proving the above lemma, we may note that taking $s > 1$ and on using the Borel–Cantelli Lemma, (4.34) implies that

$$\sup_{|a| \le K_0 v^{-1/2}(\log v)^k} v^{-1/2} |\tilde{\tilde{T}}_{vj}(a) - \tilde{T}_{vj}(a)| \to 0 \qquad \text{a.s. as} \quad v \to \infty. \tag{4.35}$$

The proof of the lemma is accomplished in several steps. First we show that for any real b, defining $T_{\nu j}(a, b)$ as similar to $\tilde{\tilde{T}}_{\nu j}(a)$ with $\hat{\beta}_\nu$ replaced by b, for $\nu \geq \nu_s^{(1)}$ (depending on s),

$$P\left\{ \sup_{|a| \leq K_0 \nu^{-1/2}(\log \nu)^k} \sup_{|b| \leq K_1 \nu^{-1/2}(\log \nu)^k} \nu^{-1/2} |T_{\nu j}(a, b) - \tilde{T}_{\nu j}(a)| > c_s^{(3)} \nu^{-\delta}(\log \nu)^{k+1} \right\} < c_s^{(4)} \nu^{-s}, \quad (4.36)$$

where K_1 is a positive constant, $c_s^{(3)}$, $c_s^{(4)}$ are positive constants depending on s. Next, in analogy to Ghosh and Sen [6, Lemma 4.1], one can show that for every s (> 0), there exist positive constants $c_s^{(5)}$ and $c_s^{(6)}$ and a positive integer ν_{s2} such that for $\nu \geq \nu_{s2}$,

$$P_{\beta=0}\{C_\nu |\hat{\beta}_{\nu j}| > c_s^{(5)}(\log \nu)^2\} \leq c_s^{(6)} \nu^{-s}. \quad (4.37)$$

Defining now $c_s^{(1)}$, $c_s^{(2)}$, and ν_s appropriately on the basis of $c_s^{(i)}$ ($i = 3, 4, 5, 6$), $\nu_s^{(1)}$ and $\nu_s^{(2)}$, one gets (4.34) from (4.35), (4.37), and (3.4). Let $H_{\nu,j,a,b}(x) = \nu^{-1} \sum_{i=1}^{\nu} u(x - (X_{ij} - a - b(c_i - \bar{c}_\nu)))$ be the sample df of $X_{ij} - a - b(c_i - \bar{c}_\nu)$'s, and let $G_{\nu,j,a,b}(x) = \nu^{-1} \sum_{i=1}^{\nu} u(x - |X_{ij} - a - b(c_i - \bar{c}_\nu)|) = H_{\nu,j,a,b}(x) - H_{\nu,j,a,b}(-x-)$ be the sample cdf for $|X_{ij} - a - b(c_i - \bar{c}_\nu)|$'s. The corresponding population cdf's are denoted respectively by $\bar{F}_{\nu,j,a,b}(x) = \nu^{-1} \sum_{i=1}^{\nu} F_{[j]}(x + a + b(c_i - \bar{c}_\nu))$, and $\bar{D}_{\nu,j,a,b}(x) = \bar{F}_{\nu,j,a,b}(x) - \bar{F}_{\nu,j,a,b}(-x-)$. Writing $\phi_{\nu j}^*(i/(\nu+1)) = a_\nu^{(j)*}(i)$ ($1 \leq i \leq \nu$, $\nu \geq 1$), one can now write

$$\nu^{-1}[T_{\nu,j}(a, b) - \tilde{T}_{\nu j}(a)] = \int_0^\infty \phi_{\nu j}^*\left(\frac{\nu}{\nu+1} G_{\nu,j,a,b}(x)\right) dH_{\nu,j,a,b}(x) - \int_0^\infty \phi_{\nu j}^*\left(\frac{\nu}{\nu+1} G_{\nu,j,a,0}(x)\right) dH_{\nu,j,a,0}(x).$$

A result analogous to Puri and Sen [11, Theorem 3.6.6] gives

$$\max_{1 \leq i \leq \nu} |\phi_{\nu j}^*(i/(\nu+1)) - \phi_j^*(i/(\nu+1))| = O(\nu^{-1/2-\delta})$$

for some $\delta > 0$, $j = 1, 2, \ldots, p$. Hence, one can write

$$\nu^{-1}[T_{\nu,j}(a, b) - \tilde{T}_{\nu j}(a)] = I_{\nu j1}(a, b) + I_{\nu j2}(a, b) + O(\nu^{-1/2-\delta}), \quad (4.38)$$

where

$$I_{\nu j1}(a, b) = \int_0^\infty \left[\phi_j^*\left(\frac{\nu}{\nu+1} G_{\nu,j,a,b}(x)\right) - \phi_j^*\left(\frac{\nu}{\nu+1} G_{\nu,j,a,0}(x)\right)\right] dH_{\nu,j,a,b}(x), \quad (4.39)$$

$$I_{\nu j2}(a, b) = \int_0^\infty \phi_j^*\left(\frac{\nu}{\nu+1} G_{\nu,j,a,0}(x)\right) d[H_{\nu,j,a,b}(x) - H_{\nu,j,a,0}(x)], \quad (4.40)$$

On integration by parts, one can write, using (3.9), (3.13), and (3.14),

$$I_{\nu j2}(a, b) = \int_0^\infty [H_{\nu, j, a, b}(x) - H_{\nu, j, a, 0}(x)] \times \phi_j^*\left(\frac{\nu}{\nu+1} G_{\nu, j, a, 0}(x)\right) \frac{\nu}{\nu+1} G_{\nu, j, a, 0}(x). \quad (4.41)$$

We shall now state a lemma. The proof follows the same line as Sen and Ghosh [13, Lemma 4.1] and Ghosh and Sen [6, Theorem 3.1]. For brevity, the details are omitted.

Lemma 4.5. *For every s (> 0), there exist two positive constants $K_s^{(1)}$ and $K_s^{(2)}$, and a positive integer ν_s^* (all of which may depend on s) such that for $\nu \geq \nu_s^*$, $k \geq 1$, and $0 < \delta < \frac{1}{4}$*

$$P\left\{ \sup_{-\infty < x < \infty} \sup_{|a| \leq K_0 \nu^{-1/2}(\log \nu)^k} \sup_{|b| \leq K_1 \nu^{-1/2}(\log \nu)^k} |H_{\nu, j, a, b}(x) - H_{\nu, j, a, 0}(x) - \bar{F}_{\nu, j, a, b}(x) + \bar{F}_{\nu, j, a, 0}(x)| > K_s^{(1)} \nu^{-1/2-\delta}(\log \nu)^k \right\} \leq K_s^{(2)} \nu^{-s}. \quad (4.42)$$

Using also the fact that $\bar{F}_{\nu,j,a,b}(x) - \bar{F}_{\nu,j,a,0}(x) = \nu^{-1} \sum_{i=1}^{\nu} [F(x + a + b(c_i - \bar{c}_\nu)) - F(x + a)] = O(\nu^{-1}(\log \nu)^{2k})$, uniformly in x, a, and $|b| \leq K_1 \nu^{-1/2}(\log \nu)^k$, one gets

$$P\left\{ \sup_{-\infty < x < \infty} \sup_{|a| \leq K_0 \nu^{-1/2}(\log \nu)^k} \sup_{|b| \leq K_1 \nu^{-1/2}(\log \nu)^k} |H_{\nu, j, a, b}(x) - H_{\nu, j, a, 0}(x)| > K_s^{(1)} \nu^{-1/2-\delta}(\log \nu)^k \right\} \leq K_s^{(2)} \nu^{-s} \quad \text{for,} \quad \nu \geq \nu_s^{**} \text{ say.} \quad (4.43)$$

Thus, by (4.41) and (4.43), one gets, by using (3.9), (3.13), and (4.14),

$$\sup_{|a| \leq K_0 \nu^{-1/2}(\log \nu)^k} \sup_{|b| \leq K_1 \nu^{-1/2}(\log \nu)^k} |I_{\nu j2}(a, b)| \leq [O(\nu^{-1/2-\delta}(\log \nu)^k)] \frac{\nu}{\nu+1} \sum_{i=1}^{\nu} K\left[1 - \frac{i}{\nu+1}\right]^{-1} = O(\nu^{-1/2-\delta}(\log \nu)^{k+1}), \quad (4.44)$$

with probability $\geq 1 - K_s^{(2)} \nu^{-s}$, for $\nu \geq \nu_s^{**}$.

Again, write

$$I_{\nu j1}(a, b) = \frac{\nu}{\nu+1} \int_0^\infty [G_{\nu, j, a, b}(x) - G_{\nu, j, a, 0}(x)] \phi_j^{*\prime}\left(\frac{\nu}{\nu+1} [\theta G_{\nu, j, a, b}(x) + (1-\theta) G_{\nu, j, a, 0}(x)]\right) dH_{\nu, j, a, b}(x) \quad (0 < \theta < 1). \quad (4.45)$$

Since $G_{\nu,j,a,b}(x) - G_{\nu,j,a,0}(x) = [H_{\nu,j,a,b}(x) - H_{\nu,j,a,0}(x)] - [H_{\nu,j,a,b}(-x) - H_{\nu,j,a,0}(-x)]$, it follows from (4.43) that

$$\sup_{-\infty < x < \infty} \sup_{|a| \le K_0 \nu^{-1/2}(\log \nu)^k} \sup_{|b| \le K_0 \nu^{-1/2}(\log \nu)^k} |G_{\nu,j,a,b}(x) - G_{\nu,j,a,0}(x)| \le K_s^{(1)} \nu^{-1/2-\delta}(\log \nu)^k$$

with probability $\ge 1 - K_s^{(2)}\nu^{-s}$ for large ν. Using arguments analogous to Sen and Ghosh [13, Theorem 4.3 (4.20)–(4.27)], one can prove now that

$$\sup_{|a| \le K_0 \nu^{-1/2}(\log \nu)^k} \sup_{|b| \le K_1 \nu^{-1/2}(\log \nu)^k} |I_{\nu j1}(a, b)| \le K_s^{(3)} \nu^{-1/2-\delta}(\log \nu)^{k+1},$$

with probability $\ge 1 - K_s^{(4)}\nu^{-s}$ for large ν. Hence, the lemma.

For proving (4.33) we need another lemma, which we prove below. For proving this lemma, we take $a_\nu^{(j)*}(i) = \phi_{\nu j}^*(i/(\nu+1)) = E\phi_j^*(U_{\nu i})$ $(1 \le i \le \nu, 1 \le j \le p)$.

Lemma 4.6. *For* $\boldsymbol{\alpha} = \boldsymbol{\beta} = \mathbf{0}$, *for every* $s > 0$, *there exist positive constants* $d_s^{(1)}$ *and* $d_s^{(2)}$ *and a positive integer* ν_{s0} *such that for* $\nu \ge \nu_{s0}$,

$$P\{|\hat{\delta}_{\nu j}| \ge d_s^{(1)}\nu^{-1/2}(\log \nu)^k\} \le d_s^{(2)}\nu^{-s}. \tag{4.46}$$

Proof. We prove only the case of $P\{\hat{\delta}_{\nu j} \ge d_s^{(1)}\nu^{-1/2}(\log \nu)^k\}$, as the other case follows similarly. Note that

$$\begin{aligned} P\{\hat{\delta}_{\nu j} \ge d_s^{(1)}\nu^{-1/2}(\log \nu)^k\} &\le P\{\hat{\delta}_{\nu j}^{(2)} \ge d_s^{(1)}\nu^{-1/2}(\log \nu)^k\} \\ &= P\{\tilde{T}_{\nu j}(d_s^{(1)}\nu^{-1/2}(\log \nu)^k) \ge 0\} \\ &= P\{\nu^{-1/2}\tilde{T}_{\nu j}(d_s^{(1)}\nu^{-1/2}(\log \nu)^k) \ge 0\}. \end{aligned} \tag{4.47}$$

It follows from Lemma 4.5 that for every $s > 0$, for large ν, with probability $\ge 1 - c_s^{(2)}\nu^{-s}$,

$$\nu^{-1/2}[\tilde{T}_{\nu j}(d_s^{(1)}\nu^{-1/2}(\log \nu)^k) - T_{\nu j}(d_s^{(1)}\nu^{-1/2}(\log \nu)^k)] \le c_s^{(1)}\nu^{-\delta}(\log \nu)^k. \tag{4.48}$$

Further, from Sen and Ghosh [6, Theorem 4.3], we have

$$\nu^{-1/2}[T_{\nu j}(d_s^{(1)}\nu^{-1/2}(\log \nu)^k) - T_{\nu j}(0)] - d_s^{(1)}(\log \nu)^k \le d_s^{(2)}\nu^{-\delta}(\log \nu)^k, \tag{4.49}$$

with probability $\ge 1 - c_s^{(2)}\nu^{-s}$, for large ν. Hence, from (4.47)–(4.49), it suffices to show that for large ν, for every $s > 0$, there exist constants $d_s^{(1)}$ and $d_s^{(2)}$ such that

$$P\{\nu^{-1/2}[T_{\nu j}(0) - \nu\mu_j^*] > d_s^{(1)}(\log \nu)^k\} \le d_s^{(2)}\nu^{-\delta}(\log \nu)^k. \tag{4.50}$$

When $\boldsymbol{\alpha} = \boldsymbol{\beta} = \mathbf{0}$, for every v, $\mathbf{R}_v^{(j)+} = (R_{v1}^{(j)+}, \ldots, R_{vv}^{(j)+})'$ is independent of $\mathbf{s}_v^{(j)} = (s(X_{1j}), \ldots, s(X_{vj}))$, where $s(u) = 2c(u) - 1$; i.e., $s(u) = 1$, 0, or -1, according as $u >, =$, or < 0. Now

$$v^{-1/2}T_{vj}(0) = v^{-1/2}\sum_{i=1}^{v}\frac{1+s(X_{ij})}{2}E\phi_j^*(U_{vR_{vi}^{(j)+}})$$

$$= \tfrac{1}{2}v^{1/2}\int_0^1 \phi_j^*(u)\,du + \tfrac{1}{2}v^{-1/2}\sum_{i=1}^{v} s(X_{ij})E\phi_j^*(U_{vR_{vi}^{(j)+}}).$$

Since (3.10) holds, from (3.13) and (3.14) the first term $= v^{1/2}\mu_j^*$. Hence, (4.49) will be proved if one can show that for large v

$$P\{v^{-1/2}T_{vjo}(0) > 2d_s^{(1)}(\log v)^k\} \le d_s^{(2)}v^{-s} \tag{4.51}$$

where

$$T_{vjo}(0) = \sum_{i=1}^{v} s(X_{ij})E\phi_j^*(U_v R_{vi}^{(j)+}), \qquad 1 \le j \le p.$$

Writing $g_v = 2d_v^{(1)}v^{1/2}(\log v)^k$, and using the Bernstein inequality, one gets

$$P\{T_{vjo}(0) - g_v\} \le \inf_{t>0} E[\exp\{t(T_{vjo}(0) - g_v)\}]. \tag{4.52}$$

Now,

$$E[\exp\{t(T_{vjo}(0) - g_v)\}] = \exp(-tg_v)E[\exp(tT_{vjo}(0))].$$

Again,

$$E[\exp(tT_{vjo}(0))] = EE\left[\prod_{i=1}^{v}\exp(ts(X_{ij})E\phi_j^*(U_v R_{vi}^{(j)+}))\,|\,\mathbf{R}_v^{(j)+}\right]$$

Using the independence of $\mathbf{s}_v^{(j)}$ and $\mathbf{R}_v^{(j)+}$ as well as the elementary inequality $\frac{1}{2}[\exp(x) + \exp(-x)] \le \exp(x^2/2)$, one gets

$$E[\exp(tT_{vjo}(0))] = E\left[\prod_{i=1}^{v}\{\tfrac{1}{2}\exp(tE\phi_j^*(U_{vR_{vi}^{(j)+}})) + \tfrac{1}{2}\exp(tE\phi_j^*(U_{vR_{vi}^{(j)+}}))\}\right]$$

$$\le E\prod_{i=1}^{v}\exp\left(\frac{t^2}{2}(E\phi_j^*(U_{vR_{vi}^{(j)+}}))^2\right)$$

$$\le E\prod_{i=1}^{v}\exp\left(\frac{t^2}{2}E\phi_j^{*2}(U_{vR_{vi}^{(j)+}})\right)$$

$$= E\exp\left(\frac{t^2}{2}\sum_{i=1}^{v}E\phi_j^{*2}(U_{vR_{vi}^{(j)+}})\right)$$

$$= E\exp\left(\frac{t^2}{2}\sum_{i=1}^{v}E\phi_j^{*2}(U_{vi})\right) = \exp\left(\frac{vt^2A_j^2}{2}\right)$$

where $A_j^2 = \int_0^1 \phi_j^{*2}(u)\,du$. Thus, from (4.52),

$$P\{T_{\nu j0}(0) > g_\nu\} \le \inf_{t>0} \exp\left(-tg_\nu + \frac{\nu t^2 A_j^2}{2}\right) = \exp\left(-\frac{g_\nu^{\,2}}{2\nu A_j^2}\right)$$
$$= \exp\left(-\frac{2d_s^{(1)^2}}{A_j^2}(\log \nu)^{2k}\right),$$

and hence, (4.50) follows. Hence, the lemma.

It follows from Lemmas 4.4 and 4.6 that for large ν, $\nu^{-1/2}[\tilde{T}_{\nu j}(\hat{\delta}_\nu) - T_{\nu j}(\hat{\delta}_\nu)] = O(\nu^{-\delta}(\log \nu)^k)$ with probability $\ge (1 - \text{const } \nu^{-s})$. Again, it follows from Sen and Ghosh [13, Theorem 4.3] that $\nu^{-1/2}[T_{\nu j}(\hat{\delta}_\nu) - T_{\nu j}(0)] + \nu^{1/2}\hat{\delta}_\nu B_j = O(\nu^{-\delta}(\log \nu)^k)$ with probability $\ge (1 - \text{const } \nu^{-s})$. Hence, with probability $\ge (1 - \text{const } \nu^{-s})$, $\nu^{-1/2}[\tilde{T}_{\nu j}(\hat{\delta}_\nu) - T_{\nu j}(0)] + \nu^{1/2}\hat{\delta}_\nu B_j = O(\nu^{-s}(\log \nu)^k)$. That is, $\nu^{1/2}\hat{\delta}_\nu B_j - \nu^{-1/2}T_{\nu j}(0) = O(\nu^{-\delta}(\log \nu)^k)$, noting that $\tilde{T}_{\nu j}(\hat{\delta}_\nu) = 0$. Inequality (4.33) now follows from Sen and Ghosh [13, Theorem 4.5]. Hence the theorem.

5. BOUNDED LENGTH (SEQUENTIAL) CONFIDENCE BANDS FOR θ

Parallel to Problems I–III of Section 4, we consider here the following three problems.

Problem I′. Confidence estimation of $\boldsymbol{\alpha}$ assuming that $\boldsymbol{\beta} = \mathbf{0}$. More specifically, we want a p-dimensional confidence rectangle for $\boldsymbol{\alpha}$ such that the length of each side $\le 2d$ ($d > 0$, preassigned) and the confidence coefficient $\ge 1 - \alpha$. This can be achieved by a direct extension of the results of Sen and Ghosh [13].

To see this, first note that under $\boldsymbol{\alpha} = \boldsymbol{\beta} = \mathbf{0}$, $\mathbf{T}_{\nu 0} = (T_{\nu 10}, \ldots, T_{\nu p0})'$ ($T_{\nu j0}$'s defined after (4.51)) has a distribution independent of F diagonally symmetric about $\mathbf{0}$. Hence, there exists a known constant $T_{\nu,\varepsilon}$ such that

$$P_{\boldsymbol{\alpha}=\boldsymbol{\beta}=\mathbf{0}}\left\{\max_{1\le j\le p} |T_{\nu j0}| \le T_{\nu,\varepsilon}\right\} = 1 - \varepsilon_\nu \to 1 - \varepsilon \qquad \text{as} \quad \nu \to \infty \tag{5.1}$$

For large ν, $\sqrt{\nu}T_{\nu,\varepsilon} \to \chi_{p,\varepsilon}^*$ where $\chi_{p,\varepsilon}^*$ is the upper $100\varepsilon\%$ point of the distribution of the maximum of $\gamma_1, \ldots, \gamma_p$ where $\boldsymbol{\gamma} = (\gamma_1, \ldots, \gamma_p)'$ is $N(\mathbf{0}, \boldsymbol{\nu})$. Define now

$$\hat{\alpha}_{L,j,\nu} = \sup\{a: T_{\nu j0}(a) > T_{\nu,\varepsilon}\}, \tag{5.2}$$
$$\hat{\alpha}_{U,j,\nu} = \inf\{a: T_{\nu j0}(a) < -T_{\nu,\varepsilon}\}, \tag{5.3}$$

where $T_{\nu j0}(a)$ is defined in the same way as $T_{\nu j0} = T_{\nu j0}(0)$, replacing X_i's by $X_i - a$'s ($1 \le j \le p$, $1 \le i \le \nu$). Then, $P_{\boldsymbol{\alpha}=\boldsymbol{\beta}=\mathbf{0}}\{\hat{\alpha}_{L,j,\nu} \le \alpha_j \le \hat{\alpha}_{U,j,\nu}\ \forall\ 1 \le j \le p\}$ $= P_{\boldsymbol{\alpha}=\boldsymbol{\beta}=\mathbf{0}}\{-T_{\nu,\varepsilon} \le T_{\nu j0} \le T_{\nu,\varepsilon}\ \forall\ 1 \le j \le p\} = 1 - \varepsilon_\nu \to 1 - \varepsilon$ as $\nu \to \infty$.

We define the stopping variable $N = N(d)$ to be the least positive integer $n\,(\geq n_0)$ such that $\max_{1 \leq j \leq p}\, (\hat{\alpha}_{U,j,n} - \hat{\alpha}_{L,n,n}) \leq 2d$. Now, using Sen and Ghosh [13, Theorem 4.3 and Lemma 5.1],

$$\sqrt{v}[T_{vj}(\hat{\alpha}_{U,j,\varepsilon}) - T_{vj}(0) + \tfrac{1}{2}\hat{\alpha}_{U,j,v} B_j] = 0(v^{-1/4}(\log v)^4) \tag{5.4}$$

with probability $\geq (1 - \text{const } v^{-s})$, for every $s > 0$, large v. Thus noting that when $a_v^{(j)*}(i) = E[\phi_j{}^*(U_{vi})]$, $1 \leq i \leq v$, $T_{vjo}(a) = 2T_{vj}(a) - \int_0^1 \phi_j{}^*(u)\,du$, for all real a, it follows from (5.3) and (5.4) that

$$-\chi_{p,\,\varepsilon}^* - \sqrt{v}\,T_{vjo}(0) + \sqrt{v}\,\hat{\alpha}_{U,\,j,\,v} B_j \to 0 \qquad \text{a.s. as} \quad v \to \infty.$$

Similarly,

$$\chi_{p,\,\varepsilon}^* - \sqrt{v}\,T_{vjo}(0) + \sqrt{v}\,\hat{\alpha}_{L,\,j,\,v} B_j \to 0 \qquad \text{a.s. as} \quad v \to \infty.$$

Thus,

$$\sqrt{v}(\hat{\alpha}_{U,\,j,\,v} - \hat{\alpha}_{L,\,j,\,v}) \to \frac{2\chi_{p,\,\varepsilon}^*}{B_j} \qquad \text{a.s. as} \quad v \to \infty.$$

Hence,

$$\max_{1 \leq j \leq p} \sqrt{v}(\hat{\alpha}_{U,\,j,\,v} - \hat{\alpha}_{L,\,j,\,v}) \to \frac{2\chi_{p,\,\varepsilon}}{\min_{1 \leq j \leq p} B_j} \qquad \text{a.s. as} \quad v \to \infty \tag{5.5}$$

It follows now from the definition of N that $\lim_{d \to 0} N(d)/s(d) = 1$ a.s., where $s(d) = \chi_{p,\varepsilon}^{*2}/d^2 \min_{1 \leq j \leq p} B_j{}^2$, and as to the rate of convergence, we can make a statement similar to (5.4). Thus, generalizing the results of Sen and Ghosh [13], we get the following theorem.

Theorem 5.1. *Under the assumptions* $F \in \mathscr{F}_p{}^0$, (2.1)–(2.3), (3.9), (3.13)–(3.14), *and* (3.17),

$$N(= N(d)) \textit{ is a nonincreasing function of } d;$$

$$N(d) < \infty \textit{ with probability } 1, \qquad EN(d) < \infty \qquad \textit{for all} \quad d > 0, \tag{5.6}$$

$$\lim_{d \to 0} N(d) = \infty \qquad \textit{a.s.}, \qquad \textit{and} \qquad \lim_{d \to 0} EN(d) = \infty$$

$$\lim_{d \to 0} N(d)/s(d) = 1 \qquad \textit{a.s.} \tag{5.7}$$

$$\lim_{d \to 0} P_{\boldsymbol{\alpha}}\{\hat{\alpha}_{L,j,N} \leq \alpha_j \leq \hat{\alpha}_{U,j,N}\ \forall\ 1 \leq j \leq p\} = 1 - \varepsilon. \tag{5.8}$$

$$\lim_{d \to 0} EN(d)/s(d) = 1. \tag{5.9}$$

We now suggest an alternate procedure for the same problem. We find a confidence region R_N for $\boldsymbol{\alpha}$ such that the maximum diameter of $R_N \leq 2d$. Our procedure is analogous to the one proposed by Srivastava [15].

We define

$$\hat{\gamma}_{jl}^{(n)} = \int_{-\infty}^{\infty}\int_{-\infty}^{\infty} \phi_j\left(\frac{n}{n+1} F_{[j]n}(x)\right)\phi_l\left(\frac{n}{n+1} F_{[l]n}(y)\right) dF_{[j,l]n}(x, y) - \mu_j \mu_l, \tag{5.10}$$

for $1 \le j \ne l \le p$ where $F_{[j]n}(x)$ and $F_{[j,l]n}(x, y)$ are the empirical df's corresponding to the true df's $F_{[j]}(x)$ and $F_{[j,l]}(x, y)$, respectively, for $j = l$, $\hat{\gamma}_{jj}^{(n)} = \gamma_{jj} = \int_0^1 \phi_j^2(u)\, du - \mu_j^2$, $1 \le j \le p$. Also, define $\hat{B}_{j,n}$ as the estimator of B_j $(1 \le j \le p)$ as in Ghosh and Sen [6, Lemma 4.2]. Define then

$$\hat{\mathbf{T}}_n = ((\hat{\tau}_{jl}^{(n)})), \qquad \hat{\tau}_{jl}^{(n)} = \hat{\gamma}_{jl}^{(n)}/\hat{B}_{j,n}\hat{B}_{l,n}, \qquad j, l = 1, \ldots, p. \tag{5.11}$$

We denote by

$$\hat{\lambda}_n = \text{max ch root of } \hat{\mathbf{T}}_n; \qquad \lambda = \text{max ch root of } \mathbf{T}, \tag{5.12}$$

where $\mathbf{T}$ is defined by (3.19); finally, $\chi_{p,\varepsilon}^2$ is defined as the upper $100\varepsilon\%$ point of the chi square distribution with p degrees of freedom. Our procedure is as follows.

Starting with an initial sample of size n_0 $(>p)$, we continue drawing observations one at a time according to a stopping time N defined by

$$N[= N(d)] = \text{smallest } n \ge n_0 \quad \text{such that} \quad \hat{\lambda}_n \le d^2 n/\chi_{p,\alpha}^2. \tag{5.13}$$

When sampling is stopped at $N = n$, construct the region R_n defined by

$$R_n = \{\mathbf{z}\colon (\hat{\boldsymbol{\alpha}}_n - \mathbf{z})'(\hat{\boldsymbol{\alpha}}_n - \mathbf{z}) \le d^2\}. \tag{5.14}$$

Then, we have the following theorem.

Theorem 5.2. *Under the assumption that* $0 < \lambda < \infty$ *and the hypothesis of Theorem* 5.1, *the results of Theorem* 5.1 *all hold for the stopping variable* $N(d)$, *defined by* (5.13), *and* R_N, *defined by* (5.14), *provided we replace* $s(d)$ *in* (5.7) *and* (5.9) *by*

$$v(d) = \chi_{p,\alpha}^2 \lambda/d^2. \tag{5.15}$$

Proof. Running down the proof of Srivastava [16], it suffices to show that $\hat{\lambda}_n \to \lambda$ a.s. as $n \to \infty$; by the Courant theorem, it thus suffices to show that

$$\hat{\mathbf{T}}_n \to \mathbf{T} \qquad \text{a.s. as} \quad n \to \infty. \tag{5.16}$$

Since $\hat{B}_{j,n}$, $j = 1, \ldots, p$, converge a.s. to B_j, $j = 1, \ldots, p$, as $n \to \infty$ (see [13]) it suffices to prove the following lemma.

Lemma 5.3. *Under* (3.4), (3.17), (3.18), *and* (3.19),

$$\hat{\gamma}_{jl}^{(n)} \to \gamma_{jl} \qquad \text{a.s. as} \quad n \to \infty \quad \text{for all} \quad 1 \le j \ne l \le p. \tag{5.17}$$

Proof. Since $\phi_j(u)$ is assumed to be nondecreasing, absolutely continuous, and square integrable inside [0, 1], by Hájek [9, Lemma 5.1], we may write for $0 < u < 1$,

$$\phi_j(u) = \phi_j^{(1)}(n) - \phi_j^{(2)}(u) + \phi_j^{(3)}(u), \tag{5.18}$$

where $\phi_j^{(1)}(n)$ is a polynomian (i.e., has bounded second derivative) and

$$\int_0^1 \{\phi_j^{(k)}(u)\}^2 \, du < \frac{\varepsilon}{2}\left[\int_0^1 \phi_j^2(u)\, du\right], \qquad k = 2, 3, \tag{5.19}$$

where $\varepsilon > 0$ is arbitrarily small. By (3.8), we may decompose the scores $a_\nu^{(j)}(i)$, $1 \le i \le \nu$, also in three parts. On the first part, involving $\phi_j^{(1)}$, a.s. convergence of $F_{[j]n}$ and $F_{[j,l]n}$ to $F_{[j]}$ and $F_{[j,l]}$ (respectively) implies the a.s. convergence of the corresponding component of $\hat{\gamma}_{jl}^{(n)}$ to that of γ_{jl}; on the other components, the Schwarz inequality and (5.19) imply that the same can be made arbitrarily small by proper choice of ε (>0). Q.E.D.

Remark. In (5.14), we could have taken a region

$$\{\mathbf{z}: (\hat{\boldsymbol{\alpha}}_n - \mathbf{z})\mathbf{A}^{-1}(\hat{\boldsymbol{\alpha}}_n - \mathbf{z}) \le d^2\},$$

where A is any given positive definite matrix. In that case, we need to define $\hat{\lambda}_n$ = max ch root of $\mathbf{A}^{-1}\mathbf{T}_n$ and λ = max ch root of $\mathbf{A}^{-1}\mathbf{T}$. The proof follows on parallel lines.

Problem II′. Confidence band for $\boldsymbol{\beta}$ treating $\boldsymbol{\alpha}$ as a nuisance parameter.

(i) *Rectangular regions.* Note that under $\boldsymbol{\beta} = \mathbf{0}$, $s_{\nu j}$'s have a completely specified distribution generated by $(n!)^p$ equally likely realizations of the ranks. Hence, there exists a known $s_{\nu,\varepsilon}$ such that

$$P_{\boldsymbol{\beta}=\mathbf{0}}\left\{\max_{1\le j\le p} |S_{\nu j}| \le S_{\nu,\varepsilon}\right\} = 1 - \varepsilon_\nu \to 1 - \varepsilon \qquad \text{as} \quad \nu \to \infty.$$

For large ν, $\sqrt{S_{\nu,\varepsilon}} \to \chi_{p,\varepsilon}^*$, the upper $100\varepsilon\%$ point of the distribution of the maximum of $\gamma_1, \ldots, \gamma_p$ where $\boldsymbol{\gamma} = (\gamma_1, \ldots, \gamma_p)'$ is $N(\mathbf{0}, \mathbf{v})$. Define now

$$\hat{\beta}_{L,j,\nu} = \sup\{b: S_{\nu j}(b) > S_{\nu,\varepsilon}\}, \qquad \hat{\beta}_{U,j,\nu} = \inf\{b: S_{\nu j}(b) < -S_{\nu,\varepsilon}\}.$$

Then,

$$P_{\boldsymbol{\beta}=\mathbf{0}}\{\hat{\beta}_{L,j,\nu} \le \beta_j \le \hat{\beta}_{U,j,\nu}\ \forall\ 1 \le j \le p\} = 1 - \varepsilon_\nu \to 1 - \eta \qquad \text{as} \quad \nu \to \infty. \tag{5.20}$$

We define the stopping variable $N = N(d)$ to be the least positive integer $n\ (\ge n_0)$ such that $\max_{1\le j\le p}(\hat{\beta}_{U,j,n} - \hat{\beta}_{L,j,n}) \le 2d$. Using Ghosh and Sen [6, Lemma 4.2], we can now prove the following theorem. The proof is omitted because of its obvious analogy to Theorem 5.1.

Theorem 5.4. *If* $F \in \mathscr{F}_p$, *then under* (2.1)–(2.3), (3.9), *and* (3.17), $N(d)$ *as defined above and the related confidence band for* $\boldsymbol{\beta}$ *satisfy the results of Theorem* 5.1 *provided we define*

$$s(d) = Q^{-1}\left(\chi_{p,\varepsilon}^{*2} \Big/ \left[d^2 \max_{1 \le j \le p} B_j^2\right]\right), \tag{5.21}$$

where $Q(n) = C_n^2$ *for* $n \ge 1$ *and is obtained by linear interpolation for non-integer* $t\,(>0)$.

(ii) *Spherical or ellipsoidal regions.* Here, we start by taking $n_0\,(\ge p)$ observations $\mathbf{X}_1, \ldots, \mathbf{X}_{n_0}$ and continue sampling one observation at a time in accordance with the stopping variable

$$N(d) = \text{smallest } n\,(\ge n_0) \quad \text{such that} \quad \lambda_n \le d^2 C_n^2 / \chi_{p,\varepsilon}^2,$$

where $\hat{\lambda}_n$ and $\chi_{p,\varepsilon}^2$ are defined in (5.12) and after that. When sampling is stopped at $N = n$, we construct the region R_n defined by

$$R_n = \{\boldsymbol{\beta} \colon (\boldsymbol{\beta}_n - \boldsymbol{\beta})'(\beta_n - \boldsymbol{\beta}) \le d^2\}, \tag{5.22}$$

where $\hat{\boldsymbol{\beta}}_n$ is defined by (4.14). Then, we have the following.

Theorem 5.5. *The conclusions of Theorem* 5.2 *hold for* $N(d)$ *and* R_n, *defined as above, provided we let*

$$v(d) = Q^{-1}(\lambda \chi_{p,\varepsilon}^2 / d^2).$$

The proof follows along the same line as in Theorems 5.1 and 5.2.

Problem III′. Confidence bands for $\boldsymbol{\theta}$. Here also, we can have either a rectangular or an ellipsoidal region for $\boldsymbol{\theta} = (\boldsymbol{\alpha}, \boldsymbol{\beta})$. We need to change $\chi_{p,\varepsilon}^{*}$ and $\chi_{p,\varepsilon}^2$ to $\chi_{2p,\varepsilon}^{*}$ and $\chi_{2p,\varepsilon}^2$, respectively, and therefore, in view of the similarity with Problems I′ and II′, the details are omitted.

REFERENCES

1. Adichie, J. N. (1967). Estimates of regression parameters based on rank tests. *Ann. Math. Statist.* **38** 894–904.
2. Albert, A. (1966). Fixed size confidence ellipsoids for linear regression parameters. *Ann. Math. Statist.* **37** 1602–1630.
3. Blum, J. R., Hanson, D. L. and Rosenblatt, J. I. (1963). On the central limit theorem for the sum of a random number of independent random variables. *Z. Wahrscheinlichkeitstheorie und Verw. Gebiete* **1** 389–393.
4. Chow, Y. S. and Robbins, H. (1965). On the asymptotic theory of fixed-width sequential confidence intervals for the mean. *Ann. Math. Statist.* **36** 457–462.
5. Ghosh, M. and Sen, P. K. (1971). Sequential confidence interval for the regression coefficient based on Kendall's tau. *Calcutta Statist. Assoc. Bull.* **20** 23–36.
6. Ghosh, M. and Sen, P. K. (1972). On bounded length confidence interval for the regression parameter based on a class of rank statistics. *Sankhyā Ser. A.* **34** 33–52.

7. Ghosh, M. and Sen, P. K. (1973). On some sequential simultaneous confidence intervals procedures. *Ann. Inst. Statist. Math.* **25** 65–75
8. Gleser, L. J. (1965). On the asymptotic theory of fixed size sequential confidence bounds for linear regression parameters. *Ann. Math. Statist.* **36** 463–467.
9. Hájek, J. (1968). Asymptotic normality of simple linear rank statistics under alternatives. *Ann. Math. Statist.* **39** 325–346.
10. Mogyorodi, J. (1967). Limit distributions for sequences of random variables with random indices. *Trans. 4th Prague Conf. Infor. Th. Statist. Dec. Fns. Random Proc.* 463–470.
11. Puri, M. L. and Sen, P. K. (1971). *Nonparametric Methods in Multivariate Analysis.* Wiley, New York.
12. Sen, P. K. (1969). On a class of rank order tests for the parallelism of several regression lines. *Ann. Math. Statist.* **40** 1668–1683.
13. Sen, P. K. and Ghosh, M. (1971). On bounded length sequential confidence intervals based on one sample rank order statistics. *Ann. Math. Statist.* **42** 189–203.
14. Sen, P. K. and Puri, M. L. (1969). On robust nonparametric estimation in some multivariate linear models. *Multivariate Analysis II* (P. R. Krishnaiah, ed.), 33–52. Academic Press, New York.
15. Srivastava, M. S. (1967). On fixed-width confidence bounds for regression parameters and mean vector. *J. Roy. Statist. Soc. Ser. B.* **29** 132–140.
16. Srivastava, M. S. (1971). On fixed-width confidence bounds for regression parameters. *Ann Math. Statist.* **42** 1403–1411.

PART V

Classification, Modelling, and Reliability

Availability Theory for Multicomponent Systems

RICHARD E. BARLOW[1,2]
UNIVERSITY OF CALIFORNIA, BERKELEY

FRANK PROSCHAN[3]
FLORIDA STATE UNIVERSITY

INTRODUCTION AND SUMMARY

In this paper we consider one of the most basic stochastic models in reliability theory, namely, the on–off process generated by failures and repairs of components in a series system. A series system of k components operates if and only if each of the k components operates. No component operates while the system is down. Furthermore, only failed components are repaired and/or replaced; repair or replacement takes a random time. Repaired components are assumed to function like new components.

There is a large literature dealing with availability, the probability that the system is functioning. However, most papers assume special repair or failure distributions (or both). Examples of this type are the paper by Gaver [7] and the paper by Obretenov *et al.* [10]. Many papers, including those just named, are concerned with parallel systems with independently operating components—that is, nonfailed components are not usually in a state of " suspended animation " during repair of a failed component, as in our model. The reader may consult the *IEEE Transactions on Reliability* as well as the *Colloquium on Reliability Theory* [4]. A model more general than ours is treated by Botez [3]. However, there is no overlap with our results.

Let X_{ir} be the duration of the rth functioning period of component i with distribution F_i assumed continuous, and mean μ_i, $i = 1, 2, \ldots, k$ (i.e., time

[1] This research has been partially supported by the Office of Naval Research under Contract N00014-69-A-0200-1036 and the National Science Foundation under grant GP-29123 with the University of California. Reproduction in whole or in part is permitted for any purpose of the U.S. Government.

[2] Present address: University of California, Berkeley.

[3] Research sponsored by the Air Force Office of Scientific Research, AFSC, USAF, under grant AFOSR-71-2058 with Florida State University.

to failure of the rth replacement for component i excluding system down times). Let D_{ir} be the duration of the repair time (or down time) for component i with distribution G_i and mean ν_i, $i = 1, 2, \ldots, k$. We assume that for $i = 1, \ldots, k$, $\{X_{ir}\}_{r=1,2,\ldots}$ and $\{D_{ir}\}_{r=1,2,\ldots}$ are mutually independent renewal processes. Note that the assumption of continuous $F_1, \ldots, F_n$ makes the probability of simultaneous failures 0.

A typical failure–repair history for such a series system might look like Fig. 1, in which component i fails at time X_{i1} and the system is down D_{i1} hours. The system again operates from time $X_{i1} + D_{i1}$ to time $X_{j1} + D_{i1}$. Component j fails at time $X_{j1} + D_{i1}$ and is replaced by time $X_{j1} + D_{i1} + D_{j1}$.

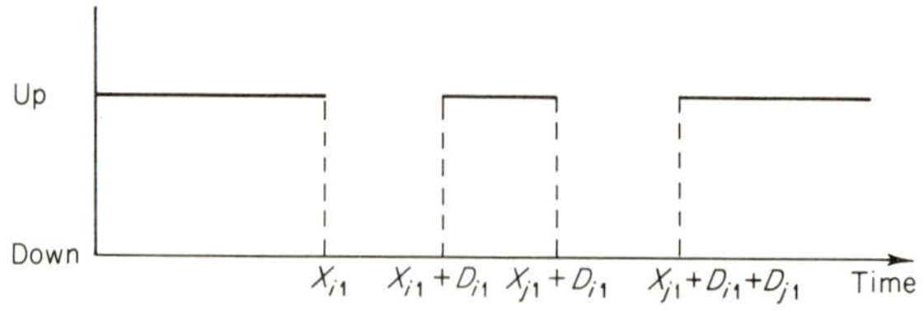

Fig. 1. History of a series system.

Let $\xi(t) = i$ if the system is down at time t due to the failure of component i ($i = 1, 2, \ldots, k$). Let $\xi(t) = 0$ if the system is operating at time t. We are interested in the limiting probability, $\lim_{t\to\infty} P[\xi(t) = i]$, that the system is in state i. The limiting system availability, $\lim_{t\to\infty} P[\xi(t) = 0]$, is of special interest.

Since only failed components are replaced with new or like-new components, the age distribution of components in the system quickly becomes mathematically very complicated. The process $\{\xi(t);\, t \geq 0\}$ has in fact no regeneration points. It is remarkable, however, that many quantities of interest are, in the limit, mathematically very simple and depend only on component mean lives and component mean repair times. The limiting average system availability, as we shall prove in Section 2, is

$$\lim_{t\to\infty} \frac{1}{t}\int_0^t P[\xi(u) = 0]\, du = \left[1 + \sum_{j=1}^{k} \frac{\nu_j}{\mu_j}\right]^{-1} \stackrel{\text{def}}{=} \pi_0, \tag{1}$$

while

$$\lim_{t\to\infty} \frac{1}{t}\int_0^t P[\xi(u) = i]\, du = \frac{\nu_i}{\mu_i}\pi_0 \stackrel{\text{def}}{=} \pi_i, \qquad i = 1, 2, \ldots, k. \tag{2}$$

These formulas are true for arbitrary failure and repair distributions. If $\lim_{t\to\infty} P[\xi(t) = i]$ exists, then it is equal to π_i. Although (1) is well known for exponential failure and exponential repair (cf. *Handbook of Reliability Engineering* [8]), a rigorous proof seems to be missing for the general case. A heuristic proof for (1) was offered by Bazovsky *et al.* [1].

Let $\tilde{N}_i(t)$ be the number of replacements of component i in time t. We show that

$$\lim_{t\to\infty} \frac{E\tilde{N}_i(t)}{t} = \frac{\pi_0}{\mu_i}, \qquad i = 1, 2, \ldots, k.$$

This result can be used in determining spare parts requirements, since $\pi_0 t/\mu_i$ will be, approximately, the number of spares of type i required in $[0, t]$.

In Section 3 we prove that the quantities

$$t^{-1/2}[\tilde{N}_i(t) - tm_i^{-1}], \qquad t^{-1/2}[\tilde{N}(t) - tm^{-1}], \qquad t^{-1/2}[U(t) - \pi_0 t]$$

are each asymptotically normal with mean 0, where

$$m_i = \mu_i \pi_0^{-1}, \qquad m^{-1} = \sum_{i=1}^{n} m_i^{-1},$$

and $U(t)$ is the system up time in $[0, t]$.

1. PRELIMINARIES

Assume that all random variables associated with the series system discussed in the introduction are defined on a probability space $(\Omega, \mathscr{A}, P)$. Recall that $\{X_{ir}(\cdot)\}_{r=1}^{\infty}$ and $\{D_{ir}(\cdot)\}_{r=1}^{\infty}$ are mutually independent renewal processes associated with component i ($i = 1, 2, \ldots, k$). We suppress the argument of random variables, $X_{ir}(w)$ evaluated at $w \in \Omega$, except where useful in Section 2. Let $S_{ir} = \sum_{r=1}^{n} X_{ir}$ and $N_i(t) = \sup\{n \mid S_{in} \le t\}$, where $S_{i0} \equiv 0$. Then $\{N_i(t);\ t \ge 0\}$ is the *renewal counting process* associated with $\{X_{ir}\}_{r=1}^{\infty}$. The following theorems are well known (cf. Feller [5] and Ross [12]).

Theorem 1.1. *Let $\{N_i(t);\ t \ge 0\}$ be a renewal counting process corresponding to $\{X_{ir}\}_{r=1}^{\infty}$, where $EX_i = \mu_i$. Then*

$$\lim_{t\to\infty} \frac{N_i(t)}{t} = \frac{1}{\mu_i} \qquad \text{a.s.} \tag{1.1}$$

and

$$\lim_{t\to\infty} \frac{EN_i(t)}{t} = \frac{1}{\mu_i} \tag{1.2}$$

where a.s. means almost surely with respect to P.

We will need the following generalization of the asymptotic normality of the renewal random variable $N_i(t)$ in Section 3. Let $\{X_r\}_{r=1}^{\infty}$ be a sequence of nonnegative random variables, *not necessarily independent nor identically distributed*, with an associated counting process $\{N(t);\ t \ge 0\}$ defined by

$$N(t) = \begin{cases} \sup\left\{n \,\middle|\, \sum_{r=1}^{n} X_r \le t\right\}, & X_1 \le t, \\ 0 & X_1 > t. \end{cases}$$

The following theorem and its corollary can be proved by the argument in Feller [6, p. 321].

Theorem 1.2

$$(\sigma^2 n)^{-1/2} \sum_{r=1}^{[n\tau]} (X_r - \mu) \xrightarrow{L} N(0, \tau) \tag{1.3}$$

as $n \to \infty$ if and only if

$$(\sigma^2 \mu^{-3} n)^{-1/2} [N(n\tau) - n\tau/\mu] \xrightarrow{L} N(0, \tau), \tag{1.4}$$

where [ ] *means greatest integer contained within the brackets, $N(\mu, \sigma^2)$ denotes the normal distribution with mean μ and variance σ^2, and $\xrightarrow{L}$ denotes convergence in law.*

Corollary 1.3. *If $\{X_r\}_{r=1}^{\infty}$ is a renewal process with $EX_r = \mu$ and* Var $X_r = \sigma^2$, *then both* (1.3) *and* (1.4) *hold.*

Billingsley [2, pp. 148–150], and Iglehart and Whitt [9] generalize Theorem 1.2 to Wiener processes on [0, 1].

2. AVERAGE SYSTEM UP TIME: ALMOST SURE RESULTS

It will be useful to study the process $\{U(t);\ t \geq 0\}$ where $U(t)$ is the system functioning time (or up time) in $[0, t]$. Similarly, let $D(t)$ be the down time in $[0, t]$, so that $U(t) + D(t) = t$. Figure 2 is a very useful representation of a series system failure history in terms of system up time $U(t)$.

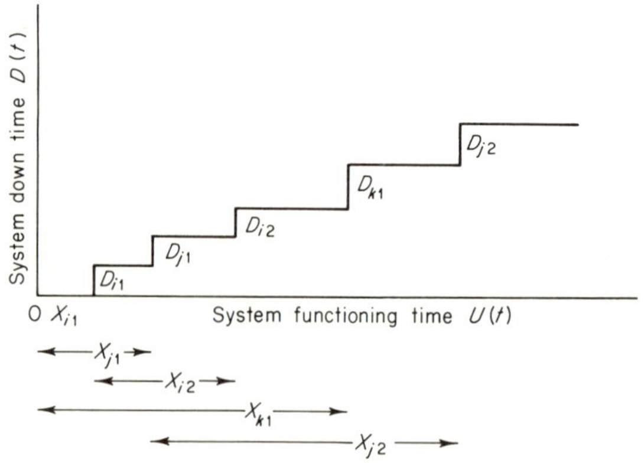

Fig. 2

Let $N_i(t)$ be the number of failures of component i in real time t. Observe that

$$\tilde{N}_i(t) = N_i[U(t)]$$

where $\{N_i(t);\, t \geq 0\}$ is the renewal counting process associated with $\{X_{ir}\}_{r=1}^{\infty}$. However, $N_i(\cdot)$ and $U(t)$ are *not* independent, since in particular $U(t) = t$ implies $N_i[U(t)] = 0$. (We assume that all components are new at $t = 0$ for definiteness. However, the limiting results are true regardless of the initial conditions.)

Lemma 2.1. *If* $0 < \mu_i < \infty$ *and* $0 \leq \nu_i < \infty$ $(i = 1, 2, \ldots, k)$, *then*

$$\lim_{t\to\infty} \frac{N_i[U(t, w), w]}{U(t, w)} = \frac{1}{\mu_i} \qquad \text{a.s.,} \tag{2.1}$$

where we have included the argument $w \in \Omega$ *to emphasize that all random variables are defined on the same probability space* (Ω, A, P).

Proof. Under the hypotheses, $U(t, w) \to \infty$ almost surely as $t \to \infty$. Since $N_i(t, w)$ and $U(t, w)$ are defined on the same probability space, (1.1) implies (2.1). ∥

Theorem 2.2. *If* $0 < \mu_i < \infty$ *and* $0 \leq \nu_i < \infty$ $(i = 1, 2, \ldots, k)$, *then*

$$\lim_{t\to\infty} \frac{U(t)}{t} = \left[1 + \sum_{j=1}^{k} \frac{\nu_j}{\mu_j}\right]^{-1} \stackrel{\text{def}}{=} \pi_0 \qquad \text{a.s.} \tag{2.2}$$

Proof. Note that

$$\sum_{i=1}^{k} \sum_{r=1}^{\tilde{N}_i(t)-1} D_{ir} \leq D(t) \leq \sum_{i=1}^{k} \sum_{r=1}^{\tilde{N}_i(t)} D_{ir}.$$

The inequality results from the fact that the system may be down at time t. Since $U(t) + D(t) = t$,

$$\frac{U(t)}{t} = \left(1 + \frac{D(t)}{U(t)}\right)^{-1} \left(1 + \sum_{i=1}^{k} \frac{1}{\tilde{N}_i(t)} \sum_{r=1}^{\tilde{N}_i(t)} D_{ir} \frac{N_i[U(t)]}{U(t)}\right)^{-1}.$$

By the strong law,

$$\frac{1}{\tilde{N}_i(t)} \sum_{r=1}^{\tilde{N}_i(t)} D_{ir} \xrightarrow[\text{a.s.}]{} \nu_i \qquad \text{as} \quad t \to \infty.$$

By Lemma 2.1,

$$\frac{N_i[U(t)]}{U(t)} \xrightarrow[\text{a.s.}]{} \frac{1}{\mu_i} \qquad \text{as} \quad t \to \infty.$$

Hence,

$$\lim_{t\to\infty} \frac{U(t)}{t} \geq \left(1 + \sum_{i=1}^{k} \frac{\nu_i}{\mu_i}\right)^{-1} = \pi_0 .$$

The reverse inequality is proved similarly.

Corollary 2.3. *Under the same conditions,*

$$\frac{EU(t)}{t} \to \pi_0 \qquad \text{as} \quad t \to \infty.$$

Proof. Since $U(t)/t \leq 1$ and $U(t)/t \to_{\text{a.s.}} \pi_0$, it follows by the Lebesgue dominated convergence theorem that

$$\frac{EU(t)}{t} \to \pi_0 \qquad \text{as} \quad t \to \infty. \quad \|$$

Corollary 2.4. *Let $D_i(t)$ be the down time for component i in* $[0, t]$. *Then under the same hypotheses,*

$$\lim_{t\to\infty} \frac{D_i(t)}{t} \underset{\text{a.s.}}{=} \frac{\nu_i}{\mu_i} \pi_0 . \tag{2.3}$$

Proof. Note that

$$\sum_{r=1}^{\tilde{N}_i(t)-1} D_{ir} \leq D_i(t) \leq \sum_{r=1}^{\tilde{N}_i(t)} D_{ir},$$

so that

$$\frac{D_i(t)}{t} \leq \frac{1}{\tilde{N}_i(t)} \left[\sum_{r=1}^{\tilde{N}_i(t)} D_{ir}\right] \frac{N_i[U(t)]}{U(t)} \frac{U(t)}{t} .$$

Hence,

$$\lim_{t\to\infty} \frac{D_i(t)}{t} \leq \frac{\nu_i}{\mu_0} \pi_0$$

by the strong law, Lemma 2.1, and Theorem 2.2.

The reverse inequality is proved similarly. Of course, it also follows from Corollary 2.4 that

$$\lim_{t\to\infty} \frac{D(t)}{t} \underset{\text{a.s.}}{=} \pi_0 \sum_{i=1}^{k} \frac{\nu_i}{\mu_i} .$$

Corollary 2.5. *Under the same hypotheses,*

$$\lim_{t\to\infty} \frac{\tilde{N}_i(t)}{t} \underset{\text{a.s.}}{=} \frac{\pi_0}{\mu_i}, \qquad i = 1, 2, \ldots, k. \tag{2.4}$$

Proof.

$$\frac{\tilde{N}_i(t)}{t} = \frac{N_i[U(t)]}{U(t)} \frac{U(t)}{t} \xrightarrow[\text{a.s.}]{} \frac{1}{\mu_i} \pi_0$$

by Lemma 2.1 and Theorem 2.2.

Corollary 2.6. *Under the same hypotheses,*

$$\lim_{t \to \infty} \frac{E\tilde{N}_i(t)}{t} = \frac{\pi_0}{\mu_i}. \tag{2.5}$$

Proof.

$$\frac{\tilde{N}_i(t)}{t} = \frac{N_i[U(t)]}{t} \leq \frac{N_i(t)}{t}.$$

By the elementary renewal theorem, (1.2),

$$\frac{EN_i(t)}{t} \to \frac{1}{\mu_i}.$$

Also, $EN_i(t) < \infty$ for all t. Hence, there exists M such that $\sup_t (EN_i(t)/t) < M$. The conclusion follows from Corollary 2.5 and the Lebesgue dominated convergence theorem. ‖

Average Availability

We call $t^{-1} \int_0^t P[\xi(u) = 0]\, du$ the *average availability* in $[0, t]$. It is a well-known property of stochastic processes that

$$t^{-1} \int_0^t P[\xi(u) = 0]\, du = \frac{EU(t)}{t}, \tag{2.6}$$

an easy consequence of Fubini's theorem. It follows from Corollary 2.3 that

$$\lim_{t \to \infty} t^{-1} \int_0^t P[\xi(u) = 0]\, du = \pi_0. \tag{2.7}$$

If $\lim_{t \to \infty} P[\xi(t) = 0]$ exists, then it can easily be shown, using (2.7), that

$$\lim_{t \to \infty} P[\xi(t) = 0] = \pi_0. \tag{2.8}$$

The limit in (2.8) will not always exist under the hypotheses of Theorem 2.2. Sufficient conditions for (2.8), for example, are that the distributions of failure times F_i be nonlattice and F_j be exponential ($j \neq i$).

Similarly, if $P[\xi(t) = i]$ exists, then

$$\lim_{t \to \infty} P[\xi(t) = i] = \pi_i. \tag{2.9}$$

System Mean Time between Failures

Each time the system is repaired, the time until next failure will of course depend on the repair history of each component. However, the average of successive up times will converge to a limit, say, μ. Likewise, the average of successive down times will converge to a limit, say, ν. To calculate these quantities, let $\tilde{N}(t) = \sum_{i=1}^{k} \tilde{N}_i(t)$ be the number of system failures in $[0, t]$.

Theorem 2.7. *If* $0 < \mu_i < \infty$ *and* $0 \le \nu_i < \infty$ $(i = 1, 2, \ldots, k)$, *then the limiting average of system up times will be a.s.*

$$\mu = \left(\sum_{i=1}^{k} \frac{1}{\mu_i}\right)^{-1}, \tag{2.10}$$

while the limiting average of system down times will be a.s.

$$\nu = \mu \sum_{i=1}^{k} \frac{\nu_i}{\mu_i}. \tag{2.11}$$

Proof. The average of system up times in $[0, t]$ will be approximately $U(t)/\tilde{N}(t)$. (The error will go to 0 a.s. as $t \to \infty$, as in previous proofs.) Since by Theorem 2.2, $\lim_{t\to\infty} U(t)/t =_{\text{a.s.}} \pi_0$, and by Corollary 2.5, $\lim_{t\to\infty} \tilde{N}_i(t)/t =_{\text{a.s.}} \pi_0/\mu_i$, it follows that

$$\mu \stackrel{\text{def}}{=} \lim_{t\to\infty} \frac{U(t)}{\tilde{N}(t)} \underset{\text{a.s.}}{=} \frac{\pi_0}{\pi_0 \sum_{i=1}^{k}(1/\mu_i)} = \left(\sum_{i=1}^{k} \frac{1}{\mu_i}\right)^{-1}.$$

The average of system down times in $[0, t]$ will be approximately

$$\sum_{i=1}^{k} \sum_{r=1}^{\tilde{N}_i(t)} \frac{D_{ir}}{\tilde{N}(t)}.$$

By Corollary 2.5 and the strong law,

$$\sum_{i=1}^{k} \frac{\tilde{N}_i(t)}{\tilde{N}(t)} \frac{1}{\tilde{N}_i(t)} \sum_{r=1}^{\tilde{N}_i(t)} D_{ir} \xrightarrow[\text{a.s.}]{} \mu \sum_{i=1}^{k} \frac{\nu_i}{\mu_i} \stackrel{\text{def}}{=} \nu. \qquad \|$$

Remark. If failure distributions were exponential, then of course $\left(\sum_{i=1}^{k} (1/\mu_i)\right)^{-1}$ would be the expected duration of a system up time each time it is up, independently of system past history. It is interesting that in the limiting sense, the average duration of a system up time is

$$\mu = \left(\sum_{i=1}^{k} \frac{1}{\mu_i}\right)^{-1}$$

for *arbitrary* failure distributions.

For a one-unit system with mean life μ and mean repair time v, the limiting system fractional up time is

$$\mu/(\mu + v). \tag{2.12}$$

For our series model, the limiting fractional system up time is, by Theorem 2.7,

$$\pi_0 = \frac{1}{1 + \sum_{i=1}^{k}(v_i/\mu_i)} = \frac{\mu}{\mu + \mu \sum_{i=1}^{k}(v_i/\mu_i)} = \frac{\mu}{(\mu + v)}, \tag{2.13}$$

where now μ is defined by (2.10) and v by (2.11). Clearly, (2.13) is the analogue of (2.12).

3. ASYMPTOTIC DISTRIBUTIONS

To obtain the asymptotic distribution of $\tilde{N}_i(t)$, $\tilde{N}(t) = \sum_{i=1}^{k} \tilde{N}_i(t)$, and $U(t)$, we must also specify the variances $\sigma_i^2 = \operatorname{Var} X_i$ and $\tau_i^2 = \operatorname{Var} D_i$ ($i = 1, 2, \ldots, k$). Let $m_i = \mu_i \pi_0^{-1}$ for $i = 1, 2, \ldots, k$. As we shall show, m_1 is approximately the mean time between failures of component i. Let * denote random variables centered at expectations. Although centered random variables will have asymptotic mean 0, the asymptotic variance is not necessarily 1. In particular, let

$$\tilde{N}_i{}^*(t) = t^{-1/2}[\tilde{N}_i(t) - tm_i^{-1}]. \tag{3.1}$$

We first state our main results before presenting proofs.

Theorem 3.1. *If* $0 < \mu_i < \infty$, $0 \le v_i < \infty$, $0 < \sigma_i^2 < \infty$, $0 \le \tau_i^2 < \infty$ *for* $i = 1, 2. \ldots, k$, *then*

$$(\tilde{N}_1{}^*(t), \tilde{N}_2{}^*(t), \ldots, \tilde{N}_k{}^*(t))$$

is asymptotically ($t \to \infty$) *multivariate normal with mean vector* 0 *and variance-covariance matrix*

$$\Sigma = (v_{ij})$$

where

$$v_{ij} = m_i^{-1} m_j^{-1} \left[\pi_0 \sum_{s=1}^{k} [v_s^2 \sigma_s^2 \mu_s^{-3} + \tau_s^2 \mu_s^{-1}] - v_i \sigma_i^2 \mu_i^{-2} - v_j \sigma_j^2 \mu_j^{-2} \right] \quad (i \ne j), \tag{3.2}$$

$$v_{ii} = v_i^2 = m_i^{-3} w_i^2, \qquad w_i^2 = \sigma_i^2 c_i^2 + \mu_i \sum_{j=1}^{k} \tau_j^2 \mu_j^{-1} + \mu_i \sum_{j \ne i} \sigma_j^2 v_j^2 \mu_j^{-3}, \tag{3.3}$$

and

$$c_i = 1 + \sum_{j \ne i} v_j \mu_j^{-1}, \qquad \pi_0 = \left[1 + \sum_{j=1}^{k} v_j \mu_j^{-1} \right]^{-1}. \tag{3.4}$$

Although immediate from Theorem 3.1, the following consequence is of sufficient practical importance to warrant explicit statement.

Corollary 3.2. *Under the conditions of Theorem* 3.1,

$$t^{-1/2}v_i^{-1}[\tilde{N}_i(t) - tm_i^{-1}]$$

is asymptotically ($t \to \infty$) $N(0, 1)$.

Theorem 3.3. *Let* $\tilde{N}(t) = \sum_{i=1}^{k} \tilde{N}_i(t)$ *and* $m^{-1} = \sum_{i=1}^{k} m_i^{-1}$. *Under the conditions of Theorem* 3.1,

$$t^{-1/2}v^{-1}[\tilde{N}(t) - tm^{-1}]$$

is asymptotically ($t \to \infty$) ($N(0, 1)$, *where*

$$v^2 = \pi_0 m^{-2} \sum_{i=1}^{k} [(m - v_i)^2 \sigma_i^2 \mu_i^{-3} + \tau_i^2 \mu_i^{-1}]. \tag{3.5}$$

Theorem 3.4. *Let* $U(t)$ *be the system up time in* $[0, t]$. *Under the conditions of Theorem* 3.1,

$$t^{-1/2}u^{-1}[U(t) - \pi_0 t]$$

is asymptotically ($t \to \infty$) $N(0, 1)$, *where*

$$u^2 = \pi_0^3 \left[\sum_{i=1}^{k} (v_i^2 \mu_i^{-3} \sigma_i^2 + \tau_i^2 \mu_i^{-1}) \right]. \tag{3.6}$$

Corollary 3.5. *Let* $D(t)$ *be the system down time in* $[0, t]$, *so that* $U(t) + D(t) = t$. *Under the conditions of Theorem* 3.1,

$$t^{-1/2}u^{-1}[D(t) - (1 - \pi_0)t] \quad \text{and} \quad t^{-1/2}[\tau_i^2 m_i^{-1} + v_i^2 w_i^2 m_i^{-3}]^{-1/2}[D_i(t) - \pi_i t]$$

are asymptotically ($t \to \infty$) $N(0, 1)$, *where* $D_i(t)$ *is the down time of component* i *in* $[0, t]$.

Application of these results to the problem of determining maintenance costs is discussed in Section 4.

Proofs of Theorems. For mathematical convenience, we will work with the random variable which is the number of *completed repairs* of component i in $[0, t]$. We do this because we want to assume that all components are new at $t = 0$. A natural cycle, then, ends with a completed repair. Asymptotically, the number of failures properly centered will have the same distribution as the number of completed repairs similarly centered. For this reason, in the remainder of this section we use $\tilde{N}_i(t)$ to mean the number of completed repairs in $[0, t]$.

Let $S_{in} = \sum_{r=1}^{n} X_{ir}$ and $S_{i0} = 0$. The time to the first completed repair of component i will then be

$$Y_{i1} = X_{i1} + D_{i1} + \sum_{j \neq i} \sum_{s=1}^{N_j(X_{i1})} D_{js},$$

where $\{N_j(x); x \geq 0\}$ is the renewal counting process associated with $\{X_{jr}\}_{r=1}^{\infty}$, as in Section 1. Similarly, the time between the $(r-1)$th and rth completed repair will be

$$Y_{ir} = X_{ir} + D_{ir} + \sum_{j \neq i} \sum_{s=N_j(S_{i,r-1})+1}^{N_j(S_{i,r})} D_{js}. \tag{3.7}$$

Let

$$Z_{in} = \sum_{r=1}^{n} Y_{ir} = S_{in} + \sum_{r=1}^{n} D_{ir} + \sum_{j \neq i} \sum_{r=1}^{N_j(S_{in})} D_{jr}. \tag{3.8}$$

Lemma 3.6. *If $0 < \mu_i < \infty$ and $0 \leq \nu_i < \infty$ $(i = 1, 2, \ldots, k)$, then*

$$\lim_{n \to \infty} n^{-1} Z_{in} \underset{\text{a.s.}}{=} \mu_i \pi_0^{-1} \overset{\text{def}}{=} m_i.$$

Proof. By the strong law, $n^{-1} S_{in} \to_{\text{a.s.}} \mu_i$ and $n^{-1} \sum_{r=1}^{n} D_{ir} \to_{\text{a.s.}} \nu_i$. Since $\mu_i > 0$, $S_{in} \to_{\text{a.s.}} \infty$ as $n \to \infty$, and since $\mu_j > 0$, $N_j(t) \to_{\text{a.s.}} \infty$ as $t \to \infty$. Hence

$$\left[\frac{1}{N_j(S_{in})} \sum_{s=1}^{N_j(S_{in})} D_{js}\right] \left[\frac{N_j(S_{in})}{S_{in}}\right] \left[\frac{S_{in}}{n}\right] \xrightarrow[\text{a.s.}]{} \nu_j \mu_j^{-1} \mu_i,$$

again by the strong law and (2.3). The conclusion of the lemma follows. ‖

The processes $\{Z_{in}; n \geq 1\}$ and $\{\tilde{N}_i(t); t \geq 0\}$ are related by

$$\tilde{N}_i(t) = \begin{cases} \max\{n \mid Z_{in} \leq t\}, & Z_{i1} \leq t, \\ 0, & Z_{i1} > t. \end{cases} \tag{3.9}$$

Since the partial sum process $\{Z_{in}; n \geq 1\}$ and the counting process $\{\tilde{N}_i(t); t \geq 0\}$ are essentially inverses of each other, it will be sufficient to determine the asymptotic normality of the partial sum process.

From (3.9) we observe that for fixed $\tau > 0$

$$m_i^{-1} n^{-1/2} \sum_{r=1}^{\tilde{N}_i(n\tau)} [Y_{ir} - m_i] \leq n^{-1/2} \left[\frac{n\tau}{m_i} - \tilde{N}_i(n\tau)\right]$$

$$\leq m_i^{-1} n^{-1/2} \sum_{r=1}^{\tilde{N}_i(n\tau)} [Y_{ir} - m_i] + m_i^{-1} n^{-1/2} Y_{i, \tilde{N}_i(n\tau)+1}, \tag{3.10}$$

so that asymptotically

$$n^{-1/2}[n\tau m_i^{-1} - \tilde{N}_i(n\tau)] \sim m_i^{-1} n^{-1/2} \sum_{r=1}^{\tilde{N}_i(n\tau)} [Y_{ir} - m_i], \tag{3.11}$$

where $\sim$ means asymptotically equivalent in distribution. On the other hand, it is well known [11] that

$$n^{-1/2} \sum_{r=1}^{N_i(n\tau)} [Y_{ir} - m_i] \sim n^{-1/2} \sum_{r=1}^{[n\tau/m_i]} [Y_{ir} - m_i] \tag{3.12}$$

since

$$\frac{N_i(n\tau)}{n\tau} \xrightarrow[\text{a.s.}]{} \frac{1}{m_i}$$

by (2.4).

It will be useful to expand $Z_{in}^* \overset{\text{def}}{=} n^{-1/2} \sum_{r=1}^{n} [Y_{ir} - m_i]$ as follows.

$$\begin{aligned} Z_{in}^* = {} & n^{-1/2}(S_{in} - n\mu_i) + n^{-1/2} \sum_{r=1}^{n} (D_{ir} - v_i) \\ & + n^{-1/2} \sum_{j \neq i} \sum_{r=1}^{N_j(S_{in})} (D_{jr} - v_j) + n^{-1/2} \sum_{j \neq i} v_j \left[N_j(S_{in}) - \frac{S_{in}}{\mu_j} \right] \\ & + n^{-1/2} \sum_{j \neq i} \frac{v_j}{\mu_j} (S_{in} - n\mu_i). \end{aligned}$$

Let $S_{in}^* = n^{-1/2}(S_{in} - n\mu_i)$ and $c_i = 1 + \sum_{j \neq i} v_j \mu_j^{-1}$. Then we can rewrite Z_{in}^* as

$$\begin{aligned} Z_{in}^* = {} & c_i S_{in}^* + n^{-1/2} \sum_{r=1}^{n} (D_{ir} - v_i) \\ & + n^{-1/2} \sum_{j \neq i} \sum_{r=1}^{N_j(S_{in})} (D_{jr} - \mu_j) + n^{-1/2} \sum_{j \neq i} v_j \left[N_j(S_{in}) - \frac{S_{in}}{\mu_j} \right]. \end{aligned}$$

Also

$$Z_{in}^* \sim c_i S_{in}^* + n^{-1/2} \sum_{j=1}^{k} \sum_{r=1}^{[n\mu_i \mu_j^{-1}]} (D_{jr} - v_j) + \sum_{j \neq i} n^{-1/2} v_j [N_j(n\mu_i) - n\mu_i \mu_j^{-1}]. \tag{3.13}$$

Note that the summands in (3.13) are now independent.

Lemma 3.7.

$$\lim_{n \to \infty} \operatorname{Var} Z_{in}^* = c_i^2 \sigma_i^2 + \mu_i \sum_{j=1}^{k} \tau_j^2 \mu_j^{-1} + \mu_i \sum_{j \neq i} \sigma_j^2 v_j^2 \mu_j^{-3} \overset{\text{def}}{=} w_i^2, \tag{3.14}$$

where $c_i = 1 + \sum_{j \neq i} v_j / \mu_j$.

Proof. Use representation (3.13) and the well-known result

$$\operatorname{Var}(n\tau)^{-1/2}[N_j(n\tau) - n\tau/\mu_j] \to \sigma_j{}^2\mu_j^{-3}\tau. \qquad \|$$

We use (3.11) and (3.12) to write

$$n^{-1/2}v_j[N_j(n\mu_i) - n\mu_i\mu_j^{-1}] \sim \frac{-v_j}{\mu_j} S^*_{j[n\mu_i\mu_j^{-1}]}. \tag{3.15}$$

Proof of Theorem 3.1. From Theorem 1.2 and representation (3.13) it is obvious that the marginal random variables $\tilde{N}_i{}^*(t)$ are asymptotically $N(0, w_i{}^2m_i^{-3})$, where $w_i{}^2$ is given by (3.14). Using (3.15) we see that

$$Z^*_{in} \sim c_i S^*_{in} - \sum_{j\neq i} v_j \mu_j^{-1} S^*_{j[n\mu_i\mu_j^{-1}]} + n^{-1/2} \sum_{j=1}^{k} \sum_{r=1}^{[n\mu_i\mu_j^{-1}]} (D_{jr} - v_j)$$

or

$$Z^*_{in} \sim \pi_0^{-1} S^*_{in} - \sum_{j=1}^{k} v_j \mu_j^{-1} S^*_{j[n\mu_i\mu_j^{-1}]} + n^{-1/2} \sum_{j=1}^{k} \sum_{r=1}^{[n\mu_i\mu_j^{-1}]} (D_{jr} - v_j). \tag{3.16}$$

From representation (3.16) it is easy to see that arbitrary linear combinations of the Z^*_{in}'s are asymptotically normal. It follows that $(\tilde{N}_1{}^*(t), \tilde{N}_2{}^*(t), \ldots, \tilde{N}_k{}^*(t))$ is asymptotically multivariate normal.

It remains to compute the variance-covariance matrix. From Theorem 1.2 and Lemma 3.7, it follows that

$$v_{ii} = v_i{}^2 = m_i^{-3} w_i{}^2,$$

as given by (3.3).

To compute $\operatorname{Cov}[\tilde{N}_i{}^*(t), \tilde{N}_j{}^*(t)]$ for $i \neq j$, recall that by (3.15)

$$\tilde{N}_i{}^*(n\tau) \sim -m_i^{-1} Z_{i[n\tau/m_i]}.$$

Hence,

$$v_{ij} \overset{\text{def}}{=} \operatorname{Cov}[\tilde{N}_i{}^*(n\tau); \tilde{N}_j{}^*(n\tau)] = m_i^{-1} m_j^{-1} \operatorname{Cov}[Z^*_{i[n\tau/m_i]}, Z^*_{j[n\tau/m_j]}]$$

for $0 \le \tau \le 1$. Using (3.16) we see that

$$Z^*_{i[n\tau/m_i]} \sim \pi_0^{-1} S^*_{i[n\tau\pi_0/\mu_i]} - \sum_{s=1}^{k} v_s \mu_s^{-1} S^*_{s[n\tau\pi_0/\mu_s]} + n^{-1/2} \sum_{s=1}^{k} \sum_{r=1}^{[n\tau\pi_0\mu_s^{-1}]} (D_{sr} - v_s).$$

It follows that for $i \neq j$,

$$\begin{aligned}\operatorname{Cov}[Z^*_{i[n\tau/m_i]}, Z^*_{j[n\tau/m_j]}] = {} & -\pi_0^{-1} \frac{v_i}{\mu_i} \operatorname{Var}(S^*_{i[n\tau\pi_0/\mu_i]}) - \pi_0^{-1} \frac{v_j}{\mu_j} \operatorname{Var}(S^*_{j[n\tau\pi_0/\mu_j]}) \\ & + \sum_{s=1}^{k} v_s{}^2 \mu_s^{-2} \operatorname{Var}(S^*_{s[n\tau\pi_0/\mu_s]}) + \sum_{s=1}^{k} \pi_0 \mu_s^{-1} \tau_s{}^2 \tau.\end{aligned}$$

Hence, letting $\tau = 1$,

$$V_{ij} = m_i m_j \left[-v_i \mu_i^{-2} \sigma_i^2 - v_j \mu_j^{-2} \sigma_j^2 + \pi_0 \sum_{s=1}^{k} [v_s^2 \mu_s^{-3} \sigma_s^2 + \tau_s^2 \mu_s^{-1}] \right]. \qquad \|$$

Proof of Theorem 3.3. From the previous proof we know that

$$\tilde{N}_n^*(n\tau) \overset{\text{def}}{=} n^{-1/2}[\tilde{N}(n\tau) - n\tau m^{-1}] \sim -\sum_{i=1}^{k} m_i^{-1} Z_{i[n\tau/m_i]}^*, \qquad 0 \le \tau \le 1.$$

From (3.17) we see that

$$\begin{aligned} \tilde{N}_n^*(n\tau) \sim &-\pi_0^{-1} \sum_{i=1}^{k} m_i^{-1} S_{i[n\tau\pi_0/\mu_i]}^* + m^{-1} \sum_{j=1}^{k} v_j \mu_j^{-1} S_{j[n\tau\pi_0/\mu_j]}^* \\ &- m^{-1} n^{-1/2} \sum_{j=1}^{k} \sum_{r=1}^{[n\tau\pi_0\mu_j^{-1}]} (D_{jr} - v_j). \end{aligned} \tag{3.18}$$

The right-hand side of (3.18) is clearly asymptotically normal with variance

$$v^2 = \pi_0 m^{-2} \sum_{j=1}^{k} [(m - v_j)^2 \mu_j^{-3} \sigma_j^2 + \tau_j^2 \mu_j^{-1}]$$

at $\tau = 1$. $\|$

Proof of Theorem 3.4. By definition of the up time, $U(t)$, in $[0, t]$,

$$\sum_{r=1}^{\tilde{N}_i(t)} X_{ir} \le U(t) \le \sum_{r=1}^{\tilde{N}_i(t)} X_{ir} + X_{i,\tilde{N}_i(t)+1}.$$

Since $\sup_{\tau \le 1}(X_{i,\tilde{N}_i(\tau n)+1})/\sqrt{n}$ converges to 0 in probability, it follows that

$$\begin{aligned} U_n^*(n\tau) &\overset{\text{def}}{=} n^{-1/2}[U(n\tau) - n\tau\pi_0] \\ &\sim n^{-1/2} \left[\sum_{r=1}^{\tilde{N}_i(n\tau)} (X_{ir} - \mu_i) + \mu_i \left[\tilde{N}_i(n\tau) - \frac{n\tau}{m_i} \right] \right] \qquad \text{for} \quad 0 \le \tau \le 1. \end{aligned}$$

As in previous proofs,

$$n^{-1/2} \mu_i [\tilde{N}_i(n\tau) - n\tau/m_i] \sim -\pi_0 Z_{i[n\tau/m_i]}^*.$$

Hence

$$U_n^*(n\tau) \sim S_{i[n\tau/m_i]}^* - \pi_0 Z_{i[n\tau/m_i]}^*.$$

Using (3.17) we see that

$$U_n^*(n\tau) \sim \pi_0 \sum_{j=1}^{k} v_j \mu_j^{-1} S_{j[n\tau\pi_0\mu_j^{-1}]}^* - \pi_0 n^{-1/2} \sum_{j=1}^{k} \sum_{r=1}^{[n\tau\pi_0\mu_j^{-1}]} (D_{jr} - v_j).$$

Clearly $U_n^*(n\tau)$ is asymptotically normal with variance

$$u^2 = \pi_0{}^3 \sum_{j=1}^{k} [v_j{}^2 \mu_j^{-3} \sigma_j{}^2 + \tau_j{}^2 \mu_j^{-1}]$$

when $\tau = 1$. ||

For application of the limiting results of Sections 2 and 3, it would be helpful to have information on rates of convergence. We suspect that convergence is relatively rapid for the a.s. results of Section 2, whereas convergence is relatively slow for the asymptotic normality results of Section 3. Further study concerning these rates of convergence should yield results of both theoretical and practical value.

4. COST OF REPAIR

First suppose that component i costs d_i *dollars to repair each time it fails* regardless of the time to complete repair. Then the cost accrued during $[0, t]$ is

$$C_1(t) = \sum_{i=1}^{k} d_i \tilde{N}_i(t),$$

and the cost per unit of time is, in the limit,

$$\lim_{t \to \infty} \frac{C_1(t)}{t} = \pi_0 \sum_{i=1}^{k} d_i \mu_i^{-1}. \tag{4.1}$$

Applying the techniques used in the proof of Theorem 3.3, we can show that

$$t^{-1/2}[C_1(t) - \pi_0 \sum_{i=1}^{k} d_i \mu_i^{-1}] \to N(0, \delta_1{}^2) \tag{4.2}$$

where

$$\delta_1{}^2 = \pi_0 \sum_{i=1}^{k} \left[d_i + \left(\sum_{j=1}^{k} d_j m_j^{-1} \right) v_i \right]^2 \mu_i^{-3} \sigma_i{}^2 + \pi_0 \left(\sum_{j=1}^{k} d_j m_j^{-1} \right)^2 \sum_{j=1}^{k} \tau_j{}^2 \mu_j^{-1}.$$

Alternatively, suppose that it costs d_i *dollars per hour of down time* for component i. Now the total cost accrued during $[0, t]$ becomes

$$C_2(t) = \sum_{i=1}^{k} d_i D_i(t)$$

where $D_i(t)$ is the down time for component i in $[0, t]$. Then

$$\lim_{t \to \infty} \frac{C_2(t)}{t} = \pi_0 \sum_{i=1}^{k} d_i v_i \mu_i^{-1} \tag{4.3}$$

by (2.7).

Theorem 4.1. *Under the conditions of Theorem* 3.1

$$t^{-1/2}\left[C_2(t) - t\pi_0 \sum_{i=1}^{k} d_i v_i \mu_i^{-1}\right] \to N(0, \delta_2{}^2) \qquad (4.4)$$

where

$$\delta_2^2 = \sum_{i=1}^{k} [\tau_i^2 m_i^{-1} + v_i^2 m_i^{-3} w_i^2]\, d_i^2$$

where $w_i{}^2$ *is given in Lemma* 3.7.

Proof.

$$n^{-1/2}\left[C_2(n\tau) - n\tau\pi_0 \sum_{i=1}^{k} d_i v_i \mu_i^{-1}\right]$$

$$\sim n^{-1/2} \sum_{i=1}^{k} \sum_{r=1}^{\tilde{N}_i(n\tau)} d_i(D_{ir} - v_i) + n^{-1/2} \sum_{i=1}^{k} v_i d_i [\tilde{N}_i(n\tau) - n\tau m_i^{-1}]$$

$$\sim n^{-1/2} \sum_{i=1}^{k} \sum_{r=1}^{n\tau m_i^{-1}} d_i(D_{ir} - v_i) - \sum_{i=1}^{k} v_i d_i m_i^{-1} Z^*_{i[n\tau m_i^{-1}]}.$$

Obviously, the normed cost function is asymptotically normal. Its variance is easily seen to be (when $\tau = 1$)

$$\delta_2{}^2 = \sum_{i=1}^{k} [m_i^{-1}\tau_i^2 + v_i^2 m_i^{-3} w_i^2]\, d_i\,. \qquad \|$$

ACKNOWLEDGMENT

We wish to thank Igor Bazovsky, who suggested the problem and obtained many of the earlier results in this area of availability.

REFERENCES

1. Bazovsky, I., MacFarlane, N. R. and Wunderman, R. (1962). Study of maintenance cost optimization and reliability of shipboard machinery. Report for ONR Contract Nonr-37400. United Control, Seattle, Washington.
2. Billingsley, P. (1968). *Convergence of Probability Measures*. Wiley, New York.
3. Botez, M. C. (1969). Sur un Modele Reparation de l'Equipment. *Cahiers Centre Études Recherche Opér.*, Institute de Statistique de l'Universite Libre de Bruxelles **11** No. 2.
4. *Colloq. Reliability Theory, Tihany, Hungary, September 1969.*
5. Feller, W. (1966). *An Introduction to Probability Theory and its Applications*, Vol. II. Wiley, New York.
6. Feller, W. (1968). *An Introduction to Probability Theory and its Applications*, 3rd ed., Vol. I. Wiley, New York.
7. Gaver, D. P. (1963). Time to failure and availability of paralleled systems with repair. *IEEE Trans. Rel.* **12** 30–38.

8. *Handbook of Reliability Engineering* (1968). NAVAIR 00-65-502. Published by Naval Ordnance Syst. Command, Washington, D.C.
9. Iglehart, D. and Whitt, W. (1969). The equivalence of functional central limit theorems for counting processes and associated partial sums. Tech. Rep. 121. Dept. of Operations Res., Stanford Univ., Stanford, California.
10. Obretenov. A., Dimitrov, B. and Uzunov, M. (1969). Study on the reliability of a system by means of stochastic processes (in Russian). *Bull. Inst. Math., Bulgaria.* **11** 159–180.
11. Renyi, A. (1957). On the asymptotic distribution of the sum of a random number of independent random variables. *Acta Math. Acad. Sci. Hungary*, **8** 193–199.
12. Ross, S. (1970). *Applied Probability Models with Optimization Applications.* Holden-Day, San Francisco, California.

Some Measures for Discriminating between Normal Multivariate Distributions with Unequal Covariance Matrices

HERMAN CHERNOFF[1]
STANFORD UNIVERSITY

1. SUMMARY AND INTRODUCTION

In a previous paper [5], a measure S was described which indicates how well one may discriminate between two normal multivariate distributions using a linear discriminant function, This measure, when applied to an example in design of experiments, led to a somewhat unexpected conclusion.

Briefly, suppose that under H_1, X is an $N(1, 1)$ random variable and Y is independently $N(1, 9)$ where $N(\mu, \sigma^2)$ represents the normal distribution with mean μ and variance σ^2. Under H_2, X is $N(-1, 9)$ and Y is independently $N(-1, 1)$. The statistician who wishes to discriminate between H_1 and H_2 is permitted n observations on X. In addition, he is given a choice between n more observations on X or n observations on Y. Applying the measure S, he is led to prefer the unbalanced choice of $2n$ observations on X to the balanced one of n observations on each X and Y. Thus it appears that if his first observations are more precise under H_1 than H_2, there is a premium on taking additional observations which are more precise under H_1 than H_2.

Is this result a consequence of restricting attention to linear discriminant functions? In this paper we extend the measure to T, which is appropriate for using the likelihood-ratio test. It is shown that using T there still is a premium for the unbalanced choice.

Section 4 contains a brief discussion of the Kullback–Leibler information numbers and how they relate to S and T.

[1] This paper was prepared with partial support from the Office of Naval Research under Contract No. N00014-67-A-0112-0051.

2. THE MEASURE S

Becker [2] suggested that

$$S = \frac{|\mu_1 - \mu_2|}{\sigma_1 + \sigma_2}$$

is a useful measure of separation between two distributions F_i with mean μ_i and variance σ_i^2, $i = 1, 2$. This measure appeared in [3], and the multivariate extension

$$S = \sup_{b \neq 0} \frac{|b'(\mu_1 - \mu_2)|}{(b'\Sigma_1 b)^{1/2} + (b'\Sigma_2 b)^{1/2}} \tag{2.1}$$

was shown to be relevant in discriminating between two multivariate normal distributions $F_i = N(\mu_i, \Sigma_i)$, $i = 1, 2$, when linear discriminant functions are used. In that case the linear discriminant function which minimizes the maximum error probability consists of selecting F_1 if

$$b_0'X > \frac{(b_0'\Sigma_1 b_0)^{-1/2}\mu_1 + (b_0'\Sigma_2 b_0)^{-1/2}\mu_2}{(b_0'\Sigma_1 b_0)^{-1/2} + (b_0'\Sigma_2 b_0)^{-1/2}}$$

where b_0 is the vector which minimizes the expression in (2.1). The corresponding error probabilities are

$$\varepsilon_1 = \varepsilon_2 = \Phi(-S).$$

If X is replaced by the sample mean of n independent observations, then the error probabilities approach zero exponentially fast in n. In fact, the error probabilities are

$$\varepsilon_{1n} = \varepsilon_{2n} = \Phi(-\sqrt{n}S) \approx (2\pi n S^2)^{-1/2} \exp(-nS^2/2). \tag{2.2}$$

The following theorem, presented in [5], is essentially a restatement of results in [1] and [6]. It may be derived by applying the method of Lagrange multipliers to the relatively simple calculation of S for the multivariate distributions $N(\mu_i, \Sigma_i)$, $i = 1, 2$, $\delta = \mu_1 - \mu_2$. We assume that the Σ_i are positive definite.

Theorem 1.

$$S^2 = t(1-t)\delta'\Sigma^{-1}\delta \tag{2.3}$$

where

$$\Sigma = t\Sigma_1 + (1-t)\Sigma_2, \tag{2.4}$$

and t is the unique solution between 0 *and* 1 *of*

$$R(t) = \delta'\Sigma^{-1}[t^2\Sigma_1 - (1-t)^2\Sigma_2]\Sigma^{-1}\delta = 0. \tag{2.5}$$

The optimal value of b is given by

$$b_0 = \Sigma^{-1}\delta \tag{2.6}$$

and is unique up to a multiplicative constant. Furthermore,

$$|b_0'\delta| = \delta'\Sigma^{-1}\delta = S^2/t(1-t) \tag{2.7}$$

and

$$t^2 b_0'\Sigma_1 b_0 = (1-t)^2 b_0'\Sigma_2 b_0 = S^2. \tag{2.8}$$

The fact that

$$\frac{dR}{dt} = 2\delta'\Sigma^{-1}\Sigma_1\Sigma^{-1}\Sigma_2\Sigma^{-1}\delta > 0 \qquad \text{for} \quad \delta \neq 0 \tag{2.9}$$

permits us to apply the Newton iterative technique where an approximation t^* to t is improved to

$$t^{**} = t^* - \left[\frac{dR(t^*)}{dt}\right]^{-1} R(t^*).$$

Theorem 1 presents S^2 as a multiple of $\delta'\Sigma^{-1}\delta$ which may be regarded as a Mahalanobis distance with respect to the weighted average Σ of the two covariance matrices Σ_1 and Σ_2. Note that if $\Sigma_1 = \Sigma_2$, the minimizing value of t is 0.5, $\Sigma = \Sigma_1 = \Sigma_2$, and $S^2 = (\delta'\Sigma^{-1}\delta)/4$.

Suppose $(X, Y)'$ is $N(\mu_i, \Sigma_i)$ under $H_i, i = 1, 2$, where $\mu_1' = (1, 1)$, $\mu_2' = (-1, -1)$, $\Sigma_1 = \left(\begin{smallmatrix}1 & 0\\ 0 & 9\end{smallmatrix}\right)$, and $\Sigma_2 = \left(\begin{smallmatrix}9 & 0\\ 0 & 1\end{smallmatrix}\right)$. Then S^2 corresponding to X is 0.25, while S^2 corresponding to (X, Y) is 0.40. Applying (2.2), the exponential rate at which the error probabilities approach zero as the sample size approaches ∞ is determined by $S^2 = 0.25$ per observation for X and $S^2/2 = 0.20$ per observation for (X, Y) if (X, Y) counts for two observations.

This illustration can be interpreted by the remark in the introduction to the effect that when using linear discriminant functions, the fact that some of the data are more precise under H_1 than under H_2 implies that there is a premium on additional data of the same sort rather than on data which are more precise under H_2 than under H_1. This remark is relevant in the *non-sequential* case, where the choice of data to be observed is to be made before any data are gathered.

Is this phenomenon due to the fact that the linear discriminant function neglects relevant information? Would it disappear if we applied the likelihood-ratio test? To answer this question we proceed to Section 3.

3. THE MEASURE T

In [3] it was shown that if the likelihood ratio is selected to minimize $\varepsilon_{1n} + \lambda\varepsilon_{2n}$ for fixed $\lambda > 0$ or to minimize $\max(\varepsilon_{1n}, \varepsilon_{2n})$ on the basis of n independent observations on X with density $f_i(x)$ under H_i, $i = 1, 2$, then

$$\lim[n^{-1} \log \varepsilon_{in}] = -I \tag{3.1}$$

where

$$I = -\log \inf_{0 \leq t \leq 1} \int f_1^{1-t}(x) f_2^{t}(x)\, dx. \tag{3.2}$$

Furthermore, the test which decides according to the sign of the logarithm of the likelihood ratio attains error probabilities satisfying (3.1). Thus comparison with (2.3) shows that $S^2/2$ is comparable to I or that T, defined by

$$I = T^2/2, \tag{3.3}$$

is comparable to S as a measure of separation between two distributions.

The following theorem characterizes T for two distinct multivariate normal distributions with positive definite matrices.

Theorem 2.

$$T^2 = \sup_{0 \leq t \leq 1} \left\{ t(1-t)\delta'\Sigma^{-1}\delta + \log \frac{|\Sigma|}{|\Sigma_1|^t |\Sigma_2|^{1-t}} \right\} \tag{3.4}$$

where

$$\Sigma = t\Sigma_1 + (1-t)\Sigma_2$$

and if the expression in braces in (3.4) *is* $H(t)$,

$$H'(t) = \delta'\Sigma^{-1}[(1-t)^2\Sigma_2 - t^2\Sigma_1]\Sigma^{-1}\delta + \log|\Sigma_1^{-1}\Sigma_2| + \mathrm{tr}[(\Sigma_1 - \Sigma_2)\Sigma^{-1}] \tag{3.5}$$

and

$$H''(t) = -2\delta'\Sigma^{-1}\Sigma_1\Sigma^{-1}\Sigma_2\Sigma^{-1}\delta - \mathrm{tr}[(\Sigma_1 - \Sigma_2)\Sigma^{-1}(\Sigma_1 - \Sigma_2)\Sigma^{-1}] < 0. \tag{3.6}$$

Before deriving this result we remark that the concavity of log determinant implies that $T^2 \geq S^2$, which is anticipated from the fact that the likelihood-ratio test is at least as good as the best linear discriminant function. This theorem leads to a computational procedure for T^2 which is very little different from that indicated for S^2 and only slightly more involved.

Proof.

$$f_1^{1-t} f_2^{t} = (2\pi)^{-k/2} |\Sigma_1|^{-(1-t)/2} |\Sigma_2|^{-t/2}$$
$$\exp\{-\tfrac{1}{2}[(1-t)(x-\mu_1)'\Sigma_1^{-1}(x-\mu_1) + t(x-\mu_2)'\Sigma_2^{-1}(x-\mu_2)]\}.$$

Completing the square, the expression in the square brackets may be written as

$$(x-\mu)'A(x-\mu) + c$$

where by matching coefficients we have

$$A = (1-t)\Sigma_1^{-1} + t\Sigma_2^{-1}, \tag{3.7a}$$

$$A\mu = (1-t)\Sigma_1^{-1}\mu_1 + t\Sigma_2^{-1}\mu_2, \tag{3.7b}$$

and

$$\mu' A\mu + c = (1-t)\mu_1'\Sigma_1^{-1}\mu_1 + t\mu_2'\Sigma_2^{-1}\mu_2. \tag{3.7c}$$

Then

$$\int f_1^{1-t}(x) f_2^{t}(x)\,dx = |\Sigma_1|^{-(1-t)/2}|\Sigma_2|^{-t/2}|A|^{-1/2}e^{-c/2}. \tag{3.8}$$

We have

$$\Sigma_1 A\Sigma_2 = \Sigma_2 A\Sigma_1 = \Sigma = t\Sigma_1 + (1-t)\Sigma_2, \tag{3.9}$$

$$\mu' A\mu = [(1-t)\mu_1'\Sigma_1^{-1} + t\mu_2'\Sigma_2^{-1}]A^{-1}[(1-t)\Sigma_1^{-1}\mu_1 + t\Sigma_2^{-1}\mu_2],$$

and

$$\begin{aligned} c = {} & (1-t)\mu_1'\,\Sigma_1^{-1}\mu_1 - (1-t)^2\mu_1'\Sigma_1^{-1}A^{-1}\Sigma_1^{-1}\mu_1 \\ & + t\mu_2'\Sigma_2^{-1}\mu_2 - t^2\mu_2'\Sigma_2^{-1}A^{-1}\Sigma_2^{-1}\mu_2 \\ & - t(1-t)[\mu_1'\Sigma_1^{-1}A^{-1}\Sigma_2^{-1}\mu_2 + \mu_2'\Sigma_2^{-1}A^{-1}\Sigma_1^{-1}\mu_1]. \end{aligned}$$

Now

$$\begin{aligned} A^{-1} &= \Sigma_1\Sigma^{-1}\Sigma_2 = \Sigma_2\Sigma^{-1}\Sigma_1, \\ \mu_1'\Sigma_1^{-1}\mu_1 &= \mu_1'\Sigma_1^{-1}(t\Sigma_1 + (1-t)\Sigma_2)\Sigma^{-1}\mu_1 \\ &= t\mu_1'\Sigma^{-1}\mu_1 + (1-t)\mu_1'\Sigma_1^{-1}\Sigma_2\Sigma^{-1}\mu_1, \\ \mu_1'\Sigma_1^{-1}A^{-1}\Sigma_1^{-1}\mu_1 &= \mu_1'\Sigma_1^{-1}\Sigma_2\Sigma^{-1}\mu_1, \\ \Sigma_1^{-1}A^{-1}\Sigma_2^{-1} &= \Sigma_1^{-1}\Sigma_1\Sigma^{-1}\Sigma_2\Sigma_2^{-1} = \Sigma^{-1}. \end{aligned}$$

By symmetry with respect to the interchange of 1 with 2 and t with $(1-t)$, it follows that

$$\begin{aligned} c &= t(1-t)[\mu_1'\Sigma^{-1}\mu_1 + \mu_2'\Sigma^{-1}\mu_2 - \mu_1'\Sigma^{-1}\mu_2 - \mu_2'\Sigma^{-1}\mu_1], \\ c &= t(1-t)\delta'\Sigma^{-1}\delta. \end{aligned} \tag{3.10}$$

Applying (3.9) and (3.10) to (3.8), we have

$$\log\left[\int f_1^{1-t}(s) f_2^{t}(x)\,dx\right] = -\tfrac{1}{2}H(t)$$

where

$$H(t) = t(1-t)\delta'\Sigma^{-1}\delta + \log\frac{|\Sigma|}{|\Sigma_1|^{t}|\Sigma_2|^{1-t}}.$$

We recall the expansions

$$(A + h\Delta)^{-1} = A^{-1} - hA^{-1}\Delta A^{-1} + h^2 A^{-1}\Delta A^{-1}\Delta A^{-1} + \cdots,$$

$$\log|A + h\Delta| = \log|A| + h\,\mathrm{tr}(A^{-1}\Delta) - \frac{h^2}{2}\,\mathrm{tr}(A^{-1}\Delta A^{-1}\Delta) + \cdots,$$

from which it follows that

$$\begin{aligned} H'(t) = {} & (1 - 2t)\delta'\Sigma^{-1}\delta - t(1-t)\delta'\Sigma^{-1}(\Sigma_1 + \Sigma_2)\Sigma^{-1}\delta \\ & + \mathrm{tr}[\Sigma^{-1}(\Sigma_1 - \Sigma_2)] + \log|\Sigma_1^{-1}\Sigma_2| \end{aligned}$$

and

$$\begin{aligned} H''(t) = {} & -2\delta'\Sigma^{-1}\delta - 2(1 - 2t)\delta'\Sigma^{-1}(\Sigma_1 - \Sigma_2)\Sigma^{-1}\delta \\ & + 2t(1-t)\delta'\Sigma^{-1}(\Sigma_1 - \Sigma_2)\Sigma^{-1}(\Sigma_1 - \Sigma_2)\Sigma^{-1}\delta \\ & - \mathrm{tr}\,\Sigma^{-1}(\Sigma_1 - \Sigma_2)\Sigma^{-1}(\Sigma_1 - \Sigma_2). \end{aligned}$$

Using

$$\Sigma^{-1} = \Sigma^{-1}(t\Sigma_1 + (1-t)\Sigma_2)\Sigma^{-1}$$

in the first term for $H'(t)$ yields (3.5). To obtain (3.6), use the above relation for the second Σ^{-1} as well as

$$\Sigma^{-1} = \Sigma^{-1}(t\Sigma_1 + (1-t)\Sigma_2)\Sigma^{-1}(t\Sigma_1 + (1-t)\Sigma_2)\Sigma^{-1}$$

in the first term in the expression for H'', yielding

$$\begin{aligned} H''(t) = {} & -2\delta'\Sigma^{-1}[t\Sigma_1\Sigma^{-1}\Sigma_2 + (1-t)\Sigma_2\Sigma^{-1}\Sigma_1]\Sigma^{-1}\delta \\ & - \mathrm{tr}[\Sigma^{-1}(\Sigma_1 - \Sigma_2)\Sigma^{-1}(\Sigma_1 - \Sigma_2)]. \end{aligned}$$

Equation (3.6) follows when we recall that $A^{-1} = \Sigma_1\Sigma^{-1}\Sigma_2 = \Sigma_2\Sigma^{-1}\Sigma_1$. From (3.7) we note that A and hence A^{-1} is positive definite. Thus the first term of H'' is negative unless $\delta = 0$. The fact that the second term of H'' is negative unless $\Sigma_1 = \Sigma_2$ can be derived from the concavity of log determinant or more directly by applying a nonsingular linear transformation which simultaneously orthogonalizes Σ_1 and Σ_2. Thus if $\Sigma_1 = R\Lambda_1 R'$ and $\Sigma_2 = R\Lambda_2 R'$ where Λ_1 and Λ_2 are diagonal positive definite matrices, the second term becomes $-\mathrm{tr}[\Lambda^{-1}(\Lambda_1 - \Lambda_2)\Lambda^{-1}(\Lambda_1 - \Lambda_2)]$ where $\Lambda = t\Lambda_1 + (1-t)\Lambda_2$. Hence $H'' < 0$ as long as the two multivariate distributions are distinct.

The algebra of this derivation could have been reduced considerably by relating $R(t)$ of Theorem 1 to the derivative of the first term of $H(t)$ and applying Theorem 1.

The expression (3.4) represents T^2 as the sum of two terms. One may be regarded mainly as a Mahalanobis distance corresponding to a weighted average of Σ_1 and Σ_2 (the weights may be close to those of Theorem 1 but will typically be different). This term essentially measures how "far" apart the means are. The second term is essentially a measure of the information contributed by the differences between the covariance matrices Σ_1 and Σ_2.

Applying the measure T^2 to independent observations on the variables X and Y of the example of Section 2 yields the following table of "separation per unit observation."

	(X_1, X_2)	(X_1, Y_1)	(Y_1, Y_2)
$S^2/2$	0.25	0.20	0.25
$T^2/2$	0.81	0.71	0.81

This indicates that even when the likelihood-ratio test is used, the phenomenon described in Section 2 remains.

4. THE KULLBACK–LEIBLER INFORMATION NUMBERS

Another measure of separation is the Kullback–Leibler information number [7] $I(F_1, F_2)$, defined by

$$I(F_1, F_2) = \int \log \frac{f_1(x)}{f_2(x)} f_1(x)\, dx. \tag{4.1}$$

For k-variate multivariate normal distributions we have

$$I(F_1, F_2) = \tfrac{1}{2}[\delta'\Sigma_2^{-1}\delta + \log|\Sigma_1^{-1}\Sigma_2| - k + \operatorname{tr}(\Sigma_2^{-1}\Sigma_1)]. \tag{4.2}$$

Here I measures the rate at which ε_{2n} approaches zero when the likelihood-ratio test is used and ε_{1n} is kept bounded away from zero and one. That is,

$$\lim_{n\to\infty} n^{-1} \log \varepsilon_{2n} = -I(F_1, F_2). \tag{4.3}$$

Thus $I(F_1, F_2)$ is comparable to $I = T^2/2$ of Section 3 and to $S^2/2$. Since ε_{2n} approaches zero more rapidly when ε_{1n} is bounded away from zero than when ε_{1n} and ε_{2n} approach zero at the same rate, it follows that

$$I(F_1, F_2) \geq T^2/2 \geq S^2/2. \tag{4.4}$$

The Kullback–Leibler numbers are additive in the sense that the information for two independent experiments is the sum of the two informations. For the illustration of Section 2, $I_X(F_1, F_2)$, the information corresponding to X is 0.877 while $I_Y(F_1, F_2)$ is 4.901. From the point of view of having ε_{2n} approach zero most rapidly when ε_{1n} is bounded away from zero, Y is more informative than X and is preferred to X whenever possible and not simply to attain a balance.

This information number also has an interpretation in terms of sequential experimentation. It is a measure of how well one can do using large-scale sequential experiments [4]. More precisely, suppose that the cost per independent observation on X is $c \to 0$, and that the choice between F_1 and F_2 is made using a Bayes sequential procedure. The risk associated with this procedure when F_1 is the true distribution is asymptotically equivalent to $(-c \log c)/I_X(F_1, F_2)$. Thus I determines how good X is for discriminating between F_1 and F_2 sequentially when F_1 is the true distribution. In view of this interpretation, it is not very surprising that one experiment is preferred to another and that there is no premium on mixing experiments when one is moderately sure of which is the correct hypothesis and experimentation is carried out sequentially.

REFERENCES

1. Anderson, T. W. and Bahadur, R. R. (1962). Classification into two multivariate normal distributions with different covariance matrices. *Ann. Math. Statist.* **33** 420–431.
2. Becker, P. (1968). *Recognition of Patterns*. Polyteknisk, Copenhagen.
3. Chernoff, H. (1952). A measure of asymptotic efficiency for tests of a hypothesis based on the sum of observations. *Ann. Math. Statist.* **23** 493–507.
4. Chernoff, H. (1959). Sequential design of experiments. *Ann. Math. Statist.* **30** 755–770.
5. Chernoff, H. (1972). The selection of effective attributes for deciding between hypotheses using linear discriminant functions. *Frontiers of Pattern Recognition*, S. Watanabe (Ed.), Academic Press, New York, pp. 55–60.
6. Clunies-Ross, C. W. and Riffenbaugh, R. H. (1960). Geometry and linear discrimination. *Biometrika* **47** 185–189.
7. Kullback, S. (1959). *Information Theory and Statistics*. Wiley, New York.

Correlation and Affinity in Gaussian Cases

KAMEO MATUSITA
THE INSTITUTE OF STATISTICAL MATHEMATICS,
TOKYO

1. INTRODUCTION

Let $\binom{X_1}{X_2}$ be a two-dimensional random variable with distribution F over R_2, and let Ω be the set of two-dimensional distributions over R_2 each of which is a direct product of one-dimensional distributions over R. Then, X_1 and X_2 are independent of each other when and only when $F \in \Omega$. Therefore, we can consider that when F comes near to Ω, the relation between X_1 and X_2 approaches independence. To express the closeness between distributions we use here the affinity between distributions. The affinity of two distributions is closely related to a distance between distributions and represents their likeness to each other (see, for example, Matusita [3]). Thus, we take up

$$\rho_{\mathrm{I}} = \max_{G \in \Omega} \rho(F, G)$$

to represent the degree of association between X_1 and X_2, where $\rho(F, G)$ denotes the affinity between F and G.[1] In previous work [2, 4] we employed this ρ_{I} to treat the problem of independence. On the other hand, when $F \in \Omega$, F becomes the direct product of the marginal distributions of X_1 and X_2. Therefore, we can also consider

$$\rho_{\mathrm{II}} = \rho(F, F_1 \times F_2)$$

as a measure of association between X_1 and X_2, where F_1 and F_2 are the marginal distributions of X_1 and X_2. These two ρ_{I}, ρ_{II} are not always equal to each other, as will be seen below.

Now, the correlation coefficient also represents an interrelation between two variates. Therefore, there naturally arises the question whether or not

[1] Suppose that the distributions F and G over the space R have density functions $p(x)$, $q(x)$ with respect to a measure m in R, respectively. Then the affinity between F and G is defined by $\rho(F, G) = \int_R [p(x)q(x)]^{1/2}\, dm$. This ρ is related to the distance $d(F, G) = \{\int_R [(p(x))^{1/2} - (q(x))^{1/2}]^2\, dm\}^{1/2}$ by $d^2(F, G) = 2(1 - \rho(F, G))$.

a relation can be found between the correlation coefficient of X_1 and X_2 and the above ρ_{I} or ρ_{II}.

In this paper we shall first give relations between ρ_{I} or ρ_{II} and the correlation coefficient, and then treat canonical correlations from our measures of association between two variates. From this it will be seen that the usual way of determining vectors which define canonical correlations can be interpreted as a maximum distance method concerning distributions.

Throughout the paper only Gaussian distributions will be considered.

2. CORRELATION COEFFICIENT AND $\rho_{\mathrm{I}}, \rho_{\mathrm{II}}$

Let $\binom{X_1}{X_2}$ be a two-dimensional random vector with Gaussian distribution $F = N(a, \Sigma)$, where

$$a = \begin{pmatrix} a_1 \\ a_2 \end{pmatrix}, \qquad \Sigma = \begin{pmatrix} \sigma_{11} & \sigma_{12} \\ \sigma_{21} & \sigma_{22} \end{pmatrix} \qquad (\sigma_{12} = \sigma_{21}).$$

The correlation coefficient of X_1 and X_2 is

$$r = \frac{\sigma_{12}}{(\sigma_{11}\,\sigma_{22})^{1/2}}.$$

Now, let Ω be the set of Gaussian distributions with mean a and covariance matrix

$$\begin{pmatrix} x & 0 \\ 0 & y \end{pmatrix} \qquad (x, y > 0).$$

Let G be a distribution of Ω. Then the affinity between F and G is calculated to be

$$\rho = 2\frac{\left|\begin{pmatrix} \sigma_{11} & \sigma_{12} \\ \sigma_{21} & \sigma_{22} \end{pmatrix}\begin{pmatrix} x & 0 \\ 0 & y \end{pmatrix}\right|^{-1/4}}{\left|\begin{pmatrix} \sigma_{11} & \sigma_{12} \\ \sigma_{21} & \sigma_{22} \end{pmatrix}^{-1} + \begin{pmatrix} x & 0 \\ 0 & y \end{pmatrix}^{-1}\right|^{1/2}}$$

$$= 2\frac{\begin{vmatrix} \sigma_{11} & \sigma_{12} \\ \sigma_{21} & \sigma_{22} \end{vmatrix}^{1/4}\begin{vmatrix} x & 0 \\ 0 & y \end{vmatrix}^{1/4}}{\begin{vmatrix} x + \sigma_{11} & \sigma_{12} \\ \sigma_{21} & y + \sigma_{22} \end{vmatrix}^{1/2}}.$$

(See Matusita [2].) Easy calculations show that this ρ becomes maximum when

$$x = [(\sigma_{11}/\sigma_{22})(\sigma_{11}\sigma_{22} - \sigma_{12}^2)]^{1/2}, \qquad y = [(\sigma_{22}/\sigma_{11})(\sigma_{11}\sigma_{22} - \sigma_{12}^2)]^{1/2},$$

or

$$x = \sigma_{11}[(1 - r^2)]^{1/2}, \qquad y = \sigma_{22}[(1 - r^2)]^{1/2},$$

where r denotes the correlation coefficient $\sigma_{12}/[(\sigma_{11}\sigma_{22})^{1/2}]$. Thus we obtain

$$\rho_1 = \max_{x,y>0} \rho(F, G) = \frac{\sqrt{2}(1-r^2)^{1/4}}{[1+(1-r^2)^{1/2}]^{1/2}}.$$

As to ρ_{II}, we have

$$\rho_{\text{II}} = 2\frac{\left|\begin{pmatrix}\sigma_{11} & \sigma_{12}\\ \sigma_{21} & \sigma_{22}\end{pmatrix}\begin{pmatrix}\sigma_{11} & 0\\ 0 & \sigma_{22}\end{pmatrix}\right|^{-1/4}}{\left|\begin{pmatrix}\sigma_{11} & \sigma_{12}\\ \sigma_{21} & \sigma_{22}\end{pmatrix}^{-1}+\begin{pmatrix}\sigma_{11} & 0\\ 0 & \sigma_{22}\end{pmatrix}^{-1}\right|^{1/2}} = \frac{2(1-r^2)^{1/4}}{(4-r^2)^{1/2}}.$$

From these relations it can immediately be seen that

1. ρ_{I} and ρ_{II} are equal to each other when and only when $r = 0$, that is, X_1 and X_2 are mutually independent;
2. ρ_{I} and ρ_{II} are both monotone-decreasing functions of r^2. When $r = 0$, $\rho_{\text{I}} = \rho_{\text{II}} = 1$, and when $r = \pm 1$, $\rho_{\text{I}} = \rho_{\text{II}} = 0$.

Although we have taken up ρ_{I} or ρ_{II} as a measure of association between X_1 and X_2, we can, of course, consider $1-\rho_{\text{I}}$, or $1-\rho_{\text{II}}$. These quantities are monotone-increasing functions of r^2; when $r = 0$, they vanish, and when $r = \pm 1$, they take on 1. Actually, Khan and Ali [1] proposed $(1-\rho_{\text{II}})^{1/2}$ as a coefficient of association between X_1 and X_2.

Further, it would be interesting to notice that the distribution in Ω which is the closest to F is not

$$N\left(a, \begin{pmatrix}\sigma_{11} & 0\\ 0 & \sigma_{22}\end{pmatrix}\right) \quad \text{but} \quad N\left(a, \begin{pmatrix}\sigma_{11}(1-r^2)^{1/2} & 0\\ 0 & \sigma_{22}(1-r^2)^{1/2}\end{pmatrix}\right).$$

Table I presents some values of r, ρ_{II}, and $-\log_e(1-\rho_{\text{II}})$ (see Matusita and Akaike [4]).

3. CANONICAL CORRELATION AND ρ_{I} OR ρ_{II}

Let $X = \binom{X_1}{X_2}$ be a k-dimensional vector variate having the Gaussian distribution $N(a, \Sigma)$, where X_1 is an s-dimensional subvector of X and X_2 is a t-dimensional subvector of X $(s + t = k)$. In this section we shall consider defining vectors of the canonical correlations between X_1 and X_2 from ρ_{I} or ρ_{II}.

TABLE I

r	ρ_{II}	$-\log_e(1-\rho_{\text{II}})$
0.0	1.	∞
0.1	0.9987	6.6454
0.2	0.9948	5.2591
0.3	0.9879	4.4145
0.4	0.9771	3.7766
0.5	0.9611	3.2468
0.6	0.9376	2.7742
0.7	0.9021	2.3238
0.8	0.8452	1.8656
0.9	0.7393	1.3444
1.0	0.	0.

Let c_1 and c_2 be s-dimensional and t-dimensional vectors, respectively. The problem we shall treat now is to find vectors c_1 and c_2 so that the interrelation between $c_1'X_1$ and $c_2'X_2$ becomes maximum. Here we employ ρ_{I} or ρ_{II}, defined in Section 1, to represent the interrelation between $c_1'X_1$ and $c_2'X_2$.

Let

$$\Sigma = \begin{pmatrix} \Sigma_{11} & \Sigma_{12} \\ \Sigma_{21} & \Sigma_{22} \end{pmatrix},$$

where Σ_{11}, Σ_{12}, Σ_{21}, Σ_{22} are $(s \times s)$, $(s \times t)$, $(t \times s)$, $(t \times t)$ matrices, respectively. Further, let

$$a = \begin{pmatrix} a_1 \\ a_2 \end{pmatrix},$$

where a_1, a_2 are s-dimensional, t-dimensional subvectors of a. Then a_1, a_2, Σ_{11}, Σ_{22} are the means and covariance matrices of X_1 and X_2, respectively. That is, X_1, X_2 are distributed as Gaussian distributions $N(a_1, \Sigma_{11})$, $N(a_2, \Sigma_{22})$. Let

$$Y_1 = c_1'X_1, \qquad Y_2 = c_2'X_2$$

where c_1, c_2 are s-dimensional, t-dimensional vectors. Then Y_1 and Y_2 are distributed as one-dimensional Gaussian distributions $F_1 = N(c_1'a_1, c_1'\,\Sigma_{11}c_1)$, $F_2 = N(c_2'a_2, c_2'\Sigma_{22}\,c_2)$, respectively. The covariance between Y_1 and Y_2 is $c_1'\Sigma_{12}\,c_2(=c_2'\Sigma_{21}c_1)$. On the other hand, $Y = \binom{Y_1}{Y_2}$ is distributed as $F = N(b, B)$ where

$$b = \begin{pmatrix} c_1'a_1 \\ c_2'a_2 \end{pmatrix} \qquad \text{and} \qquad B = \begin{pmatrix} c_1'\Sigma_{11}c_1 & c_1'\Sigma_{12}\,c_2 \\ c_2'\Sigma_{21}c_1 & c_2'\Sigma_{22}\,c_2 \end{pmatrix}.$$

Then our measures of association between Y_1 and Y_2 are

$$\rho_{\mathrm{I}}(Y_1, Y_2) = \frac{\sqrt{2}(1 - r^2)^{1/4}}{[1 + (1 - r^2)^{1/2}]^{1/2}},$$

$$\rho_{\mathrm{II}}(Y_1, Y_2) = \frac{2(1 - r^2)^{1/4}}{(4 - r^2)^{1/2}}$$

where

$$r^2 = \frac{(c_1'\Sigma_{12}c_2)^2}{c_1'\Sigma_{11}c_1 c_2'\Sigma_{22}c_2},$$

the squared correlation coefficient between Y_1 and Y_2.

Now, the maximizing of the interrelation between Y_1 and Y_2 is interpreted as making the relation between Y_1 and Y_2 as far away as possible from the independence relation, which, in turn, means the minimizing of ρ_{I} or ρ_{II}. Therefore, our problem is solved when c_1 and c_2 are found that minimize $\rho_{\mathrm{I}}(Y_1, Y_2)$ or $\rho_{\mathrm{II}}(Y_1, Y_2)$. For that, however, we have only to find c_1, c_2 that maximize r^2, since ρ_{I} (or ρ_{II}) is a monotone-decreasing function of r^2. This, however, is the usual way of finding vectors c_1 and c_2 that define canonical correlations. Therefore, our method gives the same results as the usual way does. Thus the usual way is justified from the standpoint of making the relation between two variates concerned as far away as possible from the independence relation, i.e., minimizing our measures of association ρ_{I} or ρ_{II}, which means the maximizing of the distance $(1 - \rho_{\mathrm{I}})^{1/2}$ or $(1 - \rho_{\mathrm{II}})^{1/2}$.

When a, Σ are not known, we use the sample mean and sample covariance matrix instead of them.

REFERENCES

1. Khan, A. H. and Ali, S. M. A new coefficient of association. *Ann. Inst. Statist. Math.* to be published.
2. Matusita, K. (1966). A distance and related statistics in multivariate analysis. *Multivariate Analysis* (P. R. Krishnaiah, ed.), 187–200. Academic Press, New York.
3. Matusita, K. (1967). On the notion of affinity of several distributions and some of its applications. *Ann. Inst. Statist. Math.* **19** 181–192.
4. Matusita, K. and Akaike, H. (1956). Decision rules based on the distance, for the problems of independence, invariance and two samples. *Ann. Inst. Statist. Math.* **7** 67–80.

Identification of the Structure of Multivariable Stochastic Systems

M. B. PRIESTLEY, T. SUBBA RAO and H. TONG

UNIVERSITY OF MANCHESTER
INSTITUTE OF
SCIENCE AND TECHNOLOGY

1. INTRODUCTION

We consider the problem of identifying the structure of a multivariable linear system given observations on the input and output. Our approach involves the reduction of the dimensions of the input and output vectors, and is based on the application of principal components analysis to the frequency domain representations of the input and output processes, taking into account the "loss function" associated with the control of the system under study. Some results on tests of significance for the eigenvalues of a spectral density matrix are included.

Due to the rapid development of stochastic control theory, there has been, in recent years, a considerable interest in the problem of "identification" of stochastic systems. Basically, this is the problem of constructing a model for a system given only input and output data, and no prior information regarding its physical structure. In the case of single input/single output systems almost all previous studies assume that the system is linear and time invariant, so that, in the language of linear systems theory, the problem reduces to that of estimating the system's "transfer function" and there are now well-established spectral analysis techniques developed for this problem; see, e.g., Priestley [11, 12]. In some situations we may have sufficient prior information to enable us to postulate that the system's transfer function is a rational function (of the frequency variable), with the numerator and denominator polynomials having known orders. In such cases the problem then becomes one of estimating a finite set of parameters, and this approach has been studied extensively by, e.g., Åstrom and Bohlin [3] and Box and Jenkins [4].

Consider now the case of multi-input/multi-output systems. If the system is again assumed to be linear and time invariant, we may characterize its behaviour in terms of a transfer function *matrix*, in which the (i, j)th element describes the relationship between the ith input variable and the jth output variable. In principle, the spectral analysis techniques developed for the univariate (single input/single output) case may be readily generalized to the multivariate case, and each element of the transfer function matrix may be estimated in a fairly straightforward manner—see, e.g., Parzen [9] (see also Section 4). However, in practice this "direct" approach may prove to be quite unwieldy. If nothing is known about the system (other than that it is linear and time invariant), a purely empirical method of model building may involve large numbers of input and output variables. In consequence, the transfer function matrix may be of a very large order, and even when it has been estimated it may give little insight into the basic underlying structure of the system. As an example of this situation we may refer to a recent study aimed at constructing a linear model of a large economic system. The initial investigations of this problem involved 19 input variables and 12 output variables!

It would seem desirable, therefore, to try to reduce the dimensions of both the input and output "vectors" by seeking linear transformations such that the transformed variables, whilst of lower dimensions, nevertheless characterize, as far as possible, the basic "input/output" structure of the system. Posed in this manner, there is an obvious similarity between the objective described above and that underlying the technique of "principal components analysis" as used in classical multivariate analysis. However, it is important to recall that, in the context of system identification problems, the data consist of records of multiple time series in which successive observations will, in general, be highly correlated. In addition, the natural form of the linear transformations (referred to above) will correspond to "filters" involving "lagged" terms of each time series. The key step in overcoming these difficulties is, as noted by Brillinger [5, 6], to transform the whole problem into the *frequency domain*. The "lagged" time-domain relationships then immediately transform into classical "regression" relationships between the "Fourier components" of the various time series. Moreover, we may further exploit the orthogonality of these "Fourier components" at different frequencies, so that our model becomes amenable to the techniques of classical multivariate analysis.

In this paper, we discuss the applications of principal components analysis to the problems of (a) reducing the dimension of the input variables, and (b) reducing the dimension of the output variables, using criteria which arise specifically from the "control" aspects of the problem. We argue that the appropriate criteria for (a) and (b) are quite different; in particular, in case

(b) one takes account of the "loss function" associated with the system under study.

Before discussing these problems further, we first recall two well-known results in classical multivariate analysis.

2. TWO RESULTS IN PRINCIPAL COMPONENTS ANALYSIS

In our study of multivariable control systems we shall make use of the following two results, both of which are discussed in detail by Rao [16]. The presentation given below follows that of Rao's paper.

Preliminary Result (A)

Let $\mathbf{X} = (X_1, X_2, \ldots, X_p)'$ be a column vector representing a zero-mean $[p]$ random variable, and let $\mathbf{\Sigma} = E[\mathbf{XX}']$ denote the variance–covariance matrix of $\mathbf{X}$. Consider the quantity

$$\mathscr{L} = E[\mathbf{X}'\mathbf{QX}],$$

where $\mathbf{Q}$ is a given $p \times p$ symmetric positive definite matrix. (To motivate later applications we will refer to $\mathscr{L}$ as a "*loss function.*") We now wish to choose $q\ (\leq p)$ vectors, $\mathbf{L}_1, \mathbf{L}_2, \ldots, \mathbf{L}_q$, which are orthonormal with respect to $\mathbf{Q}^{-1}$ and are such that if $Y_i = \mathbf{L}_i'\mathbf{X}_i$ $(i = 1, \ldots, q)$, then $\sum_{i=1}^{q} \operatorname{var}(Y_i)$ gives the best approximation to $\mathscr{L}$ over the class of all q linear combinations of $\mathbf{X}$ of the above form. This is equivalent, essentially, to the problem of finding the principal components of $\mathbf{X}$ when $\mathbf{X}$ belongs to an "oblique space" with "metric" $\mathbf{x}'\mathbf{Qx}$. The solution [16] may be shown to be given by choosing

$$\mathbf{L}_i = \mathbf{Q}_i \qquad (i = 1, \ldots, q),$$

where $\mathbf{Q}_1, \mathbf{Q}_2, \ldots, \mathbf{Q}_q$ are the first q eigenvectors of $\mathbf{\Sigma}$ with respect to $\mathbf{Q}^{-1}$, i.e., are the first q vectors associated with the determinantal equation

$$|\mathbf{\Sigma} - \lambda\mathbf{Q}^{-1}| = 0. \tag{2.1}$$

The eigenvectors $\mathbf{Q}_i$ thus satisfy the equations

$$\mathbf{\Sigma Q}_i = \lambda_i \mathbf{Q}^{-1}\mathbf{Q}_i, \qquad i = 1, \ldots, q, \tag{2.2}$$

where $\lambda_1 \geq \lambda_2 \geq \cdots \geq \lambda_q \geq 0$, are the first q eigenroots of Eq. (2.1). The $\mathbf{Q}_i$, as defined above, may be chosen to be orthonormal with respect to $\mathbf{Q}^{-1}$, i.e.,

$$\mathbf{Q}_i'\mathbf{Q}^{-1}\mathbf{Q}_j = 0, \qquad i \neq j,$$

and

$$\mathbf{Q}_i'\mathbf{Q}^{-1}\mathbf{Q}_i = 1 \qquad (\text{all } i).$$

It then follows immediately that

$$E[Y_i Y_j] = \mathbf{Q}_i'\boldsymbol{\Sigma}\mathbf{Q}_j = \lambda_j \mathbf{Q}_i'\mathbf{Q}^{-1}\mathbf{Q}_j = \begin{cases} \lambda_j, & i = j, \\ 0, & i \neq j. \end{cases} \tag{2.3}$$

Thus, the $\{Y_i\}$ are orthogonal variables, with $\operatorname{var}(Y_i) = \lambda_i$ $(i = 1, \ldots, q)$, and the result may be summarized in the form

$$\sum_{i=1}^{q} \operatorname{var}(Y_i) = \sum_{i=1}^{q} \lambda_i \leq \sum_{i=1}^{p} \lambda_i = E[\mathbf{X}'\mathbf{Q}\mathbf{X}] = \mathscr{L} \tag{2.4}$$

(with equality if $q = p$).

Rao [16] suggests that the adequacy of the "fit" of the Y_i may be tested by considering the statistic

$$\Lambda = \frac{\lambda_1 + \lambda_2 + \cdots + \lambda_q}{\lambda_1 + \lambda_2 + \cdots + \lambda_p}. \tag{2.5}$$

It should be pointed out that the problem considered by Rao [16] is, in fact, the "sample version" of the one described above, and corresponds to the situation where one has n independent observations on $\mathbf{X}$, say $\mathbf{X}_1$, $\mathbf{X}_2$, ..., $\mathbf{X}_n$. In place of the "loss function" $\mathscr{L}$, Rao considers the sum of squares of the distances between all pairs of points, which with the metric $\mathbf{x}'\mathbf{Q}\mathbf{x}$, is given by

$$\mathscr{D}_X = \sum_{i=1}^{n} \sum_{j=1}^{n} (\mathbf{X}_i - \mathbf{X}_j)'\mathbf{Q}(\mathbf{X}_i - \mathbf{X}_j). \tag{2.6}$$

The corresponding "distance measure" for the transformed variables

$$\mathbf{Y} = (\mathbf{L}_1'\mathbf{X}, \mathbf{L}_2'\mathbf{X}, \ldots, \mathbf{L}_q'\mathbf{X})'$$

is given by (with $\mathbf{L}_1 \cdots \mathbf{L}_q$ as orthonormal axes of the $[q]$ space)

$$\mathscr{D}_Y = \sum_{i=1}^{n} \sum_{j=1}^{n} (\mathbf{Y}_i - \mathbf{Y}_j)'(\mathbf{Y}_i - \mathbf{Y}_j). \tag{2.7}$$

Now it is easily shown that

$$\mathscr{D}_X = 2n \operatorname{trace}(\hat{\boldsymbol{\Sigma}}^{(X)}\mathbf{Q}), \tag{2.8}$$

$$\mathscr{D}_Y = 2n \operatorname{trace}(\hat{\boldsymbol{\Sigma}}^{(Y)}) \tag{2.9}$$

where $\hat{\boldsymbol{\Sigma}}^{(X)}$, $\hat{\boldsymbol{\Sigma}}^{(Y)}$ are the sample dispersion matrices of $\mathbf{X}_1, \ldots, \mathbf{X}_n; \mathbf{Y}_1, \ldots, \mathbf{Y}_n$, respectively.

Rao [16] then shows that the transformation which makes $\mathscr{D}_Y$ the best approximation to $\mathscr{D}_X$ is given by setting $\mathbf{L}_i \equiv \mathbf{Q}_i$ $(i = 1, \ldots, q)$. The analogy with the "loss function" problem is apparent on noting that

$$\mathscr{L} = E[\mathbf{X}'\mathbf{Q}\mathbf{X}] = \operatorname{trace}(\boldsymbol{\Sigma}\mathbf{Q}). \tag{2.10}$$

Thus, replacing $\hat{\boldsymbol{\Sigma}}$ by $\boldsymbol{\Sigma}$, we obtain

$$\mathscr{D}_X \sim 2n\mathscr{L}, \qquad \mathscr{D}_Y \sim 2n \sum_{i=1}^{q} \text{var}(Y_i).$$

(Equivalently, we may say that $(1/2n^2)\mathscr{D}_X$ is the "sample version" of $\mathscr{L}$.)

Preliminary Result (B)

Let $\mathbf{Z}(m \times 1)$ be a vector of "instrumental" random variables associated with the vector random variable $\mathbf{X}(p \times 1)$. (Note that $\mathbf{Z}$ may include some or all of the elements of $\mathbf{X}$.) We wish to determine q linear combinations of $\mathbf{Z}$, say $Y_i = \mathbf{M}_i'\mathbf{Z}$ $(i = 1, \ldots, q)$, such that the "predictive efficiency" of $\mathbf{Y}$ for $\mathbf{X}$ is a maximum over all suitable choices of the $\mathbf{M}_1, \ldots, \mathbf{M}_q$. Writing $\mathbf{M} = (\mathbf{M}_1, \mathbf{M}_2, \ldots, \mathbf{M}_q)$, $Y = (Y_1, Y_2, \ldots, Y_q)'$, the above transformations may be written as

$$\mathbf{Y} = \mathbf{M}' \cdot \mathbf{Z} \tag{2.11}$$

Now let the variance-covariance matrix of $(\mathbf{X}, \mathbf{Z})$ be denoted by

$$\mathbf{C}_{XZ} = \begin{pmatrix} \boldsymbol{\Sigma} & \boldsymbol{\theta} \\ \boldsymbol{\theta}' & \boldsymbol{\Gamma} \end{pmatrix}, \qquad \text{say.}$$

Then the variance-covariance matrix of $(\mathbf{X}, \mathbf{Y})$ is given by

$$\mathbf{C}_{XY} = \begin{pmatrix} \boldsymbol{\Sigma} & \boldsymbol{\theta}\mathbf{M} \\ \mathbf{M}'\boldsymbol{\theta}' & \mathbf{M}'\boldsymbol{\Gamma}\mathbf{M} \end{pmatrix}, \tag{2.12}$$

and the variance-covariance matrix of the "residual" (after fitting the least-squares regression of $\mathbf{X}$ on $\mathbf{Y}$) is given by

$$\mathbf{C}_R = \boldsymbol{\Sigma} - \boldsymbol{\theta}\mathbf{M}(\mathbf{M}'\boldsymbol{\Gamma}\mathbf{M})^{-1}\mathbf{M}'\boldsymbol{\theta}'. \tag{2.13}$$

Rao [16] suggests two alternative measures of (lack of) predictive efficiency of $\mathbf{Y}$ for $\mathbf{X}$, namely,

$$\text{(a)} \quad \text{trace}\, \mathbf{C}_R \qquad \text{and} \quad \text{(b)} \quad \|\mathbf{C}_R\|,$$

where $\| \ \|$ denotes the Euclidean norm.

If we adopt the measure (a), then we choose $\mathbf{M}$ so as to minimize trace $\mathbf{C}_R$, i.e., to maximize

$$\text{trace}[\boldsymbol{\theta}\mathbf{M}(\mathbf{M}'\boldsymbol{\Gamma}\mathbf{M})^{-1}\mathbf{M}'\boldsymbol{\theta}']. \tag{2.14}$$

Assuming, without loss of generality, that the $\mathbf{M}_i$ are chosen to be orthonormal with respect to $\boldsymbol{\Gamma}$, i.e., that $\mathbf{M}_i'\boldsymbol{\Gamma}\mathbf{M}_j = \delta_{ij}$, it may be shown that (2.14) is maximized by choosing $\mathbf{M}_1, \mathbf{M}_2, \ldots, \mathbf{M}_q$ as the first q eigenvectors of the matrix $(\boldsymbol{\theta}'\boldsymbol{\theta})$ with respect to $\boldsymbol{\Gamma}$. Thus, the $\mathbf{M}_i$ are given by

$$\boldsymbol{\theta}'\boldsymbol{\theta}\mathbf{M}_i = \lambda_i \boldsymbol{\Gamma}\mathbf{M}_i, \qquad i = 1, \ldots, q, \tag{2.15}$$

where $\lambda_1 \geq \lambda_2 \geq, \ldots, \geq \lambda_q$ are the first q eigenroots of the determinantal equation

$$|\boldsymbol{\theta}'\boldsymbol{\theta} - \lambda\boldsymbol{\Gamma}| = 0. \tag{2.16}$$

3. MULTIVARIABLE LINEAR SYSTEMS

A multi-input/multi-output linear system, with an additive "noise disturbance" at the output stage, may be described schematically as in Fig. 1. Here, there are

m inputs	$\{U_1(t), \ldots, U_m(t)\} = \mathbf{U}'(t)$, say,
p "true" outputs	$\{X_1(t), \ldots, X_p(t)\} = \mathbf{X}'(t)$, say,
p "noise" components	$\{\zeta_1(t), \ldots, \zeta_p(t)\} = \boldsymbol{\zeta}'(t)$, say,
p "measured" outputs	$\{Y_1(t), \ldots, Y_p(t)\} = \mathbf{Y}'(t)$, say.

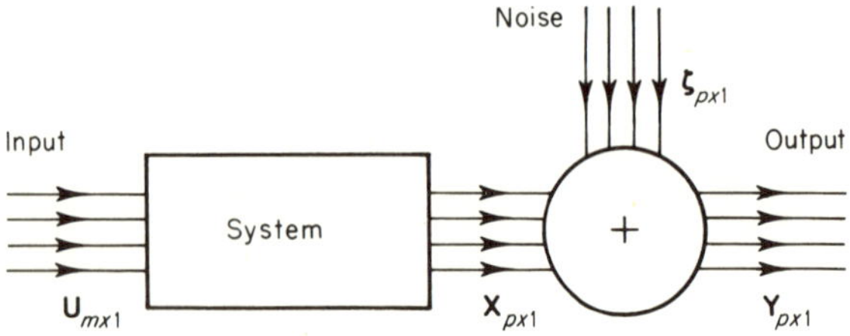

Fig. 1

If the system is linear, time invariant, and physically realizable, then the processes $\mathbf{U}(t)$, $\mathbf{X}(t)$, $\mathbf{Y}(t)$, $\boldsymbol{\zeta}(t)$ satisfy relationships of the form

$$X_i(t) = \sum_{u=0}^{\infty} \{a_{i1}(u)U_1(t-u) + \cdots + a_{im}(u)U_m(t-u)\}$$

or,

$$X_i(t) = \sum_{j=1}^{m} \sum_{u=0}^{\infty} a_{ij}(u)U_j(t-u), \qquad i = 1, \ldots, p, \tag{3.1}$$

and

$$Y_i(t) = X_i(t) + \zeta_i(t), \qquad i = 1, \ldots, p. \tag{3.2}$$

Assuming now that all the processes are zero-mean and weakly stationary, and that the time parameter t is discrete (say, $t = 0, \pm 1, \pm 2, \ldots$), we may introduce the following spectral representations:

$$U_k(t) = \int_{-\pi}^{\pi} e^{i\omega t}\, dZ_U^{(k)}(\omega), \qquad k = 1, \ldots, m, \tag{3.3}$$

$$X_k(t) = \int_{-\pi}^{\pi} e^{i\omega t}\, dZ_X^{(k)}(\omega), \qquad k = 1, \ldots, p, \tag{3.4}$$

$$\zeta_k(t) = \int_{-\pi}^{\pi} e^{i\omega t}\, dZ_\zeta^{(k)}(\omega), \qquad k = 1, \ldots, p, \tag{3.5}$$

$$Y_k(t) = \int_{-\pi}^{\pi} e^{i\omega t}\, dZ_Y^{(k)}(\omega), \qquad k = 1, \ldots, p. \tag{3.6}$$

Recall that, in each of the above representations, $Z(\omega)$ is a process with orthogonal increments, and we have, e.g.,

$$E[|dZ_U^{(k)}(\omega)|^2] = f_{U_k U_k}(\omega)\, d\omega, \qquad \text{each } k, \tag{3.7}$$

$$E[dZ_U^{(k)}(\omega)\, dZ_U^{*(l)}(\omega)] = f_{U_k U_l}(\omega)\, d\omega, \qquad \text{each } k, l, \tag{3.8}$$

where $f_{U_k U_k}(\omega)$ is the *spectral density function* of $U_k(t)$, and $f_{U_k U_l}(\omega)$ is the *cross-spectral density function* between $U_k(t)$ and $U_l(t)$, with similar results for each of the other processes. (We assume, of course, that the various density functions exist for all ω.) The *spectral density matrix* of, e.g., the process $\mathbf{U}(t)$ is then defined by

$$\mathbf{F}_{UU}(\omega) = \{f_{U_k U_l}(\omega)\} \qquad (k, l = 1, \ldots, m), \tag{3.9}$$

and the *cross-spectral density matrix* between, e.g., $\mathbf{Y}(t)$ and $\mathbf{U}(t)$, is defined by

$$\mathbf{F}_{YU}(\omega) = \{f_{Y_k U_l}(\omega)\} \qquad (k = 1, \ldots, p, \quad l = 1, \ldots, m). \tag{3.10}$$

Using the spectral representations (3.3)–(3.6), we may rewrite Eqs. (3.1), (3.2) in the *frequency domain form* as follows:

$$dZ_X^{(i)}(\omega) = A_{i1}(\omega)\, dZ_U^{(1)}(\omega) + \cdots + A_{im}(\omega)\, dZ_U^{(m)}(\omega), \qquad i = 1, \ldots, p, \tag{3.11}$$

where, for each j, k ($j = 1, \ldots, p$, $k = 1, \ldots, m$),

$$A_{jk}(\omega) = \sum_{u=0}^{\infty} a_{jk}(u) e^{-iu\omega}. \tag{3.12}$$

$A_{jk}(\omega)$ is called the *transfer function* between the kth input, $U_k(t)$, and the jth output, $X_j(t)$. The $(p \times m)$ matrix $\mathbf{A}(\omega)$, defined by

$$\mathbf{A}(\omega) = \{A_{jk}(\omega)\}, \tag{3.13}$$

is called the *transfer function matrix* of the complete system.

Writing $d\mathbf{Z}_U(\omega) = \{dZ_U^{(1)}(\omega), \ldots, dZ_U^{(m)}(\omega)\}'$ (with a similar notation for the other processes), we may write (3.11) and (3.2) in the more concise forms

$$d\mathbf{Z}_X(\omega) = \mathbf{A}(\omega)\, d\mathbf{Z}_U(\omega), \tag{3.14}$$

$$d\mathbf{Z}_Y(\omega) = d\mathbf{Z}_X(\omega) + d\mathbf{Z}_\zeta(\omega), \tag{3.15}$$

or

$$d\mathbf{Z}_Y(\omega) = \mathbf{A}(\omega)\, d\mathbf{Z}_U(\omega) + d\mathbf{Z}_\zeta(\omega). \tag{3.16}$$

If we assume now that the "noise disturbance," $\zeta(t)$, is *uncorrelated* with the input process, $\mathbf{U}(t)$, i.e., $E[\zeta_i(s)U_j(t)] = 0$, all i, j, s, t, then it follows that $d\mathbf{Z}_\zeta(\omega)$ is uncorrelated with $d\mathbf{Z}_U(\omega')$ (all ω, ω'). (This is a reasonable assumption in the case of "open-loop" systems—see Priestley [11].) In this case, it follows immediately from (3.16) that (using definitions of the form (3.7), (3.8)),

$$\mathbf{F}_{YU}(\omega) = \mathbf{A}(\omega)\mathbf{F}_{UU}(\omega). \tag{3.17}$$

Equation (3.17) is basic in the frequency domain approach to the identification of linear systems. If the spectral matrices $\mathbf{F}_{UU}(\omega)$, $\mathbf{F}_{YU}(\omega)$ are known, a priori, then $\mathbf{A}(\omega)$ may be determined explicitly from

$$\mathbf{A}(\omega) = \mathbf{F}_{YU}(\omega)\mathbf{F}_{UU}^{-1}(\omega). \tag{3.18}$$

4. IDENTIFICATION OF THE SYSTEM STRUCTURE

In practice, the transfer function matrix, $\mathbf{A}(\omega)$, is unknown, as are the spectral matrices $\mathbf{F}_{YU}(\omega)$, $\mathbf{F}_{UU}(\omega)$, and we have to infer the structure of the system from sample records of observations on $\mathbf{U}(t)$ and $\mathbf{Y}(t)$, say from $t = 1, \ldots, T$. We may, of course estimate each of the elements of $\mathbf{F}_{YU}(\omega)$, $\mathbf{F}_{UU}(\omega)$ from the sample records, using the standard "spectral window" technique; see, e.g., Parzen [9]. We would then obtain estimates $\hat{\mathbf{F}}_{YU}(\omega)$, $\hat{\mathbf{F}}_{UU}(\omega)$, of $\mathbf{F}_{YU}(\omega)$, $\mathbf{F}_{UU}(\omega)$, respectively, and a natural estimate of $\mathbf{A}(\omega)$ would be given by

$$\hat{\mathbf{A}}(\omega) = \hat{\mathbf{F}}_{YU}(\omega)\hat{\mathbf{F}}_{UU}^{-1}(\omega). \tag{4.1}$$

However, as explained previously, such a "direct" approach may prove unwieldy. In many cases the total set of input variables may be "too large" in the sense that the output variables could be "explained" in terms of linear relationships on just a few linear combinations of the input variables. Similarly, the set of output variables may be "too large" in the sense that the "control" of the system to some set of desired operating conditions may be achieved by concentrating attention on just a few linear combinations of the p output variables.

Thus, given no prior information on the transfer function matrix, $\mathbf{A}(\omega)$, we may consider two stages of the reduction of variables:

(a) reduction of the dimension of the output vector, and
(b) reduction of the dimension of the input vector.

Once we have reduced the dimensions of the input and output vectors, we may then estimate the (transformed) transfer function matrix, *which will now have reduced dimensions*, by applying the technique underlying Eq. (4.1) to the transformed input and output variables.

We now consider these two stages in more detail.

5. REDUCTION OF THE DIMENSION OF THE OUTPUT VECTOR

As noted previously, the criteria which we propose for reducing the dimensions of the input and output vectors are quite different. In the context of control systems, the objective is usually to control the output so that it follows, as closely as possible, some prescribed form. Suppose (without loss of generality) that the objective is to control the output vector so that it remains as close as possible to *zero* (the "set point"). We then introduce a "loss function," $\mathscr{L}$, which measures the cost of deviations of the output vector from its "set point." Suppose now that we have a *quadratic loss function*, of the form,

$$\mathscr{L} = E[\mathbf{Y}'(t)\mathbf{Q}\mathbf{Y}(t)] \tag{5.1}$$

where $\mathbf{Q} = \{q_{jk}\}$ is a given (symmetric) positive definite matrix. (Note that if $\mathbf{Y}(t)$ is weakly stationary, then $\mathscr{L}$ is independent of t.) If we wish to reduce the dimension of $\mathbf{Y}(t)$ by transforming to a new output variable, $\mathbf{W}(t)$, it is reasonable, therefore, to demand that the transformed variable, $\mathbf{W}(t)$, should characterize, as far as possible, the important features of the systems output which we wish to control. Thus, our objective is to find q ($\leq p$) linear combinations of the $\{Y_i(t)\}$ which are orthogonal, and whose total variances approximate as closely as possible the loss function $\mathscr{L}$.

Now, rewriting (5.1) in terms of its frequency domain representation, we have

$$\begin{aligned}
\mathscr{L} &= E[\mathbf{Y}'(t)\mathbf{Q}\mathbf{Y}(t)] \\
&= \sum_j \sum_k q_{jk} E[Y_j(t) Y_k^*(t)] \qquad \text{(treating } \mathbf{Y}(t) \text{ as a complex-valued variable)} \\
&= \sum_j \sum_k q_{jk} \int_{-\pi}^{\pi} \int_{-\pi}^{\pi} e^{i\omega t} e^{-i\omega' t} E[dZ_Y^{(j)}(\omega)\, dZ_Y^{*(k)}(\omega')] \\
&= \sum_j \sum_k q_{jk} \int_{-\pi}^{\pi} E[dZ_Y^{(j)}(\omega)\, dZ_Y^{*(k)}(\omega)]
\end{aligned}$$

(remembering that the $dZ_Y^{(i)}(\omega)$ are cross-orthogonal over ω). Hence, we may write

$$\mathcal{L} = \int_{-\pi}^{\pi} E[d\mathbf{Z}_Y^*(\omega)\mathbf{Q}\, d\mathbf{Z}_Y(\omega)]. \tag{5.2}$$

We may now consider the integrand of the right-hand side of (5.2) *separately*, for each ω. Thus, we write

$$\mathcal{L} = \int_{-\pi}^{\pi} \mathcal{L}(\omega)\, d\omega \tag{5.3}$$

where

$$\mathcal{L}(\omega) = E[d\mathbf{Z}_Y^*(\omega)\mathbf{Q}(d\omega^{-1})\, d\mathbf{Z}_Y(\omega)]. \tag{5.4}$$

Now consider q linear combinations of the $\{Y_i(t)\}$, say, $W_1(t), \ldots, W_q(t)$, where

$$W_i(t) = \sum_{j=1}^{p} \sum_{u=-\infty}^{\infty} p_{ij}(u) Y_j(t-u), \qquad i = 1, \ldots, q. \tag{5.5}$$

In frequency domain terms (5.5) may be written (using an obvious notation) as

$$dZ_W^{(i)}(\omega) = \mathbf{P}_i(\omega)\, d\mathbf{Z}_Y(\omega), \qquad i = 1, \ldots, q, \tag{5.6}$$

where

$$\mathbf{P}_i'(\omega) = \{p_{i1}(\omega), \ldots, p_{ip}(\omega)\}$$

and

$$p_{jk}(\omega) = \sum_{u=-\infty}^{\infty} p_{jk}(u)e^{-iu\omega}, \qquad j = 1, \ldots, q; \quad k = 1, \ldots, p. \tag{5.7}$$

Writing $\mathbf{P}(\omega) = \{P_1(\omega), \ldots, \mathbf{P}_q(\omega)\}$, Eq. (5.6) becomes

$$d\mathbf{Z}_W(\omega) = \mathbf{P}'(\omega) d\mathbf{Z}_Y(\omega). \tag{5.8}$$

Recall now that we wish to choose the $\{p_{ij}(u)\}$ so that $\sum_{i=1}^{q} \text{var}\{W_i(t)\}$ gives the best approximation (over the class of all q linear orthogonal transformations of $Y_i(t)$) to $\mathcal{L} = E[\mathbf{Y}'(t)\mathbf{Q}\,Y(t)]$. But, we have

$$\text{var}\{W_i(t)\} = \int_{-\pi}^{\pi} E[|dZ_W^{(i)}(\omega)|^2], \qquad i = 1, \ldots, q.$$

Hence, the problem may be formulated equivalently in the form: *for each* ω, choose q vectors $\mathbf{P}_1(\omega), \ldots, \mathbf{P}_q(\omega)$ so that $\sum_{i=1}^{q} \text{var}\{dZ_W^{(i)}(\omega)\}$ gives the best approximation to $\mathcal{L}(\omega) = E[d\mathbf{Z}_Y^{*\prime}(\omega)\mathbf{Q}(d\omega^{-1})\, d\mathbf{Z}_Y(\omega)]$.

Using "preliminary result (A)," we find that, for each ω, the optimal choice of $\mathbf{P}_i(\omega)$ is the ith eigenvector of Σ, the variance-covariance matrix of

$d\mathbf{Z}_Y(\omega)$, with respect to $[\mathbf{Q}(d\omega^{-1})]^{-1} = \mathbf{Q}^{-1}d\omega$. But $E[dZ_Y^{(i)}(\omega)\, dZ_Y^{*(j)}(\omega)] = f_{Y_iY_j}(\omega)\, d\omega$, hence,

$$\mathbf{\Sigma} \equiv \mathbf{F}_{YY}(\omega)\, d\omega, \tag{5.9}$$

where $\mathbf{F}_{YY}(\omega)$ is the spectral density matrix of $\mathbf{Y}(t)$. Thus, $\mathbf{P}_i(\omega)$ *is the ith eigenvector of* $\mathbf{F}_{YY}(\omega)$ *with respect to* $\mathbf{Q}^{-1}$, corresponding to the determinantal equation

$$|\mathbf{F}_{YY}(\omega) - \lambda_i \mathbf{Q}^{-1}| = 0. \tag{5.10}$$

Once $\mathbf{P}_i(\omega)$ is determined from (5.10), the corresponding time-domain transformation, corresponding to the $\{p_{jk}(u)\}$, is given by inverting (5.7), yielding

$$p_{jk}(u) = \frac{1}{2\pi}\int_{-\pi}^{\pi} p_{jk}(\omega)e^{i\omega u}\, d\omega. \tag{5.11}$$

6. REDUCTION OF THE DIMENSION OF THE INPUT VECTOR

Consider now the problem of reducing the dimension of the input vector, $\mathbf{U}(t)$. Since we are trying to describe the behaviour of the system in terms of linear relationships, it seems reasonable to seek r $(\leq m)$ linear combinations of $U_1(t), \ldots, U_m(t)$, say, $V_1(t), \ldots, V_r(t)$, so that the linear predictive efficiency of $V_1(t), \ldots, V_r(t)$ for the output vector is a maximum. However, at this stage we have already removed "redundant" elements from the original output $\mathbf{Y}(t)$, so that it seems appropriate now to consider the predictive efficiency of $\mathbf{V}(t)$ for the transformed output, for $\mathbf{W}(t)$.

Let

$$V_j(t) = \sum_{k=1}^{m} \sum_{u=-\infty}^{\infty} m_{jk}(u)U_k(t-u), \qquad j = 1, \ldots, r. \tag{6.1}$$

or, in frequency domain terms (with an obvious notation),

$$dZ_V^{(j)}(\omega) = \mathbf{M}_j'(\omega)\, d\mathbf{Z}_U(\omega), \qquad j = 1, \ldots, r, \tag{6.2}$$

where

$$\mathbf{M}_j'(\omega) = \{m_{j1}(\omega), \ldots, m_{jm}(\omega)\}$$

and

$$m_{jk}(\omega) = \sum_{u=-\infty}^{\infty} m_{jk}(u)e^{-i\omega u}. \tag{6.3}$$

Writing $\mathbf{M}(\omega) = \{\mathbf{M}_1(\omega), \ldots, \mathbf{M}_r(\omega)\}$ (6.2) may be written as

$$d\mathbf{Z}_V(\omega) = \mathbf{M}'(\omega)\, d\mathbf{Z}_U(\omega). \tag{6.4}$$

Now consider a linear "regression" of $W_j(t)$ on $V_1(t), \ldots, V_r(t)$ of the form

$$W_j(t) = \sum_{k=1}^{r} \sum_{u=-\infty}^{\infty} b_{jk}(u) V_k(t-u) + \text{residual}. \tag{6.5}$$

The mean square error is given by

$$M = E\left[W_j(t) - \sum_{k=1}^{r} \sum_{u=-\infty}^{\infty} b_{jk}(u) V_k(t-u)\right]^2 \tag{6.6}$$

$$= E\left[\int_{-\pi}^{\pi} e^{i\omega t}\left\{dZ_W^{(j)}(\omega) - \sum_{k=1}^{r} B_{jk}(\omega)\, dZ_V^{(k)}(\omega)\right\}\right]^2, \tag{6.7}$$

where

$$B_{jk}(\omega) = \sum_u b_{jk}(u) e^{-iu\omega}.$$

Thus,

$$M = \int_{-\pi}^{\pi} E\left[\left| dZ_W^{(j)}(\omega) - \sum_{k=1}^{r} B_{jk}(\omega)\, dZ_V^{(k)}(\omega)\right|^2\right]. \tag{6.8}$$

Hence, the problem of choosing $\mathbf{V}(t)$ in terms of linear predictive efficiency for $\mathbf{W}(t)$ is equivalent to the problem of choosing $d\mathbf{Z}_V(\omega)$ in terms of linear predictive efficiency for $d\mathbf{Z}_W(\omega)$. Returning to Eq. (6.4), and using "preliminary result (B)" we see that, *for each* ω, *the optimal choice of* $\mathbf{M}_j(\omega)$ *is the jth eigenvector of* $\boldsymbol{\theta}'(\omega)\boldsymbol{\theta}(\omega)$ *with respect to* $\boldsymbol{\Gamma}(\omega)$, where $\boldsymbol{\theta}(\omega)$ is the covariance matrix between $d\mathbf{Z}_W(\omega)$ and $d\mathbf{Z}_U(\omega)$, and $\boldsymbol{\Gamma}(\omega)$ is the variance-covariance matrix of $d\mathbf{Z}_U(\omega)$. Thus,

$$\begin{aligned}\boldsymbol{\theta}(\omega) &= E[d\mathbf{Z}_W(\omega)\, d\mathbf{Z}_U^{*\prime}(\omega)] \\ &= \mathbf{P}'(\omega) E[d\mathbf{Z}_Y(\omega)\, d\mathbf{Z}_U^{*\prime}(\omega)] \qquad \text{using (5.8)}, \\ &= \mathbf{P}'(\omega)\mathbf{F}_{YU}(\omega)\, d\omega, \end{aligned} \tag{6.9}$$

and

$$\boldsymbol{\Gamma}(\omega) = E[d\mathbf{Z}_U(\omega)\, d\mathbf{Z}_U^{*\prime}(\omega)] = \mathbf{F}_{UU}(\omega)\, d\omega. \tag{6.10}$$

Hence, $\mathbf{M}_j(\omega)$ is the jth eigenvector of

$$\mathbf{F}'_{YU}(\omega)\mathbf{P}(\omega)\mathbf{P}'(\omega)\mathbf{F}_{YU}(\omega) \qquad \text{w.r. to} \quad \mathbf{F}_{UU}(\omega),$$

corresponding to the determinantal equation

$$|\mathbf{F}'_{YU}(\omega)\mathbf{P}(\omega)\mathbf{P}'(\omega)\mathbf{F}_{YU}(\omega) - \lambda_j \mathbf{F}_{UU}(\omega)| = 0. \tag{6.11}$$

Once the $\{\mathbf{M}_j(\omega)\}$ are determined from (6.11), the corresponding time-domain transformations $\{m_{jk}(u)\}$ are found by inverting (6.3), yielding

$$m_{jk}(u) = (1/2\pi) \int_{-\pi}^{\pi} m_{jk}(\omega) e^{iu\omega}\, d\omega \tag{6.12}$$

7. PRACTICAL PROBLEMS

It should be noted that, in applying the techniques discussed in Sections 6 and 7 to practical problems, the following points arise.

(a) So far, we have allowed the transformations of both the output and input variables to be "two-sided," in the sense that in (5.5) the summation over u in the right-hand side extends from $-\infty$ to ∞. In other words, we have allowed $\mathbf{W}(t)$ to depend on both past and *future* values of $\mathbf{Y}(t)$. In practice, it may be desirable to restrict $\mathbf{W}(t)$ to be a function of past values only of $\mathbf{Y}(t)$—with a similar restriction for the relationship between $\mathbf{V}(t)$ and $\mathbf{U}(t)$. Under these restrictions we are led to considering "constrained" optimisation problems of the type familiar in the Wiener–Kolmogorov prediction theory since, in effect, we are now restricting the $\{\mathbf{P}_i(\omega)\}$ and $\{\mathbf{M}_j(\omega)\}$ to belong to the class of "backward transforms." The solution of these constrained problems may possibly involve Wiener–Hopf type analysis.

(b) In practice, the spectral matrices $\mathbf{F}_{YY}(\omega)$, $\mathbf{F}_{YU}(\omega)$, $\mathbf{F}_{UU}(\omega)$ will, in general, be unknown, and will have to be estimated numerically from observations on $\mathbf{U}(t)$ and $\mathbf{Y}(t)$. Although, as noted previously, there are standard methods available for estimating these matrices, they can be evaluated only at a *discrete* set of frequencies, say $\omega_1, \omega_2, \ldots, \omega_K$. In turn, this means that the vectors $\{\mathbf{P}_i(\omega)\}$, $\{\mathbf{M}_j(\omega)\}$, can be evaluated only at the same discrete set of frequencies, and the corresponding time-domain transformations, $\{p_{jk}(u)\}$, $\{m_{jk}(u)\}$ will have to be determined by computing the "discrete" Fourier transforms of $\hat{\mathbf{P}}_i(\omega)$, $\hat{\mathbf{M}}_j(\omega)$ rather than using (5.11), (6.12) directly.

Both these points require further consideration, and we hope to investigate them with the aid of some numerical studies.

8. TESTS OF SIGNIFICANCE OF EIGENVALUES

So far, we have not discussed the problem of choosing appropriate values for q (the reduced dimension of the output vector) and r (the reduced dimension of the input vector). However, it was noted in Section 2 that, in the context of classical multivariate analysis, the analogous problem could be studied via the statistic (2.5). As a first step toward investigating this problem further, we present in this and the following sections some general results concerning the eigenvalues of an estimated spectral density matrix.

First we note that, in Section 5, we were concerned with the eigenvalues of a spectral matrix $\mathbf{F}$ with respect to a matrix $\mathbf{Q}^{-1}$. However, since the matrix $\mathbf{Q}$ is real positive definite, we may write $\mathbf{Q}^{-1} = \mathbf{U}^2$ for some real symmetric positive definite $\mathbf{U}$. The roots of the equation $|\hat{\mathbf{F}} - \lambda\mathbf{Q}^{-1}| = 0$ are then identical to those of $|\mathbf{U}^{-1}\hat{\mathbf{F}}\mathbf{U}^{-1} - \lambda\mathbf{I}| = 0$. It follows that if $\hat{\mathbf{F}} = \sum_\alpha \mathbf{Z}_\alpha \mathbf{Z}_\alpha'^*$, say ($\mathbf{Z}_\alpha$'s being independent), is distributed as complex Wishart, then $\mathbf{U}^{-1}\hat{\mathbf{F}}\mathbf{U}^{-1} = \sum (\mathbf{U}^{-1}\mathbf{Z}_\alpha)(\mathbf{U}^{-1}\mathbf{Z}_\alpha)'^{*}$ is also distributed as complex Wishart because the independence of $\{\mathbf{Z}_\alpha\}$ implies that of the $\{\mathbf{U}^{-1}\mathbf{Z}_\alpha\}$. Hence, without loss of generality, we may study the distribution of the eigenvalues of a complex Wishart matrix.

Let $\mathbf{X}'(t) = (X_1(t), X_2(t), \ldots, X_p(t))$ denote a general p-dimensional zero-mean, real, stationary Gaussian time series, with spectral density matrix $\mathbf{F}(\omega)$. Let $\lambda_1(\omega) > \lambda_2(\omega) > \cdots > \lambda_p(\omega) > 0$ be the roots of the equation $|\mathbf{F}(\omega) - \lambda(\omega)\mathbf{I}| = 0$. Suppose $\mathbf{X}(t)$ $(t = 1, 2, \ldots, T)$ is a sample of size T. Define the *cross periodogram* by

$$I_{jk}(\theta_m) = Z_j(\theta_m)Z_k^*(\theta_m)$$

where

$$Z_j(\theta_m) = (2\pi T)^{-1/2} \sum_{t=1}^{T} X_j(t)e^{i\theta_m t},$$

and

$$\theta_m = \frac{2\pi m}{T}, \qquad m = 0, 1, 2, \ldots, \left[\frac{T}{2}\right].$$

A well-known form of estimate of $f_{jk}(\omega)$ is obtained by averaging the cross periodogram ordinates in the neighbourhood of the point ω, giving

$$\hat{f}_{jk}(\omega) = (2n+1)^{-1} \sum_{m=-n}^{n} I_{jk}(\omega + \theta_m)$$

(where n is a "bandwidth" parameter). The estimated spectral matrix is then

$$\hat{\mathbf{F}}(\omega) = \{\hat{f}_{jk}(\omega)\} = (2n+1)^{-1}\mathbf{A}(\omega)$$

where $\mathbf{A}(\omega) = \{A_{jk}(\omega)\}$ with $A_{jk}(\omega) = \sum_{m=-n}^{n} I_{jk}(\omega + \theta_m)$. Alternatively, we can write

$$\hat{\mathbf{F}}(\omega) = \frac{1}{n'} \sum_{m=1}^{n} \mathbf{Z}(\omega' + \theta_m)\mathbf{Z}^{*\prime}(\omega' + \theta_m)$$

where $n' = 2n + 1$, $\omega' = \omega - [2\omega(n+1)]/T$ and

$$\mathbf{Z}'(\theta) = (Z_1(\theta), Z_2(\theta), \ldots, Z_p(\theta)).$$

The joint distribution of the "distinct" elements of $\mathbf{A}(\omega)$ is known as the complex Wishart distribution [7] and is denoted by $W_p^{\,c}(\mathbf{F}(\omega), n)$. Let $l_1(\omega)$, $l_2(\omega), \ldots, l_p(\omega)$ be the roots of the equation

$$|\hat{\mathbf{F}}(\omega) - l(\omega)\mathbf{I}| = 0$$

and assume $l_1(\omega) > l_2(\omega) > \cdots > l_p(\omega) > 0$ for all ω. The roots $l_1(\omega), \ldots, l_p(\omega)$ are thus the sample estimates of $\lambda_1(\omega), \ldots, \lambda_p(\omega)$, respectively.

Consider a test of the hypothesis $H_1: \lambda_1(\omega) = \lambda_2(\omega) = \cdots = \lambda_p(\omega) = \lambda(\omega)$ (where $\lambda(\omega)$ is unknown). It is well known that testing the equality of all the eigenvalues of $\mathbf{F}(\omega)$ is equivalent to testing that $\mathbf{F}(\omega)$ is a diagonal matrix with all its diagonal elements equal. This test is known as a "sphericity test" [1]. We now briefly present the extension of this test in the case of a complex Wishart distribution (see Priestley *et al.* [13, 14] for further details and also Pillai and Nagarsenki [10]). The likelihood ratio criterion L for testing H_1 is

$$L = \left[\frac{\prod_{i=1}^{p} l_i^{1/p}}{\sum (l_i/p)}\right]^{pn'} \tag{8.1}$$

and the rth moment of L is

$$E(L^r) = p^{pn'r} \frac{\Gamma(n'p)}{\prod_{i=1}^{p}\Gamma(n'+1-i)} \frac{\prod_{i=1}^{p}\Gamma[n'(1+r)+(1-i)]}{\Gamma(n'p(1+r))}.$$

We can now make use of the approximation suggested by Box [1, Th. 8.6.1] to obtain the following asymptotic distribution of L;

$$\Pr\{-2\rho \log_e L \le z\} \doteqdot \Pr\{\chi_f{}^2 \le z\}$$

where

$$\rho = 1 - \frac{1+2p^2}{6n'p} \qquad \text{and} \qquad f = p^2 - 1.$$

In other words, under H_1 the statistic

$$2\rho n'\left\{-\log_e|\mathbf{A}| + p \log_e\left(\frac{\operatorname{tr}\mathbf{A}}{p}\right)\right\}$$

is, for large n', approximately distributed as a central χ^2 with f degrees of freedom.

To test the equality of all eigenvalues at several frequencies $(\omega_1, \omega_2, \ldots, \omega_M)$, say (spaced sufficiently wide apart), we consider the criterion

$$\bar{L} = \prod_{l=1}^{M} L(\omega_l).$$

where $L(\omega_l)$ is given by (8.1). It can be shown that, under certain assumptions about the frequencies $(\omega_1, \omega_2, \ldots, \omega_M)$, the set $\{\mathbf{A}(\omega_1), \mathbf{A}(\omega_2), \ldots, \mathbf{A}(\omega_M)\}$ is approximately a set of independent complex Wishart matrices (for details refer to Wahba [17, 18]). Since $\bar{L}$ is the product of independent likelihood ratio criteria, we can use a χ^2 approximation for $-2 \log_e \bar{L}$. By equating the first two moments, we can show that $-2\rho \log \bar{L}$ is approximately distributed as a χ^2 with Mf degrees of freedom.

9. TESTING THE EQUALITY OF $\lambda_{k+1}(\omega) = \lambda_{k+2}(\omega) = \cdots = \lambda_p(\omega) = \lambda(\omega)$

Consider a test of the hypothesis H_2: $\lambda_{k+1}(\omega) = \cdots = \lambda_p(\omega) = \lambda(\omega)$, where $\lambda(\omega)$ is unknown. Without loss of generality we assume $\mathbf{F}(\omega)$ is a diagonal matrix. Following Lawley [8], we partition $\mathbf{F}$, $\hat{\mathbf{F}}$ (suppressing ω), as follows:

$$\mathbf{F} = \begin{pmatrix} \Lambda & 0 \\ 0 & \lambda I \end{pmatrix}, \qquad \hat{\mathbf{F}} = \begin{pmatrix} \hat{\mathbf{F}}_{11} & \hat{\mathbf{F}}_{12} \\ \hat{\mathbf{F}}_{21} & \hat{\mathbf{F}}_{22} \end{pmatrix}, \qquad \Lambda = \begin{pmatrix} \lambda_1 & 0 & 0 \\ 0 & \lambda_2 & 0 \\ & \cdots & \\ 0 & & \lambda_k \end{pmatrix}.$$

Define $\delta\hat{F} = \hat{\mathbf{F}} - \mathbf{F}$. $\delta\hat{f}_{ij} = \hat{f}_{ij} - f_{ij}$, $\delta l_i = l_i - \lambda_i$. It can be shown that (for details refer to Priestley *et al.* [13])

$$l_r = \lambda_r + \delta\hat{f}_{rr} - \delta\hat{f}_{rr} \sum_l{}' \frac{|\delta\hat{f}_{rl}|^2}{(\lambda_r - \lambda_l)^2}$$
$$+ \sum_l{}' \frac{|\delta\hat{f}_{rl}|^2}{(\lambda_r - \lambda_l)} - \sum_l{}' \sum_m{}' \frac{\delta\hat{f}_{rl}\,\delta\hat{f}_{lm}\,\delta\hat{f}_{mr}}{(\lambda_r - \lambda_l)(\lambda_m - \lambda_r)} + O(\delta^4) \qquad (9.1)$$

where $\sum'$ indicates that the summation is to be over all values from 1 to k except r. To construct a test of H_2, it can be shown that, under H_2,

$$2\rho' n' \left\{ -\log_e |\hat{\mathbf{F}}_{22}| + q \log_e \left(\frac{\operatorname{tr} \hat{\mathbf{F}}_{22}}{q} \right) \right\}$$

is approximately distributed as a χ^2 with f' degrees of freedom where

$$f' = q^2 - 1, \qquad \rho' = 1 - \frac{1 + 2q^2}{6n'q}, \qquad q = p - k.$$

From (9.1) we can obtain (see also Brillinger [6])

$$E(l_r(\omega_k)) = \lambda_r(\omega_k) + \frac{1}{n'} \sum_l{}' \frac{\lambda_r(\omega_k)\lambda_l(\omega_k)}{(\lambda_r(\omega_k) - \lambda_l(\omega_k))} + O\left(\frac{1}{n'^2}\right),$$

$$E(l_r(\omega_k) - \lambda_r(\omega_k))^2 = \frac{\lambda_r^2(\omega_k)}{n'} + O\left(\frac{1}{n'^2}\right),$$

and

$$\operatorname{cov}(l_r(\omega_k), l_s(\omega_k)) = O\left(\frac{1}{n'^2}\right).$$

10. ASYMPTOTIC THEORY FOR THE DISTRIBUTION OF EIGENVALUES

Let us denote the eigenvectors corresponding to the eigenvalues $\lambda_1(\omega), \ldots, \lambda_p(\omega)$ of $\mathbf{F}(\omega)$ by $\gamma_1(\omega), \ldots, \gamma_p(\omega)$, respectively, so that writing $\Gamma(\omega) = \{\gamma_1(\omega), \gamma_2(\omega), \ldots, \gamma_p(\omega)\}$, we have

$$\mathbf{\Gamma}^{*\prime}\mathbf{F}\mathbf{\Gamma} = \begin{pmatrix} \lambda_1 & & \mathrm{O} \\ & \lambda_2 & \\ \mathrm{O} & & \lambda_p \end{pmatrix} = \mathbf{\Lambda}, \qquad \text{say,} \tag{10.1}$$

where we have suppressed the argument ω in (10.1) for simplicity of notation. Following Anderson [2], we put

$$\mathbf{U} = \frac{1}{\sqrt{n'}}(\mathbf{\Gamma}^{*\prime}\mathbf{A}\mathbf{\Gamma} - n'\mathbf{\Lambda}) \tag{10.2}$$

and observe that, asymptotically, $\operatorname{vec}\mathbf{U} \sim N_{p^2}^c(\mathbf{0}, \mathbf{\Sigma}_u)$ where $\operatorname{vec}\mathbf{U}$ denotes the column vector obtained from $\mathbf{U}$ by placing its columns under one another successively and $\mathbf{\Sigma}_u = \operatorname{diag}[\lambda_1{}^2, \lambda_1\lambda_2, \ldots, \lambda_1\lambda_p, \ldots, \lambda_p{}^2]$.

Consider the null hypothesis

$$\begin{aligned} H_3\colon \lambda_1 = \lambda_2 = \cdots = \lambda_{q_1} &= \nu_1, \\ \lambda_{q_1+1} = \cdots = \lambda_{q_1+q_2} &= \nu_2, \\ &\cdots \\ \lambda_{p-q_r+1} = \cdots = \lambda_p &= \nu_r, \end{aligned}$$

where $\nu_1 > \nu_2 > \cdots > \nu_r$ are (unknown) constants. Employing standard arguments we may deduce that the likelihood criterion L for testing the subhypothesis

$$H_4\colon \lambda_{q_1+\cdots+q_{k-1}+1} = \cdots = \lambda_{q_1+\cdots+q_k} = \nu_k$$

(where ν_k is an unknown constant) is

$$\left[\prod_{j\in M_k} d_j \Big/ \left(q_k^{-1}\sum_{j\in L_k} d_j\right)^{q_k}\right]^{n'},$$

where $M_k = \{q_1 + q_2 + \cdots + q_{k-1} + 1, \ldots, q_1 + q_2 + \cdots + q_k\}$ and $d_1 > d_2 > \cdots > d_p$ are the sample estimates of $\lambda_1, \ldots, \lambda_p$, i.e., are the eigenvalues of the matrix $(1/n')\mathbf{A}$. It can then be shown (see Priestley *et al.* [14]) that, under H_4,

$$-2\log L \doteqdot \frac{1}{2\nu_k{}^2}\left[2\sum_{\substack{i<j\\ i,j\in M_k}} |u_{ij}|^2 + \sum_{i\in M_k} |u_{ii}|^2 - \frac{1}{q_k}\left|\sum_{i\in M_k} u_{ii}\right|^2\right]$$

so that asymptotically,

$$-2 \log L \sim \chi^2_{f'} \qquad \text{where} \qquad f' = {q_k}^2 - 1.$$

This result thus confirms that given in Section 9.

REFERENCES

1. Anderson, T. W. (1958). *An Introduction to Multivariate Statistical Analysis.* Wiley, New York.
2. Anderson, T. W. (1963). Asymptotic theory for principal component analysis, *Ann. Math. Statist.* **34** 122–148.
3. Åstrom, K. J. and Bohlin, T. (1966). Numerical identification of linear dynamic systems from normal operating records. *Proc. IFAC Symp. Theory Self-Adaptive Control Syst., September 1965.* Plenum, New York.
4. Box, G. E. P. and Jenkins, G. M. (1970). *Time Series, Forecasting, and Control.* Holden-Day, San Francisco, California.
5. Brillinger, D. R. (1964). A frequency approach to the techniques of principal components, factor analysis, and canonical variates in the case of stationary time series. *Roy. Statist. Soc. Conf., September 1964.* (Held in Cardiff, U.K.)
6. Brillinger, D. R. (1969). The canonical analysis of stationary time series. *Multivariate Analysis II* (P. R. Krishnaiah, ed.). Academic Press, New York.
7. Goodman, N. R. (1963). Statistical analysis based on a certain multivariate complex Gaussian distribution. *Ann. Math. Statist.* **35** 152–176.
8. Lawley, D. N. (1956). Tests of significance for the latent roots of covariance and correlation matrices. *Biometrika* **43** 128–136.
9. Parzen, E. (1966). On empirical multiple time series analysis. *Proc. Fifth Berkeley Symp. Math. Statist. Prob.* Univ. of California Press, Berkeley.
10. Pillai, K. C. S. and Nagarsenker, B. N. (1971). On the distribution of the sphericity test criterion in classical and complex normal populations having unknown covariance matrices. *Ann. Math. Statist.* **42** 764–767.
11. Priestley, M. B. (1969). Estimation of transfer functions in closed loop stochastic systems. *Automatica* **5** 623–632.
12. Priestley, M. B. (1971). Fitting relationships between time series. *I.S.I. Session, 38th, Washington, August 1971.*
13. Priestley, M. B., Subba Rao, T. and Tong, H. (1972). Tests of significance for the eigenvalues of a spectral matrix. Tech. Rep. No. 32. Dept. of Math., U.M.I.S.T., June 1972.
14. Priestley, M. B., Subba Rao, T. and Tong, H. (1972). Testing the equality of K hermitian matrices. Tech. Rep. No. 33. Dept. of Math., U.M.I.S.T., June 1972.
15. Priestley, M. B., Subba Rao, T. and Tong, H. (1972). Asymptotic Distribution of Eigenvalues of a Sample Spectral Density Matrix. Tech. Rep. No. 34. Dept. of Math., U.M.I.S.T., June 1972.
16. Rao, C. R. (1964). The use and interpretation of principal component analysis in applied research. *Sankhyā Ser. A* **26** 329–358.
17. Wahba, G. (1968). On the distribution of some statistics useful in the analysis of jointly stationary time series. *Ann. Math. Statist.* **39** 1849–1862.
18. Wahba, G. (1971). Some tests of independence for stationary multivariate time series. *J. Roy. Statist. Soc. Ser. B* **33** 153–166.

An Information Function Approach to Dimensionality Analysis and Curved Manifold Clustering[1]

J. N. SRIVASTAVA
COLORADO STATE UNIVERSITY

Let there be s "independent" variables $X_1, \ldots, X_s$. Let $\{X\}$ be a set of values of the $(s \times 1)$ vector $X' = (X_1, \ldots, X_s)$, and let the N elements of $\{X\}$ be $(x_{1r}, x_{2r}, \ldots, x_{sr})$ $(r = 1, \ldots, N)$. For fixed r, let $y_{jr} = \phi_j(x_{1r}, x_{2r}, \ldots, x_{sr}) + \varepsilon_{jr}$ $(j = 1, \ldots, l)$, where ϕ_j are certain "fairly smooth" continuous functions, and the ε_{jr} are random variables with zero mean. The ε_{jr} are assumed mutually independent and also independent of the X's; in particular, the ε_{jr} could all be constant (equal to zero). Let $\{Y\}$ be the set having the N elements $(y_{1r}, y_{2r}, \ldots, y_{lr})$, $(r = 1, \ldots, N)$. We assume that $l \geq s$. The problem considered in this paper is: Given only the set $\{Y\}$, and assuming that the integer s, the set $\{X\}$, and the functions ϕ_j are unknown, what can we say about these? A method (particularly for "estimating" s) based on Shannon's information function is suggested. The results of certain Monte Carlo studies conducted in this connection are discussed. Certain associated problems are considered.

1. INTRODUCTION

For ease of discussion, we first elaborate the problem outlined in the preceding summary. The set $\{Y\}$ may be considered as a set of N points in an l-dimensional Euclidean space E_l with axes, say, $Y_1, Y_2, \ldots, Y_l$. Suppose, for the moment, that all $\varepsilon_{jr} = 0$. Then, clearly, the points $\{Y\}$ lie on an s-dimensional surface. Thus s may be called the ("intrinsic") "dimensionality" of the "data" points $\{Y\}$.

Since the set $\{Y\}$ is generated using the set $\{X\}$, we need to say something about the set $\{X\}$. First, of course, the elements of $\{X\}$ are vectors with real coordinates, so that $\{X\}$ is a set of points (which may perhaps be repeated)

[1] This research was supported by AFOSR grant 72-2364.

in an s-dimensional Euclidean space E_s. Actually, in live statistical situations, the X_i could be random variables. Since the ϕ_j are, to a large extent, at our disposal, we can without much loss of generality assume the X_i to be independent random variables uniformly distributed in the interval [0, 1]. For our Monte Carlo studies, we have occasionally generated the X_i, and hence $\{X\}$, by assuming them to have the uniform distribution as above. However, occasionally, in order to reduce the difficulties arising due to sampling fluctuations, we generated $\{X\}$ by letting X_i take the $(t + 1)$ values $0, 1/t, 2/t, \ldots, t/t = 1$ for some positive integer t, so that $N = (t + 1)^s$.

We said earlier that the functions ϕ_j are "fairly smooth." In this paper, we shall not try to attach a precise meaning to this concept, which hopefully would be attempted in future work. Broadly, from an intuitive viewpoint, what we mean is that these functions are not too crinky; there are few ups and downs, and the functions (though they may be nonlinear) are continuous and have only a relatively "few" maxima and minima in the range of interest. As will be seen later on, in the Monte Carlo studies we have used polynomials; trigonometric, exponential, and logarithmic functions; and mixtures of these.

It is thus clear that we are postulating that when the ε's are all zero, the set of points $\{Y\}$ lies on a "smooth" s-dimensional surface inside the l-dimensional Euclidean space E_l. The question is, given only the points $\{Y\}$, and with no knowledge of s, $\{X\}$, or the ϕ_j, can we determine these? The main problem considered in this paper is the determination of s, which would of course be the first problem to be considered in any attempt to solve the general problem.

Given the N points $\{Y\}$, there would be an infinite number of surfaces, or "curved manifolds," on which the points could lie. Of course, the assumption of "smoothness" would tend to cut down the number of such surfaces. Still, however, it is clear that it would be impossible (by any technique) to find the nature of the surface unless the "sample size" N is large enough, so that all the "curves" and "gradations" of the surface are "well represented."(The reader may have noticed that many words or phrases are set in quotation marks. The reason is that we intentionally do not want to define the corresponding concepts too precisely. Very precise definitions, at this stage, may hinder cultivations of intuitive thought which seems to be more essential from the data-analytic viewpoint.)

Obviously, our best hope of retrieving s or the ϕ_j arises when the set $\{Y\}$ represents well the "main" features of the underlying unknown curved manifold, and all the ε's are zero. Now, when the ε's are not zero, but rather are random variables, the intrinsic surface and its dimensionality might still be retrievable if the range of variation or the variance of the random variables is small compared to the "curvatures" of the surface.

It is clear that the problem presented above is a generalization of the factor analysis problem. In the latter, the functions ϕ_j, though not completely known, are still known to the extent that they are linear functions of the x's. The generalization here is twofold: first, nonlinearity could be present; and second, the nature of nonlinearity is not known. It may be emphasized that both factor analysis and principal component analysis can be considered as techniques in the general area of cluster analysis, the clusters there being closer to linear manifolds. Thus, in general, all such problems may be considered as relating to curved manifold clustering.

There are obvious applications of this class of questions to data analysis. Indeed, in scientific research where an attempt is being made to uncover natural laws, the data gathered would often lie on such manifolds, if the errors were negligible. If errors are present, as may happen when many unaccounted factors with small effects are present, the original functional relationship will get distorted to a more probabilistic picture. The question arises, how to collect data (both when errors are and are not present) so that the "natural laws" can be more easily discoverable. This is the "design" question, and we will not consider it here. However, we may add that the present work may be a necessary step toward handling that question.

Mention should be made of multidimensional scaling. Here, a class of problems similar to ours is considered, except that instead of starting with a set of points $\{Y\}$ in E_l, we are usually given the set of $\binom{N}{2}$ distances between each pair of points. In the last decade, because of its great importance (particularly in psychology), this area has been very active.

The paper of Shepherd and Carroll [18] also considers problems similar to ours. However, the methods in the present paper are different, and so far as the author is aware, completely new. The basic idea behind the new method is to divide up E_l by a certain rectangular lattice, find the number of points of $\{Y\}$ in each cell, and look at the entropy of the resulting distribution. The aim of this paper is to present a few techniques associated with this method, and to look at the potentialities of the method through Monte Carlo studies. This has resulted in the opening up of certain promising avenues, theoretical work along which would be illuminating. Such work will be presented in other communications. (The problems in this paper were suggested to the author from the viewpoint of possible applications to meterology.)

2. ENTROPY IN THE DISCRETE AND CONTINUOUS CASES

Consider a random variable A which takes values $a_1, a_2, \ldots, a_n$ with probabilities $p_1, p_2, \ldots, p_n$, respectively. Without loss of generality, we shall assume $p_i > 0$ for all i. Then, the entropy of this scheme, denoted by $H(A)$, is given by

$$H(A) = -\sum_{i=1}^{n} p_i \log p_i. \tag{1}$$

The entropy, as defined above for the discrete case, can be considered as a measure of "uncertainty" in the value that A assumes.

Now, let B be another discrete random variable which takes values b_1, $b_2, \ldots, b_m$ with probabilities $p_1', p_2', \ldots, p_m'$, respectively. Also, let $p_{\alpha\beta} = \text{Prob}\{B = b_\beta \mid A = a_\alpha\}$. Then $H(B) = -\sum p_i' \log p_i'$, is the (unconditional) entropy of B. Consider the scheme, consisting of the pair of events (a_α, b_β), for all α and β. The entropy in this bivariate case is given by

$$H(A, B) = -\sum_{\alpha=1}^{n} \sum_{\beta=1}^{m} (p_\alpha p_{\alpha\beta}) \log(p_\alpha p_{\alpha\beta}). \tag{2}$$

Again, let $H_\alpha(B)$ denote the conditional entropy of B given $A = a_\alpha$, so that $H_\alpha(B) = -\sum_{\alpha=1}^{n} p_{\alpha\beta} \log p_{\alpha\beta}$. Let $H_A(B)$ denote the average conditional entropy of B, i.e., $H_A(B) = \sum_\alpha p_\alpha H_\alpha(B)$. Hence, we have the well-known formula

$$H(A, B) = H(A) + H_A(B). \tag{3}$$

Also, $H(B) \geq H_A(B)$, with equality holding when A and B are independent. If $H(A, B) = H(A)$, then $H_A(B) = 0$, which can be interpreted to mean that the uncertainty in B given the value of A is zero. Thus the more B is "dependent" on A, the closer are $H(A, B)$ and $H(A)$. Indeed, if there is a one–one relationship between A and B, then $H(A, B) = H(A)$. For example, suppose A is always positive, and $B = A^2$. Then $m = n$, $p_{\alpha\beta} = 0$, if $\alpha \neq \beta$, and $p_{\alpha\alpha} = 1$, so that $H(A, B) = H(A) = H(B)$. In other words, the uncertainty in A, B and (A, B) is the same.

This nice property, however, does not hold under the present definition of entropy in the continuous case, which we now consider. For simplicity, throughout this paper, we shall limit ourselves in the continuous case to random variables with absolutely continuous distributions. Let A^* be such a random variable, with probability density $f(a)$. Then, following (say) Renyi [13], we define the entropy

$$H(A^*) = -\int_{-\infty}^{\infty} f(a) \log f(a)\, da. \tag{4}$$

This entropy is not generally invariant under one-to-one transformations, or even scale transformations. Thus, if $B^* = \theta A^*$, where θ (>0) is a constant, then it can be easily checked that

$$H(B^*) = H(\theta A^*) = H(A^*) + \log \theta. \tag{5}$$

Now, one would normally think that the uncertainty associated with A^* should be the same as that associated with θA^*, but in fact it is not so under the present definition. Now, let $F(a)$ be the distribution function of A^*. Consider the transformation $U = F(a)$. Then θU (where $\theta > 0$) has the uniform distribution in the interval $(0, \theta)$, and $H(U) = \log \theta$. (Incidentally, this shows that in the continuous case, random variables can have negative or zero entropy, as well.) If entropy were to be invariant under one-to-one transformations, then U and A^* should have the same entropy.

Now, let $A_1, A_2, \ldots, A_p$ be (continuous) random variables, and let $F_i(a_i)$ be the distribution function of A_i. Let $U_i = F_i(a_i)$, $i = 1, \ldots, n$. Then we introduce a new concept, called "normalized entropy" of the random variables, which is given by $H^*(A_1, \ldots, A_p)$, where

$$H^*(A_1, \ldots, A_p) = H(U_1, \ldots, U_p)$$
$$= -\int g(u_1, \ldots, u_p) \log g(u_1, \ldots, u_p)\, du_1 \cdots du_p, \quad (6)$$

where $g(u_1, \ldots, u_p)$ is the joint probability density of the random variables $U_1, U_2, \ldots, U_p$. Clearly, for $p = 1$, we have $H^*(A_1) = H^*(U_1) = 0$, so that the normalized entropy of any two univariate random variables (absolutely continuous with respect to Lebesgue measure, say) is the same. The normalized entropy is thus essentially a multivariate concept. It is also clear that the normalized entropy of $(A_1, \ldots, A_p)$ as defined above is invariant under transformations of the variables A_i to B_i, such that (for $i = 1, \ldots, p$) B_i is a one-to-one function of A_i. The following important relation between the standard and normalized entropies can be easily established.

Theorem 1. *We have*

$$H^*(A_1, \ldots, A_p) = H(A_1, \ldots, A_p) - \sum_{i=1}^{p} H(A_i) \leq 0. \quad (7)$$

To given an example, let $A_1^0, \ldots, A_p^0$ have the p-variate normal distribution with covariance matrix Σ and correlation matrix R, and let the variance of A_i^0 be σ_{ii}. Then

$$H^*(A_1^0, \ldots, A_p^0) = \frac{1}{2}\log|\Sigma| - \frac{1}{2}\sum_{i=1}^{p} \log \sigma_{ii} = \frac{1}{2}\log|R|.$$

When $A_1, \ldots, A_p$ are independent, then $H^*(A_1, \ldots, A_p) = 0$. The quantity H^* does measure the amount of uncertainty; a large value (in the algebraic sense) of H^* corresponds to a higher degree of uncertainty. Thus $H^* = -1$ and -40, respectively, correspond to a (relatively) higher and a lower degree of uncertainty.

The introduction of H^* is not meant to suggest that the standard measure H is less useful. Both have their importance. We can say that, in a sense, the measure H behaves as if *all* (real) variables are to be measured to the *same* number of decimal places. Thus if the variables A and B correspond to measuring the lengths of objects in centimeters and millimeters, respectively, then (numerically) $B = 10A$, and $H(B) = H(A) + \log 10$. The ((log 10) units of) extra information under B can be explained as follows. If lengths are measured to k decimal places, irrespective of the unit of measurement, then obviously, expressing the length in millimeters would give more information than using centimeters. The measure H^* behaves as if any variable A is measured to a number of decimal places large enough so that upon transformation to the uniform distribution, we shall obtain values correct to, say, k decimal places (where k is fixed).

How is a measure of uncertainty (like H^*) connected with the dimensionality analysis under curved manifold clustering? To see this, consider an example. Take two (continuous) random variables A and B. If A and B are independent (i.e., the dimensionality is 2, the maximum possible in this case), the uncertainty about the knowledge of one (say, A) given the other (B) is maximum; in this case, we find that $H^*(A, B)$ has the large value 0. On the other hand, if B is "dependent" on A, then $H^*(A, B)$ will be less. Thus, intuitively, the lower the dimensionality, the less should be the uncertainty.

3. UNCERTAINTY AND DIMENSIONALITY

First, we introduce the concept [13] of the information of order α, which is used later on. For the discrete case, let there be a (discrete) random variable Y_d, which takes distinct values with probabilities $p_1, p_2, \ldots$, respectively. Then we define

$$I_1(Y_d) \equiv H(Y_d) = -\sum_k p_k \log p_k \tag{8}$$

$$I_\alpha(Y_d) = (1-\alpha)^{-1} \log\Big(\sum_k p_k^{\alpha}\Big), \qquad \alpha \neq 1, \quad \alpha > 0. \tag{9}$$

The quantity $I_\alpha(Y_d)$ is called the information of order α, and we have $\lim_{\alpha \to 1} I_\alpha(Y_d) = I_1(Y_d)$.

Next, let $Y_1, Y_2, \ldots, Y_l$ have a l-dimensional absolutely continuous distribution with probability density function $f(Y_1, \ldots, Y_l)$. For $\alpha > 0$, let

$$I_1(Y_1, \ldots, Y_l) = -\int f(Y_1, \ldots, Y_l) \log f(Y_1, \ldots, Y_l)\, dY_1 \cdots dY_l, \tag{10}$$

$$I_\alpha(Y_1, \ldots, Y_l) = (1-\alpha)^{-1} \log \int_{-\infty}^{\infty} \{f(Y_1, \ldots, Y_l)\}^\alpha\, dY_1 \cdots dY_l. \tag{11}$$

Then $I_\alpha(Y_1, \ldots, Y_l)$ is the information of order α regarding $(Y_1, \ldots, Y_l)$, and again, $\lim_{\alpha \to 1} I_\alpha = I_1$.

Next, let Y_{jm} ($j = 1, \ldots, l$; m a positive integer) be discrete random variables defined by $Y_{jm} = [mY_j]/m$, where the Y_j are as above for (10) and (11), and where $[\theta]$ denotes the largest integer less than or equal to θ. Then, we have, from Renyi [13]

Theorem 2. *If $I_\alpha(Y_{1m}, \ldots, Y_{lm})$ is finite, we have*

$$\lim_{m \to \infty} \frac{I_\alpha(Y_{1m}, \ldots, Y_{lm})}{\log m} = l, \tag{12}$$

$$\lim_{m \to \infty} [I_\alpha(Y_{1m}, \ldots, Y_{lm}) - l \log m] = I_\alpha(Y_1, \ldots, Y_l). \tag{13}$$

The above theorem shows that there is a connection between dimensionality l and information $I_\alpha(Y_{1m}, \ldots, Y_{lm})$, at least for the case where the variables $Y_1, \ldots, Y_l$ have an l-dimensional absolutely continuous distribution. The question arises: Can the above relation between l and I_α be generalized and exploited to help solve the problem of estimating s that we set forth in Section 1?

To consider this, let us examine the situation more closely. The variable Y_{jm} is a discrete approximation to Y_j, being constant over intervals of length $1/m$. To approximate l using (12), one would need to know $f(Y_1, \ldots, Y_l)$, which in applications would be unknown. Also, the approximation "error" on the right-hand side of (13) depends upon the scale (etc.) of measurement of the Y_j, and may be large compared to $\log m$, even for moderately large values of m. If a sample of size N from the distribution of $(Y_1, \ldots, Y_l)$ is given, then even if N were large and m relatively small, there would (in general) be too many cells in which the space of $(Y_{1m}, \ldots, Y_{lm})$ gets divided, and hence a rather thin distribution of frequency in the various cells. All these facts would militate against Theorem 2 taking effect. In what follows, we still try to develop techniques to exploit the central ideas of this theorem.

Consider $Y_1, \ldots, Y_l$ again. Let $F_j(y_j)$ be the density of Y_j, and let the mth quantiles of Y_j be $-\infty = \theta_{j0}, \theta_{j1}, \ldots, \theta_{jm} = \infty$. Let

$$p(i_1, \ldots, i_l) = \text{Prob}\{\theta_{j, i_j} < Y_j \le \theta_{j, i_j+1}; j = 1, \ldots, l\}.$$
$$i_j = 0, 1, \ldots, m-1; j = 1, \ldots, l. \tag{14}$$

We have divided the space of the l variables into m^l cells. Note that

$$\sum_{\substack{j=1 \\ j \neq j'}}^{l} \sum_{i_j=0}^{m-1} p(i_1, i_2, \ldots, i_l) = \frac{1}{m} \qquad \text{for} \quad j' = 1, \ldots, l. \tag{15}$$

Now, let $Y_1, \ldots, Y_l$ correspond to the Y's introduced in Section 1. Consider the sample $\{Y\}$ of N points. For this sample, let $\hat{\theta}_{jr}$ $(j = 1, \ldots, l; r = 0, \ldots, m)$ be the sample mth quantiles for Y_j, and let $\hat{p}(i_1, \ldots, i_l)$ be the estimated probabilities, using the sample $\{Y\}$ and the quantiles $\hat{\theta}_{jr}$. Let

$$Q_{m1} = -\sum_{j=1}^{l} \sum_{i_j=0}^{m-1} [p(i_1, \ldots, i_l)] \log[p(i_1, \ldots, i_l)], \tag{16}$$

$$\hat{Q}_{m1} = -\sum_{j=1}^{l} \sum_{i_j=0}^{m-1} [\hat{p}(i_1, \ldots, i_l)] \log[\hat{p}(i_1, \ldots, i_l)]. \tag{17}$$

Thus, Q_{m1} is the same as I_1 in (8), where p_k are replaced by the $p(i_1, \ldots, i_l)$. Similarly, for $\hat{Q}_{m1}$. Also, in the same way, one can define for $\alpha > 0$, $\alpha \neq 1$, the quantities $Q_{m\alpha}$ and $\hat{Q}_{m\alpha}$ using (9) instead of (8). One can prove

Theorem 3. *Let $U_j = F_j(Y_j)$; that is, U_j is the transform of Y_j to uniform $j = 1, \ldots, l$. Then if $Y_1, \ldots, Y_l$ have an absolutely continuous distribution, we have*

$$\lim_{m\to\infty} (Q_{m\alpha} - l \log m) = I_\alpha(U_1, \ldots, U_l). \tag{18}$$

Notice that, unlike in (13), the right-hand side of (18) does not depend on the scale of the Y's, etc.

What about a lower dimensionality, say, s? There are some results. For example, we can easily prove

Theorem 4. (i) *Suppose $Y_1, \ldots, Y_s$ have an s-dimensional absolutely continuous distribution. Suppose also that Y_α ($\alpha = s+1, \ldots, l$) is a monotonic function of Y_β for some β with $1 \leq \beta \leq s$. Let $Q_{m\alpha}$ be calculated for the whole set $\{Y_1, \ldots, Y_l\}$. Then*

$$\lim_{m\to\infty} (Q_{m\alpha} - s \log m) = I_\alpha(U_1, \ldots, U_s) \tag{19}$$

where the U_j are defined as before.

(ii) *Suppose Y_1 has an absolutely continuous symmetric distribution with median* 0, *and let $Y_2 = Y_1^2$. Then if $Q_{m\alpha}$ below refers to (Y_1, Y_2), we have*

$$\lim_{m\to\infty} (Q_{m\alpha} - (1) \log m) = \log 2. \tag{20}$$

The above theorems are still, however, not directly helpful, since $Q_{m\alpha}$ (which cannot be calculated from $\{Y\}$) is used rather than $\hat{Q}_{m\alpha}$, and furthermore, since they refer to the case when m is large. With increasing m, the number of cells m^l grows very fast, and even large values of N may, in general, not give good estimates $\hat{p}(i_1, \ldots, i_l)$. Indeed, because of the latter fact, we must consider the smallest possible values of m, such as even 2 or 3. In this connection, we can establish the following interesting result.

Theorem 5. *Let the Y_j and U_j be defined as in Theorem 3. Then for every m we have*

$$l \leq (\log m)^{-1}[Q_{m\alpha} - I_\alpha(U_1, \ldots, U_l)]. \tag{21}$$

Returning now to the original problem of finding s using $\{Y\}$ and in view of the above facts, it was decided to look at the quantities $(\hat{Q}_{m\alpha}/\log m)$. As the Monte Carlo examples of Section 5 indicate, this does give a fair insight into the problem. Before going into these examples, we discuss certain other useful associated techniques.

4. SOME AUXILIARY TECHNIQUES

Consider again the curved manifold (say, S) of dimensionality s whose nature we want to explore with the help of the set of points $\{Y\}$. Sometimes, particularly when the ε's are nonzero and large, some of the points in $\{Y\}$ may lie too far away from S. In such cases it may be useful to omit certain points in order to get a better picture of the essential features of S. Again, in general, it may help to increase N by generating some new points (using the old set $\{Y\}$). Certain techniques for such purposes are now presented. Throughout, $\hat{P}$ denotes the l-dimensional array of numbers $\hat{p}(i_1, \ldots, i_l)$, with m^l cells.

4.1. δ-Deletion

We shall illustrate this for the case $l = 2$ or 3. The generalization for large l is easy. When $l = 2$, $\hat{P}$ is an $(m \times m)$ matrix with row and column sums equal to $(1/m)$, so that $(m\hat{P})$ is doubly stochastic. By Birkhoff's theorem, $\hat{P}$ belongs to a convex polyhedron whose vertices are permutation matrices. In other words, there exist permutation matrices $E_1, \ldots, E_k$, say, such that

$$\hat{P} = \delta_1 E_1 + \delta_2 E_2 + \cdots + \delta_k E_k. \tag{22}$$

Such a decomposition is not unique. A certain decomposition which is meaningful in the present situation, and which can be made unique by using conventions, is now described. Let $\gamma_{11} = \max_{i_1} \min_{i_2} p(i_1, i_2)$, $\gamma_{12} = \max_{i_2} \min_{i_1} p(i_1, i_2)$, $\gamma_1 = \max(\gamma_{11}, \gamma_{12})$. Let γ_1 be in the cell (i_{11}^*, i_{12}^*), where i_{11}^* and (next) i_{12}^* are a minimum. Next, let

$$\gamma_{21} = \max_{i_1 \neq i_{11}^*} \min_{i_2 \neq i_{12}^*} p(i_1, i_2), \quad \gamma_{22} = \max_{i_2 \neq i_{12}^*} \min_{i_1 \neq i_{11}^*} p(i_1, i_2), \quad \gamma_2 = \max(\gamma_{21}, \gamma_{22}),$$

and let γ_2 be in the cell (i_{21}^*, i_{22}^*), where i_{21}^* (and then) i_{22}^* are a minimum. Again, let

$$\gamma_{31} = \max_{i_1 \neq i_{11}^*, i_{21}^*} \min_{i_2 \neq i_{12}^*, i_{22}^*} p(i_1, i_2), \quad \gamma_{32} = \max_{i_2 \neq i_{12}^*, i_{22}^*} \min_{i_1 \neq i_{11}^*, i_{21}^*} p(i_1, i_2),$$

$$\gamma_3 = \max(\gamma_{31}, \gamma_{32})$$

and suppose γ_3 occurs in cell (i_{31}^*, i_{32}^*), where i_{31}^*, i_{32}^* are a minimum (in that order). Next, let

$$\gamma_{41} = \max_{i_1 \neq i_{11}^*, i_{21}^*, i_{31}^*} \min_{i_2 \neq i_{12}^*, i_{22}^*, i_{32}^*} p(i_1, i_2),$$

and so on. By continuing this procedure, we get the numbers $\gamma_1, \gamma_2, \ldots, \gamma_l$. Let $\delta_1 = \max(\gamma_1, \gamma_2, \ldots, \gamma_m)$. Then, one can check that

$$\hat{P} = \delta_1^* E_1 + P_1 \tag{23}$$

where E_1 is a permutation matrix, and a multiple of P_1 is doubly stochastic. The above procedure is then repeated, and P_1 expressed as $P_1 = \delta_2{}^*E_2 + P_2$, and so on. Finally, by continuing the above we obtain

$$\hat{P} = \delta_1^* E_1 + \cdots + \delta_k^* E_k \tag{24}$$

for some k, where E_i are permutation matrices, and $\delta_1^* \le \delta_2^* \le \cdots \le \delta_k^*$. Now choose $\delta > 0$, and suppose k' is such that $\delta_1^* \le \delta_2^* \le \cdots \le \delta_{k'}^* \le \delta$, and $\delta_{k'+1}^* > \delta$. Let $P_0{}^* = \sum_{i>k'} \delta_i{}^* E_i$, and let $P_0 = \theta_0 P_0{}^*$, where θ_0 is chosen so that the sum of all elements in P_0 equals 1. Then, one could work with P_0 instead of P for the computation of the $Q_{m\alpha}$; this process is called δ-deletion. Each matrix E_i may be interpreted as representing a "direction" of variation, and the above process eliminates directions where the probability is small.

When $l = 3$, we proceed similarly. Let

$$\gamma_{11} = \max_{i_1} \min_{i_2, i_3} p(i_1, i_2, i_3), \qquad \gamma_{12} = \max_{i_2} \min_{i_1, i_3} p(i_1, i_2, i_3),$$
$$\gamma_{13} = \max_{i_3} \min_{i_1, i_2} p(i_1, i_2, i_3),$$

$\gamma_1 = \max(\gamma_{11}, \gamma_{12}, \gamma_{13})$, and suppose $(i_{11}^*, i_{12}^*, i_{13}^*)$ is the cell where γ_1 occurs where i_{11}^*, i_{12}^*, and i_{13}^* are minimum (and in that order). Next, let

$$\gamma_{21} = \max_{i_1 \ne i_{11}^*} \min_{i_2 \ne i_{12}^*,\, i_3 \ne i_{13}^*} p(i_1, i_2, i_3),$$

etc., and get γ_2. Thus, as before, we shall obtain $\gamma_1, \gamma_2, \ldots, \gamma_m$. Let $\delta_1{}^* = \max(\gamma_1, \gamma_2, \ldots, \gamma_m)$. Then $\hat{P} = \delta_1^* E_1^* + P_1$, where E_1^* is a three-dimensional "permutation array" with m^3 cells out of which m cells have a 1 and the rest have 0, and such that if cells (i_1, i_2, i_3) and (i_1', i_2', i_3') both have a 1, then $i_1 \ne i_1', i_2 \ne i_2', i_3 \ne i_3'$. It is seen, then, that a multiple of P_1 (like $\hat{P}$) is also stochastic, and so by repeating the above procedure, we get $P_1 = \delta_2^* E_2^* + P_2$, where $E_2{}^*$ is also a permutation array. Hence, continuing, we get $\hat{P} = \delta_1^* E_1^* + \delta_2^* E_2^* + \cdots + \delta_h^* E_h^*$, for some h, where $\delta_1{}^* \le \delta_2{}^* \le \cdots \le \delta_h{}^*$. Again, δ-deletion can be done as before.

4.2. Scraping

Consider $\hat{P}$ again. Let $\hat{p}(i_1, \ldots, i_l) = n(i_1, \ldots, i_l)/N$, where $n(i_1, \ldots, i_l)$ is an integer. Then, if $n(i_1, \ldots, i_l) = 1$, eliminate that point of $\{Y\}$ which falls in the cell $(i_1, \ldots, i_l)$; do this for all cells. Thus some points of $\{Y\}$ will get "scraped" out. Suppose N_1 points are left; call the new set of points $\{Y\}_1$. Starting with the set $\{Y\}_1$, one can again obtain the mth quantiles, etc., and new frequencies $n^{(1)}(i_1, \ldots, i_l)$, say, using which one can calculate a new value of $Q_{m\alpha}$, say, $Q_{m\alpha}^{(1)}$. This is called scraping of order 1. One can now perform a scraping of order 1 on the set $\{Y\}_1$; this will be called scraping of order 2 on $\{Y\}_1$, and so on.

Scraping seems particularly useful when N is large and m is small. In a sense, it reveals the "main" part of the surface S.

4.3. Increasing N

Consider again $\{Y\}$, whose points are $(y_{1r}, \ldots, y_{lr})$, $r = 1, \ldots, N$. Let $\omega_{rr'}^2 = \sum_{j=1}^{l}(y_{jr} - y_{jr'})^2$, $r \neq r'$, $r, r' = 1, \ldots, N$. Choose a number $\delta > 0$ and a positive integer k. If $\omega_{rr'}^2 \leq \delta$, generate $(k-1)$ additional points $(z_{srr'1}, \ldots, z_{srr'l}) = (s/k)(y_{1r}, \ldots, y_{lr}) + [1 - (s/k)](y_{1r'}, \ldots, y_{lr'})$, $s = 1, 2, \ldots, (k-1)$; do this for all r, r'. Add the set of all points so generated to the old set $\{Y\}$ to get a set of points, say, $\{Y\}^{(1)}$. The process can be repeated on $\{Y\}^{(1)}$ with a new δ and k, if desired; and so on.

The above procedure has been found particularly useful when N is small, but small values of m are not revealing enough. When N is quite large, the above, though not necessary, may still be advantageous.

5. MONTE CARLO STUDIES

Consider the set $\{Y\}$ again. We shall continue the notation developed in the summary and Section 1. In this section, we present many examples of functions ϕ_j, etc., and the values of $Q_{m\alpha}$ obtained to give us an idea of how the quantity

$$\hat{S}_m = [Q_{m\alpha}/\log m], \tag{25}$$

where $[\theta]$ denotes the largest integer less than or equal to θ, would behave as an estimate of s. We first give the values of l, s, ϕ_j (in terms of $x_1, \ldots, x_s$), etc., for each case.

Example 5.1. $l = 4$, $s = 1$, $\phi_1 = x_1$, $\phi_2 = 2x_1$, $\phi_3 = x_1^2$, and $\phi_4 = x_1^3$.

Example 5.2A. $l = 4$, $s = 2$, $\phi_1 = \sin^{-1}\sqrt{x_1}$, $\phi_2 = \sin^{-1}\sqrt{x_2}$, $\phi_3 = \tan^{-1}[\frac{1}{2}(x_1 + x_2)]$, and $\phi_4 = x_1^2 + x_2^2$.

Example 5.2B. $l = 6$, $s = 2$; ϕ_1, ϕ_2, ϕ_3, ϕ_4 are the same as in Example 5.2A; $\phi_5 = 3x_1^2 + \exp(x_2)$, and $\phi_6 = 2\log(1 + x_1) + 3x_2^4$.

Example 5.3. $l = 6$, $s = 2$; $\phi_1 = 2x_1$, $\phi_2 = x_1^2$, $\phi_3 = x_1^3$, $\phi_4 = 2x_1 + x_2$, $\phi_5 = x_1^2 + x_2^4$, and $\phi_6 = x_1^2 + x_1^3 + x_2^4$.

Example 5.4A. $l = 5$, $s = 3$; $\phi_1 = x_1 + 2x_2 + 3x_3$, $\phi_2 = x_1^2 + 2x_2^2 + 4x_3^2$, $\phi_3 = (1 + x_1)(1 + x_2)(1 + x_3)$, $\phi_4 = \exp(x_1) + \exp(2x_2) + \exp(x_3)$, and $\phi_5 = \log(1 + x_1) + \sin(2\pi x_2) + \cos(2\pi x_3)$.

Example 5.4B. $l = 8$, $s = 3$; ϕ_1, ϕ_2, ϕ_3 are the same as in Example 5.4A; $\phi_4 = \exp(x_1) + \exp(2x_1) + \exp(3x_1)$, $\phi_5 = \log(1 + x_1) + \sin(2\pi x_1) + \cos(2\pi x_2)$, $\phi_6 = (x_1^2 + x_2^4)/(1 + x_2 + x_3)$, $\phi_7 = (x_1^3 + x_3^3)/(1 + x_1 + x_2)$, and $\phi_8 = \sin(2\pi x_1) + x_2^2 + x_3^2$.

Example 5.5A. $l = 7$, $s = 4$; ϕ_1, ϕ_2, ϕ_3, ϕ_4, ϕ_5 are the same as ϕ_1, ϕ_2, ϕ_3, ϕ_7, and ϕ_8, respectively, of Example 5.4B; $\phi_6 = x_1 + x_2^2 + x_3 + x_4^3$, and $\phi_7 = x_1 x_4 + x_2 x_3 + x_4^2$.

Example 5.5B. $l = 8$, $s = 4$; $\phi_1, \ldots, \phi_7$ are as in Example 5.5A, and ϕ_8 the same as ϕ_6 of Example 5.4B.

Example 5.5C. $l = 9$, $s = 4$; $\phi_1, \ldots, \phi_8$ are as in the last example, and ϕ_9 the same as ϕ_5 of Example 5.4B.

Example 5.5D. $l = 10$, $s = 4$; $\phi_1, \ldots, \phi_9$ are as in the last example, and ϕ_{10} the same as ϕ_4 of Example 5.4B.

Example 5.6. $l = 3$, $s = 2$; $\phi_1 = \cos(2\pi x_1)$, $\phi_2 = \cos(2\pi x_2)\sin(2\pi x_1)$, and $\phi_3 = \sin(2\pi x_2)$, $\sin(2\pi x_1)$.

The result of the Monte Carlo studies are presented in Table I, which we now explain. The entry in the column headed X is S if the points $\{X\}$ are systematically chosen as explained in Section 1; it is R if we choose each x_{ir} (independently and) randomly in (0, 1). The column N shows the sample size; a plus sign indicates that scraping of order 1 is done and a minus sign implies that the number of points has been increased by the methods of Section 4.3. The suffix m in $Q_{m\alpha}$ is omitted. Also, in certain rows, certain cells are left blank; the entry in such cases is identical with the last row above it which does have a nonblank in this column. The column entitled ε has 0 in it if all ε's are taken to be zero for that case; the entry Ua (where a is some real number) indicates that each ε was randomly chosen, being uniformly distributed over the interval $(-a, a)$. Similarly, the entry Na in the column for ε indicates that ε are chosen randomly, being distributed as normal with mean 0 and variance a.

Most of the studies conducted were exploratory in nature. The functions ϕ_j were chosen rather arbitrarily to produce variety. Only Example 5.6, on the sphere, was studied in detail and more systematically. The results presented are representative of the general trend.

The estimator $\hat{s}_{m1}$ of (25) seems to do not too badly in most cases, except when s is large, in which case it overestimates s by 1 in many cases. It seems that for large l, one should use a relatively large m as well as a large value of α, say, $\alpha = 10$. Unfortunately, our first set of subroutines to handle various computations are not written in the most efficient way possible. In the near future, we hope to have a much more efficient computer program. Studies in

TABLE I

Values of $Q_{m\alpha}$ in the Monte Carlo Studies

Ex.	l	s	N	X	ε	m	$\log_e m$	Q_1	Q_2	Q_3	Q_5
5.2A	4	2	625	S	0	5	1.61	3.97	3.77	3.65	3.51
					U0.1			4.54	4.20	3.95	3.67
5.2B	6				0	3	1.10	3.21	2.84	2.64	2.45
						4	1.39	3.94	3.56	3.31	3.05
						5	1.61	4.43	4.10	3.89	3.65
						10	2.30	5.62	5.40	5.20	4.91
5.3					0	2	0.69	1.55	1.18	1.08	1.02
					U0.01			1.69	1.23	1.11	1.04
					U1.0	3	1.10	5.08	4.52	4.06	3.57
					0	5	1.61	3.48	3.06	2.79	2.53
					U0.01			3.68	3.21	2.92	2.63
					U0.5	10	2.30	6.41	6.39	6.36	6.22
					U1.0			6.42	6.42	6.41	6.37
					N1.0			6.43	6.43	6.42	6.42
5.4A	5	3	1331		U0.5	2	0.69	2.56	2.04	1.83	1.66
						10	2.30	6.90	6.59	6.06	5.33
					0	5	1.61	5.25	4.63	4.11	3.60
5.4B	8				0	2	0.69	3.60	3.27	3.10	2.90
						3	1.10	4.84	4.41	4.10	3.75
5.5A	7	4	2401			2	0.69	2.50	2.18	2.05	1.94
5.5B	8							3.92	3.42	3.20	3.00
						3	1.10	5.95	5.34	4.95	4.58
5.5C	9				0	2	0.69	3.82	3.36	3.11	2.88
					N0.25			4.38	3.82	3.54	3.28
5.5D	10				0			4.36	3.89	3.62	3.35
					U0.5			5.34	4.84	4.55	4.23
5.6	3	2	25	S	N0.05	4	1.39	3.22	3.22	3.22	3.22
			52−					3.04	2.84	2.68	2.49
			25	R	N0.05			2.87	2.77	2.68	2.54
			106−					2.93	2.70	2.56	2.43
			82−	S				2.86	2.68	2.55	2.40
			25	R	0			2.87	2.77	2.68	2.54
			196−					2.82	2.59	2.47	2.34
			25					2.98	2.88	2.77	2.58
			424−					2.66	2.44	2.30	2.12
			215−	S		6		3.66	3.53	3.45	3.38
			25	R				3.00	2.94	2.88	2.78
			250−					3.36	3.19	3.08	2.94
			25					3.05	3.00	2.95	2.84
			424−					3.16	2.94	2.82	2.69
			64		U1.0	3	1.10	2.84	2.66	2.50	2.27
			42+					2.52	2.32	2.14	1.91
			64		0	6	1.79	3.62	3.24	2.84	2.43

the directions suggested by the present results (particularly with larger m) will then be carried out, and reported separately. This project also includes a simultaneous application of "weaving" techniques (Section 4.3), and those of "smoothing" (Section 4.2 or 4.1). Weaving in order to amplify N seems to be useful in most situations. Smoothing is particularly useful when ε's are random and possibly large.

We have created more problems than we have answered.

REFERENCES

1. Degerman, R. (1970). Multidimensional analysis of complex structure: Mixtures of class and quantitative variation. *Psychometrika* **35** 475–491.
2. Dempster, A. P. (1971). An overview of multivariate data analysis. *J. Multivariate Anal.* **1** 316–345.
3. Good, I. J. (1950). *Probability and the Weighing of Evidence.* Griffin, London.
4. Guttman, L. (1968). A general nonmetric technique for finding the smallest coordinate space for a configuration of points. *Psychometrika* **33** 469–506.
5. Hamdan, M. A. and Tsokos, C. P. (1971). An information measure of association in contingency tables. *Information and Control* **19** 174–179.
6. Khinchin, A. I. (1957). *Mathematical Foundations of Information Theory.* Dover, New York.
7. Klahr, D. (1969). A Monte Carlo investigation of the statistical significance of Kruskal's nonmetric scaling procedure. *Psychometrika* **34** 319–330.
8. Kruskal, J. B. (1964). Multidimensional scaling by optimizing goodness of fit to a nonmetric hypothesis. *Psychometrika* **29** 1–27.
9. Kruskal, J. B. (1964). Nonmetric multidimensional scaling: A numerical method. *Psychometrika* **29** 115–129.
10. Kullback, S. (1959). *Information Theory and Statistics.* Wiley, New York.
11. Osborne, D. K. (1970). Further extensions of a theorem of dimensional analysis. *J. Mathematical Psychology* **7** 236–242.
12. Rao, C. R. (1965). *Linear Statistical Inference and its Application.* Wiley, New York.
13. Renyi, A. (1970). *Probability Theory.* North-Holland Publ. Amsterdam, and Amer. Elseveir, New York.
14. Ruspini, E. H. (1969). A new approach to clustering. *Information and Control* **15** 22–32.
15. Schönemann, P. N. (1970). On metric multidimensional unfolding. *Psychometrika* **35** 349–366.
16. Shannon, C. E. (1956). The bandwagon. *IRE Trans. Information Theory* **2** 3.
17. Shepard, R. N. (1966). Metric structures in ordinal data. *J. Mathematical Psychology* **3** 287–315.
18. Shepard, R. N. and Carroll, J. D. (1965). Parametric representation of nonlinear data structures. *Multivariate Analysis* (P. R. Krishnaiah, ed.). Academic Press, New York.
19. Young, F. W. (1970). Nonmetric multidimensional scaling: Recovery of metric information. *Psychometrika* **35** 455–473.

Nonlinear Iterative Partial Least Squares (NIPALS) Modelling: Some Current Developments

HERMAN WOLD

DEPARTMENT OF STATISTICS
UNIVERSITY OF GÖTEBORG
GÖTEBORG, SWEDEN

1. INTRODUCTION AND SUMMARY

1.1. Let me quote from the report, published in 1966, where the NIPALS approach was introduced.[1] "The validity and possible advantages of the [NIPALS] approach have to be judged from case to case. At this early stage of development it is too early to venture a broad appraisal. It is safe to say however that the [NIPALS] procedures, thanks to their simplicity and their adaptability to the mathematical structure of the various models, open up new vistas in the domain of nonlinear estimation." The present report, drawing from the experiences that have emerged in the theoretical and applied work with NIPALS modelling, focuses on the potential use of the NIPALS approach in the recent many-faceted development of path models in the social sciences. Three general types of models are considered, denoted L, L^*, and L^{**}, and distinguished by which categories of the explanatory variables are directly or indirectly observed. The models have a background in two central lines of evolution in econometrics (directly observed variables) and the behavioural sciences (latent, or indirectly observed, variables). In another perspective of multivariate analysis, the models belong under two broad lines of evolution, from high degrees of information in the modelling to lower degrees of information.

In the NIPALS approach, models L, L^*, and L^{**} are specified in terms of predictors (conditional expectations). This makes a nonlinear model structure if one or more of the explanatory variables are indirectly observed. Earlier results carry over from econometrics to provide consistent estimation of models L by ordinary least squares (OLS) regression, and of models L^* by the fix-point (FP) method and related NIPALS procedures. In models L^{**}

[1] Wold [30, p. 413].

the great diversification in the model design makes a challenge to the estimation techniques. Four case studies (Section 5) illustrate the flexibility of NIPALS estimation when it comes to models L^{**}.

The following general views on the reach and limitation of the NIPALS approach are stated on the joint basis of theoretical considerations and applied work:

(i) Being specified in terms of predictors, NIPALS models make for clear-cut coordination of the conceptual specification of the model and its operative use for purposes of causal analysis and/or prediction;

(ii) In comparison with other approaches, and in particular the maximum likelihood method, NIPALS is often more general, and typically so since it works with a smaller number of zero intercorrelation assumptions between residuals and variables. Hence the NIPALS approach makes for models that give a closer fit to the given observations, as is reflected in successful applications to real-world data.

(iii) NIPALS estimation procedures are sequences of OLS regressions, and are therefore easy to program and handle on the computer.

A draft of this report was presented at the Bell Laboratories when I was serving as visiting consultant in June 1972. I am indebted to Dr. R. Gnanadesikan, head of the statistics department, and several associates for valuable comments on the theory and application of NIPALS modelling. Specifically, the discussion with D. J. Kruskal on model versus estimation aspects of NIPALS is gratefully acknowledged.

The numerical results in Section 5 draw from current team work projects at the Dept. of Statistics, University of Göteborg. In Section 5.1 the computer calculations have been performed by P. Högberg, in Section 5.3 by B. Areskough, in Section 5.4 by P. Nilsson. Further results of the various projects will be published elsewhere.

2. WHAT IS NIPALS MODELLING?

2.1. NIPALS modelling is a design for the linearization of models that are nonlinear in the parameters. The design is an *ad hoc* combination of (i) model specification in terms of causal and/or predictive relations, and (ii) parameter estimation.

To specify, we consider a nonlinear model M. Let A be a subset of the parameters of M, and let M_A be the model obtained by keeping all parameters of M fixed except those in A. We shall say that M_A is a *submodel* of M. In forming a submodel M_K, model M may possibly be subject to a suitable transformation.

With regard to (i), NIPALS modelling involves two characteristic features:

2.1.1. The parameters are suitably grouped, say in three groups A, B, C with parameters

$$\alpha_i\,(i = 1, \ldots, r); \qquad \beta_j(j = 1, \ldots, s); \qquad \gamma_k\,(k = 1, \ldots, t) \tag{1}$$

in such manner that the corresponding submodels M_A, M_B, M_C are linear in their free parameters. Typically, a submodel, say M_A, takes the form

$$Y = \alpha_1 X_1 + \cdots + \alpha_r X_r + \varepsilon \tag{2}$$

The variables $X_i\,(i = 1, \ldots, r)$ and Y are functions that may involve any variables of model M and any of its parameters except those in group A. The residual ε is the deviation between Y and its systematic part $\alpha_1 X_1 + \cdots + \alpha_r X_r$. It may well be that the variables X_i and Y are vector variables.

2.1.2. For each submodel M_K ($K = A$, B, or C), say M_A, relation (2) is specified as a predictor for its left-hand variable Y. That is, the systematic part of Y is assumed to be the conditional expectation of Y for given $X_i\,(i = 1, \ldots, r)$; in symbols,

$$E(Y \mid X_1, \ldots, X_r) = \alpha_1 X_1 + \cdots + \alpha_r X_r \tag{3}$$

With regard to (ii), NIPALS modelling involves a third characteristic feature:

2.1.3. In accordance with 2.1.1, the estimation is performed iteratively by the standard technique known as *relaxation*.[2] In accordance with 2.1.2, each step of the iteration involves the application of OLS regression to a submodel M_K ($K = A$, B, C) to obtain proxy estimates of its parameters. Writing

$$\alpha, \beta, \gamma \qquad \text{and} \qquad a, b, c \tag{4}$$

for the parameters (1) and their NIPALS estimates in vector form, let

$$a^{(1)}, b^{(1)}, c^{(1)}; \ldots ; a^{(s)}, b^{(s)}, c^{(s)}; \ldots \tag{5}$$

denote the consecutive parameter proxies given by the iterative procedure. The proxies $a^{(1)}$, $b^{(1)}$ that start the iterative procedure are specified arbitrarily or by some *ad hoc* device, and the proxies $a^{(s)}$, $b^{(s)}$, $c^{(s)}$ in the sth round are obtained by OLS regression in accordance with submodels M_A, M_B, and M_C, using $b^{(s-1)}$, $c^{(s-1)}$ and $c^{(s-1)}$, $a^{(s)}$, and $a^{(s)}$, $b^{(s)}$ as fixed parameter proxies, respectively. Finally, if the iterative procedure converges, the limiting values

$$a = \lim_{s\to\infty} a^{(s)}; \qquad b = \lim_{s\to\infty} b^{(s)}; \qquad c = \lim_{s\to\infty} c^{(s)} \tag{6}$$

are the NIPALS estimates of the parameters α, β, γ.

[2] See, for example, Varga [26].

2.2. Comments

2.2.1. To repeat, the relaxation device 2.1.3 as used separately is a standard technique in numerical analysis. It is the combination with predictor specification and OLS regression that is specific to NIPALS modelling.
2.2.2. In the predictor specification 2.1.2 the asymmetric modelling of the variables Y and X_i is a key feature. The asymmetry makes for clear-cut coordination of the conceptual definition of the model M and its operative use for purposes of cause–effect analysis and/or prediction. The systematic part in formula (2) provides a least squares prediction (3) of Y in terms of the X_i's. Relations (2)–(3) may or may not be causal; that is, they may or may not include the hypothesis that the X_i's are causal factors that influence the effect variable.[3]
2.2.3. By this report I take the opportunity to change the name of the device from NIPALS estimation to NIPALS modelling. This is a better name since the approach involves not only parameter estimation but also model specification. From the outset the specification of submodels in terms of predictors was a key feature of the NIPALS approach.[1]
2.2.4. OLS regression as applied to a predictor (2)–(3) provides parameter estimates, say

$$a_1, \ldots, a_r \tag{7}$$

that under general conditions are consistent in the large-sample sense.[4] It is sufficient for consistency that the first- and second-order sample moments of the variables Y, X_i $(i = 1, \ldots, r)$ tend to the corresponding population moments as the sample size increases indefinitely, and that the population has no collinearity among the variables X_i. The salient point of the argument is that OLS regression parameters are continuous functions of the population moments of first and second order.

The argument carries over to iterative OLS regression procedures. Hence under general conditions of regularity the NIPALS estimates (6) are consistent.
2.2.5. While the flexibility of NIPALS modelling shows off in a diversified scatter of applications, most of the theoretical and applied work has thus far dealt with two types of models, namely, interdependent (ID) systems and principal components. The NIPALS modelling of ID systems will be in focus in Sections 4.2–4.3. As to the NIPALS estimation of principal components, a key advantage is that the data need not be complete.[5] For another thing, the NIPALS estimation procedure extends without change to the

[3] To link up with the author's earlier work on the definition and meaning of causal concepts, see [32].

[4] Wold [28].

[5] Christoffersson [8].

approach known as Single Value Decomposition.[6] These two features combine to make for a broad range of applications of principal components modelling. (See, for example, Christoffersson [8], Wold and Sjöström [34].)
2.2.6. Dr. Joseph Kruskal has posed the interesting question whether an explicit definition can be given for the class of nonlinear models that constitutes the scope of NIPALS modelling. NIPALS modelling is highly flexible, allowing the combined use of several devices, including parameter grouping and relaxation; auxiliary transformation of the model; and modelling the predictors in terms of indirectly observed variables and other hypothetical constructs. Hence I see NIPALS modelling as an open ended array of models with unlimited complexity in the combined use of several devices. Illustrations and case studies of these various devices, rather than a closed general formula, will provide a feeling for the scope of NIPALS modelling.

3. LOW INFORMATION VERSUS HIGH INFORMATION MODELLING

3.1. In the social sciences over the last decade there has been a forceful development of causal model building by the use of path models.[7] A new and most significant feature is that econometric methods have been incorporated in the modelling, a merger that brings together widely different lines of evolution and thereby opens up new vistas and broadens the scope of the approach. The main incentive for the present paper is to explore these developments from the point of view of NIPALS modelling. For background purposes, we shall in this section take a look at the information aspects of model building, and in particular two lines of evolution from low information to high information modelling.

3.2. Speaking broadly, evolution in science proceeds in stages of increasing knowledge, from primitive models that involve rudimentary or relatively little information, to advanced models that incorporate plenty of detailed and well-established information. Chart 3.2 marks the distinction between descriptive and explanatory knowledge, and indicates two lines of evolution from low information to high information models.[8] Some illustrations, mainly from the social sciences, are briefly spelled out.

3.2.1. Univariate measurement provides no explanation, and is low in descriptive knowledge, whether the information is obtained under controlled

[6] Eckart–Young [10b]. In fact, the NIPALS model for the principal components remains the same, but the data input is a matrix of raw data (time series or cross section data) and not the corresponding correlation matrix. See also Carroll–Chang [6].

[7] See the recent readers edited by Blalock [4] and Costner [9].

[8] For discussions that led to the 2×2 classification in Chart 3.2 I am indebted to my son Svante W.

experimental conditions or by sampling surveys. Classification of uni- or multivariate measurements is low in explanatory information and is usually on the low side in description. Graph theory is somewhat more advanced in information, but is on the whole on the low side with regard to both description and explanation.

3.2.2. Controlled experiments of the stimulus–response type provide a high degree of explanation, inasmuch as they yield reproducible knowledge in the form of cause–effect relationships. Typically, they provide one parameter for each stimulus variable, other causal influences being eliminated, either by keeping them constant or by the wellknown device of randomization. Hence the model involves only some few parameters and is therefore on the low side as regards description. Similarly, cause-effect models based on nonexperimental data are high in information and low in description, although the assessment of such relations by multiple regression analysis provides some information about the influence of factors that it is not the primary purpose of the model to investigate. Typical cases in point are the econometric relations of consumer demand as influenced by market price and consumer income, and the Cobb–Douglas production functions with capital and labour as explanatory variables.

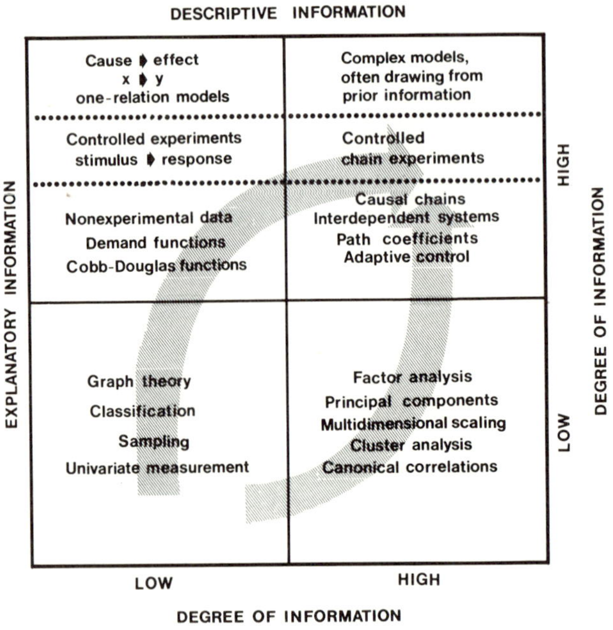

Chart 3.2

3.2.3. Through the two left-hand quadrants of Chart 3.2 now considered goes a line of evolution to the upper right-hand quadrant. Here the models are high in both explanation and description, inasmuch as they are complex structures that have high aspirations of explanatory power, and provide plenty of descriptive knowledge by their many parameters. Models in the nature of causal flows take a central place. The models may derive their explanatory information from controlled experiments; cases in point are the chain experiments of organic chemistry.[9] They may also be nonexperimental, their explanatory power being based on prior information in the form of theoretical argument or nonexperimental evidence. At this advanced stage of evolution we find the demographic forecasting models with their detailed classification after age, sex, and other relevant factors. Econometrics has reached this stage by the multirelational models of Tinbergen (causal chains) and Haavelmo (interdependent systems) and their followers.[10]

3.2.4. A second line of evolution from the lower left to the upper right goes through the lower right-hand quadrant. Typical of this quadrant are principal component (PC) models which are high in description thanks to their many parameters, while they often are low in explanatory aspiration. Factor analysis (FA) models are closely related to PC models, both using the device of hypothetical variables indirectly observed as latent variables or constructs. FA and PC models introduce asymmetry between the variables, and thereby an explanatory element in the modelling. Here as in the other quadrants there are no sharp lines of demarcation relative to adjacent quadrants. In psychology and education the FA and PC models usually are based on experimental data; this makes for reproducible features in the factors and components, bringing the model forward in the evolution to the upper-right quadrant.

Part of the problem in PC and FA modelling is to assess the appropriate number of dimensions of the model. Reference is made to multidimensional scaling and other methods for the assessment of dimensionality.[11] As a preparatory stage for PC and FA modelling, the methods of dimension analysis belong under a relatively early phase of the second line of evolution in model building.

Along the second line of evolution, at an early phase near the lower-left quadrant, we find correlation coefficients, association coefficients, χ^2 analysis for the testing of independence between variables, and many related ap-

[9] For example, Woodward's synthesis of reserpine [35] is a chain experiment with no less than 22 links.

[10] Tinbergen [22, 23], Haavelmo [12]. For a review with emphasis on information aspects, see Meissner [18].

[11] For the treatment of a problem of multidimensional scaling by the use of NIPALS modelling, see Carroll and Chang [6].

proaches in modelling. The symmetric treatment of the variables in these approaches is symptomatic of the low degree of explanatory aspirations. Canonical correlations mark a somewhat higher stage of description.

Partial correlation coefficients introduce asymmetry among the variables, and thereby make a step in the direction of explanation. Further up on the second line of evolution, the "path models" of Sewall Wright take the form of systems of partial correlations (or, as emphasized by John Tukey, partial regressions).[12] In the genetic applications of Wright the path models are distinctly explanatory in their combined use of nonexperimental data and prior theoretical arguments; accordingly, they have been placed in the upper-right quadrant.

3.2.5. Sociology, for a long time hampered by the anticausal ban of Russell [20], is now the scene of an explosion of causal model building. In sociological models in the form of causal flows the two lines of evolution in Chart 3.2 run together, on the one hand the causal chain and interdependent systems of econometrics, on the other hand the path models of genetics. In the most recent developments, furthermore, the sociological models make use of indirectly observed variables, a key feature in factor analysis and principal components and some econometric models.[13] This melting pot of new developments offers plenty of opportunities to NIPALS modelling.

4. CAUSAL FLOW MODELS IN ECONOMETRICS AND THE BEHAVIOURAL SCIENCES

4.1. Graphic Illustrations; Notation

Charts 4.1–4.6 illustrate models that will be in focus. Some comments with a view to bridging the terminological differences between econometric and path models follow.

Charts 4.1 and 4.3–4.6 illustrate path models. Chart 4.2 is of econometric type (Tinbergen's arrow scheme) and illustrates the same model as Chart 4.1. All through, the arrows indicate causal and/or predictive relations. Endogenous variables are variables to which one or more arrows are directed. Exogenous variables are variables from which one or more variables are directed, but to which no variable is directed. Predetermined variables are either exogenous or lagged endogenous variables. In the graphs of path models, directly observed variables are represented by squares, indirectly observed

[12] Wright [36]; Tukey [24].

[13] Specific reference is made to the socio-economic modelling by Adelman and Morris [1], a merger of econometric models and factor analysis that makes a significant breakthrough into the upper-right quadrant.

variables by circles.[14] The notations y, y^* refer to a directly observed variable and the corresponding indirectly observed variable.

In econometric models illustrated by Tinbergen's arrow scheme the data are time series, and for each time point there is one observation for each variable. Path model graphs, for example, Chart 4.1, do not specify the individual observations. To put it otherwise, in path models the relations between the variables usually are static; Tinbergen's arrow schemes usually refer to dynamic models. When Chart 4.1 is translated to the arrow scheme in Chart 4.2 the model remains static, as indicated by the solid arrows. If x is replaced by y_1 lagged by one time unit, as indicated by the broken arrows, the model becomes dynamic.

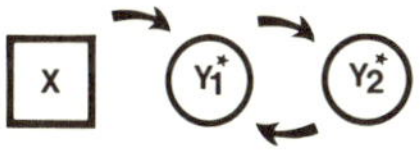

Chart 4.1

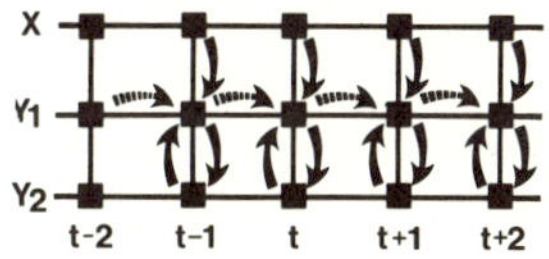

Chart 4.2

4.1.1. *Models L, L*, L**.*[15] For a unified treatment of the merging lines in econometric and path modelling we shall consider three models where the explanatory variables are specified alternatively as directly or indirectly observed.

The models are assumed to have n endogenous variables $Y = (y_1, \ldots, y_n)$ and m predetermined variables $Z = (z_1, \ldots, z_m)$. To avoid constant terms in the various relations, all variables are assumed to be measured from their mean.

We use the same notations $\mathrm{B} = |\beta_{ik}|$, $\Gamma = |\gamma_{ik}|$ for the parameters of all models L–L^{**} although in general they are numerically different. Estimated values of B, Γ are denoted B, G.

Lest models L–L^{**} take an endogenous variable to be explained by itself, we assume that in matrix $\mathrm{B} = |\beta_{ik}|$ all elements β_{ii} in the main diagonal are zero.

4.1.2. *Behavioural Relations versus Identities.*[16] In this brief exposition we shall focus on systems where all relations of the structural form are behavioural. This is the typical case for models in sociology and the behavioural sciences. In econometric models the structural form often includes one or more identities; that is, deterministic (residual-free) relations with numerically specified

[14] This usage is adopted from Jöreskog [15].

[15] For models L and L^* the exposition leans heavily on Mosbaek and Wold [19].

[16] For a treatment from the present point of view, see Ågren [2b].

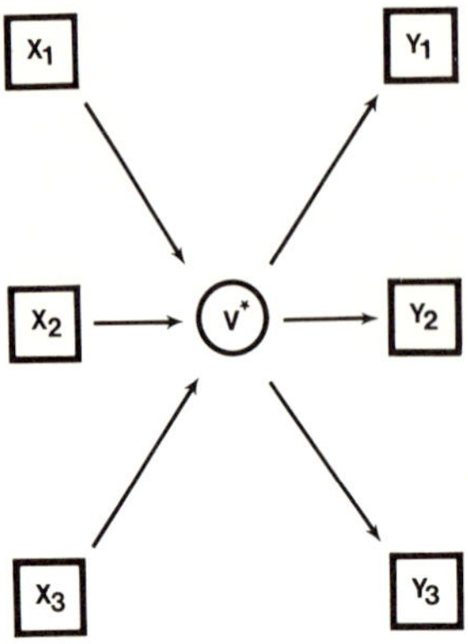

Chart 4.3

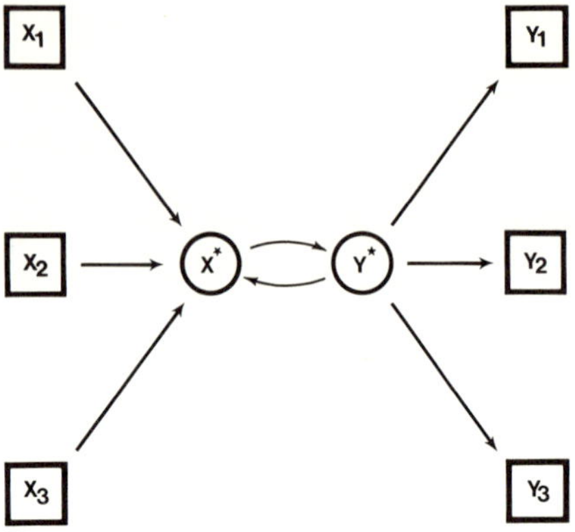

Chart 4.4

coefficients. Thus unless explicitly stated otherwise we shall assume that there are no identities in the system under consideration.

4.1.3. *Causal Chain* (*CC*) *versus Interdependent* (*ID*) *Systems.* This distinction, first defined for models L, carries over to models L^* and L^{**}: In CC-type models, but not in ID-type models, the endogenous variables y_i can be ordered so that matrix B is *subdiagonal*; that is, $\beta_{ik} = 0$ for all $k > i$. For example, Charts 4.1–4.2 and 4.4 illustrate ID systems, Charts 4.3 and 4.5–4.6 CC systems.

In ID systems, accordingly, it is the coefficients β_{ik} with $k > i$ that make for interdependence. Speaking broadly, the interdependence is weak if all coefficients β_{ia} with $\beta_{ia}\beta_{ai} \neq 0$ are numerically small.

4.1.4. *The Parity Principle.*[17] This name refers to the well-known principle that if a model has p parameters at free disposal, these can as a rule be determined so that the model satisfies p—and not more than p—restrictions or assumptions. For example, multiple OLS regression honours the parity principle, inasmuch as the regression residual has zero intercorrelation with each explanatory variable (including the constant "variable"), and there is one parameter for each explanatory variable.

4.2. Model L: All Variables Are Directly Observed[15]

Using vector notation, we define L as follows,

$$\text{Structural form:} \qquad Y = \mathrm{B}Y + \Gamma Z + \delta \tag{1}$$

$$\text{Reduced form:} \qquad Y = \Omega Y + \varepsilon \tag{2a}$$

$$\text{with} \quad \Omega = [I - \mathrm{B}]^{-1}\Gamma \quad \text{and} \quad \varepsilon = [I - \mathrm{B}]^{-1}\delta \tag{2b, c}$$

[17] Wold [33], Chapter 9.4(b). The parity principle rules with more force in deterministic than in stochastic models.

4.2.1. To prepare for the definition of CC and ID systems, we write down the corresponding predictor specifications, giving for the structural form (1)

$$E(Y \mid Y, Z) = \mathrm{B}Y + \Gamma Z \tag{3}$$

and for the reduced form (2a–c)

$$E(Y|Z) = \Omega Z = [I - \mathrm{B}]^{-1}\Gamma Z \tag{4a, b}$$

In words, (3) assumes that for each endogenous variable y_i the systematic part of the ith structural relation gives the conditional expectation of y_i, for known values of the variables y_j, z_k that occur in the right-hand member of the same ith relation. In (4a, b) it is assumed that the systematic part of the ith relation of the reduced form is the conditional expectation of y_i, given the variables z_k that occur in the same ith relation.

Comment. The specifications (3) and (4a,b) are incompatible, inasmuch as (4a, b) in general is not valid if (3) holds true, and conversely.

4.2.2. *Causal Chain* (*CC*) *versus Interdependent* (*ID*) *Systems of Type L.* CC and ID systems are often specified with regard to the residuals and their correlation properties. The following brief review emphasizes the connection with the predictor specifications (3)–(4) of linear CC and ID systems.

(i) The following assumption is typical for CC systems: In each structural relation the residual is uncorrelated with the variables y_j, z_k that are explanatory in the same relation. The corresponding predictor specification is (3); as is readily seen, the correlation assumptions are simple implications of (3).

(ii) For CC systems, on the predictor specification (3) or the correlation assumptions (i), OLS regression as applied to the relations of the structural form (1) gives consistent parameter estimates B, G. Formula (2b) then gives consistent estimates for the parameters Ω of the reducted form, say

$$W = [I - B]^{-1}G \tag{5}$$

Well to note, OLS regression directly applied to the reduced form (2) of a CC system in general will *not* give consistent estimates for the parameters Ω.

(iii) In CC systems, the structural form (1) can be used in accordance with (3) for purposes of predictive and causal inference. If part of the structural form is used for prediction in terms of given values for some of the variables Y, Z, the given values are constrained by the relations

$$Y = [I - B]^{-1}GZ \tag{6}$$

Subject to supplementary conditions, the reduced form (2a) of a CC system can be used for prediction in accordance with (5) but the residuals (prediction errors) in general will be larger than in the structural form.

(iv) The following assumption is classical for ID systems: In the ith structural relation ($i = 1, \ldots, n$) the residual δ_i is uncorrelated with all predetermined variables z_k ($k = 1, \ldots, m$). The corresponding predictor specification is (4); as is readily seen, the correlation assumptions are simple implications of (4a).[4]

Comment. By the comment in 4.2.1, the predictor specification (4) of the reduced form of ID systems rules out the predictor specification (3) of the structural form. To spell out, the classical specification (iv) of ID systems in general implies

$$E(Y \mid Y, Z) \neq \mathrm{B}Y + \Gamma Z \tag{7}$$

(v) Corollaries to (4a b) and (7): For classical ID systems, OLS regression gives consistent parameter estimates when applied to the reduced form, but in general not when applied to the structural form.

Reference is made to the estimation methods that have been devised for ID systems under the classical assumptions, including the FIML (full information maximum likelihood), LIML (limited information maximum likelihood), TSLS (two-stage least squares) and 3SLS (three-stage least squares) methods.[18] All of these methods begin by estimating the parameters Ω of the reduced form by OLS regression, proceeding in various ways to the estimation of the parameters B, Γ, This is the famous identification problem of classical ID systems.

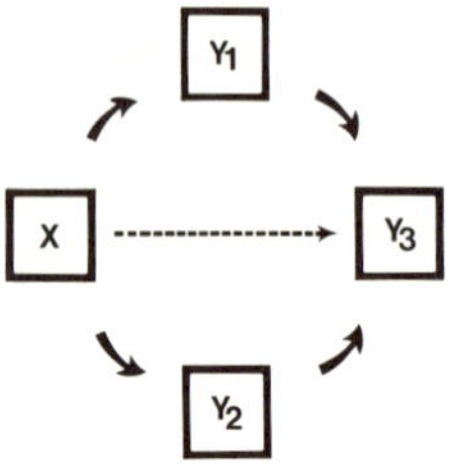

Chart 4.5

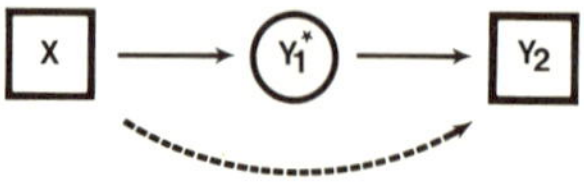

Chart 4.6

(vi) In view of (7) and the first para of (v), what is the operative meaning of the structural form of an ID system?[19] We shall see that an operative answer to this much-discussed question is provided by an interpretation in terms of models L^*.

Comment. Clearly, the correlation assumptions (i) of CC systems honour the parity principle 4.1.4, whereas the assumptions (iv) of classical ID systems will violate the parity principle as soon as the system is overidentified—and

[18] For an exposition from the present point of view, see Lyttkens [19, Chapter 11].
[19] Specific reference is made to the symposium by Christ *et al.* [7].

this is the customary case in practice. An illustration is the one-loop ID system with all $\beta_{ik} = 0$ except for

$$\beta_{1n} \neq 0 \qquad \text{and} \qquad \beta_{i,\,i-1} \neq 0 \qquad (i = 2, \ldots, n) \tag{8}$$

Then if there are n predetermined variables, all different, and just one occurring in each structural relation, we see that there are $2n$ parameters B, Γ, whereas the classical assumptions (iv) require n^2 zero intercorrelations between residuals and predetermined variables. This violates the parity principle as soon as the loop system has three or more relations.

In the subsequent L^* approach to ID systems, we shall obtain not only an operative interpretation of the structural form, but also, as a by-product, a generalization of the classical assumptions (iv) that honours the parity principle.

4.3. Model L^*: Each Explanatory Variable Is the Systematic Part of the Corresponding Directly Observed Variable.[15]

The definition of model L^* is, in structural form,

$$Y = \mathrm{B} Y^* + \Gamma Z + \varepsilon \qquad \text{with} \qquad Y = Y^* + \varepsilon \tag{9a,b}$$

The reduced form takes the same form (2a-c) as for model L, now with the same residual ε in (9) as in (2). Hence

$$Y^* = [I - \mathrm{B}]^{-1} \Gamma Z \tag{10}$$

If we use (9b) to rewrite (9a) as in (1), the ensuing residuals are again related by (2c). Thus far, accordingly, there is only a formal difference between models L and L^*.

4.3.1. *Classic ID versus General ID (GEID) systems of Type L^*.* The classic ID assumptions 4.2.2 (iv) of model L carry over without change to classic ID systems of type L^*, again giving the predictor (4), and again (7) but not (3).

In GEID systems, the residual ε_i of the ith structural relation ($i = 1, \ldots, n$) is assumed to be uncorrelated with the variables Y^* and Z that occur in the ith relation.[20]

Under general regularity conditions GEID systems in structural form can be specified as predictors, giving

$$E(Y \mid Y^*, Z) = \mathrm{B} Y^* + \Gamma Z \tag{11}$$

The reduced form can be used for prediction on the basis of (4a,b) or (11), subject to the same residual ε as for (9b). If part or the whole of the structural form is used for prediction on the basis of (11), in terms of given values for Y^* and Z, the given values are constrained by (10).

Turning to estimation procedures, we note that (11) does not permit us

[20] Wold [29]; see also Mosbaek and Wold [19].

to apply OLS regression directly to the structural form (9), inasmuch as Y^* is not observed and must be assessed by the estimation procedure.

(i) *Estimation of classic ID systems.* The classic methods 4.2.2 (v) for the estimation of ID systems of type L carry over to type L^* by using (9b) to rewrite (9a) as (1), and returning to (9a) after the estimation. The TSLS method is directly suited for model L^*, since it first uses OLS regression to estimate Y^* from the reduced form, carries the estimated Y^* into the structural form (9a), and then estimates (9a) by OLS regression.

(ii) *Estimation of GEID systems.* Reference is made to the fix-point (FP) method, a NIPALS procedure devised for the estimation of classic ID and GEID systems.[20] The FP method has been improved and extended in several ways. The family of FP methods includes the FFP (Fractional FP) method of A. Ågren, the RFP (Recursive FP) method of L. Bodin and the PFP (Parametric FP) method of Lyttkens.[21] The various FP methods, all of which are iterative, may be valid under different conditions, but they are equivalent in the sense that if two of the iterative procedures converge, these will converge to the same limiting estimates.

(iii) *The special case of CC systems of type L^*.* Here the FP estimation reduces to n consecutive OLS regressions, one for each relation of the structural form, and each step involving a substitution of Y^* as estimated by the previous OLS regressions. Such a system will in general have larger residuals than the corresponding CC system of type L; the argument is here the same as in 4.2.2 (iii).

(iv) The GEID-FP approach has been designed with a view to avoiding two weaknesses of the classic estimation methods 4.2.2 (v). We note:

1. In the reduced form (2) of an ID system each relation will as a rule contain all of the predetermined variables Z. Hence OLS estimation of the reduced form, which is the first phase of all classic methods, is marred by collinearities when there are many predetermined variables, as is the case in medium size or large systems. The FP method works all the time in the structural form, where each relation as a rule contains only some few parameters, and thereby avoids the collinearity trouble.[20]

2. When the structural parameters B, Γ are not uniquely determined by the parameters Ω of the reduced form (overidentified ID systems, which is the typical case in practice), classic ID systems violate the parity principle 4.1.4. The FP method, on the other hand, honours the parity principle.[22]

[21] Ågren [2a-b], Bodin [5a-b], Lyttkens [33, Chapter 13.1].

[22] The FP method shows up favourably in comparative studies of different estimation methods as applied to real-world data; see Bergström [3a, Table 17.1.4; 3b]. It would seem that the parity principle gives a plausible explanation of the fine performance of the FP method. Note that these applications deal with ID systems where the structural form contains one or more identities among the behavioural relations; cf. 4.1.2.

(v) This last statement requires some qualification for ID and GEID systems that contain identities (see 4.1.2). The FP family of modelling methods carries over to such systems with little or no formal change.[20–22] If there is only a small number of identities, the parity principle will be less violated by the GEID than by the classic ID assumptions, but if there are many identities the situation is reversed. Reference is made to the IIV (iterative instrument variables) estimation method by E. Lyttkens, which has been designed with particular view to ID systems with many identities, and which differs from FP and NIPALS modelling in being symmetric with regard to the endogenous variables Y.[23]

4.4. Models L^{**}: All Explanatory Variables May Be Indirectly Observed

The definition of model L^{**} is, in structural form

$$Y = \mathrm{B}Y^* + \Gamma Z^* + \Delta \tag{12}$$

where Y^* is defined as for models L^*, while Z^* denotes the vector of predetermined variables z_i or z_i^*, each of which either is directly observed or is latent and observed by means of one or more indicators. As for models L^*, solving for Y^* gives the reduced form,

$$Y = [I - \mathrm{B}]^{-1}Z^* + \Delta \tag{13}$$

with the same residual in both forms; that is,

$$Y = Y^* + \Delta \tag{14}$$

The predictor specifications for the two forms read

$$E(Y|\, Y^*, Z^*) = \mathrm{B}Y^* + \Gamma Z^*, \tag{15}$$

$$E(Y|Z^*) = [I - \mathrm{B}]^{-1}\Gamma Z^* \tag{16}$$

and are subject to much the same interpretation as for (4) and (11).

Models L^{**} take many forms. In what follows, the aim of the exposition is not broad coverage, but rather to explore some few of the new types of problems that arise. Specifically, we shall be concerned with models L^{**} where the design involves one or both of the following features:

(i) Latent variables with one or more indicators;

(ii) "Split Variable Modelling," that is, the systematic part of a causal variable and the remaining part are not necessarily assumed to have the same causal influence.

[23] Lyttkens [17; also cf. 2.2.2.].

The limited purpose of Sections 4.4 and 5 must be emphasized. Section 5 gives four case studies to illustrate the devices (i)–(ii), and we shall conclude the present section with some general remarks to link up with the review of Models L and L^* in Sections 4.1–4.3.

4.4.1. *Models L^* and L^{**}; Some Estimation Aspects*

(i) The maximum likelihood (ML) method is of general scope in NIPALS modelling, as for other types of model. Specific reference is made to the important work of K. G. Jöreskog, who has used the Fletcher–Powell approach to develop computer techniques for the estimation of broad classes of covariance structures.[24]

(ii) The FP method does not carry over directly from L^* to L^{**} and has to be combined with other devices to cope with the indirectly observed Z^* that occurs in (12). We proceed to discuss the situation by way of case studies (Section 5), showing how the NIPALS approach can be adapted to estimate some few models quoted from the current literature. The emphasis is on the operative interpretation of the models and on the flexibility of the NIPALS approach.

(iii) *Ågren's convergence principle.* For the FP procedure (and similarly for FFP), Ågren derives an approximate convergence criterion on the argument that the iterative procedure, if convergent, will in the limit behave as the well-known iterative procedure for solving with respect to y^* the equation system $y^* = \mathrm{B}y^* + \Gamma Z$ that involves matrices B, Γ which do not vary in the course of the iterations. Although approximate, Ågren's criterion is very useful in applied work. Hence it can be expected that Ågren's argument, which we shall call Ågren's convergence principle, is of wide scope and extends to other NIPALS procedures under consideration.

(iv) The exposition in Sections 4.1–4.3 of FP and related NIPALS approaches provides a diversified array of special cases if seen in conjunction with the case studies in Sections 5.1–5.4. In conclusion, this array of special cases will now be commented upon with regard to the general relationship between model specification, estimation procedure, and other phases of NIPALS modelling.

Once a NIPALS model is specified in accordance with 2.1.1–2.1.2, the basic design of the corresponding estimation procedure can be inferred, each predictor being translated into an OLS regression, and the graphical path pattern of the model indicating the consecutive order of the regressions. It is then straightforward matter to program the NIPALS estimation procedure for the computer. The further developments will depend very much on the degree of complexity of the model. For very simple models the NIPALS

[24] Jöreskog [15], which paper includes references to his earlier related work.

procedure can be replaced by algebraic solutions for the parameter estimates.[25] The more complex the model, the more of the numerical analysis will be computer work. Test runs of the computer program will show whether the procedure performs to satisfaction, or whether refinements are needed to make for better convergence and/or more speed in the computations.[21] As to standard errors and small-sample bias, the classical techniques using Taylor developments may carry over to provide large-sample approximations.[26] Again, the complexity of the resulting formulas increases rapidly with the complexity of the model. Hence "rough-and-dirty" methods come to the fore, and in particular it would seem that John Tukey's jackknife is promising in applications to NIPALS estimation procedures.

5. CASE STUDIES IN NIPALS MODELLING

5.1. Haldane's Breathing Experiments upon Himself (1905)

Reference is made to the model, illustrated in Chart 4.1, and the statistical analysis presented by Turner and Stevens [25]. Their model being of type L^*, we quote it in the notation (4.1):

$$y_1 = \gamma_{10} + \gamma_{11} z_1 + \beta_{12} y_2^* + \varepsilon_1, \qquad y_2 = \gamma_{20} + \beta_{21} y_1^* + \varepsilon_2 \quad \text{(1a,b)}$$

where z_1 is the percentage of CO_2 inhaled, y_1 the percentage of CO_2 in the blood, and y_2 the depth of breathing in cm^3. We spell out the vectors of the variables and the parameter matrices,

$$Y = \begin{bmatrix} y_1 \\ y_2 \end{bmatrix}; \quad Z = \begin{bmatrix} 1 \\ z_1 \end{bmatrix}; \quad \mathrm{B} = \begin{bmatrix} 0 & \beta_{12} \\ \beta_{21} & 0 \end{bmatrix}; \quad \Gamma = \begin{bmatrix} \gamma_{10} & \gamma_{11} \\ \gamma_{20} & 0 \end{bmatrix}. \quad \text{(2a,d)}$$

Relation (1a) is underidentified. Turner and Stevens avoid this difficulty by assuming, on the basis of prior argument,

$$\gamma_{11} (= g_{11}) = 1. \quad \text{(3a)}$$

Using FIML estimation (see 4.2.2), Turner and Stevens obtain

$$b_{12} = -0.002764; \qquad b_{21} = 977.1. \quad \text{(3b,d)}$$

We shall see that a slight change in Turner–Stevens' model will suffice to avoid the prior assumption (3a).

5.1.1. Let us modify relation (1a) by introducing

$$\varepsilon_2 = y_2 - y_2^* \quad \text{(4)}$$

[25] For a case in point, see Lyttkens [33, Chapter 14].

[26] Lyttkens [16].

as a third explanatory variable, say

$$y_1 = \gamma_{10} + \gamma_{11} z_1 + \beta_{12} y_2^* + \mu_{12} \varepsilon_2 + \varepsilon_1 \tag{5}$$

Further let M be the model obtained by combining (5) with (1b) and by specifying both relations as predictors, giving

$$E(y_1 \mid z_1, y_2^*, \varepsilon_2) = \gamma_{10} + \gamma_{11} z_1 + \beta_{12} y_2^* + \mu_{12} \varepsilon_2 \tag{6a}$$

$$E(y_2 \mid y_1^*) = \gamma_{20} + \beta_{21} y_1^* \tag{6b}$$

Model M is a case of SVD (Split Variable Design) modelling in the sense of 4.4(ii). We see that Model M reduces to (1a, b) if

$$\mu_{12} = 0 \tag{7}$$

On the other hand, if

$$\mu_{12} = \beta_{12} \tag{8}$$

the relations (5) and (6a) reduce to an ordinary regression,

$$y_1 = \gamma_{10} + \gamma_{11} z_1 + \beta_{12} y_2 + \varepsilon_1 \tag{9a}$$

with

$$E(y_1 \mid z_1, y_2) = \gamma_{10} + \gamma_{11} z_1 + \beta_{12} y_2 \tag{9b}$$

Let M_1 be the SVD model defined by (9a, b) in conjunction with (1b) and (6b). We see that M_1 can be consistently estimated by two OLS regressions. First, OLS as applied to (9a) gives estimates for the parameters involved and for the systematic part y_1^*. Second, using the estimated y_1^*, OLS is applied to (1b). The following OLS estimates have been obtained for Model M_1,

$$g_{11} = 0.51; \qquad b_{12} = -0.0009; \qquad b_{21} = 916.04 \tag{10}$$

5.1.2 *Comments.*

(i) Although Model M_1 is rather similar to Turner–Stevens' model (1a, b), there are marked differences in the numerical results. It is a question whether Haldane's data contains enough information to confirm Turner-Stevens' model, but I do not wish to argue in favour of M_1 as an alternative hypothesis.

(ii) It will be noted that the same device does not work if applied to the model, say M_2, obtained by remodelling (1b) instead of (1a). In fact, if we assume

$$y_2 = \gamma_{20} + \gamma_{21} y_1 + \varepsilon_2 \tag{11}$$

and apply OLS regression, the resulting systematic part y_2^* will be exactly linear in y_1; hence OLS regression as applied to (1a) will give $g_{11} = 0$, which obviously is a nonsense result. This is not an accidental happening; we shall come back to this point in Comment (iv).

(iii) Model M_1 (or M_2) can be seen as a NIPALS hybrid between a causal chain and an interdependent system. The limited purpose of section 5.1.1 is to show that it may suffice to form such a hybrid in order to remove under-identification. Turner-Stevens' model was chosen for illustration because of its very simple structure, and not with a view to make more than a formal comparison. Cases of SVD modelling with more subject-matter substance will follow in Section 5.2.

(iv) The SVD model M formed by (5) and (1b) in conjunction with (6a, b) has provided a unified view of the two models under consideration, namely Turner–Stevens' model (1a, b), which is an ID system, and the hybrid (9a, b) between a causal chain and an ID system. Turner–Stevens' system (1a, b) is underidentified, and it will be noted that the SVD approach (5) in itself did not help to make the system (1a, b) identified. This comment is of general scope, and it is clear that for the SVD approach to be of relevance in an ID system, the system must be overidentified. Again with reference to Section 5.2, other areas where SVD modeling may be useful are ordinary (unirelational) regression analysis, and causal chain systems.

5.2. Milton Friedman's "Permanent Income" Hypothesis[27]

We shall consider three models (12)–(14), all of which refer to consumer demand for n goods ($i = 1, \ldots, n$),

$$y_{it} = \gamma_{io} + \gamma_{i1} x_{it} + \gamma_{i2} z_t + \varepsilon_{it}, \tag{12}$$

$$y_{it} = \gamma_{io} + \gamma_{i1} x_{it} + \gamma_{i2} z_t^* + \varepsilon_{it}, \tag{13}$$

$$y_{it} = \gamma_{io} + \gamma_{i1} x_{it} + \gamma_{i2} z_t^* + \mu_{i2} z_t^{**} + \varepsilon_i, \tag{14}$$

The models are specified as predictors; to spell out for (12),

$$E(y_{it} | x_{it}, z_t) = \gamma_{io} + \gamma_{i1} x_{it} + \gamma_{i2} z_t \tag{15}$$

The models are logarithmic transforms of demand functions with constant elasticities elasticities γ_{i1}, γ_{i2}, μ_{i2}; again spelling out for (12),

$$d_{it} = c p_{it}^{\gamma_{i1}} q_t^{\gamma_{i2}} (1 + \Delta_i) \tag{16}$$

[27] Friedman [11].

The variables are specified as follows.

$y_{it} = \log d_{it}$ is consumer demand for the ith good;

$x_{it} = \log p_{it}$ is the price of the ith good;

$z_t = \log q_t$ is consumer income;

$z_t^* = \log q_t^*$ is consumer income, the permanent part;

$z_t^{**} = \log \dfrac{q_t}{q_t^*}$ is consumer income, the transient part.

The relations

$$q = (q^*)(q/q^*); \qquad z = z^* + z^{**} \tag{17a,b}$$

show that if

$$\mu_{i2} = \gamma_{i2} \qquad \text{or} \qquad \mu_{i2} = 0, \tag{18a,b}$$

model (14) reduces to (12) or (13), respectively.

5.2.1. Model (12) is a system of ordinary demand functions. By the argument in 2.2.4, OLS regression will provide consistent estimates for the demand elasticities γ_{i1}, γ_{i2}.

5.2.2. Model (13) is specified in accordance with Friedman's hypothesis of permanent income. Reference is made to Singh and Drost [21], who estimate a model of this type by two different NIPALS procedures. The model assumes permanent income to depend upon past incomes by a scheme of distributed lags,

$$z_t^* = \sum_{s=1}^{h} \lambda^s z_{t-s} \tag{19}$$

5.2.3. Model (14) develops Friedman's hypothesis by including the transient income z^{**} as a separate explanatory variable. The application of model (14) to different commodities allows us to explore whether (18a) or (18b) is valid, or, more generally, whether $\mu_{i2} \neq \gamma_{i2}$. The ith commodity may be said to satisfy Friedman's hypothesis in a restrictive sense if the empirical findings support the hypothesis (18b), and in a more general sense if the date indicate

$$0 < \mu_{i2} < \gamma_{i2} \qquad \text{or} \qquad \gamma_{i2} < \mu_{i2} < 0 \tag{20}$$

Reference is made to Willassen [27] for applications of (14) and related models. and for further theoretical analysis of the approach.

5.3. Multiple Causes and Multiple Indicators of an Unobservable Variable

Chart 4.3, from Hauser and Goldberger [14], refers to the following problem: Determine the variable v^* so that it coincides with x_1, x_2, x_3 as closely as possible, and at the same time explains as much as possible of y_1, y_2, y_3.

Hauser and Goldberger solve the problem by linear matrix algebra, Eqs. (45)–(57). In his contribution to this symposium, Jöreskog [15] gives a maximum likelihood solution, using Fletcher and Powell's iterative method of osculatory descent.

The Appendix to Hauser and Goldberger's paper gives a treatment in terms of canonical correlations, with reference to Duncan, Watts, and Blalock. Here follow two versions of a related approach, using NIPALS procedures designed in accordance with Charts 4.3–4.4.

5.3.1. We restate the problem as referring to m variables $x_1, \ldots, x_m$ and n variables $y_1, \ldots, y_n$, which are normalized to zero means and unit variances. The latent variable v^* is assumed to be linear in the variables x_i, y_k, say

$$v^* = \gamma\left(\sum_i \alpha_i x_i + \sum_k \beta_k y_k\right) \tag{21}$$

where γ is a normalization constant that gives v^* unit variance.

In accordance with Chart 4.3, we assume that v^* is linear in the variables $x_1, \ldots, x_m$,

$$v^* = \sum_i \alpha_i x_i + \varepsilon \tag{22}$$

and that each y_k is linear in v^*,

$$y_k = \beta_k v^* + \varepsilon_k; \qquad k = 1, \ldots, n \tag{23}$$

Relations (22)–(23) are submodels in the sense of 2.1, and in accordance with 2.1.2 they will be specified as predictors, that is

$$E(v^* \mid x_1, \ldots, x_m) = \sum_i a_i x_i \tag{24}$$

$$E(y_k \mid v^*) = \beta_k v^* \tag{25}$$

Let a_i, b_k, g, v^* denote the NIPALS estimates for the unknowns α_i, β_k, γ, v^*. Starting with an arbitrary set of coefficient proxies, say

$$a_i^{(1)} = b_k^{(1)} = 1 \qquad (i = 1, \ldots, m; k = 1, \ldots, n) \tag{26}$$

we describe the calculations in the sth step of the procedure on the basis of the proxy $v^{(s-1)}$ obtained in the previous step. First, the proxies $a_i^{(s)}$ are obtained by the multiple OLS regression

$$v^{(s-1)} = \sum_i a_i^{(s)} x_i + e^{(s)} \tag{27}$$

Second, the proxy $b_k^{(s)}$ is for each $k = 1, \ldots, n$ obtained by the simple OLS regression

$$y_k = b_k^{(s)} v^{(s-1)} + e_k^{(s)} \tag{28}$$

Finally, we form

$$v^{(s)} = g^{(s)}\left(\sum_i a_i^{(s)} x_i + \sum b_k^{(s)} y_k\right) \tag{29}$$

where $g^{(s)}$ is determined so as to give $v^{(s)}$ unit variance.

5.3.2. In the following table, the first line quotes Hauser–Goldberger's estimates of the coefficients α_i, β_k, normalized in accordance with (21).

	α_1	α_1	α_3	β_1	β_2	β_3
H–G	0.353	0.108	0.486	0.129	0.278	0.187
(i)	0.155	0.086	0.242	0.346	0.422	0.334
(ii)	0.211	0.095	0.314	0.213	0.448	0.301

Line (i) shows the corresponding results for the NIPALS model (21)–(25) estimated by the NIPALS procedure (26)–(29). Line (ii) shows the results for a NIPALS model designed in accordance with Chart 4.4, having two latent variables which are linear forms of the variables x_i and y_k, respectively.

Comment. As always in NIPALS modelling, the two NIPALS models are specified as predictors, and thereby allow an immediate interpretation in predictive and/or causal terms. It will further be noted that the two NIPALS models, but not Hauser–Goldbergers's model, honour the parity principle 4.1.4.

5.4. A Problem of Path Coefficients

Path models as initiated and developed by Wright [36] are causal chain systems of type L, where all variables and residuals are normalized to zero mean and unit variance, and the ensuing coefficients for variables and residuals are called path coefficients (usually denoted by p_{ab}). The linear operations of product moments and substitution of variables give the fundamental rule that a total path coefficient between a primary factor and an effect is the sum of compound path regressions connecting the primary factor and the effect.[28] In Chart 4.5, for example, the fundamental rule gives

$$r_{14} = p_{12} r_{24} + p_{13} r_{34}. \tag{30}$$

In overidentified systems such relations are more or less approximate, since compound path regressions do not necessarily coincide with the corresponding partial regression coefficients. Such a deviation is indicated by the broken arrow in Chart 4.5, and similarly in Chart 4.6.

[28] Wright [36]; also cf. Tukey [24], Turner and Stevens [25], and Duncan [10a].

Focusing on the path model behind Chart 4.6, we pose the problem of constructing a variable

$$y_2^* = \alpha_1 y_1 + \alpha_2 y_2 + \alpha_3 y_3 \tag{31}$$

which, when substituted for y_2 in the model, makes the broken arrow vanish. That is, we require

$$r(y_1, y_3 \mid y_2^*) = 0 \tag{32a}$$

or, which is the same,

$$r(y_1, y_3) = r(y_1, y_2^*)r(y_2^*, y_3) \tag{32b}$$

5.4.1. The following NIPALS procedure is proposed for the construction of y_2^*.

The start: Form an arbitrary linear combination of the variables y_i, say

$$y_2^{(1)} = c_1 \sum_{i=1}^{3} y_i \tag{33a}$$

where c_1 is determined so as to give $y_2^{(1)}$ unit variance, in symbols

$$\frac{1}{N} \sum_i (y_2^{(1)})^2 = 1 \tag{33b}$$

where N denotes the number of observations.

The steps from $2s - 1$ to $2s$ and $2s + 1$:

(i) Form the OLS regression

$$y_1 = b_2^{(2s)} y_2^{(2s-1)} + b_3^{(2s)} y_3 + e^{(2s)} \tag{34a}$$

and calculate

$$y_2^{(2s)} = c_{2s}(b_2^{(2s)} y_2^{(2s-1)} + b_3^{(2s)} y_3) \tag{34b}$$

where c_{2s} is determined so as to give $y_2^{(2s)}$ unit variance.

(ii) Form the OLS regression

$$y_3 = b_2^{(2s+1)} y_2^{(2s)} + b_1^{(2s+1)} y_1 + e^{(2s+1)} \tag{35a}$$

and calculate

$$y_2^{(2s+1)} = c_{2s+1}(b_2^{(2s+1)} y_2^{(2s)} + b_1^{(2s+1)} y_1) \tag{35b}$$

where c_{2s+1} is determined so as to give $y_2^{(2s+1)}$ unit variance.

5.4.2. For illustration, let y_1, y_2, y_3 be three variables with unit variance and intercorrelations ρ. Then

$$y_2^* = y_1 + \alpha y_2 + y_3 \tag{36a}$$

with

$$\alpha = \sqrt{4\rho^2 + 2\rho} - 2\rho \tag{36b}$$

will satisfy the condition (32b), as is readily verified .

With data input corresponding to $\rho = \frac{1}{2}$, the NIPALS procedure (33)–(35) has given $\alpha = \sqrt{2} - 1$, in agreement with (36b).

REFERENCES

1. Adelman, I. and Morris, C. T. (1967). *Society, Politics and Economic Development. A Quantitative approach.* Baltimore, Maryland.

2a. Ågren, A. (1970). Fractional fix-point (FFP) estimation. *Interdependent Systems. Structure and Estimation* (E. J. Mosbaek and H. Wold *et al.*), Chapter 3.6. North-Holland Publ., Amsterdam.

2b. Ågren, A. (1972). Extensions of the fix-point method. Theory and applications. Doctoral Thesis, Univ. of Uppsala, Uppsala, Sweden.

3a. Bergström, R. (1969). NIPALS and related methods for ID systems applied to Monte Carlo and real world data. Wold-Lyttkens (eds.) [33, Chapter 17].

3b. Bergström, R. (1973). Studies of iterative methods for the estimation of interdependent systems, especially the FP and IIV methods. Doctoral Thesis, Univ. of Uppsala, Uppsala, Sweden, to be published.

4. Blalock, Jr., H. M., ed. (1971). *Causal Models in the Social Sciences.* Aldine, Chicago, Illinois and Atherton, New York.

5a. Bodin, L. (1970). Recursive fix-point (RFP) estimation. *Interdependent Systems. Structure and Estimation* (E. J. Mosbaek and H. Wold, *et al.*), Chapter 3.7. North-Holland Publ., Amsterdam.

5b. Bodin, L. (1973). Recursive fix-point estimation. Theory and applications. Doctoral Thesis, Univ. of Uppsala, Uppsala, Sweden, to be published.

6. Carroll, J. D. and Chang, Y. (1970). Analysis of individual differences in multidimensional scaling, with an *N*-way generalization of Eckart–Young decomposition. *Psychometrika* **35** 283–320.

7. Christ, C. F., Hildreth, C., Liu, T.-Ch. and Klein, L. R. (1960). A symposium on simultaneous equation estimation. *Econometrica* **28** 835–871.

8. Christoffersson, A. (1970). The one component model with incomplete data. Doctoral Thesis, Univ. of Uppsala, Uppsala, Sweden.

9. Costner, H. C., ed. (1971). *Sociological Methodology, 1971.* Jossey-Bass, San Francisco, California.

10a. Duncan, O. T. (1966). Path analysis: Sociological examples. *Amer. J. Sociology* **72** 1–16; also Blalock [4, pp. 115–138].

10b. Eckart, C. and Young, G. (1936). The approximation of one matrix by another of lower rank. *Psychometrika* **1** 211–218.

11. Friedman, M. (1957). *A Theory of the Consumption Function.* Princeton Univ. Press, Princeton, New Jersey.

12. Haavelmo, T. (1943). The statistical implications of a system of simultaneous equations. *Econometrica* **11** 1–12.

13. Haldane, J. and Priestley, J. (1905). Lung ventilation. *J. Physiol.* **32** 225–266.

14. Hauser, R. M. and Goldberger, A. S. (1971). The treatment of unobservable variables in path analysis. *Sociological Methodology, 1971* (H. C. Costner, ed.), 81–117. Jossey-Bass, San Francisco, California.

15. Jöreskog, K. G. (1973). Analysis of covariance structures. *Multivariate Analysis-III* (P. R. Krishnaiah, ed). Academic Press, New York, pp. 263–285.

16. Lyttkens, E. (1970). Some asymptotic formulas for standard errors and small-sample bias of FP estimates. *Interdependent Systems. Structure and Estimation* (E. J. Mosbaek and H. Wold, *et al.*), Chapter 3.9. North-Holland Publ., Amsterdam.

17. Lyttkens, E. (1970). Iterative instrumental variables (IIV) estimation. *Interdependent Systems. Structure and Estimation* (E. J. Mosbaek and H. Wold, *et al.*), Chapter 11.4. North-Holland Publ., Amsterdam.
18. Meissner, W. (1969). Ökonometrische Modelle. Rekursivität versus Interdependenz aus der Sicht der Kybernetik. Duncker and Humblot, Berlin.
19. Mosbaek, E. J. and Wold, H., with contr. by Lyttkens, E., Ågren, A., and Bodin, L., (1970). *Interdependent Systems. Structure and Estimation.* North-Holland Publ., Amsterdam.
20. Russell, B. (1914). On the notion of cause, with application to the free-will problem. *Our Knowledge of the External World*, 214–246. Norton, New York.
21. Singh, B. and Drost, H. (1971). An alternative econometric approach to the permanent income hypothesis: An international comparison. *Rev. Econom. and Statist.* **53** 326–334.
22. Tinbergen, J. (1937). *An Econometric Approach to Business Cycle Problems.* Hermann, Paris.
23. Tinbergen, J. (1940). Econometric business cycle research. *Rev. Econom. Studies* **7** 73–90.
24. Tukey, J. W. (1954). Causation, regression and path analysis. *Statistics and Mathematics in Biology* (O. Kempthorne, T. A. Bancroft, J. W., Gowen and J. L. Luch, eds.). Iowa State Univ. Press, Ames.
25. Turner, M. E. and Stevens, C. D. (1959). The regression analysis of causal paths. *Biometrics* **15** 236–258; also Blalock [4, pp. 75–99].
26. Varga, R. S. (1962). *Matrix Iterative Analysis.* Prentice-Hall, Englewood Cliffs, New Jersey.
27. Willassen, Y. (1972). Three econometric applications of NIPALS modelling. *Sem. Econom. College France, Paris, November 1972.*
28. Wold, H. (1963). On the consistency of least squares regression. *Sankhyā Ser. A* **25** No. 2, 211–215.
29. Wold, H. (1965). A fix-point theorem with econometric background, I–II. *Ark. Mat.* **6** 209–240.
30. Wold, H. (1966). Nonlinear estimation by iterative least squares procedures. *Research Papers in Statistics; Festschrift for J. Neyman* (F. David, ed.), 411–444. Wiley, New York.
31. Wold, H. (1969). Nonexperimental statistical analysis from the general point of view of scientific method. *Bull. Inst. Internat. Statist.* **52** No. 1, 391–424.
32. Wold, H. (1969). Mergers of economics and philosophy of science. A cruise in deep seas and shallow waters. *Synthese* **20** 427–482.
33. Wold, H. and Lyttkens, E., eds. (1969). Nonlinear iterative partial least squares (NIPALS) estimation procedures. *Bull. Inst. Internat. Statist.* **53** 29–51.
34. Wold, S. and Sjöström, M. (1972). Statistical analysis of the Hammett equation. I. Methods and model calculations. *Chemica Scripta* **2** 49–55.
35. Woodward, R. B. *et al.* (1958). The total synthesis of reserpine. *Tetrahedron* **2** 1–57.
36. Wright, S. (1934). The method of path coefficients. *Ann. Math. Statist.* **5** 161–215.

Titles of Contributed Papers

R. Askey (University of Wisconsin): Radial Characteristic Functions

S. K. Badhe (Sandoz-Wander, Inc.): Multivariate Generalization of Behrens-Fisher-Welch V Statistics and Their Distributions for Fixed Values of Unknown Ratios of Variances

S. K. Basu (University of North Carolina): Density Versions of Multivariate Central Limit Theorems

K. N. Berk (Illinois State University): Consistent Autoregressive Spectral Estimates

M. W. Browne (Institute for Personal Research, S. Africa, and Educational Testing Service, Princeton): Generalized Least Squares Estimators in the Analysis of Covariance Structures.

S. K. Chatterjee (University of North Carolina and University of Calcutta): Rank Procedures for Certain Multivariate Mixture and Classification Problems

E. J. Dudewicz and S. R. Dalal (University of Rochester): Multiple Comparisons with a Control when Variances are Unequal

P. Ghosh (Federal City College, Washington): A Note on Multivariate Symmetrical and Unimodal Distributions

F. Alberto Grunbaum (Courant Institute of Mathematical Sciences): An Inverse Problem for Gaussian Processes

G. Kaufman (Massachusetts Institute of Technology) **and S. J. Press** (University of Chicago): Bayesian Factor Analysis

C. F. Kossack (University of Georgia): Applications of Multivariate Analysis

A. M. Kshirsagar (Texas A & M University) **and S. F. Musket** (Southern Methodist University): A Note on the Discrimination in the Case of Zero Mean Differences

C. Lin (University of Southern California) **and P. D. Berger** (Boston University): A Method for Evaluating Multivariate Normal Probability Integrals

A. M. Mathai (McGill University, Canada): Exact Non-Null Distributions of a Class of Test Statistics having Certain Structures

S. F. Musket (Sun Oil Company): An Evaluation of Kendall's Order-Statistics Method for Discriminant Analysis and Related Studies

R. L. Obenchain (Bell Telephone Laboratories): Linear Reduction of Dimensionality Using the Principles of Affine Commutativity and Invariance.

S. R. Paranjape and C. Park (Miami University, Oxford, Ohio): On Multi-Parameter Wiener Processes

W. J. Park (Wright State University): On Strassen's Version of the Law of the Iterated Logarithm for a Multi-Parameter Gaussian Process

F. J. Schuurmann and V. B. Waikar (Miami University, Oxford, Ohio): Tables for the Power Function of Roy's Test for $\Sigma = I$ Against Two-Sided Alternatives

M. Siotani (Kansas State University and Aerospace Research Laboratories): Distributions of the Ranges in the Multivariate Case